Introduction to protein architecture

Introduction to protein architecture

The structural biology of proteins

Arthur M. Lesk
University of Cambridge

OXFORD
UNIVERSITY PRESS

OXFORD
UNIVERSITY PRESS

Great Clarendon Street, Oxford OX2 6DP

Oxford University Press is a department of the University of Oxford.
It furthers the University's objective of excellence in research,
scholarship, and education by publishing worldwide in

Oxford New York

Athens Auckland Bangkok Bogotá Buenos Aires Calcutta Cape Town
Chennai Dar es Salaam Delhi Florence Hong Kong Istanbul Karachi
Kuala Lumpur Madrid Melbourne Mexico City Mumbai Nairobi Paris
São Paulo Singapore Taipei Tokyo Toronto Warsaw

with associated companies in Berlin Ibadan

Published in the United States by Oxford University Press Inc., New York

First published 2001

British Library Cataloguing in Publication Data

Data available

Library of Congress Cataloguing in Publication Data

Data available

ISBN 0 19 850474 8 (Pbk)

3 5 7 9 10 8 6 4 2

Typeset by J & L Composition Ltd, Filey, N. Yorks
Printed
on acid-free paper
by Giunti Industrie Grafiche, Italy

Preface

One of the most important scientific legacies that our generation will leave to its successors is the detailed information about biological sequences and structures now being determined and archived. Full-genome sequencing projects have provided complete information about numerous bacterial and viral genomes, and the complete DNA sequences of three eukaryotes: yeast, the nematode worm *Caenorhabditis elegans,* and the fruit fly *Drosophila melanogaster*. The human genome should be 90% completed at about the date at which this book is published, and finished within a year or two afterwards. The number of protein and nucleic acid structures determined by X-ray crystallography and Nuclear Magnetic Resonance spectroscopy is also growing at a rapid rate. The Protein Data Bank contains over 10 000 sets of coordinates. Systematic structure determinations of *all* proteins from selected organisms—projects called structural genomics—are underway.

A new field, bioinformatics, has grown up around these data. It has captured the interest of many scientists for its intellectual challenges, its potential for useful applications, and its promising scope for careers.

There are many books on proteins, written from many different points of view. In this one the emphasis is on protein architecture, on proteins as three-dimensional patterns. Rutherford said, 'All science is either physics or stamp collecting'. This book embodies my reply, that the study of proteins includes the best qualities of both. It is intended for use as a textbook for advanced undergraduates and beginning graduate students, and as a reference for workers in the field.

Three types of problems appear at the ends of chapters. Exercises are short and straightforward applications of material in the text. Problems also require no information not contained in the text, but require lengthier answers or in some cases calculations. The third category, 'weblems' (word suggested by V.I. Lesk), require access to the World Wide Web, preferably through a graphical interface. Weblems are designed to give readers practice with the tools required for further study and research in the field.

The web site for this book is at

www.oup.co.uk/best.textbooks/biochemistry/protein

The author is grateful to many colleagues. During my recent professional activities I have been indebted to the entire community of protein crystallographers and NMR spectroscopists who have generated the data with which I work, and to the University of Cambridge and the Medical Research Council Laboratory of Molecular Biology for facilities and support. For specific help I thank E. Abola, C.R. Barker, A.J. Barrett, A.C. Bloomer, R. Bazzo, C.-I. Brändén, G. Bricogne, R.W. Carrell, C. Chothia, F. Cohen, J. Deisenhofer, P.R. Evans, M. Gait, M.B. Gerstein, E. Gherardi, N. Grant, L. Holm, K.C. Holmes, T.S. Horsnell, T.J.P. Hubbard, E.Y. Jones, T.A. Jones, A. Lenton, E.L. Lesk, M.E. Lesk, V.E. Lesk, V.I. Lesk, A.G.W. Leslie, G. Lingley, L. Lo Conte, P. Margiotta, V. Morea, A.G. Murzin, B. Nall, B. Pashley, A. Pastore, X.-Y. Pei, M.F. Perutz, T. Rabbitts, N. Rawlings, R.J. Read, F.W. Roberts, G.D. Rose, W. Rypniewski, G. Schertler, P.B. Sigler, J.M. Smith, M. Teeter, I.M. Tomlinson, A. Tramontano, G. Vriend, A.G. Weeds, J. Sussman, J.E. Walker and J.C. Whisstock. I thank the staff of Oxford University Press for their skills and patience in producing this book.

Cambridge A.M.L.
June 1999

Dedicated to A. Pastore and A. Tramontano

Dedicated to A. Pastore and A. Tarantino

Contents

Sections marked with an asterisk may be skipped on a first reading.

CHAPTER 1

The photosynthetic reaction centre: protein structure in a microcosm

The reaction centre from the purple bacterium *Rhodopseudomonas viridis* is the site of the initial step in the capture of light energy in photosynthesis. The reaction centre is a membrane-bound complex of four proteins, binding 14 low molecular weight cofactors. These cofactors include the chromophores that absorb the excitation energy that is converted to electrochemical potential—or redox energy—across the membrane.

In the early 1980s, reports that reaction centres formed regular two-dimensional arrays *in vivo* made crystallographers very, very interested. For any object to form a crystal, it must possess a definite and fixed structure. And if it does, that structure can be determined. Echoes of the original excitement felt by J.D. Bernal in 1934 upon seeing the X-ray diffraction pattern of crystalline pepsin, and realizing the implication that someday a complete atomic model of a protein would be revealed:

> The wet crystals gave individual X-ray reflections, which were rather blurred owing to the large size of the crystal unit cell, but which extended all over the films to spacings of about 2 Å. That night [in late April 1934], Bernal, full of excitement, wandered around the streets of Cambridge, thinking of the future and of how much it might be possible to know about the structure of proteins if the photographs he had just taken could be interpreted in every detail. (Hodgkin and Riley 1968.)

From 1934 it took 25 years to determine a complete protein structure—J.C. Kendrew, M.F. Perutz, and colleagues announced the structures of myoglobin in 1959 and haemoglobin in 1960. But by 1982, when the first three-dimensional crystals of the reaction centre were obtained, X-ray crystallographic technique had much advanced. The archives of protein structure (the Protein Data Bank) contained 150 structures, including myoglobin and haemoglobin, enzymes such as lysozyme and chymotrypsin, the hormone insulin, and fragments of antibodies. Very satisfying—and yet cells are not dilute solutions of non-interacting proteins like the figures in an L.S. Lowry painting. To understand the integration of the activities of proteins, it is necessary to go to higher levels of structure.

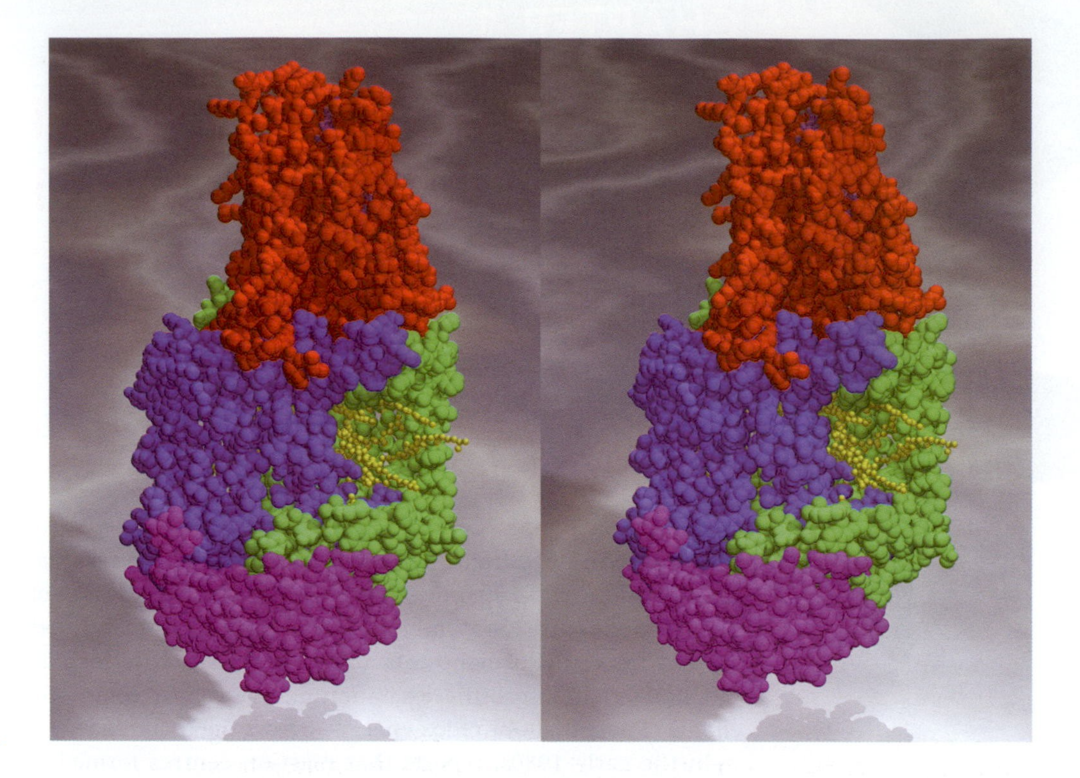

Fig. 1.1a

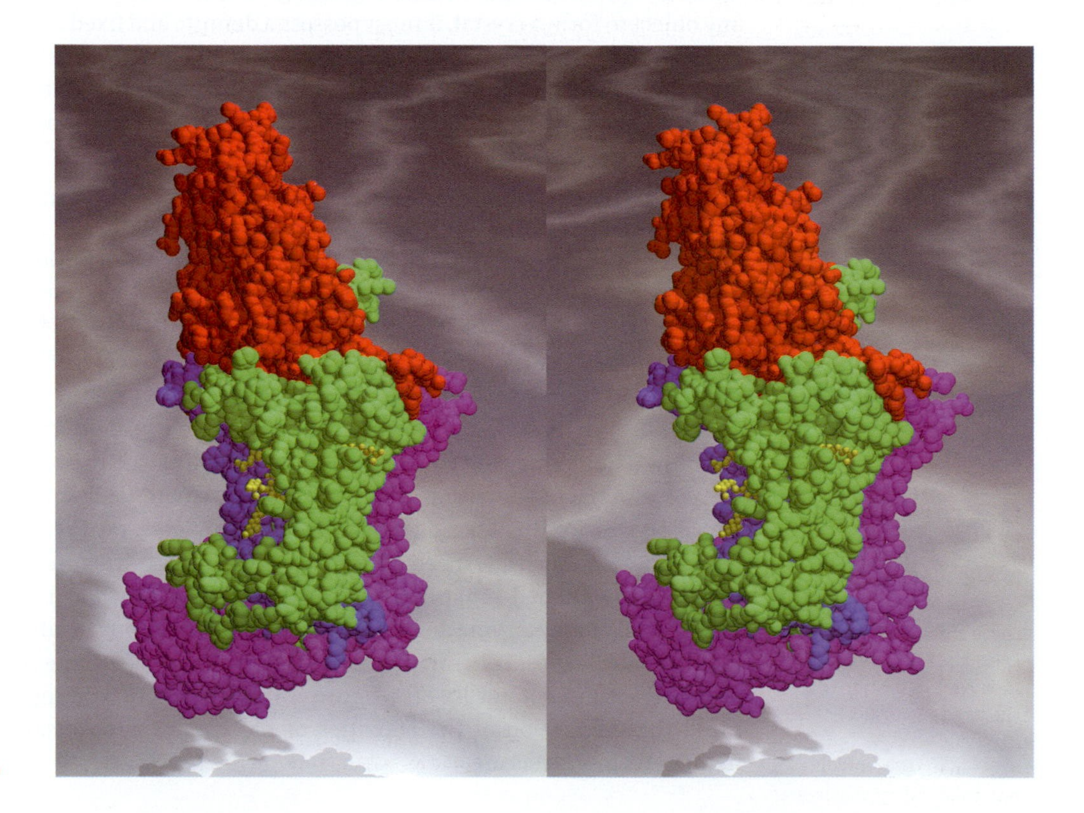

Fig. 1.1b

Fig. 1.1 (Facing page) The photosynthetic reaction centre from *Rhodopseudomonas viridis*, solved by J. Deisenhofer, H. Michel and R. Huber [1PRC]. (1PRC is the designation of the coordinate set in the Protein Data Bank, described in the next chapter). This figure shows an 'all-atom' representation. Parts (a) and (b) show two orientations, at right angles to each other. The four proteins are distinguished by colour: light, blue; medium, green; heavy, purple; cytochrome, orange.

Macromolecular complexes, such as the reaction centre, are the first step to higher levels of structure. Comparing the material available for the first edition of this book with what is known today, not only are there many more structures—the Protein Data Bank currently contains over 10 000 entries—but structures of substantially greater complexity have been determined. In addition to the photosynthetic reaction centre, large macromolecular complexes for which atomic coordinates are available include the proteasome (responsible for controlled degradation of proteins), the GroEL–GroES chaperonin complex (which catalyses the folding of proteins), the nucleosome (a DNA–protein complex that underlies chromosome structure), the F1 fragment of ATPase (an enzyme that uses the energy of a proton gradient across a membrane to synthesize ATP), the subunits of the ribosome (the protein-synthesizing apparatus) and numerous virus structures.

As a macromolecular complex, the reaction centre was a 'breakthrough' structure. It was also the first to reveal the structure of a membrane-bound protein. Membrane proteins were—and to some extent still remain—outside the mainstream, because of the difficulty of preparing them in a state suitable for structural analysis.

The reaction centre illustrates all the essential principles of protein conformation. Let us look at it and take it apart.

The reaction centre from *Rhodopseudomonas viridis*

Figure 1.1 shows a representation of all atoms of the reaction centre except hydrogens, in 'front and side' views.* It is approximately what one could see in an atomic-resolution microscope, assuming that different proteins

* Readers may find a stereo viewer useful. The most common type comes with a stand appropriate for material lying flat on a table. Lorgnettes, held directly in front of the eyes, are more comfortable for reading a book. (Available from Agar Scientific Ltd, Stansted, Essex CM24 8DA, U.K. or Pelco, 4595 Mountain Lakes Boulevard, Redding, CA 96003, U.S.A.)

Many readers will become adept at seeing stereo unaided. The trick is to develop an unusual control of the eye muscles. In ordinary vision, when an object moves away from us we diverge our eyes *and* refocus for the greater distance. We are accustomed to coupling these muscular adjustments automatically. To view stereo pictures, we must diverge our eyes, but keep the focus appropriate for an object that is relatively near. A useful initial goal in learning to do this is to diverge the eyes without refocusing—imagine looking through the page at a distant object behind it—to produce a view in which each eye is receiving only its proper image, but these images are out of focus. Then try to focus them. (You will see *three* images—two from each eye, with the two in the centre overlapping; it is the centre one that will appear in stereo.)

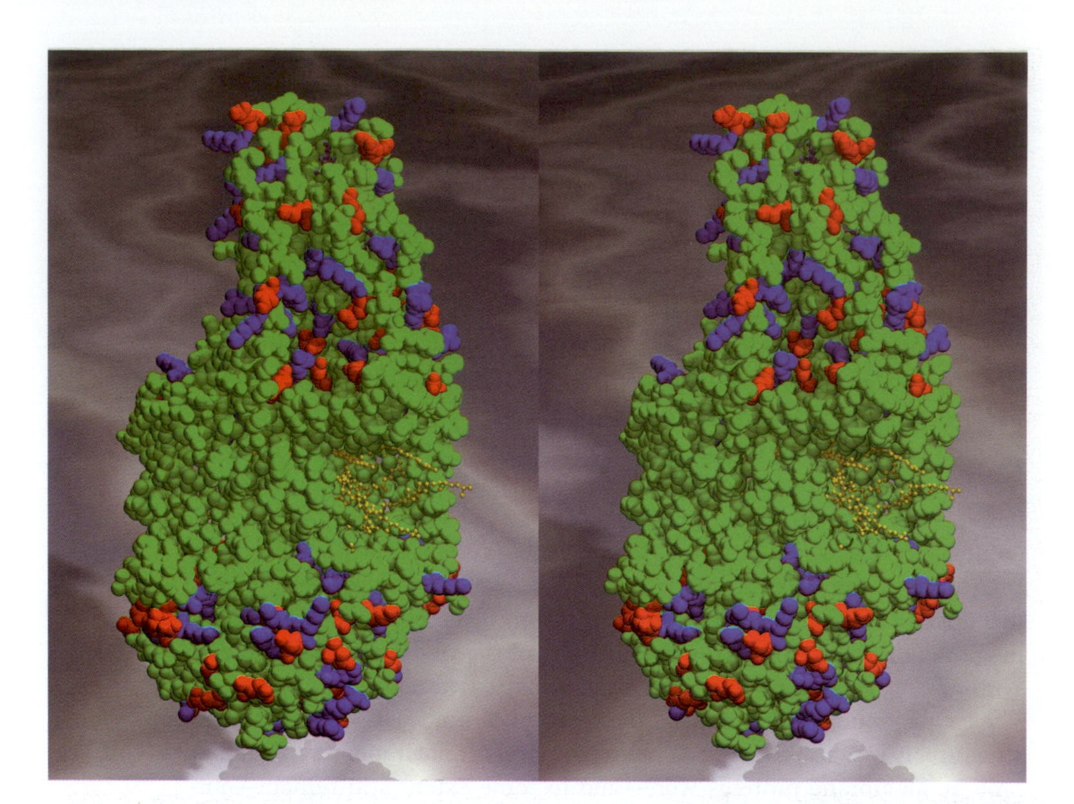

Fig. 1.2 The reaction centre coloured by residue charge (positive, blue; negative, red; neutral, green). The 'green belt' around the waist of the molecule corresponds to the membrane-spanning segment.

could be stained different colours. The assembly as a whole has dimensions 72 Å × 72 Å × 132 Å. The chromophores are embedded in the two central subunits and are only partly visible. The four proteins are the light (L, in blue), medium (M, in green), and heavy (H, in purple) subunits, and a cytochrome (orange).

Pictures showing all atoms give a general impression of the layout of the subunits in the complex, but fail to give a clear depiction of the architecture. For example, it would be difficult to identify helices in Figure 1.1. But these pictures do make the important point that proteins are compact, densely-packed assemblies of atoms. Think of the formation of the native state of a protein as a process of intramolecular crystallization. A more sparsely distributed grouping of atoms would not be stable.

Before discussing other representations of the reaction centre, consider one more all-atom picture. Figure 1.2 shows the structure with an alternative colour coding. Positively charged residues are coloured blue, negatively charged residues red, the others green. The wide horizontal swathe across the centre of the molecule that is entirely green (except for the cofactors) corresponds to the region that traverses the membrane. The proteins present surfaces of different character to the aqueous environment, at top and bottom, and to the membrane. The surfaces of the H subunit and the cytochrome are

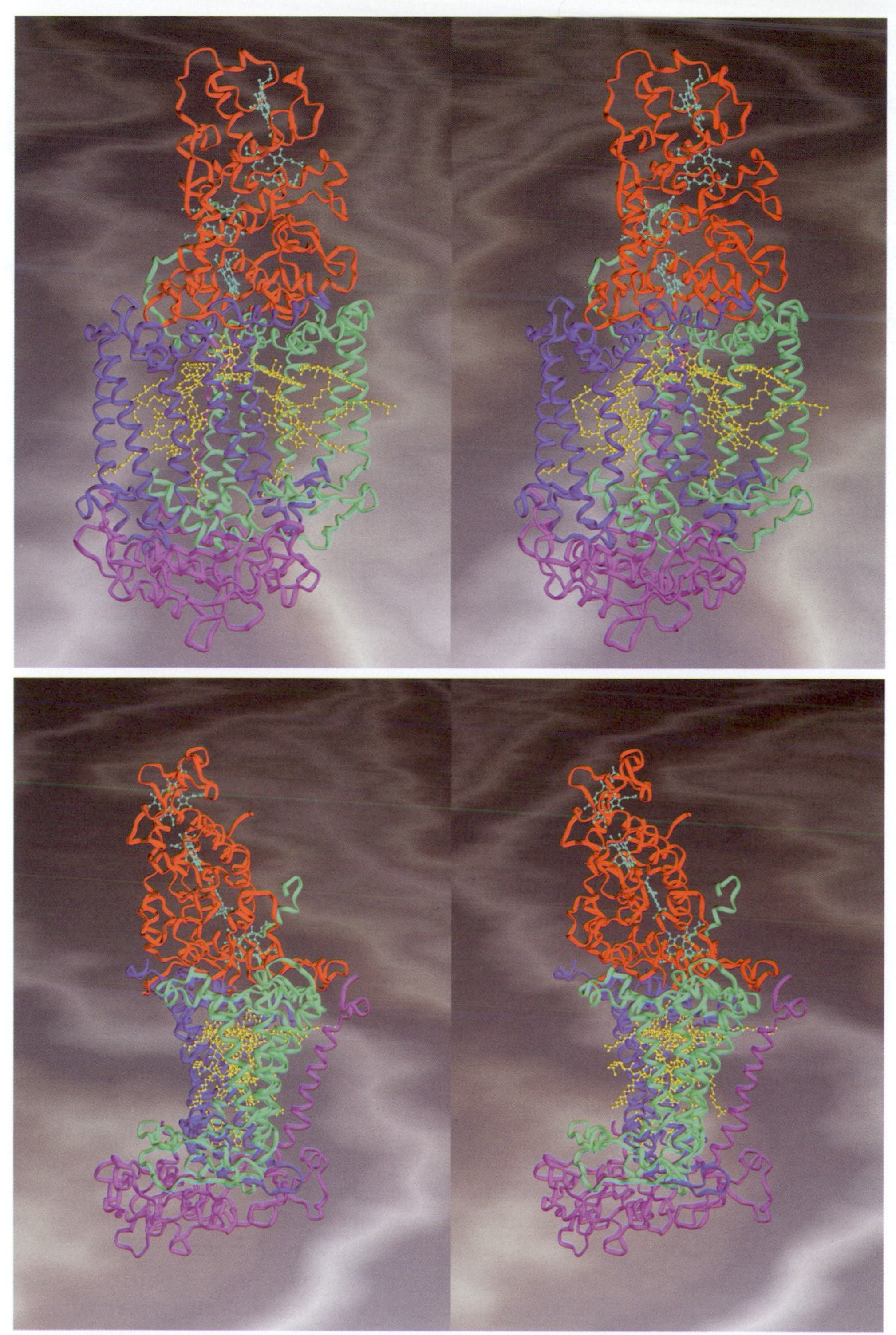

Fig. 1.3 Simplified representation of the reaction centre, in same orientations as in Figure 1.1. The course of the chain is shown by a smooth curve. The cofactors are shown in full detail, and are visible here to a much greater extent than in Figures 1.1 and 1.2.

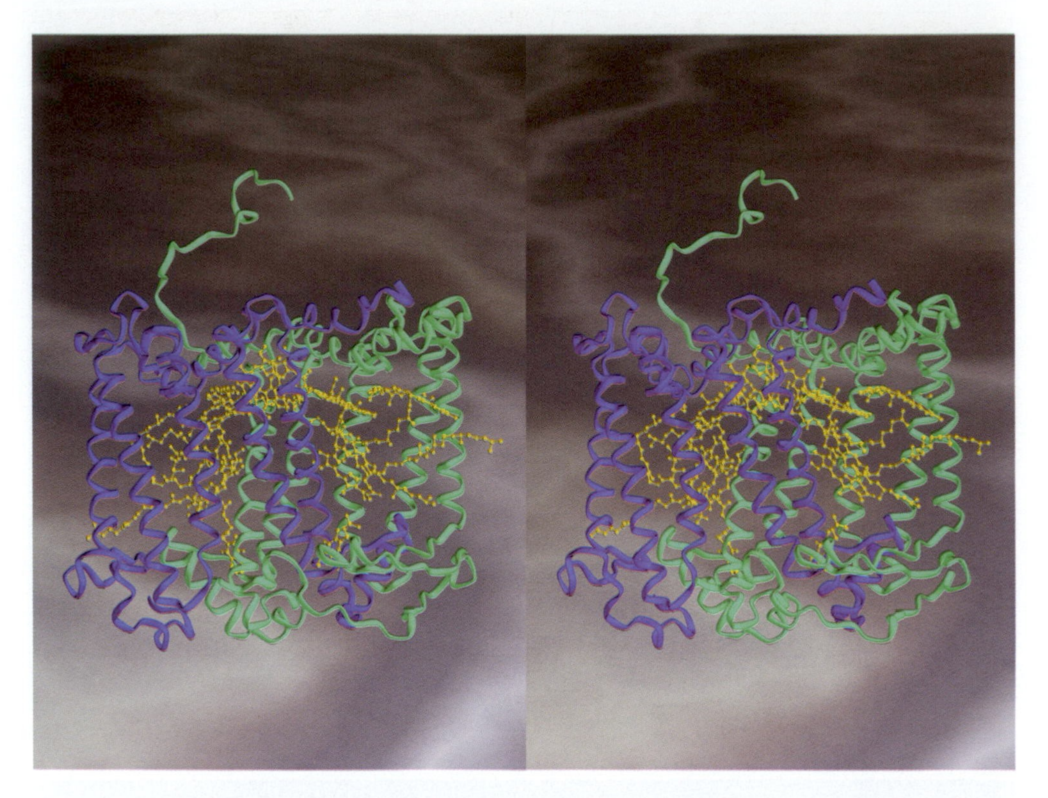

Fig. 1.4 Simplified representation, membrane-spanning subunits L and M only.

typical of proteins in aqueous environments: a tossed salad of charged, polar and uncharged residues. The surfaces presented to the lipid environment of the membrane are devoid of charged residues. As we shall see, the interiors of all proteins resemble these membrane-exposed surfaces in physicochemical character. Protein interiors contain primarily uncharged residues, some polar ones, and never—well, hardly ever—charged ones.

For a more intelligible representation of the architecture of the reaction centre, we must be more selective about the detail that we display. In Figure 1.3 the protein is simplified by running a smooth curve along the backbone; cofactors remain in full atomic detail. This picture is far more perspicuous. The cofactors are all visible. It is possible to recognize features of the folding pattern, such as helices.

Isolating the L and M subunits (Figure 1.4), we see another aspect of the structure that would be difficult to identify from the all-atom picture. There is symmetry in the structure, in the spatial relation between the two subunits. To say that there a symmetry means two things: first, that the objects related by symmetry are similar in structure; second, that there is a simple geometric relationship between them. Of course exact symmetry requires that the objects be identical; here we are dealing with an approximation to

Fig. 1.5 (Facing page) (a) Helices from the H subunit and (b) sheet from cytochrome subunit of the reaction centre.

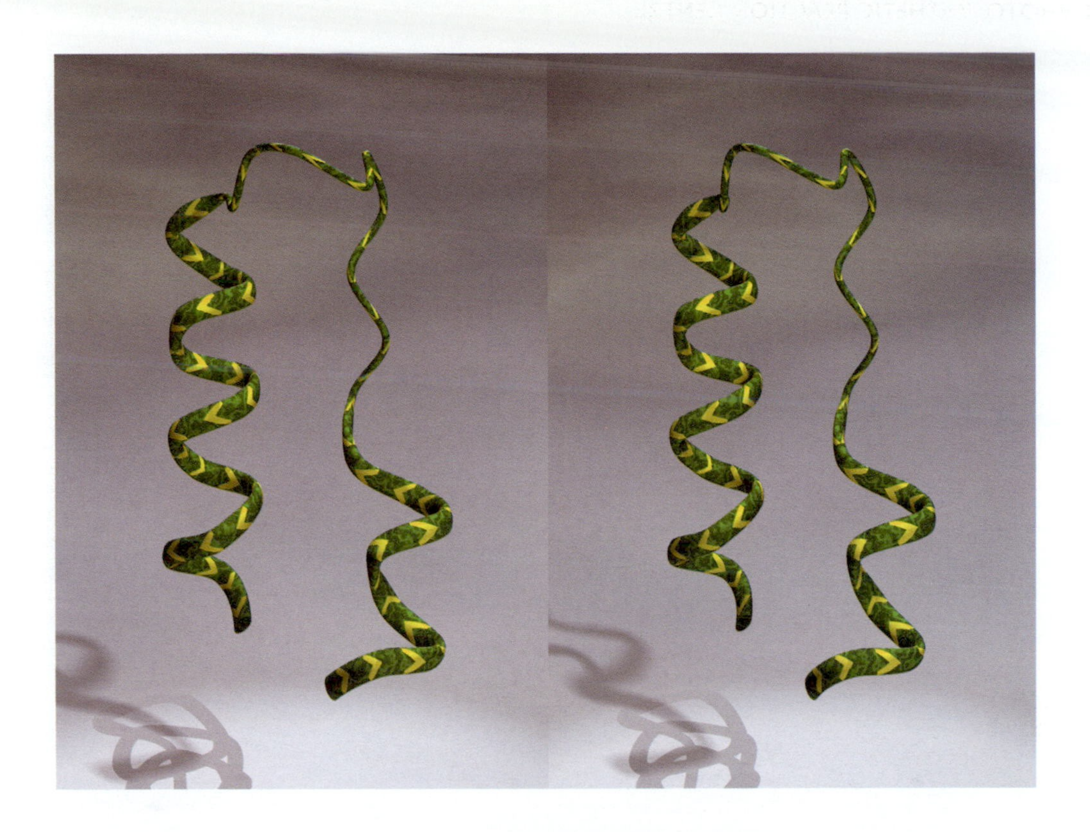

Fig. 1.5a

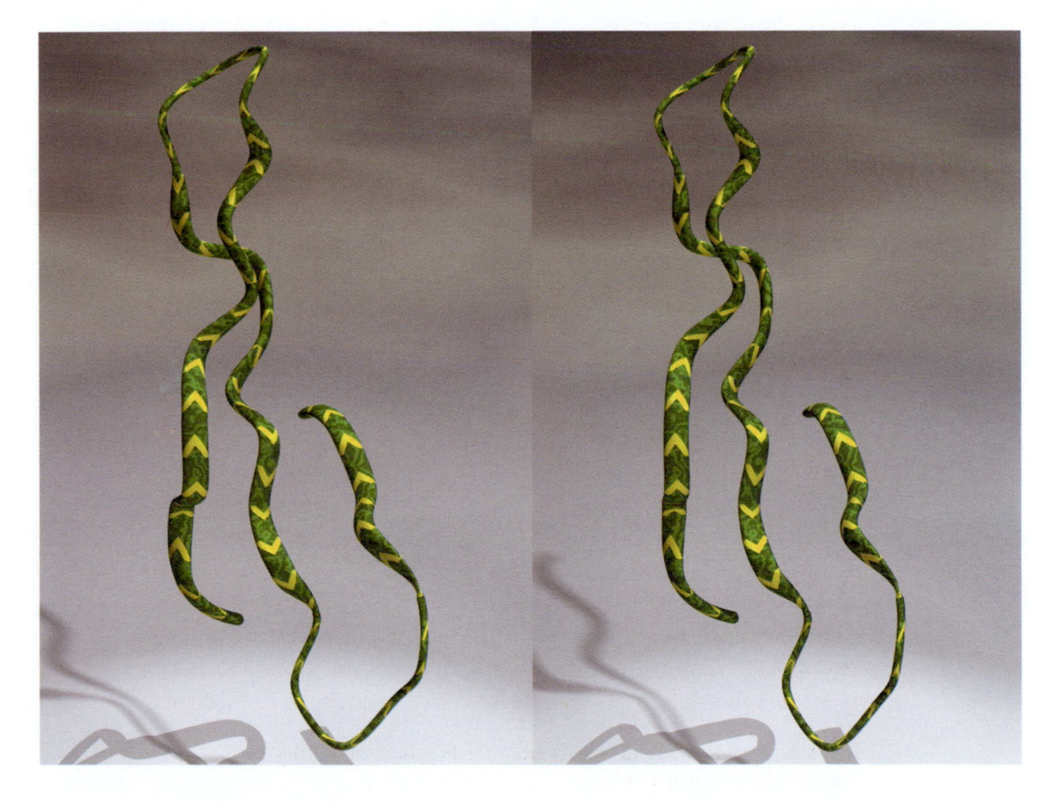

Fig. 1.5b

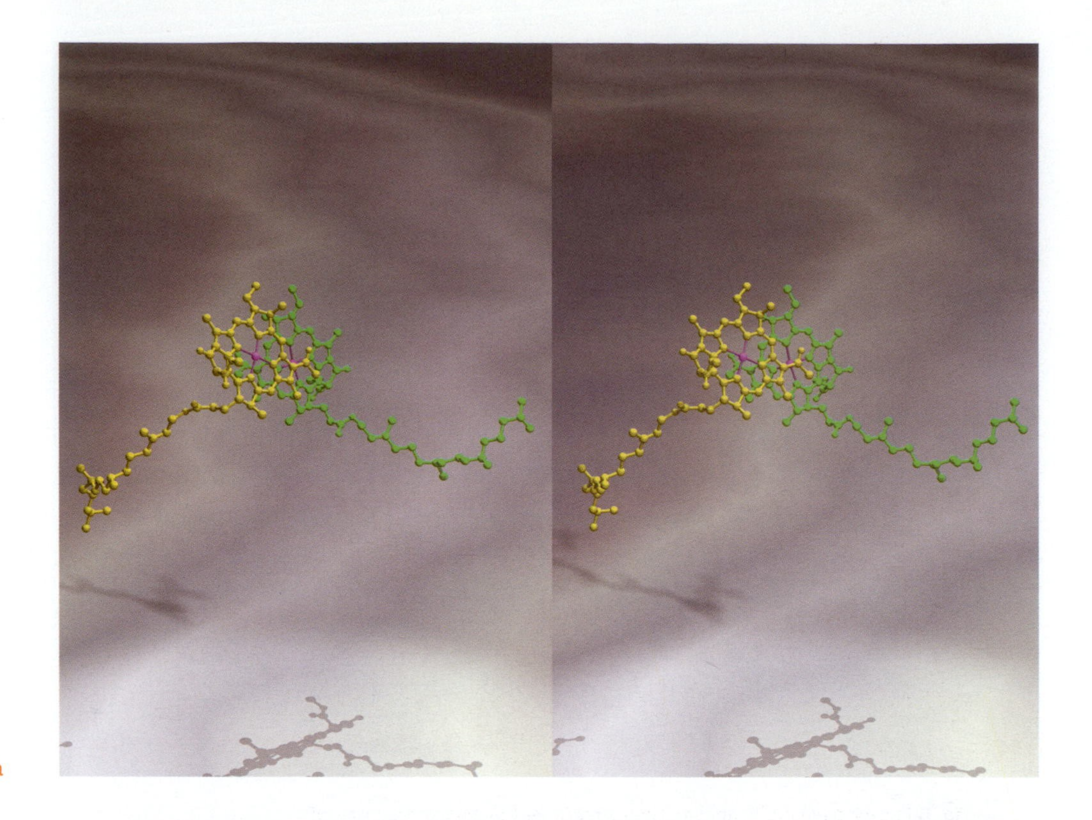

Fig. 1.6a

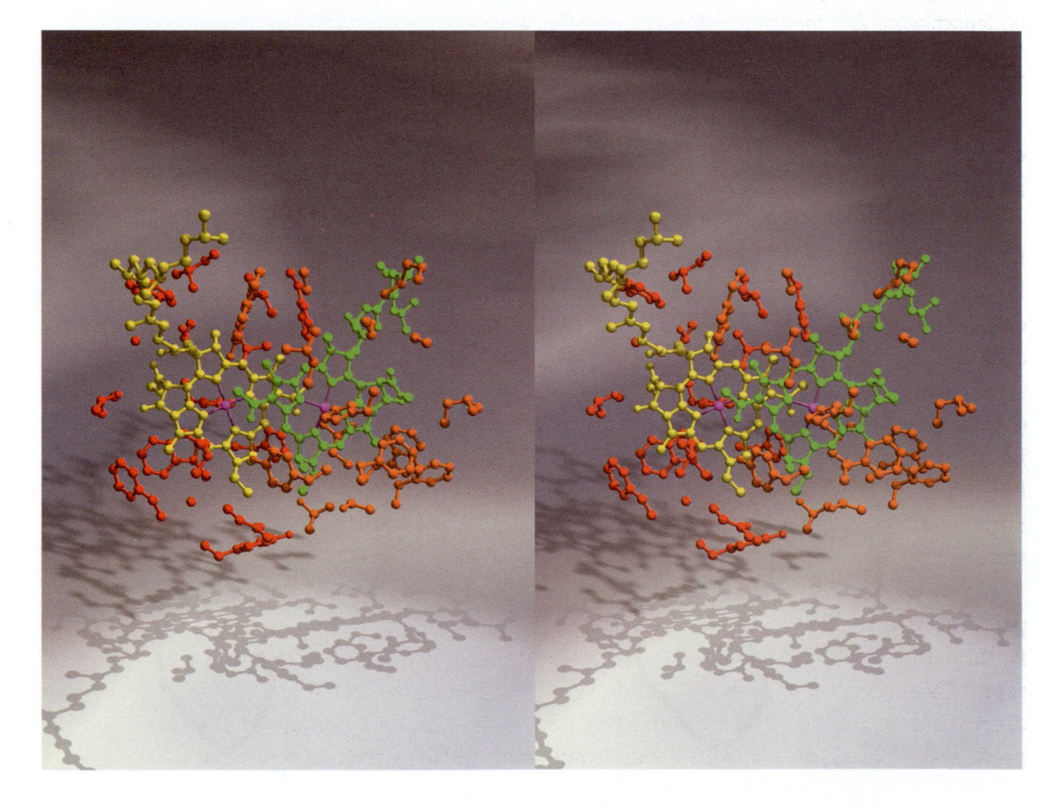

Fig. 1.6b

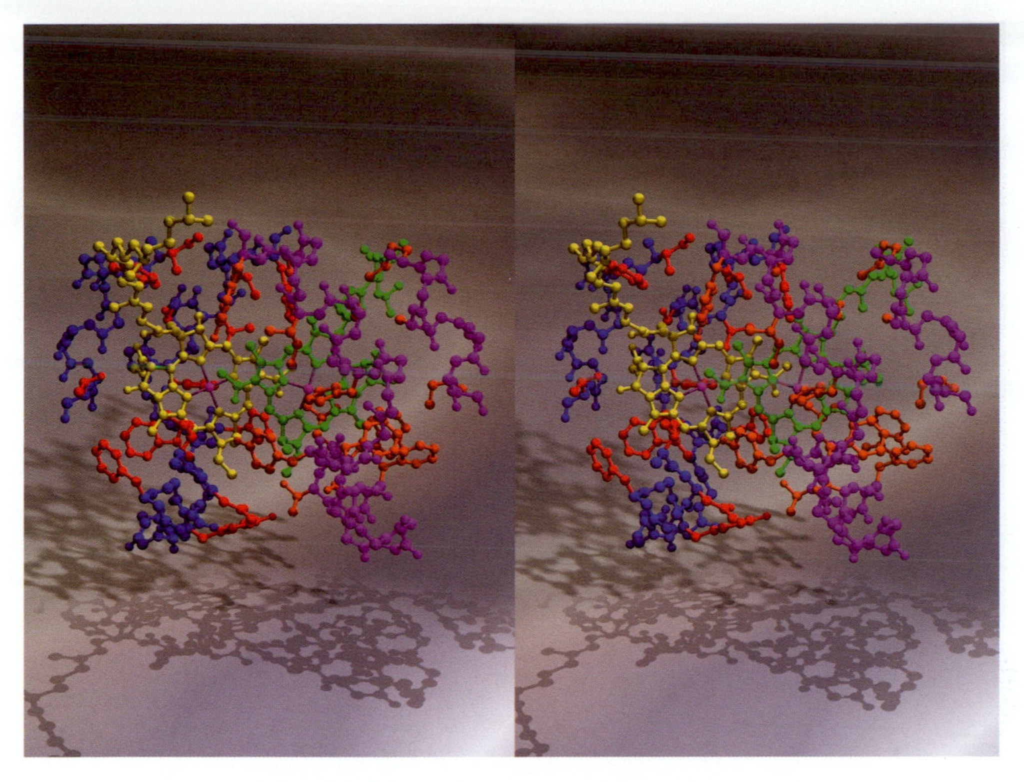

Fig. 1.6c

Fig. 1.6 (a, b Facing page and c above) The environment of the 'special pair of bacteriochlorophylls', the site of initial photochemical events in photosynthesis. (a) The special pair alone, (b) the special pair plus the sidechains that interact with it, (c) the special pair plus the sidechains and backbone of the residues that interact with it.

exact symmetry. Can you see the spatial relationship? It is a rotation by approximately 180° around a vertical axis parallel to the page.

Helices are prominent landmarks in the L and M subunits, and appear also in the other subunits of the reaction centre and in many other proteins (see Figure 1.5a; the chevrons make it possible to follow the direction of the chain). They are all 'right-handed' helices, a consequence of the fact that amino acids in proteins have the L configuration. A second recurrent structural pattern, called a *sheet*, appears in the cytochrome subunit (Figure 1.5b). The sheet is constructed of several segments that interact laterally. It is pleated like an accordion. Like most sheets in proteins, it is twisted and rumpled—very different from the sheets on a newly-made bed. Short connecting segments called *turns* or *loops* link the strands of the sheet shown here.

Helices and sheets are like 'Lego' pieces, standard units of structure that proteins can assemble in different ways. They were predicted by Linus Pauling and Robert Corey in 1950 from model building. Upon reading their report, Max Perutz went into his laboratory, and photographed the X-ray diffraction pattern of a horse hair. The keratin in the hair contained the predicted helices! Later he and his colleagues found them to be the basis of

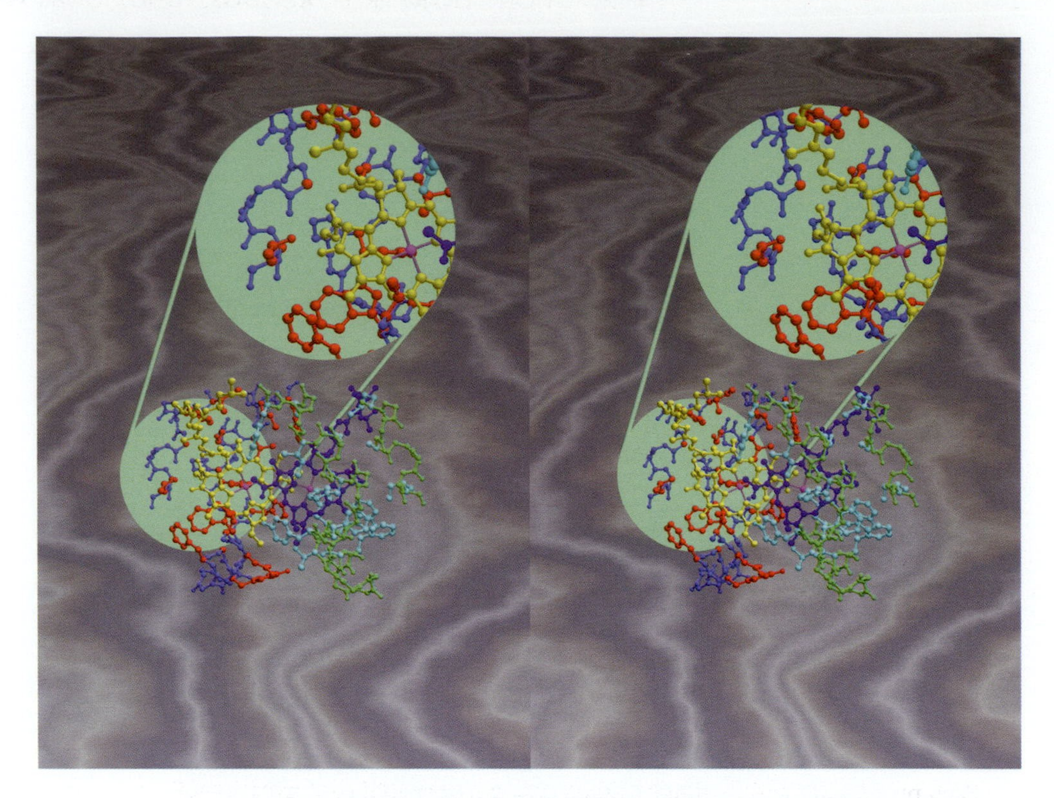

Fig. 1.7 A 'blow-up' of the region around the special pair to show, in a single picture, *both* the immediate environment of one bacteriochlorophyll and its structural context.

the structures of myoglobin and haemoglobin, the first globular protein structures determined. Some other proteins are also built of helices; some are built of sheets; others contain combinations of both. Helices are common in the transmembrane segments of many membrane proteins such as the L and M subunits of the reaction centre.

With no representation of the sidechains, Figures 1.3 and 1.4 give the impression that proteins are light and airy—which we know to be untrue, from Figures 1.1 and 1.2. To address questions about intimate details of intramolecular interactions, we can draw a more complete picture, but need to choose the right representation. Pictures that show the pattern of chemical bonds portray the topology or connectivity of a structure more clearly than all-atom pictures: Kekulé's famous dream, in which he visualized benzene as a ring of six snakes, is an apt example. The price paid is the loss of the space-filling or packing attributes of the assembly of atoms.

As an example, at the heart of the reaction centre is the 'special pair' of bacteriochlorophylls, to which excitation energy is transferred, and which is the site of the ionization—the initial step in electron transfer. Of course the structure and environment of the special pair is crucial, and we want to understand the nature of the interactions between the proteins and these ligands. Figure 1.6 shows the special pair and its environment. The final picture is a complex one, so we build it up in stages. Figure 1.6a shows the special pair on its own. Each bacteriochlorophyll contains a large ring structure—the chlorin ring—and a long phytyl tail. The magnesium atoms at the centres of the chlorin rings are in magenta. Figures 1.6b and 1.6c show the contacts of the chlorin rings with the protein (the contacts to the phytyl tail are not shown). Figure 1.6b shows the sidechains only; Figure 1.6c shows the backbone as well. Note that most of the contacts are made to sidechains; the folding pattern of the backbone places them in proper position to form these contacts.

Figure 1.6 showed the details of the binding site of the special pair. But—important as these details are—we are seeing the carpentry of protein structure and not the architecture: We have lost the global aspects; we have lost the structural context of the special pair, its place in the grand scheme. Is there any way to show both the details and the whole, the *gestalt*? Figure 1.7 gives one approach. It combines, in one picture, a focus on the environment of one bacteriochlorophyll of the special pair with a wider view of its structural context.

This presentation of different pictures of the reaction centre has followed the steps of scientists who dissect and analyse protein structures. The complex is built of individual proteins, each a compact 'intramolecular crystal' of atoms. They fit together snugly to impart a fixed and definite structure to the complex. They present surfaces of different physicochemical character to the aqueous environment and to the membrane.

Each protein has an individual folding pattern, or spatial conformation of its chain. No one structure can do more than hint at the variety of folds in the known proteins, but we have already seen enough to be able to imagine how different spatial combinations of helices and sheets can create a panoply of patterns.

We see also, in the relation between the L and M subunits, the process of evolution at work. Divergence of gene sequences, through mutation, leads to divergence of structure. True, the structures of the L and M subunits are similar; many features of the folding pattern remain intact; the divergence of the structures is conservative. This means that only a relatively small degree of specialization in these structures was required for adequate selective advantage. The close similarity does not imply that these proteins have reached the limits of possible structural divergence. One thing that studies of evolution have taught us is that the cumulative effect of many small changes can be quite grand indeed.

G.E. Hutchinson wrote of 'The ecological theatre and the evolutionary play.' Translated into our terms, the laws of physics and chemistry and the functional properties of life processes provide the theatre—that is, the constraints, the 'rules of the game'. Protein evolution is the performance. Some features of protein structures are what they are because the laws of physics and chemistry would not allow them to be otherwise: if you hire a cast of L amino acids, expect right-handed α-helices. Some of what we see is necessitated by the mechanism of evolution: a protein structure that required a unique amino acid sequence—so that any mutation would destroy it—would not be observed in nature because evolution could never find it. (If any mutation would destroy the structure, it could not have any precursor.) And some of what we see is historical accident. It is by no means easy to sort out these effects. The creative tension among them will pervade and animate our discussions.

Conclusions

In this chapter we have used the reaction centre to illustrate a variety of features of protein structure. We have explored different methods of representing structures, and seen that we must finely tune the contents and complexity of the picture to the question we want to ask.

We have begun to appreciate the hierarchical nature of protein architecture. Amino acids are formed of atoms. Amino acids join up to form linear polypeptide chains. Regions of the polypeptide chain within proteins form paradigm structures such as helices and strands of sheet. Helices and sheets assemble into compact structures. Individual proteins assemble into macromolecular complexes.

With the completion of genome sequencing projects, such as yeast, *Caenorhabditis elegans*, *Drosophila melanogaster* and human, we shall have fully detailed information about the blueprints of organisms. Structural work is proceeding also, and it will not be long before we have complete information about individual protein structures in simple organisms. (We can already assign structures to approximately half of the proteins in *E. coli*.) The next logical step will be to assemble this information into higher-order structures.

For a greater challenge still, recognize that genome sequences and protein structures give us *static* information only, while life is a dynamic process. Goethe wrote that 'Architecture is frozen music.' It is a long way from the architecture of a piece of steak to molecular mechanisms of consciousness and emotions. But, if love be the music of food, play on!

Recommended reading and references

Branden, C.-I. and Tooze, J. (1998). *Introduction to protein structure*, (2nd edn). Garland Publishing Co., New York.

Casey, D. (1992). *Primer on molecular genetics*. U.S. Department of Energy. Available at: http://www.ornl.gov/TechResources/Human_Genome/publicat/primer/intro.html

Deisenhofer, J. and Michel, H. (1989). The photosynthetic reaction center from the purple bacterium *Rhodopseudomonas viridis*. *Science* **245**, 1463–73.

Fersht, A. (1998). *Structure and mechanism in protein science: a guide to enzyme catalysis and protein folding*. W.H. Freeman & Co., New York.

Hodgkin, D.C. and Riley, D.P. (1968). Some ancient history of protein X-ray analysis. In: *Structural chemistry and molecular biology*, (A. Rich and N. Davidson, ed.). W.H. Freeman & Co., San Francisco, pp. 16–28.

Judson, H.F. (1978). *The eighth day of creation: makers of the revolution in biology*. Simon and Schuster, New York.

Lesk, A.M. (2000). The unreasonable effectiveness of mathematics in molecular biology. *The Mathematical Intelligencer* **22**, 28–37.

Matthews, B.M. (1997). Recent transformations in structural biology. *Methods in Enzymology* **276**, 3–10.

Exercises and problems

Exercises

1.1. On a photocopy of Figure 1.1a, draw a 25 Å wide horizontal strip that indicates the position of the membrane. (The height of the entire complex in Figure 1.1a is 132 Å.)

1.2. From an inspection of Figures 1.3 and 1.4, which pairs of subunits of the reaction centre make contacts with each other?

1.3. On a photocopy of Figure 1.3, indicate with highlighter as many helical regions as you can find.

1.4. Referring to Figure 1.4, how many transmembrane helices do you see in the L and M subunits?

1.5. On a photocopy of Figure 1.4, number the helices in one molecule. Then number the helices in the other molecule so that helices related by symmetry have the same numbers.

Problems

1.1. The amino acid sequences of the L and M chains of the *R. viridis* reaction centre have the following optimal alignment:

```
                  10                   20                30        40
L        A-LLSFERKYRVRGGTL-----------IGGDLFDFW--------VGPYFVGFFGVSAIF
         |   ::  : ::||  :           :|  ::::|        :|| ::|  |::|:
M        ADYQTIYTQIQARGPHITVSGEWGDNDRVGKPFYSYWLGKIGDAQIGPIYLGASGIAAFA
                 10        20        30        40        50        60

                 50        60               70        80        90
L        FIFLGVSLIGYAASQGPTWDP-------FAISINPPDLKYGLGAAPLLEGGFWQAITVCA
         |   :: :| :  :    :||       | ::: ||  :||:|  || :||:|    :
M        FGSTAILIILFNMAAEVHFDPLQFFRQFFWLGLYPPKAQYGMGIPPLHDGGWWLMAGLFM
                 70        80        90       100       110       120

             100       110       120       130       140       150
L        LGAFISWMLREVEISRKLGIGWHVPLAFCVPIFMFCVLQVFRPLLLGSWGHAFPYGILSH
           :: || :|    :| ||:| |:   | : ||:   :  ::| |:|||::: |:||  |
M        TLSLGSWWIRVYSRARALGLGTHIAWNFAAAIFFVLCIGCIHPTLVGSWSEGVPFGIWPH
                130       140       150       160       170       180

             160       170       180       190       200
L        LDWVNNFGYQYLNWHYNPGHMSSVSFLFVNAMALGLHGGLILSVANPGDG-------DKV
         :||:: |: :| |::| | |  |::| :  :: :: ||: ||:||  |         |:
M        IDWLTAFSIRYGNFYYCPWHGFSIGFAYGCGLLFAAHGATILAVARFGGDREIEQITDRG
                190       200       210       220       230       240

          210       220       230       240       250       260
L        KTAEHENQYFRDVVGYSIGALSIHRLGLFLASNIFLTGAFGTIASGPFWTRGWPEW---W
          ::|:   ::| ::|::    |:|| | |::  :::::: | : :| | : :|  |
M        TAVERAALFWRWTIGFNATIESVHRWGWFFSLMVMVSASVGILLTGTF-VDNWYLWCVKH
                250       260       270       280        290

             270
L        GWWLDIPFWS--------------   273
         |   | | :
M        GAAPDYPAYLPATPDPASLPGAPK   323
       300       310       320
```

Here the | indicates identical residues in the two sequences, and : indicates a conservative substitution (that is, substitution of one amino acid by another with similar size and charged or polar character). A '-' indicates a deletion. (a) How many positions in this alignment contain identical residues in both sequences? (b) How many positions in this alignment contain identical residues or conservative substitutions in both sequences? (c) What percent of positions (out of a total of 273) contain identical residues? (d) If at least 25% identical residues is taken to be the threshold for expecting two proteins to be homologous—that is, related by evolution, descended from a common ancestor—do the L and M chains of the *R. viridis* appear to be homologous?

1.2. The charged amino acids are arginine (R) and lysine (K) (positively charged) and aspartic acid (D) and glutamic acid (E) (negatively charged). Using a highlighter, indicate all stretches in the sequences of the L and M proteins (see Problem 1.1) that contain at least 8 successive uncharged amino acids. Compare the number you find to the number of transmembrane helices.

1.3. Photocopy Figure 1.2. Determine the areas occupied on the page of the regions above and below the membrane-spanning region (see Exercise 1.1). Count the number of positive residues and the number of negative residues in these regions. Assume that each positively charged residue has a single positive charge and that each negatively charged residue has a single negative charge (in units of e^-). (a) Is there a net positive or negative charge in either of these regions? (b) Estimate the density of the *absolute value* of the charge. (These are only estimates because you can only measure the area of the 'silhouette' of a three-dimensional object.)

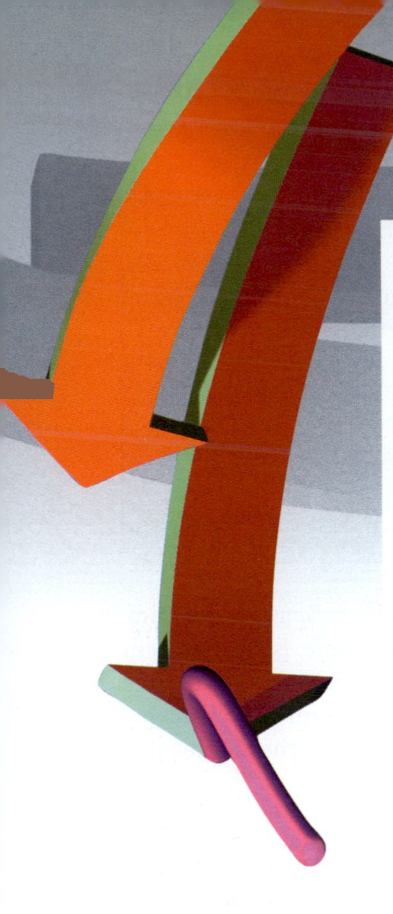

CHAPTER 2

In vivo, in vitro, in silicio

Why study proteins?

In the drama of life on a molecular scale, proteins are where the action is.

Proteins are fascinating molecular devices. They play a variety of roles in life processes: there are structural proteins (viral coat proteins, molecules of the cytoskeleton, epidermal keratin); catalytic proteins (the enzymes); transport and storage proteins (haemoglobin, myoglobin, ferritin); regulatory proteins (including hormones and many proteins that control genetic transcription); and proteins of the immune system and the immunoglobulin superfamily, including proteins involved in cell–cell recognition and signalling.

Until recently, molecular biologists have studied individual proteins, learning their secrets one by one. However, the most profound significance of proteins lies in their collective properties as a class of biomolecules, because these properties include all their potential, as well as their actual, characteristics. Proteins have an underlying chemical unity; they have the ability to organize themselves in three dimensions; and the system that produces them can create inheritable structural variations, conferring the ability to evolve. Recent developments permit study of the entire spectrum of proteins of an organism.

Genomics

Felix Frankfurter wrote that 'the constitution of the United States is most significantly not a document but a stream of history'. This is also a perfect characterization of genomes, which contain not only snapshots of individual organisms but also reflections of the development of life.

Whole-genome sequencing projects have transformed molecular biology. It is not merely that the amount of sequence and structural data is increasing at spectacular rates (exponentially, with a doubling time of ~ 1 year). For the first time we have *complete* information about the genomes of about 50 prokaryotes, and, among higher organisms, Yeast, the nematode

Whole-genome sequencing projects

Genome	Length (base pairs)	Date of completion
Epstein–Barr virus	0.172×10^6	1984
Bacterium (*Escherichia coli*)	4.8×10^6	1997
Yeast (*Saccharomyces cerevisiae*)	14.4×10^6	1996
Nematode worm (*Caenorhabditis elegans*)	100×10^6	1998
Fruit fly (*Drosophila melanogaster*)	180×10^6	1999
Human (*Homo sapiens*)	3300×10^6	2001

worm *Caenorhabditis elegans*, and the fruit fly *Drosophila melangaster* are complete and others are on the way. The human genome is 90 per cent sequenced, and the target date for its completion is 2001. (See http://www.ebi.ac.uk/genomes/info.html) Before, we were limited to studying selected available examples. The new results have laid bare the choices that nature has made.

These developments have spawned new disciplines, indeed, new industries, that draw upon many scientific disciplines, including biology, physics, chemistry, computer science and mathematics. *Bioinformatics* is the systematization of the data into structured and interlinked computer data banks, and the development of tools for access to these data. The results support research and development of methods to draw structural inferences from sequence data and *vice versa*. Computing plays an essential role in this enterprise.

The Internet and World Wide Web are essential for the distribution of data and the development of information-retrieval tools. One goal of this book is to introduce readers to a set of useful web sites, and practice in using them.

The three-dimensional counterpart of genome sequence determination, the solution of all the protein *structures* in an organism, is known as *structural genomics*. When we know the sequences and structures of all macromolecules in a yeast cell, we will have a complete but *static* knowledge of the components of a living object. The next step will be to learn about how the components are assembled and how their individual activities are integrated, called the *proteome project*. R.J. Simpson has compared the determination of genome sequences and protein structures to providing a roster of orchestral instruments; and the understanding of life to the appreciation of their integrated activities in a concert performance.

Although structural data are not as complete as sequence data, detailed atomic structures are now available for over 10 000 proteins. These structures reveal the variety of spatial patterns that nature produces in this family of molecules, which is the focus of this book. They support a number of fascinating and useful scientific endeavours, including:

1. **Interpretations of the mechanisms of function of individual proteins.** The catalytic activity of an enzyme, such as the serine proteinase

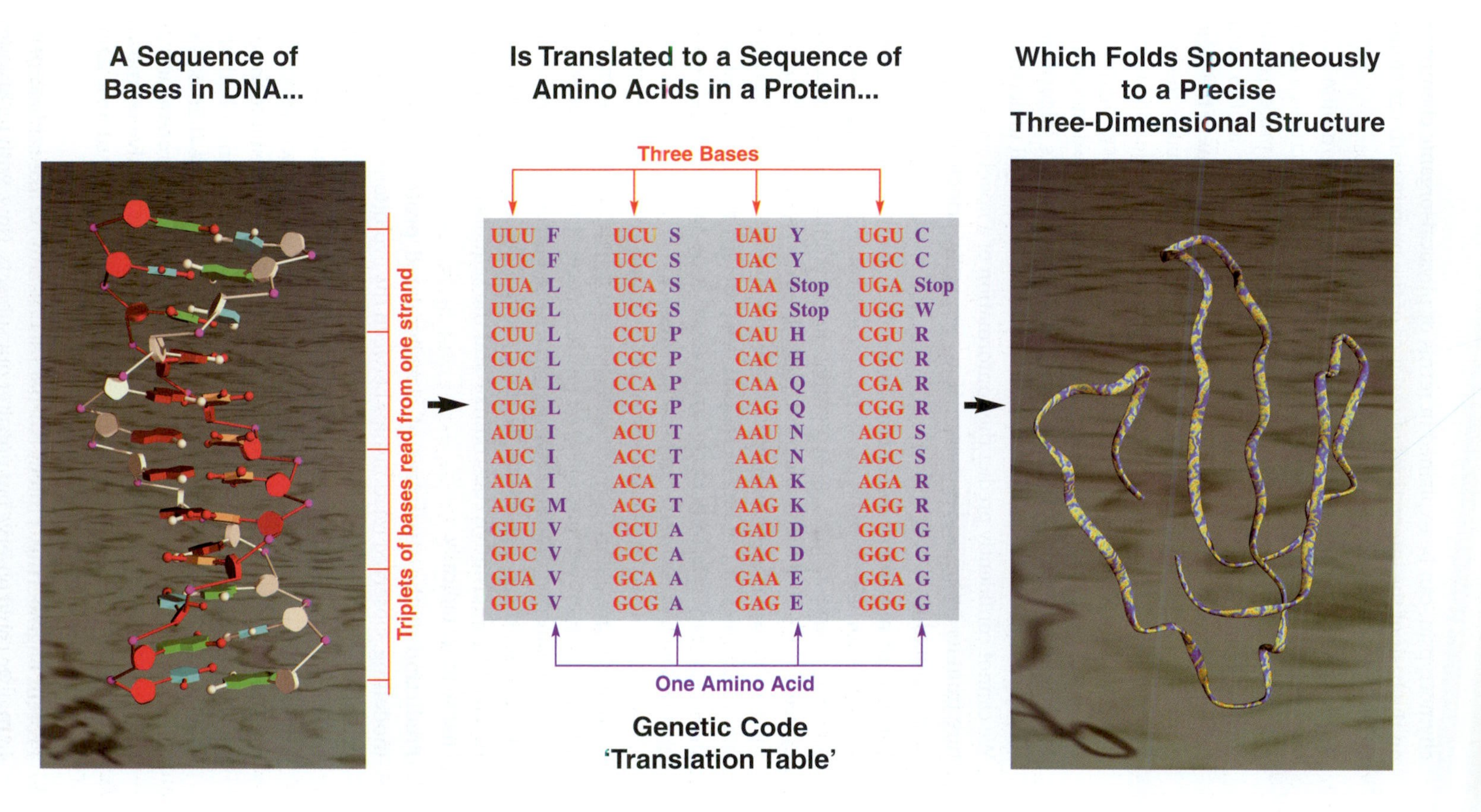

A Sequence of Bases in DNA...
Triplets of bases read from one strand
Is Translated to a Sequence of Amino Acids in a Protein...
Three Bases
UUU F UCU S UAU Y UGU C
UUC F UCC S UAC Y UGC C
UUA L UCA S UAA Stop UGA Stop
UUG L UCG S UAG Stop UGG W
CUU L CCU P CAU H CGU R
CUC L CCC P CAC H CGC R
CUA L CCA P CAA Q CGA R
CUG L CCG P CAG Q CGG R
AUU I ACU T AAU N AGU S
AUC I ACC T AAC N AGC S
AUA I ACA T AAA K AGA R
AUG M ACG T AAG K AGG R
GUU V GCU A GAU D GGU G
GUC V GCC A GAC D GGC G
GUA V GCA A GAA E GGA G
GUG V GCG A GAG E GGG G
One Amino Acid
Genetic Code
'Translation Table'
Which Folds Spontaneously to a Precise Three-Dimensional Structure

Fig. 2.1 Expression of gene sequences as three-dimensional structures of proteins. The three-dimensional structure is implicit in the amino acid sequence. An interesting paradox: The translation of DNA sequences into amino acid sequences is very simple to describe logically; it is specified by the genetic code. The process of folding of the polypeptide chain into a precise three-dimensional structure is very difficult to describe logically. However, translation requires the immensely complicated machinery of the ribosome, tRNAs and associated enzymes; but protein folding occurs spontaneously!

chymotrypsin, can be explained in terms of physical–organic chemistry, on the basis of the interactions of residues of the protein with the atoms around the scissile bond (see Chapter 6).

2. **Approaches to the 'protein folding' problem.** The amino acid sequences of proteins dictate their three-dimensional structures. Under physiological conditions of solvent and temperature, most proteins fold spontaneously to an active native state; that is, their amino acid sequence dictates their three-dimensional structure. In terms of the logic of life, the folding of proteins is the point at which nature makes the leap from the one-dimensional information stored in the genetic code to the three-dimensional world we inhabit: nucleotide sequence enciphers amino acid sequence; amino acid sequence encodes three-dimensional conformation (Figure 2.1). It cannot be said that nature's 'algorithm' relating amino acid sequence to protein structure is yet well understood. We cannot confidently predict the conformation of a novel protein structure from its amino acid sequence. However, many principles of protein architecture have become clear.

3. **Patterns of molecular evolution.** There are several families of protein structures for which we know dozens or even hundreds of amino acid sequences, and at least 20 structures; for example, the globins, the cytochromes c, and the serine proteinases. It has been possible to analyse the mechanism of evolution, in that we can observe the structural and functional roles of the sets of residues that are strongly conserved and those that vary relatively freely, and can describe the structural consequences of changes in the amino acid sequence.

   Mutations, insertions and deletions in the amino acid sequence perturb protein conformation. However, selection can impose constraints on the structure to preserve function. As a result, a 'core' of the structure tends to be well-conserved during evolution. When proteins evolve with changes in function, these constraints on the structure are relaxed—or rather, replaced by alternative constraints—and the sequences and structures change more radically.

4. **Prediction of the structures of closely-related proteins: homology modelling.** Observed relationships between the evolutionary divergence of amino acid sequence and divergence of protein structure in families of homologous proteins make possible the prediction of protein structure from amino acid sequence, in favourable cases. Suppose we determine the DNA sequence of a gene, and discover that its translated amino acid sequence is related to one or more proteins for which we know both the sequence and the structure. As a rule of thumb, if no more than 60% of the residues have changed between the unknown, target, protein, and its nearest relative of known structure, then the nearest relative of known structure will provide a reasonable quantitative model for almost all of the unknown protein.

5. **Protein engineering.** Protein biochemists used to be like astronomers, in that we could observe but not alter our subjects. Now, with techniques

of genetic engineering, it is possible to design and test modifications of known proteins, and to design novel ones. Potential applications include:

(a) Modifications to probe mechanisms of function, such as the method of 'alanine scanning'—the separate conversion of each amino acid in a protein to alanine, or even the *allumwandlung**—the conversion to each of the other 19—to identify residues important for structure or function, or the construction of 'mermaid haemoglobin': a hybrid of mammalian and fish chains to test theories of the Bohr effect (the dependence of oxygen affinity on pH).

(b) Attempts to enhance thermostability, for example, by introducing disulphide or salt bridges, or by optimizing the choice of amino acids that pack the core. Thermostable enzymes would have industrial application as ingredients in laundry products to permit higher wash temperatures.

(c) Clinical applications, such as the transfer of the active site from a rat antibody to a human antibody framework, to produce a molecule that retains therapeutic activity in humans but reduces the side effects arising from the patient's immunological response against the rat protein. This usefully extends the range of possible therapeutic antibodies, because rats but not humans can raise antibodies against many human tumours.

(d) Modifying antibodies to give them catalytic activity. Two features of all enzymes are: the ability to bind substrate specifically, and the juxtaposition of bound substrate with residues that catalyse a chemical change (by stabilizing the transition state of a reaction). Immunoglobulins provide the binding and discrimination; the challenge to the chemist is to introduce the catalytic function.

6. **Drug design.** There are many proteins specific to pathogens that we want to inactivate. Knowing the structure of the AIDS proteinase, or the neuraminidase of influenza virus, it should be possible to design molecules that will bind tightly and specifically to an essential site on these molecules, to interfere with their function.

Now that so many protein structures have been solved, a new scientific speciality has grown up around them. Questions of structure, function and evolution can be addressed by interrogating the molecules themselves, examining the positions of individual atoms. We can analyse, we can classify, and we can in some cases predict. We are thereby gaining some insight into the design principles of this fascinating class of molecules.

Protein structure and conformation

Proteins are polymers containing a *backbone* or *main chain* of repeating units—the peptides—with a *sidechain* attached to each (Figure 2.2). They are

* A chess problem in which in different variations a pawn must change upon promotion to all other possible pieces.

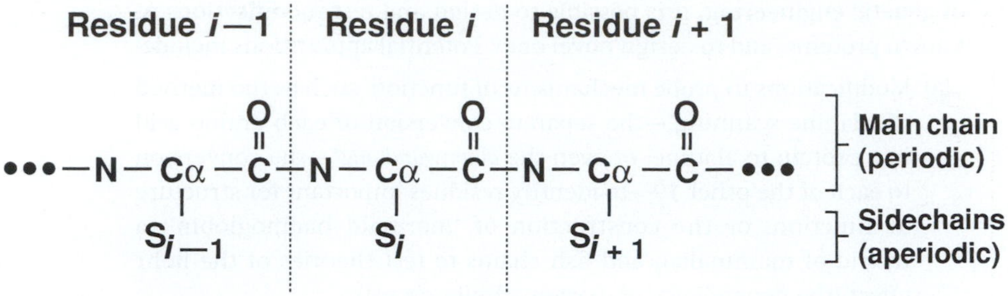

Residue $i-1$ | Residue i | Residue $i+1$

$$\cdots-\mathrm{N}-\mathrm{C}\alpha-\overset{\displaystyle \mathrm{O}}{\overset{\|}{\mathrm{C}}}-\mathrm{N}-\mathrm{C}\alpha-\overset{\displaystyle \mathrm{O}}{\overset{\|}{\mathrm{C}}}-\mathrm{N}-\mathrm{C}\alpha-\overset{\displaystyle \mathrm{O}}{\overset{\|}{\mathrm{C}}}-\cdots$$

Main chain (periodic)

Sidechains (aperiodic): S_{i-1}, S_i, S_{i+1}

Fig. 2.2 Formation of the polypeptide chain. In each protein there is a common repetitive main chain and an individual sequence of sidechains S_{i-1}, S_i, S_{i+1}, etc.

> **Nomenclature**
>
> The atoms in the main chain of each residue are denoted N, Cα, C and O. The sidechain is attached to the Cα. Sidechain atoms are identified by the chemical symbol and successive letters from the Greek alphabet, proceeding out from Cα. Thus the sidechain of methionine has atoms Cα, Cβ, Cγ, Sδ, Cε.

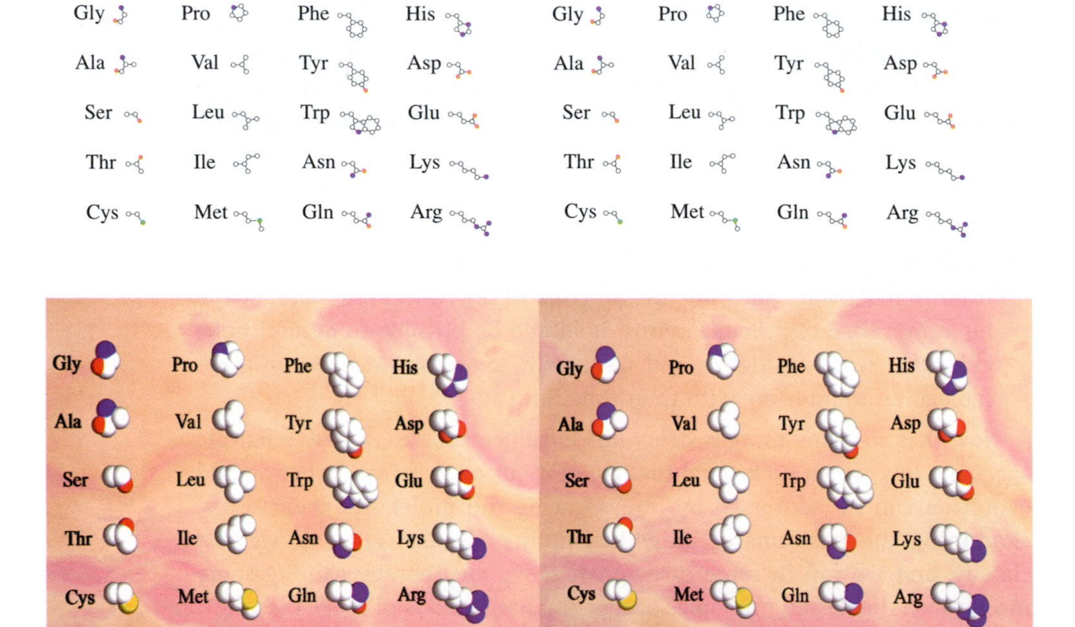

Fig. 2.3 The 20 canonical amino acids, shown in (a) ball-and-stick and (b) space-filling representations. Grey or black, carbon; red, nitrogen; blue, oxygen; yellow, sulphur. This is the cast of characters that play all the different roles in different proteins.

The amino acids and their three-letter and one-letter codes

Glycine	**Gly**	**G**	Valine	Val	V	Tyrosine*	Tyr	Y	Histidine*	His	H
Alanine	**Ala**	**A**	Isoleucine	Ile	I	Methionine	Met	M	Aspartic acid*	Asp	D
Serine*	**Ser**	**S**	Leucine	Leu	L	Tryptophan*	Trp	W	Glutamic acid*	Glu	E
Threonine*	**Thr**	**T**	Proline	Pro	P	Asparagine*	Asn	N	Lysine*	Lys	K
Cysteine	Cys	C	Phenylalanine	Phe	F	Glutamine*	Gln	Q	Arginine*	Arg	R

Black: small Green: medium-sized and large hydrophobic Red: acidic, negatively charged

***Can form sidechain hydrogen bond or salt bridge** Magenta: polar Blue: basic, positively charged

analogous to strings of Christmas tree lights—the wire is like the repetitive backbone, and the order of colours of the bulbs variable. The unique sequence of the sidechains gives each protein its individual characteristics.

The amino acids

Natural proteins contain a basic repertoire of 20 amino acids (Figure 2.3 and Box at top of this page). All except glycine have an asymmetric carbon, and proteins contain the L isomer. Proline is special because its sidechain is linked to the backbone by ring closure. Some proteins do contain amino acids outside the canonical set of 20, but these are produced by chemical modification after the protein is synthesized, or by introduction of a selenocysteine during translation, as in glutathione peroxidase. Ions, small organic ligands and even water molecules are also integral parts of many protein structures.

The amino acids vary in size, hydrogen-bonding potential and charge

The 20 sidechains vary in their physico-chemical properties. These characteristics determine how serious a perturbation of a structure occurs when, as proteins evolve, one amino acid replaces another.

Some sidechains are electrically neutral: because of the thermodynamically unfavourable interaction of hydrocarbons with water, they are called 'hydrophobic' residues (see Box: The hydrophobic effect). Other sidechains are polar: asparagine and glutamine contain amide groups; serine, threonine and tyrosine contain hydroxyl groups. Polar sidechains, like main chain

The hydrophobic effect

The *hydrophobicity* of an amino acid is a measure of the thermodynamic interaction between the sidechain and water. Hydrocarbon sidechains such as those of leucine and phenylalanine interact unfavourably with water, as does the oil in salad dressing. Just as oil–water mixtures separate spontaneously into two phases, there is a tendency for the hydrophobic sidechains to sequester themselves in the interior of a protein, away from contact with water. This *hydrophobic effect* provides an important component of the driving force for protein folding. Its significance was predicted by W.J. Kauzmann shortly before the first X-ray crystal structures of proteins confirmed it.

Amino acid hydrophobicity scale*

2.25	Trp
1.80	Ile
1.79	Phe
1.70	Leu
1.54	Cys
1.23	Met
1.22	Val
0.96	Tyr
0.72	Pro
0.31	Ala
0.26	Thr
0.13	His
0.00	Gly
−0.04	Ser
−0.22	Gln
−0.60	Asn
−0.64	Glu
−0.77	Asp
−0.99	Lys
−1.01	Arg

* Fauchère, J. and Pliška, V. (1983). Hydrophobic parameters π of amino acid side chains from the partitioning of N-acetyl-amino-acid amides. *Eur. J. Med. Chem.* **18**, 369–75.

peptides, can participate in hydrogen bonding. Other sidechains are charged: aspartic acid and glutamic acid are negatively charged; lysine and arginine are positively charged. The charged atoms occur at or near the ends of the relatively long and flexible sidechains; the atoms proximal to the backbone are non-polar. Two sidechains with positive and negative charge can approach each other in space to form a 'salt bridge'.

Protein folding

Any possible conformation of the polypeptide chain of a protein places different sets of residues in proximity. The interactions of the sidechains and main chain, with one another and with the solvent and with ligands, determine the energy of the conformation. Proteins have evolved so that one folding pattern of the chain produces a set of interactions that is significantly more favourable than all others. This corresponds to the native state.

Proof that protein structure is dictated by the amino acid sequence alone is based on experiments first carried out by C. Anfinsen, who showed that the denaturation of ribonuclease—the breakup of the native structure by heat or urea—was reversible. If the denatured molecules are returned to normal conditions of temperature and solvent, both structure and enzymatic activity return.

Formation of the native state is a *global* property of the protein. In most cases, the entire protein (or at least a large part) is necessary for stability. This is because many of the stabilizing interactions involve parts of the protein that are very distant in the polypeptide chain, but brought into spatial proximity by the folding.

Proteins are only marginally stable, and achieve stability only within narrow ranges of conditions of solvent and temperature. Overstep these boundaries, and proteins lose their definite compact structure, and even their helices and sheets, and take up states with disorder in the backbone conformation and few if any specific interactions among residues: the pieces of the jigsaw puzzle have been pulled apart and deformed. The free energy of stabilization of proteins under ordinary conditions is typically only about 20–60 $kJ \cdot mol^{-1}$ (5–$15\ kcal \cdot mol^{-1}$). The various homeostatic mechanisms, by which the internal environment of our bodies is maintained at a relatively constant temperature, salt concentration and acidity, are essential to health if only because proteins lose the effectiveness of their function if any of these conditions vary beyond limits which are in many cases quite narrow.

Protein folding pathways

It is not enough for the native conformation of a protein to be stable; the protein must be able to find it, in a short time, starting from a denatured state characterized by a random population of unfolded conformations, A 'folding pathway' is built into the structure by natural selection, just as the native state is. The nature of the pathway is elusive, because it is rare that it is possible to trap intermediates.

Protein structures depend on a variety of chemical forces for their stability and for their affinity and specificity for ligands

1. **Covalent and coordinate chemical bonds.** Many proteins contain covalent chemical bonds in addition to those of the polypeptide backbone and the sidechains. Disulphide bridges between cysteine residues are quite common. Figure 2.4 shows the small protein crambin, which contains three disulphide bridges. Disulphide bridges can also link different polypeptide chains, as in insulin (Figure 2.5) and in immunoglobulins.

   Metal ions are integral parts of the structures of many proteins. Figure 2.5 shows the off-axial zinc-binding site of pig insulin. Figures 2.6a and 2.6b show a 'Zinc-finger' protein, Zif268, in which the ion is coordinated directly by sidechains. In other cases, the metal is not bound directly to the protein, but is part of a larger ligand. The Mg^{2+} ion of bacteriochlorophyll, in the reaction centre illustrated in Chapter 1, is an example. Another is sperm whale myoglobin, binding an iron-containing haem group (Figure 2.6c). Cytochrome c is another haem protein but has a very different fold, and binds haem in a different way (Figure 2.6d).

2. **Hydrogen bonding.** Polar atoms in proteins make hydrogen bonds to water in the unfolded state. In the folded state, the hydrogen-bonding potential of atoms buried in the interior of the protein must somehow be

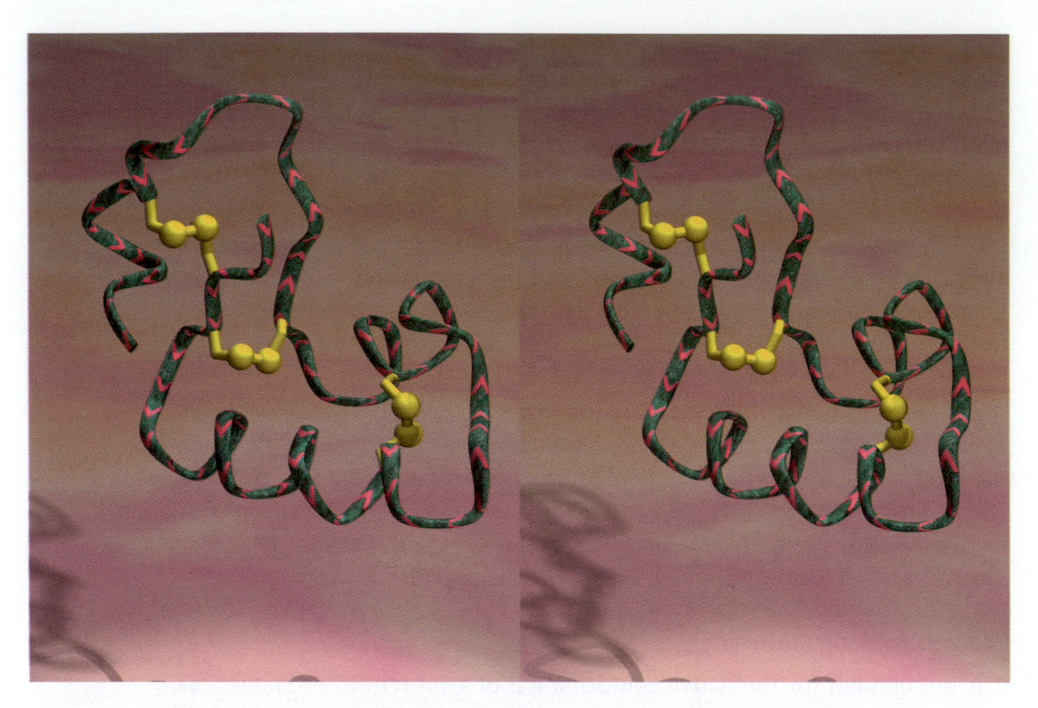

Fig. 2.4 Crambin [1CRN]. The double-lollipops represent disulphide bridges.

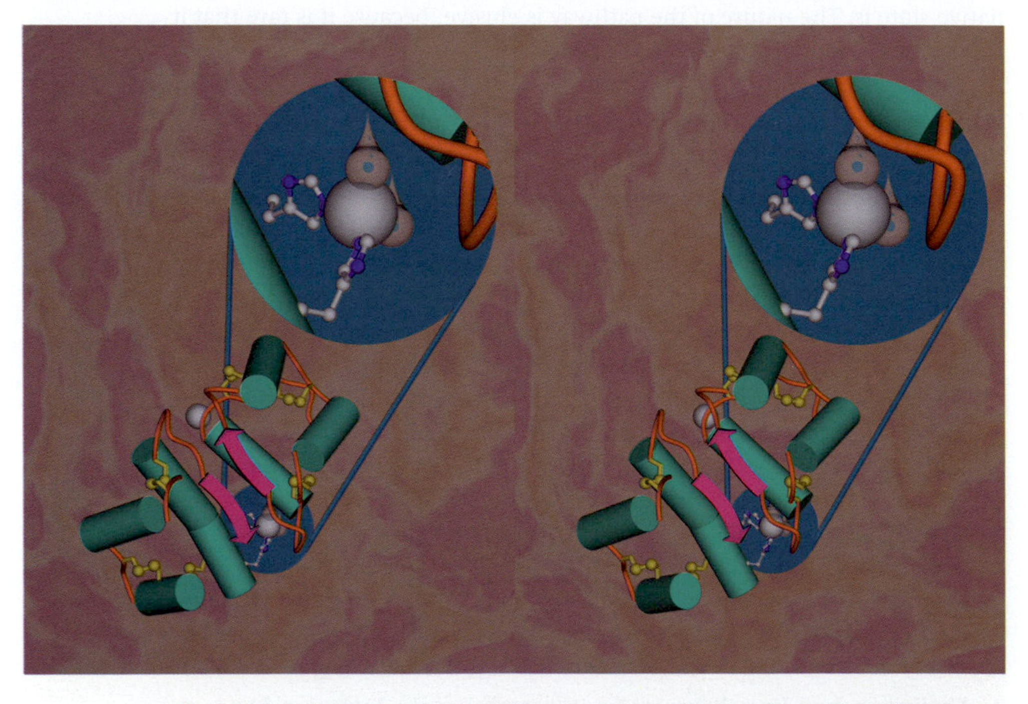

Fig. 2.5 The off-axial zinc-binding site in the 4-Zn insulin dimer [1ZNI]. The off-axial zinc-binding site is 'blown up'. This zinc ion is coordinated by two histidine residues and two water molecules, represented somewhat fancifully by teardrops.

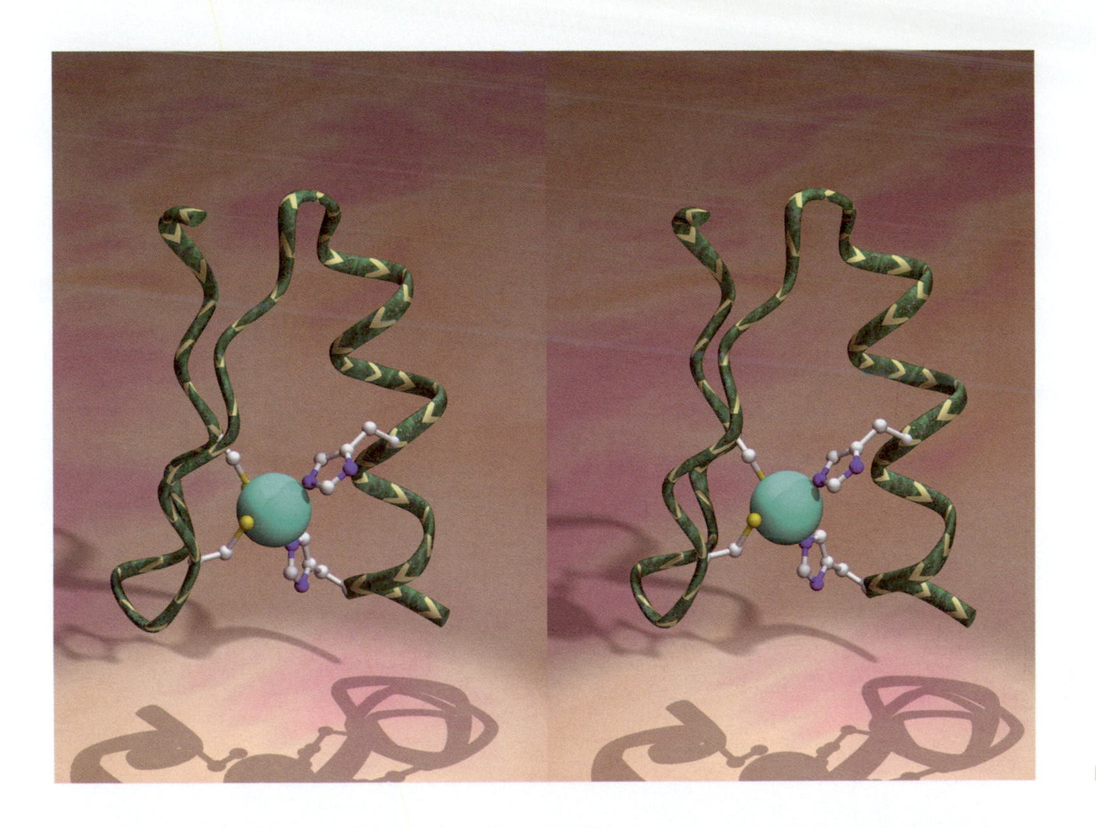

Fig. 2.6a

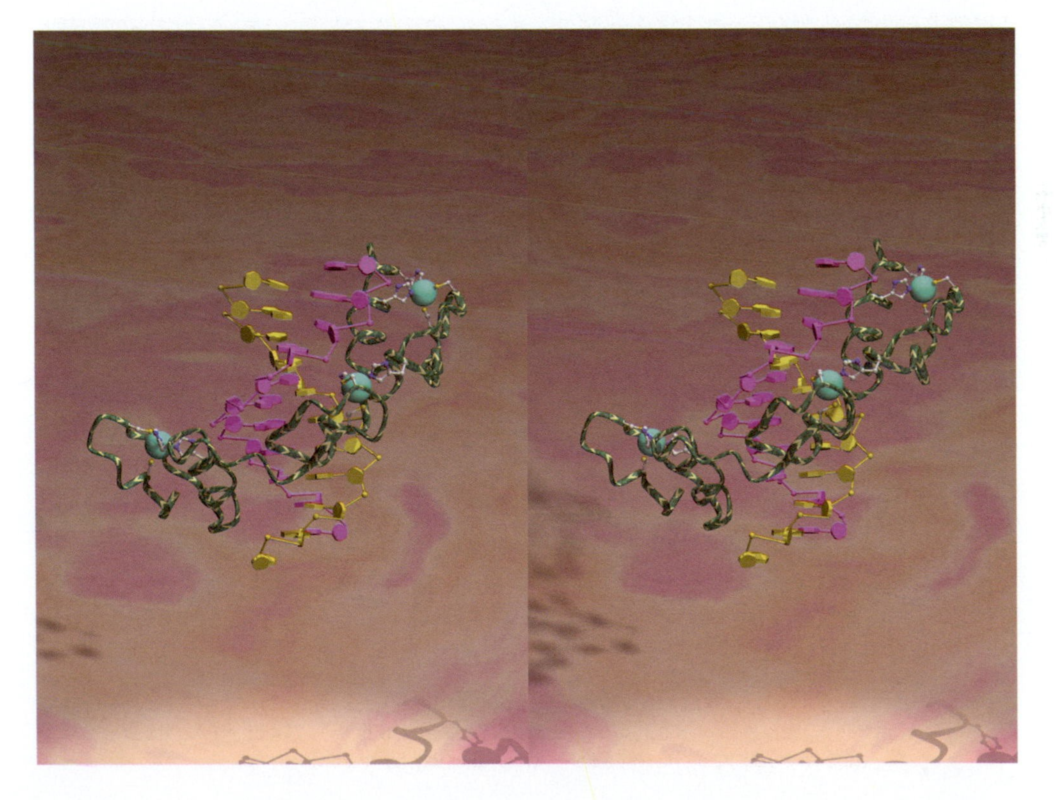

Fig. 2.6b

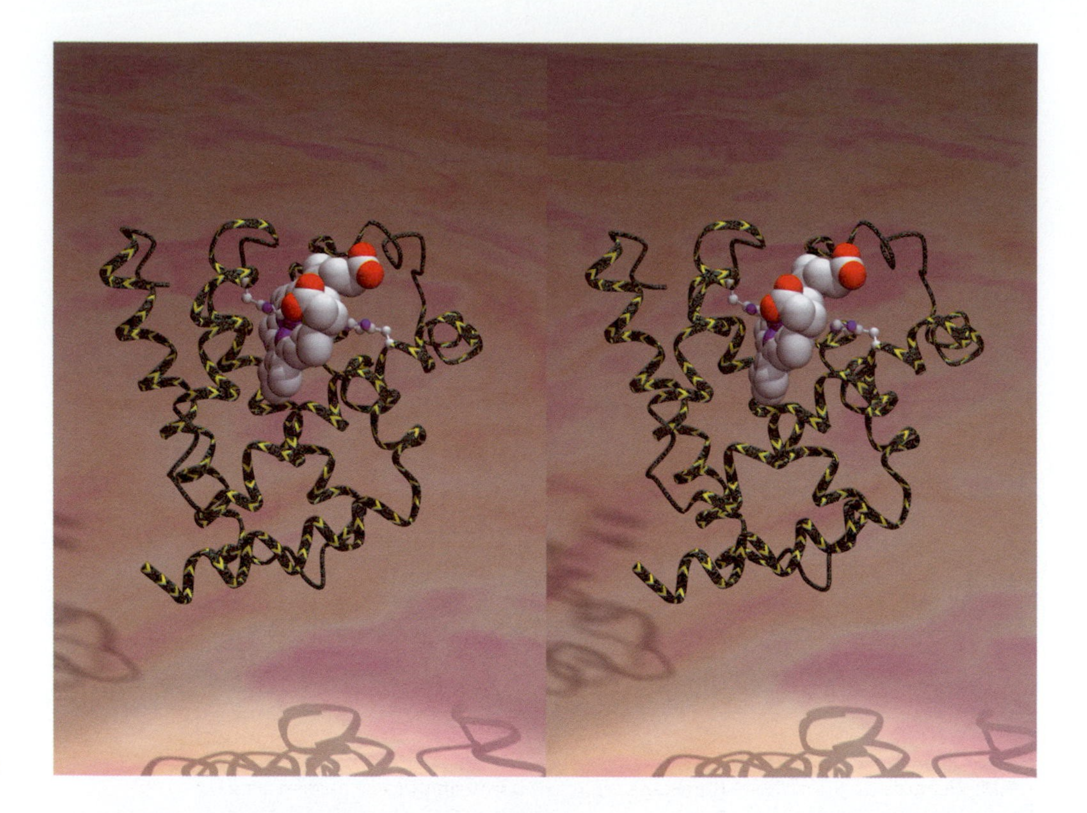

Fig. 2.6c

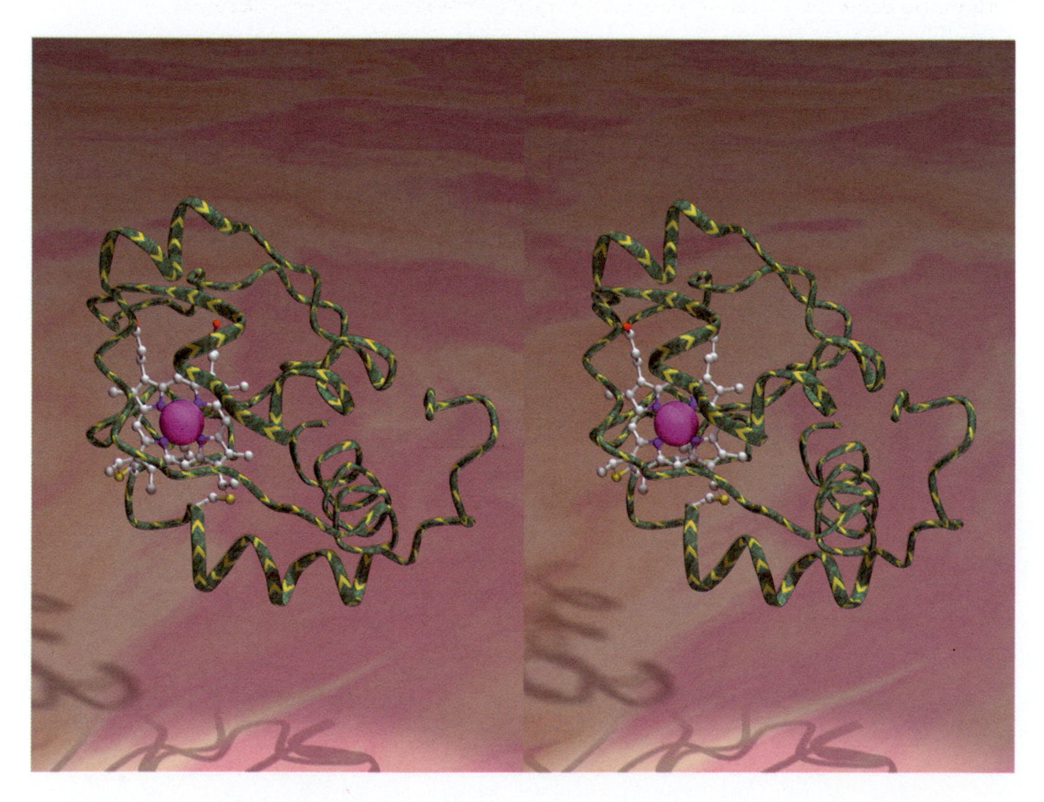

Fig. 2.6d

Fig. 2.6 (Facing and previous page) Some proteins contain metal ions as integral parts of their structure, either bound directly to amino acids or within organic cofactors. (a) One domain from 'zinc-finger' peptide Zif268 [1AAY]; (b) three zinc-finger domains from Zif268, binding DNA; haem-containing proteins: (c) sperm whale myoglobin [1MBD]; (d) rice cytochrome c [1CCR].

satisfied. The main chain, containing peptide groups, *must* pass through the interior, and some polar sidechains are also buried. They thereby lose their interactions with water. To recover the energy, buried polar atoms form protein–protein hydrogen bonds. The standard secondary structures—helices and sheets—achieve the formation of hydrogen bonds by the main chain atoms.

Hydrogen bonds also contribute to the binding of cofactors and substrates. Figure 2.7 shows the binding of NAD to a domain of horse liver alcohol dehydrogenase. NH groups protruding from the end of one of the helices form hydrogen bonds to phosphate groups of the cofactor.

3. **The hydrophobic effect.** For proteins that take on their native states in aqueous environments, hydrophobic residues congregate in the interior and charged residues are on the surface. The *accessible surface area* of the

Accessible and buried surface area

F.M. Richards first developed the geometric approach to analysis of protein structures. He defined the area of molecular surface accessible to a water molecule (modelled as a sphere 1.4 Å in radius) and the packing density of the atoms in protein interiors, and, with B. Lee, wrote the first computer programs to calculate them. C. Chothia extensively applied accessible surface area calculations to rationalize the hydrophobic contribution to the thermodynamics of protein folding and interactions. Observed regularities include:

1. The basic calibration. Each $Å^2$ of buried surface area contributes 105 J (25 cal) of free energy of stabilization.
2. The accessible surface area (A.S.A.) of monomeric proteins of up to about 300 residues varies as the $\frac{2}{3}$ power of the molecular weight M: A.S.A. = $11.1\ M^{\frac{2}{3}}$.
3. The formation of oligomeric proteins from monomers buries an additional 1000–5000 $Å^2$ of surface. Lower values characterize proteins for which the monomer structure is stable in isolation; higher values characterize proteins in which association must stabilize the structure of the monomers as well as the complex.
4. Nature of the buried area. The average solvent-accessible surface of monomeric proteins—the protein *exterior*—is ~58% non-polar (hydrophobic), ~29% polar, and ~13% charged. The average buried surface of monomeric proteins—the protein *interior*—is ~60% non-polar (hydrophobic), ~33% polar, and ~7% charged. Many people who have read or heard about the hydrophobic effect expect the large *buried* hydrophobic surface but are surprised at how large the *exterior* hydrophobic area is. In fact, the main 'take-home message' about the difference between the surface and the interior is that proteins almost never bury charged groups.

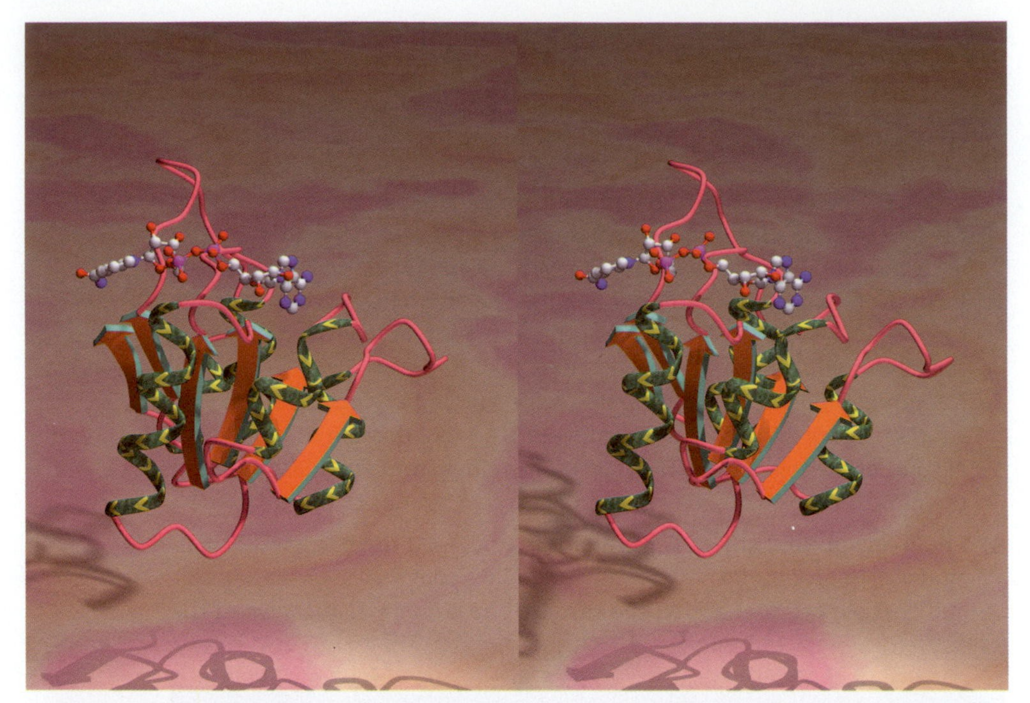

Fig. 2.7a

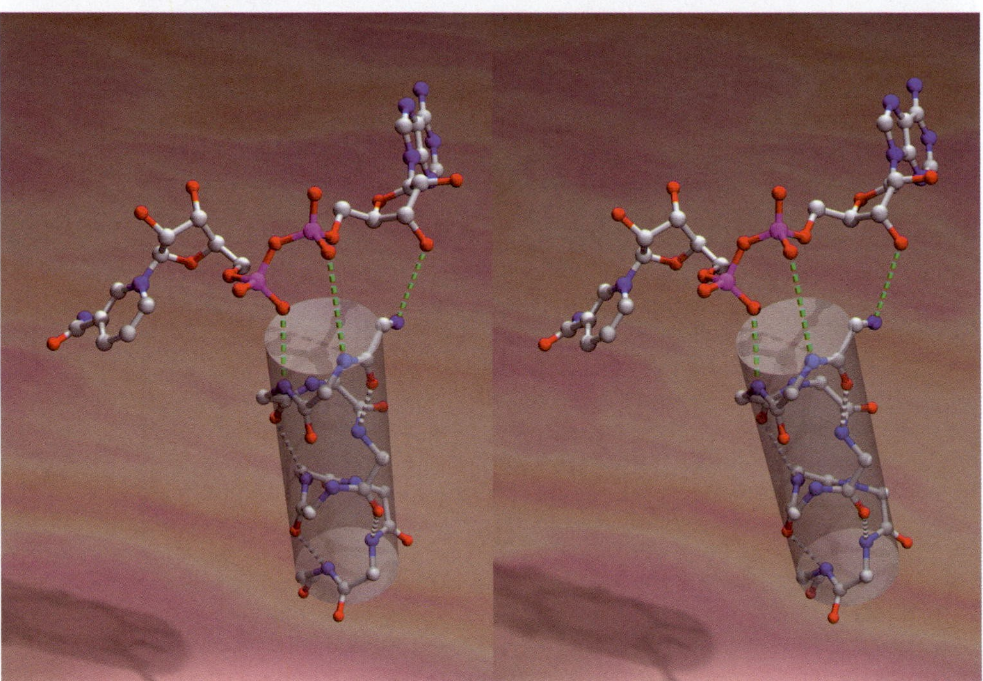

Fig. 2.7b

Fig. 2.7 NAD binding by horse liver alcohol dehydrogenase [6ADH]. (a) NAD binds to a domain containing a central parallel β-sheet flanked by helices. An 'NAD-binding domain' with this folding pattern appears in many dehydrogenases and in other proteins that bind related molecules. (b) The region around the first helix, showing hydrogen bonding between the N–H groups in the last turn of the helix and the phosphate oxygens.

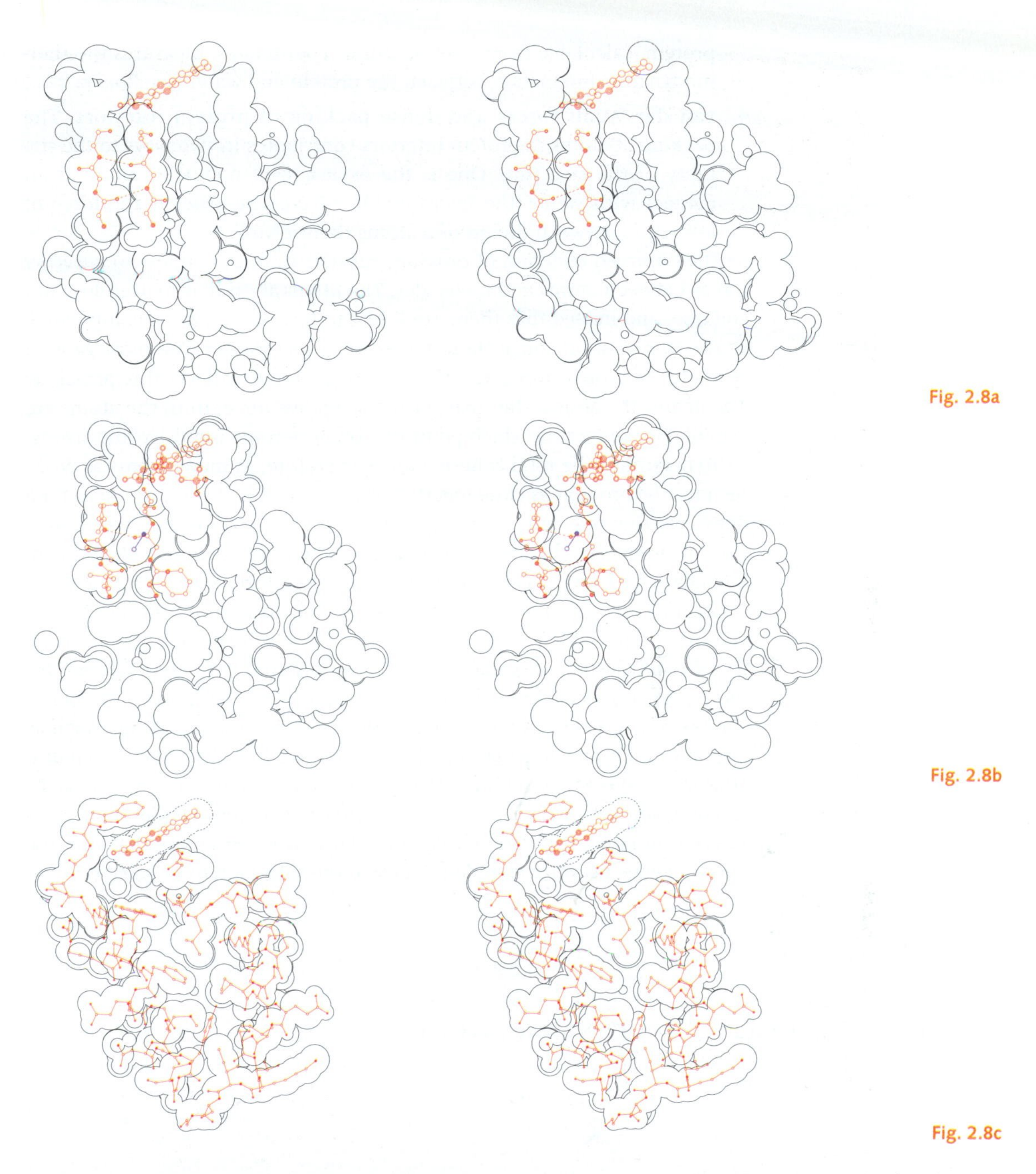

Fig. 2.8 Serial sections through the β-sheet of flavodoxin [5NLL] show the dense packing of protein interiors. Flavodoxin contains a five-stranded β-sheet with two helices packed against each side, and the prosthetic group FMN. Each drawing shows three serial sections through van der Waals envelopes of the atoms. The sections are cut 1 Å apart. The prosthetic group is shown in red, at the top of the picture. (a) Sections passing through the mean plane of the sheet. The main chains of three residues from each of two adjacent strands of sheet are also shown in red. (b) Sections passing through the sidechains above the sheet (in this orientation). The main chain and sidechains of three residues from each of two adjacent strands of sheet are also shown in red. The distal atoms of a methionine, from a helix packing against the sheet (shown in blue), insert between sidechains of residues on the two adjacent strands. (c) Sections passing through two helices packed against the sheet (below the sheet in this orientation). In this drawing only the flavin ring of the FMN is shown.

protein, calculable from a set of atomic coordinates, measures the thermodynamic interaction between the protein and water (see Box, p. 27).

4. **van der Waals forces and dense packing of protein interiors.** The packing of atoms in protein interiors contributes in two ways to the stability of the structure. One is the exclusion of non-polar atoms from contact with water (the hydrophobic effect). The other is the force of attraction between the protein atoms themselves.

The observed cohesion of ordinary substances bears witness to *attractive* forces between atoms and molecules. The observation that matter does not collapse, and indeed that there are limits to how far it can be compressed, shows that at short range these forces must be *repulsive.* The most general type of interatomic force, the van der Waals force, reflects this principle: The nearer the atoms, the stronger the attractive force, until the atoms are actually 'in contact', at which point the forces become repulsive and strong.

To maximize the total cohesive force, therefore, as many atoms as possible must be brought as close together as possible. It is the requirement for a dense packing that imposes a requirement for *structure* on the protein interior. It produces a jigsaw-puzzle-like fit of interfaces between elements of secondary structure packed together in protein interiors (see Figure 2.8).

Such a complementarity of opposing surfaces is also responsible for the specificity with which many enzymes bind their substrates. Chymotrypsin, trypsin and elastase are related proteinases with different specificities: chymotrypsin cleaves adjacent to large flat non-polar sidechains, such as phenylalanine; trypsin cleaves next to long positively charged sidechains, such as arginine and lysine; and elastase cleaves next to small non-polar sidechains. How is this specificity achieved? Each proteinase contains a crevice near its catalytic site into which the sidechain adjacent to the scissile bond must insert. The size, shape and charge distribution of the atoms that line the pocket create a surface complementary to the sidechain to be selected.

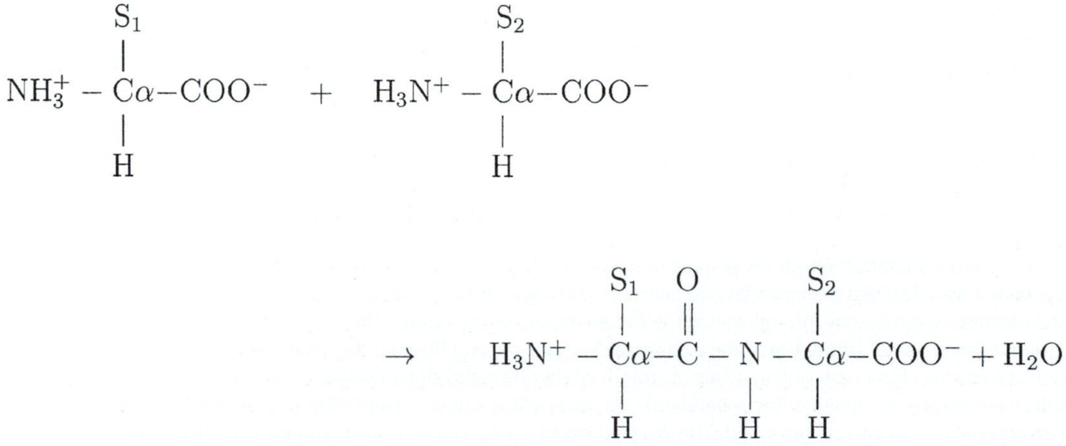

$$\begin{array}{ccc} & S_1 & \\ & | & \\ NH_3^+ - & C\alpha - COO^- & \\ & | & \\ & H & \end{array} \quad + \quad \begin{array}{ccc} & S_2 & \\ & | & \\ H_3N^+ - & C\alpha - COO^- & \\ & | & \\ & H & \end{array}$$

$$\rightarrow \quad \begin{array}{c} \\ \\ H_3N^+ - \\ \\ \\ \end{array}\begin{array}{c} S_1 \\ | \\ C\alpha \\ | \\ H \end{array}\begin{array}{c} O \\ \| \\ -C- \\ \\ \\ \end{array}\begin{array}{c} \\ \\ N \\ | \\ H \end{array}\begin{array}{c} S_2 \\ | \\ - C\alpha \\ | \\ H \end{array}\begin{array}{c} \\ \\ -COO^- + H_2O \\ \\ \\ \end{array}$$

Fig. 2.9 Formation of the peptide bond.

Conformation of the polypeptide chain

The condensation of amino acids produces a polypeptide chain, with the backbone atoms linked through the peptide bond (Figure 2.9).

The folding pattern of the polypeptide chain can be described in terms of angles of internal rotation around the bonds in the main chain (Figure 2.10). The bonds in the polypeptide backbone between the N and Cα, and between the Cα and C, are single bonds. Internal rotations around these bonds are

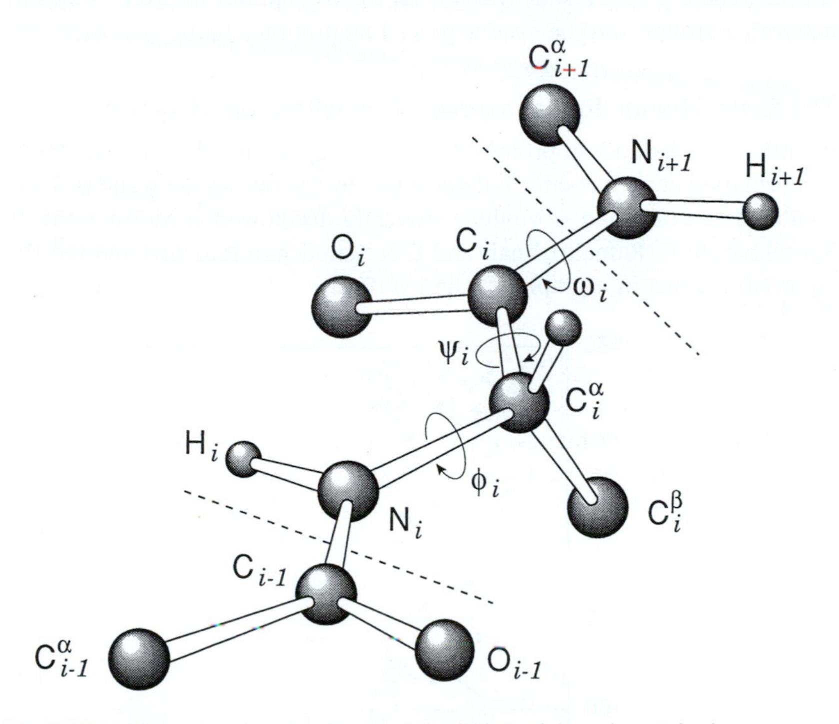

Fig. 2.10 Conformational angles describing the folding of the polypeptide chain. Geometrically, the main chain of a protein is a succession of points in space: N–Cα–C–N–Cα–C (the carbonyl oxygens are not in the main chain itself although they are often loosely accorded 'honorary' main chain status). To a good approximation, the bond lengths and angles—the distances between every two successive points, and the angles between every three successive points—are constant. The degrees of freedom of the chain then involve four successive atoms, and consist of rotations in which the first three atoms are held fixed, and the fourth atom is rotated around the bond linking the second and third. (An ordinary paper clip easily unwinds to give a chain with four vertices, and is a useful object with which to practise.) By convention, the *cis* conformation is 0°, and a positive angle corresponds to looking down the bond between atoms 2 and 3, and rotating the distant atom in a clockwise direction.

The entire conformation of the protein can be described by these angles of internal rotation. Each set of four successive atoms in the mainchain defines an angle. In each residue *i* (except for the N- and C-termini) the angle ϕ_i is the angle defined by atoms Cα (of residue *i* – 1)–N–Cα–C, and the angle ψ_i is the angle defined by atoms N–Cα–C–N (of residue *i* + 1). ω_i is the angle around the peptide bond itself, defined by the atoms Cα–C–N (of residue *i* + 1)–Cα (of residue *i* + 1). ω is restricted to be close to 180° (*trans*) or 0° (*cis*). Conformational angles of sidechains are called χ_1, χ_2, . . ., proceeding out along the sidechain. For example, χ_2 of a methionine residue is the dihedral angle defined by atoms Cα, Cβ, Cγ, Sδ. (See http://www.chem.qmw.ac.uk/iupac/misc/biop.html for full details.)

not restricted by the electronic structure of the bond, but only by possible steric collisions in the conformations produced. In contrast, the peptide bond itself has partial double-bond character, with restricted internal rotation. This was first recognized by L. Pauling. The peptide group occurs in *cis* and *trans* forms, with the *trans* isomer the more stable. For all the amino acids except proline, the energy difference between *cis* and *trans* states is very large. For proline, the energy difference is only about 5 $kJ{\cdot}mol^{-1}$ (1.2 $kcal{\cdot}mol^{-1}$). As a result, virtually all the *cis* peptides in proteins appear between a proline and the residue preceding it in the chain.

The Sasisekharan–Ramakrishnan–Ramachandran diagram

Because most residues in proteins have *trans* peptide bonds, the main chain conformation of each residue is determined by the two angles ϕ and ψ. Some combinations of ϕ and ψ produce sterically disallowed conformations. V. Sasisekharan, C. Ramakrishnan and G.N. Ramachandran first plotted the 'allowed' regions in a graph of ϕ and ψ (Figure 2.11). There are two main

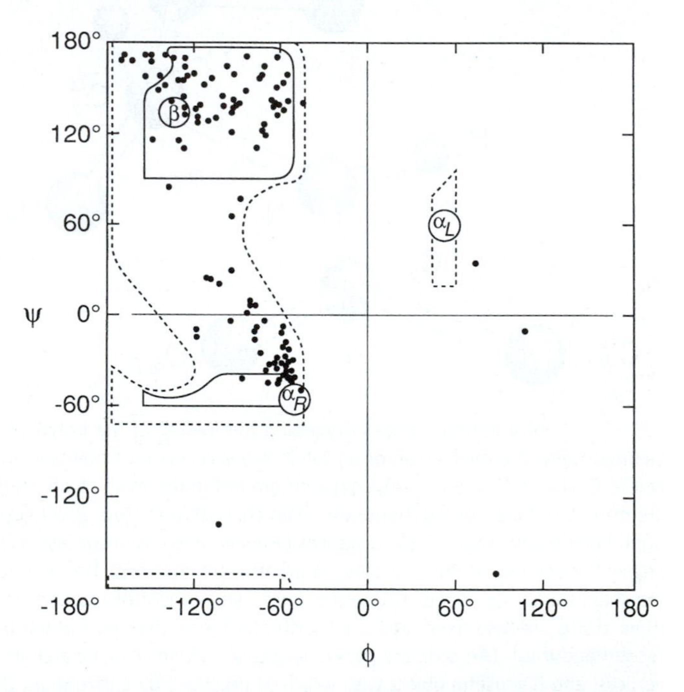

Fig. 2.11 Sasisekharan–Ramakrishnan–Ramachandran diagram: The regions of sterically 'allowed' values of ϕ and ψ, assuming the *trans* conformation: $\omega = 180°$. Broken lines enclose regions estimated to include the maximum tolerable limits of steric strain. The diagram indicates the conformations of three recurrent conformational patterns: the right handed α-helix, α_R; the β-strand, β; and the left-handed α-helix, α_L. (Residues in the α_L conformation in proteins are primarily glycines, with a few asparagines.) Charts of this type were first developed by V. Sasisekharan, C. Ramakrishnan and G.N. Ramachandran. C. Schellman found similar results independently, at about the same time. Also plotted on this diagram is the distribution of conformational angles for the residues in a typical protein solved at high-resolution and well-refined: ribonuclease A [7RSA]. Notice that most but not all of the residues have conformations within the allowed regions.

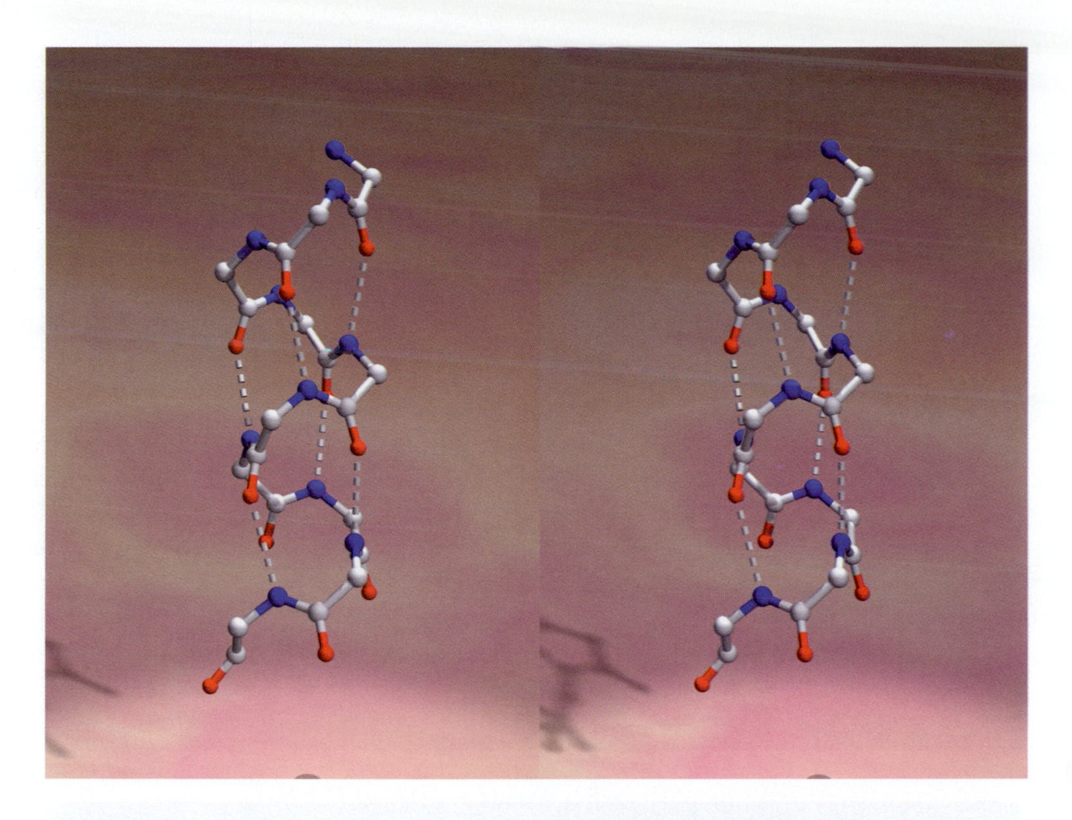

Fig. 2.12a

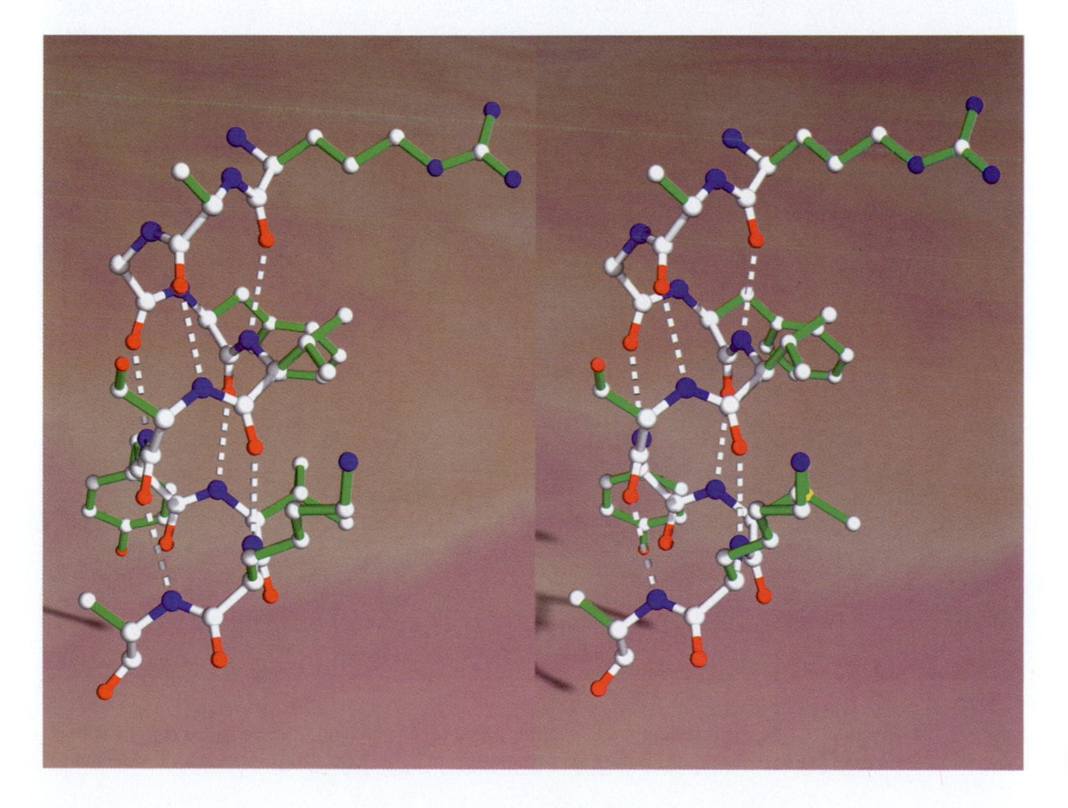

Fig. 2.12b

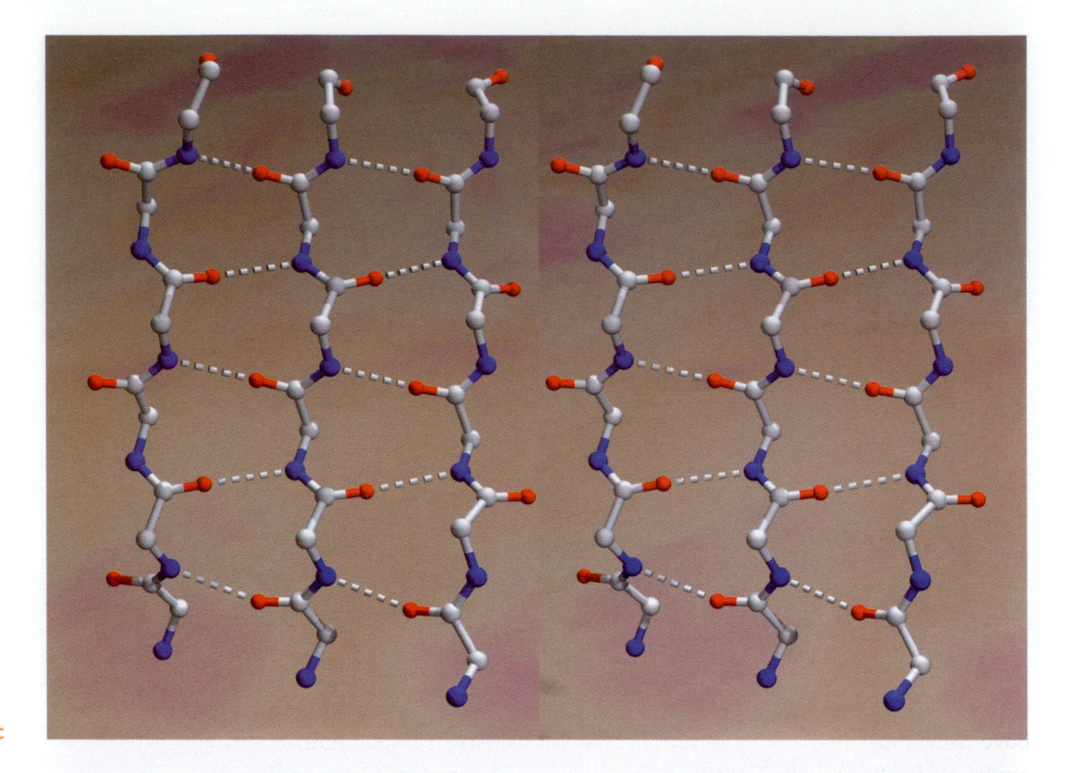

Fig. 2.12c

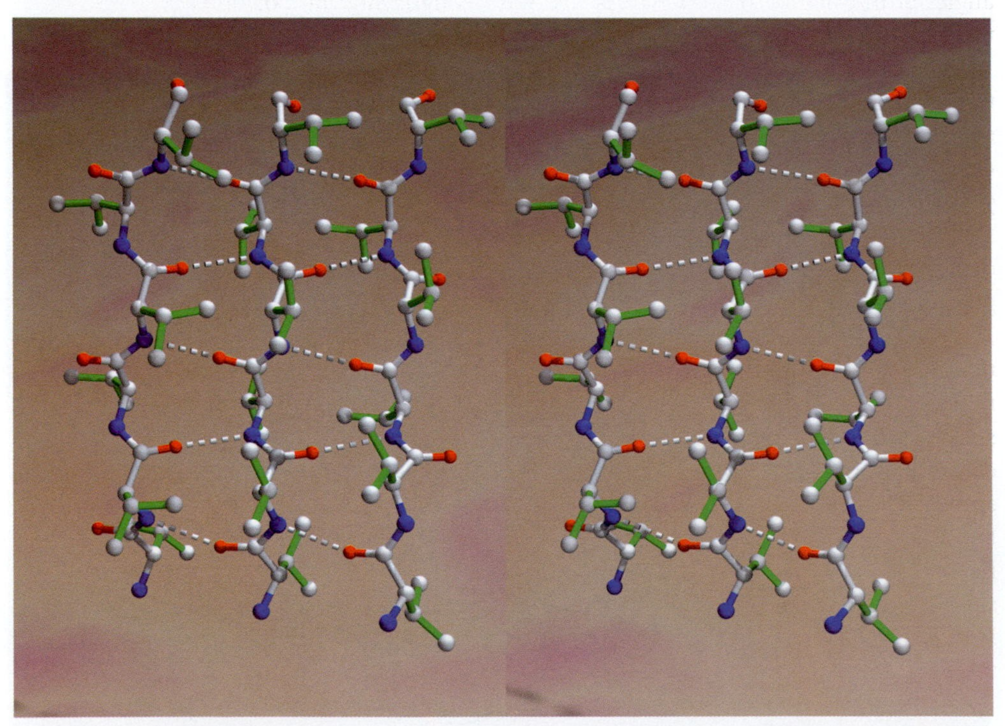

Fig. 2.12d

Fig. 2.12 (a, b Preceding page, c, d this page) (a) Main chain of α-helix. (b) α-helix with sidechains. (c) Main chain of three β-strands forming a sheet. (d) β-strands with sidechains.

allowed regions, one around $\phi = -57°$, $\psi = -47°$ (denoted α_R) and the other around $\phi = -125°$, $\psi = +125°$ (denoted β) with a 'neck' between them. The mirror image of the α_R conformation, denoted α_L, is allowed equally for glycine residues only, as it must be because glycine is achiral.

The two major allowed regions correspond to the two major types of secondary structure: helix and sheet. A stretch of residues, all with conformations in the α_R region, would form a helix (Figures 2.12a and 2.12b). The helix is right-handed, like the threads of an ordinary bolt. (A helix formed by a stretch of residues in the α_L region would form the corresponding left-handed helix.) In the β region, the chain is nearly fully extended. A stretch of residues, all with conformations in the β region, would form a strand as found in a sheet (Figure 2.12c and 2.12d).

It is probably no coincidence that the same conformations that correspond to low-energy states of individual residues also permit the formation of structures with extensive mainchain hydrogen bonding. The two effects can thereby cooperate to lower the energy of the native state.

Sidechain conformation

Sidechain conformations are also described by angles of internal rotation, denoted χ_1 up to χ_5, working out along the sidechain. Different sidechains have different numbers of degrees of freedom. An Arg sidechain has five angles of internal rotation, whereas Gly and Ala (ignoring the positions of the methyl hydrogens) have none at all.

The conformations of any sidechain corresponding to different combinations of values of the χ angles are called *rotamers.* Is there any equivalent for sidechains of the Sasisekharan–Ramakrishnan–Ramachandran plot, that describes their allowed or preferred conformations? In general, the values of χ_1—the rotation around the Cα–Cβ bond—tend to cluster around 60° (g−), 180° (t), and 60° (g+), to avoid eclipsed conformations. In particular cases the distributions may be very highly skewed indeed. For valine, the conformation $\chi_1 \sim 180°$ is populated almost exclusively.

Most sidechains have a relatively small repertoire of preferred conformations. Statistical analysis of patterns of conformational angles in well-determined protein structures produce *rotamer libraries,* collections of preferred sidechain conformations. The local backbone structure influences sidechain conformation, because placement of the backbone and Cβ atoms creates specific loci for potential steric collision. Secondary structure can thereby bias the distribution of sidechain rotamers. For example, $\chi_1 \sim +60°$ (g−) is hardly ever observed for any residue in a helix, except for serine, the sidechain OH of which can form a hydrogen bond to the C=O of the preceding residue. *Backbone-dependent rotamer libraries* specify the sidechain conformational states preferred for different backbone conformations.

Rotamer libraries are useful for modelling, because one need consider only a small, discrete set of possible conformations. Also useful for modelling related proteins is the observation that as proteins evolve, sidechain conformation tends to be conserved. That is, homologous residues in related proteins tend to have similar sidechain conformations, even when they are mutated. The reason is that sidechains pack against their neighbours. Each sidechain fits into a 'cage' created by neighbouring sidechains. When a residue is changed by mutation, the new sidechain, in order to fit into the old cage, tends to adopt a conformation similar to that of the sidechain it replaces.

The known protein structures

Protein structure historically considered

Pasteur was one of the first scientists to think in three-dimensional terms about the structure of molecules in general and biological molecules in particular. Readers will know of his classic experiment in separating racemic tartaric acid, by manual selection of crystals of different shape. Pasteur recognized that the different crystal forms reflect a difference in underlying chemical constitution. Later, in studying the fermentation of tartaric acid, he observed that enzymes also were selecting only one form, discriminating between the two kinds of tartaric acid molecules on the basis of their three-dimensional molecular structure.

As the structural approach to organic chemistry developed in the late nineteenth century, the analysis of proteins progressed. Fischer showed that proteins were formed from amino acids, linked into a linear chain. (That the amino acids in proteins have a fixed order was not demonstrated until the determination of the amino acid sequence of insulin by F. Sanger, 50 years later.) Fischer also provided a structural interpretation of the selectivity of enzymes—their ability to discriminate among very similar molecules, as Pasteur had observed in the fermentation of tartaric acid. In 1894 Fischer wrote:

> Only with a similar geometrical structure can molecules approach each other closely, and thus initiate a chemical reaction. To use a picture, I should say that the enzyme and substrate must fit each other like a lock and key . . .

But locks have more complicated structures than keys. The detailed determination of the molecular structures of proteins had to await developments in crystallography.

Protein crystals had been known since at least 1840, when haemoglobin crystals were observed. However, the determination of detailed molecular structures from crystallography became possible only in 1912, with the discovery of X-ray diffraction. In 1913, while the Braggs were solving the structure of simple minerals such as rock salt, Nishikawa and Ono took X-ray diffraction photographs of silk, and concluded, qualitatively, that the material must contain some ordered structure at the molecular level. Subsequently, the study of X-ray diffraction patterns of protein fibres was pursued with great energy. It is interesting that the structures of hair and silk contain, in simplified forms, archetypical features of globular protein structures: helices and sheets. The study of fibre diffraction stimulated considerable ingenuity in building models of these molecules. This was a good rehearsal for the challenge of the first globular proteins to be solved by X-ray crystallography—myoglobin and haemoglobin, followed by lysozyme and chymotrypsin—which presented much more complex and difficult problems.

Other types of experiments were revealing that protein molecules have an internal ordered structure (distinct from the orderly arrangement of molecules in crystals). Mild heating could reversibly disrupt the structure without breaking the primary chemical bonds themselves. This is called denaturation: the coagulation of albumin when an egg is boiled is a common example. (It also illustrates the tendency of denatured proteins to aggregate.) In 1929, Hsien Wu concluded that the native form of the protein is a compact ordered structure stabilized by weak interactions, and that denaturation produces a disorganized, unfolded structure. A.E. Mirsky and L. Pauling reached similar conclusions, and identified hydrogen bonds as an important component of the forces that stabilize protein structure.

When Bernal took the first X-ray photograph of a crystalline globular protein in 1934, he started the process that led to the determination of the structures of myoglobin, haemoglobin, and first a trickle and then a cascade of crystal structures of other proteins. For many years, X-ray crystallography was the only source of detailed macromolecular structures. A rival appeared in the 1980s, when K. Wüthrich and R.R. Ernst developed methods for solving protein structures by Nuclear Magnetic Resonance (NMR) spectroscopy. Protein structure solution by NMR is now a thriving field.

X-ray crystallography and NMR are the two major techniques for determining the structures of proteins, accounting for all but a very few of the known structures. The determination of the structure of bacteriorhodopsin by R. Henderson and colleagues established electron diffraction as a method for solving the structures of membrane proteins that form regular two-dimensional arrays. Neutron diffraction has in a few cases added details of crucial hydrogen atoms and the unambiguous identification of water molecules.

Methods of structure determination

X-ray crystallography

The X-ray structure determination of a protein begins with its isolation, purification and crystallization. If a suitable crystal is placed in an X-ray beam, diffraction will be observed, arising from the regular microscopic arrangement of the molecules in the crystal. In principle, the diffraction pattern specifies the electron density of the matter in the unit cell of the crystal. A model of the protein can be built into a display of the electron density. But the experimental data are incomplete—we can measure only the *intensities* of the diffracted rays. To compute the electron density we need more information, about the *phases* of the diffracted rays. There are several ways in practice to solve the 'phase problem':

1. Classically, the method of *isomorphous replacement* determines phases by combining diffraction data from a native crystal with data from other

crystals containing the same protein packed in the same way, but modified by the addition of a heavy atom.

2. Many new proteins are similar to proteins of known structure. It is possible to 'find' the unknown molecule in the unit cell of the crystal by placing a known relative in different positions and orientations. The related structure then provides approximate phases. This is called the method of *molecular replacement*. It currently accounts for about half the newly-determined structures.
3. Certain atoms absorb as well as scatter X-rays, with effects on the diffraction pattern that contain phase information. Solution of crystal structures from measurements of the variation of the intensity distribution in the diffraction pattern over a range of wavelengths is called the method of *Multiwavelength Anomalous Dispersion*, or MAD. It requires a tunable X-ray source, available at a synchrotron.
4. Knowledge of general features of electron density distributions in crystals—for instance, that they must always be positive, and have certain statistical properties—permits calculation of phases directly from the experimental data. Such *direct methods* have solved the structures of small molecules for many years, and have now become useful in protein crystallography.

The equipment and technique of data collection, and software for data reduction and structure solution, are now integrated into a mature and effective technology. The rate-limiting step is getting good crystals, which remains in the realm of intelligently-guided black magic.

Model building and refinement

With phases good enough to compute and display a rough electron density, the crystallographer builds a model of the structure into it. This model and the experimental data furnish the input to a computational process of refinement: by a least-squares fitting procedure the atoms are shifted about to maximize the agreement between the model and the measurements. The refinement program also ensures that the structure has proper stereochemistry. In addition to the atomic coordinates, a thermal parameter or 'B-factor' is reported, to indicate the relative mobility of an atom.

In effect, the B-factor measures the size of the atomic peak in the electron density, and can in practice include the effects of disorder as well as vibration. Typically, long sidechains on the protein surface are mobile—like 'the countless chuckles of the waves of the sea'. In contrast, for small molecules, which form very well ordered crystals, the effective sizes of the atoms do arise primarily from thermal vibrations. Indeed, by collecting data at low temperature—by 'freezing out the vibrations'—it is even possible to determine changes in atomic electron density distributions arising from the formation of chemical bonds. However, for proteins it is appropriate to regard B-factors merely as empirical parameters, and the reader is cautioned to think twice before venturing to enquire too deeply into the multitude and variety of sins that they may in the most unfavourable cases cover.

How accurate are the structures?

There are a number of fairly reliable indicators of the accuracy of the atomic coordinates in published protein structures. Some are derived in the process of the structure determination and reflect experimental observations; others are inherent in the atomic coordinates themselves.

The *resolution* of an X-ray structure determination is a measure of how much data was collected. The more data, the more detailed the features in the electron-density map to be fitted, and, of course, the greater the ratio of number of observations to the number of atomic coordinates and B-factors to be determined. Resolution is expressed in Å, with a *lower* number signifying a higher resolution.

Experience has shown that the confident determination of different structural features is dependent on different thresholds of resolution.

Confidence in structural features of proteins determined by X-ray crystallography

(These are *rough* estimates, and depend strongly on the quality of the data.)

Structural feature	Resolution				
	5 Å	3 Å	2.5 Å	2.0 Å	1.5 Å
Chain tracing	—	Fair	Good	Good	Good
Secondary structure	Helices fair	Fair	Good	Good	Good
Sidechain conformations	—	—	Fair	Good	Good
Orientation of peptide planes	—	—	Fair	Good	Good
Protein hydrogen atoms visible	—	—	—	—	Good

Some unusually favourable protein crystals permit data collection to ultrahigh resolution—1.5 Å and below. There are several advantages to collecting very high resolution data, if the crystal permits it.

1. It may be possible to resolve static disorder: some sidechains or even regions of main chain may occupy two or more distinct conformations, which may be blurred together at low resolution, but which can be identified and refined independently. The results provide a picture of protein microstates.
2. The locations of water molecules are better defined.
3. Stereochemical details, such as deviations from planarity of the peptide bond, can be accurately determined. Only at very high resolution can refinement be carried out independent of any force field to impose correct stereochemistry. This permits tests of and improvements to these force fields. Some proteins contain genuine large deviations from stereochemical norms; only at very high resolution can one be confident that these are not experimental errors.
4. Refinement can produce good enough phases for protein hydrogen atoms to become visible in electron density maps.

5. At superhigh resolution (~0.5 Å), one can even measure electron densities in chemical bonds, vital for understanding the mechanisms of certain enzymes.

Still another advantage is the applicability of direct phasing methods, which in favourable cases can produce a model quickly and correctly given an accurate set of experimental data at 1.3 Å resolution or better!

Measures of structural quality

A report of an X-ray structure determination will include a statistic called the *R-factor,* a measure of how well the model reproduces the experimental intensity data. Other things being equal, the lower the R-factor the better the structure. The R-factor is a fraction expressed as a percentage; R = 0% would be an impossible ideal case—no disorder, no experimental error; R = 60% for a collection of atoms placed randomly in the unit cell of the crystal. A typical, well-refined protein structure, based on 2.0 Å resolution data, will have an R-factor of less than 20%. A related quantity, the *free R-factor,* measures the agreement between the model and a subset of the experimental data withheld during the refinement. The free R-factor give an unbiased measure of the agreement.

Having seen enough well-determined protein structures to know what they should look like, it is possible to subject atomic coordinate sets to a scrutiny independent of the experimental data. Good protein structures:

1. are compact, as measured by their surface area and packing density;
2. have hydrogen bonds of reasonable geometry, with few hydrogen bonds 'missing' in places where they would be expected, e.g. secondary structures;
3. have a distribution of backbone conformational angles in a Sasisekharan–Ramakrishnan–Ramachandran diagram confined almost entirely to the allowed regions (see Figure 2.11).

Nuclear Magnetic Resonance

Nuclear Magnetic Resonance (NMR) spectra measure the energy levels of the magnetic nuclei in atoms. These levels are sensitive to the environment of the atom. From effects transmitted between atoms bonded to each other, which affect the precise frequency of the signal from an atom (the *chemical shift*), NMR can determine the values of conformational angles. In particular, chemical shifts can define secondary structures. From interactions through space between nonbonded atoms <5 Å apart (the Nuclear Overhauser Effect or NOE), NMR can identify pairs of atoms close together in the structure, including those not close together in the sequence. The information about atoms distant in the sequence but close together in space is crucial; without it there would be no way to assemble individual regions into the correct overall structure.

In protein structure determination by NMR, the experimental data show a set of peaks corresponding to the interactions between pairs of atoms. Most work is done with protons, but other nuclei such as ^{13}C and ^{15}N also give signals and can identify additional pairs of neighbouring atoms. (Chemical shifts of C and N atoms in histidine sidechains can reveal the protonation and tautomeric state of the ring.) Conversely, selective deuteration will bleach out certain signals and simplify the spectrum. The first task of the spectroscopist is to assign the spectrum—that is, to correlate the peaks with the amino acids in the sequence. Once this is done, the data provide a set of distance constraints—that is, a list of atoms that are close together in space. This specifies the secondary structure, and gives indications of tertiary interactions.

Mathematics tells us that from a complete and exact set of interatomic distances, we could calculate the structure directly. But NMR spectroscopy gives an approximate and incomplete set of interatomic distances. The ambiguity in the structure that they define is correspondingly increased. As with X-ray crystallography, it is a combination of the fit to the experimental data together with force fields to enforce proper stereochemistry, that enter the computations that produce sets of atomic coordinates for the structure.

Proteins in solution are not subject to the constraints of crystal packing, and are expected to flounce around somewhat. Indeed, typically the result of an NMR experiment is a set of similar but not identical structures (often ~15–20 of them), all of which are comparably consistent with the combination of experimental data and stereochemical restraints. This may genuinely reflect structural dynamics; or may arise from insufficient distance measurements linking parts of the structure, rendering long-range spatial relationships uncertain. Fortunately it is possible to decide between these alternatives experimentally.

X-ray crystallography and NMR spectroscopy each have advantages and disadvantages. The main advantage of NMR is that it is not necessary to produce crystals; this is sometimes a severe impediment to X-ray structure determination. NMR gives us a window into protein dynamics, on the time scale of about 10^{-9}–10^{-6} seconds. A disadvantage of NMR is the limit on the size of a protein that can be studied. Speaking roughly, a 50–100 residue protein is a piece of cake. A 100–300 residue protein is solvable without problems by using suitable strategies of isotopic labelling (e.g., uniform double and triple labelling—with ^{2}H, ^{13}C, and ^{15}N—or selective labeling of different amino acids). Recent new technical developments might allow NMR to solve much larger structures, although much research must still be done to bring this to fruition.

An X-ray crystal structure certainly provides more precise values of atomic coordinates than does NMR; whether they are more accurate depends on how strongly you wish to argue that the structures of proteins seen in crystals are artificially constrained. The observation of enzymatic activity in the crystalline state proves that the essentials of the native state are present. Comparisons of X-ray structure determinations of the same protein in different crystals, with different intermolecular interactions, suggests that the perturbation of protein structures by the crystal environment is usually small, and localized to the regions in which the molecules are in contact in the crystal.

Low temperature electron microscopy (cryo–EM)

What are the prospects for extension of structural studies to very large aggregates, which may be difficult to prepare as single crystals suitable for X-ray diffraction? Electron microscopy of specimens at liquid-nitrogen temperatures has revealed structures of assemblies in the range M_r = 500 000 to 4 × 10^8, 100–1500 Å in diameter, such as the hepatitis B virus core shell or the clathrin coat. These results are not at atomic resolution—one sees 'blobs' rather than individual atoms. However, at 3–4 Å resolution, achievable in some cases, one can begin to recognize features, such as helices and sheets. Particularly exciting is the idea of combining electron microscopy with high-resolution X-ray crystal structures of the component proteins separately. By determining the positions of the individual proteins within the aggregate, the high-resolution structures can be assembled into a full atomic model of the entire complex.

The Protein Data Bank

The Protein Data Bank (PDB) is the collection of publicly available structures of proteins, nucleic acids, and other biological macromolecules. On January 4, 2000 the PDB contained 11 401 entries which can be classified into:

1. protein structures, which may include cofactors, substrates, inhibitors or other ligands including nucleic acids,
2. oligonucleotide or nucleic acid structures,
3. carbohydrate structures determined by X-ray fibre diffraction (note that some of the protein structures contain oligosaccharide moieties),
4. hypothetical models of protein structures (the PDB no longer collects models).

The PDB also contains purely bibliographic entries corresponding to structures not deposited, and, for some proteins, experimental X-ray diffraction measurements from which the structures were determined. A separate database maintained by G. Gilliland stores conditions of protein crystallization.

There is a certain amount of duplication among the PDB entries. In some cases there has been a redetermination of a structure, crystallized under different conditions, or re-solved at higher resolution. The structures of some proteins have been determined independently by X-ray crystallography and NMR. In some cases the structure of a protein has been determined in different states of ligation: free, and with one or more ligands. In other cases the structures of very closely related molecules have been determined. This makes it difficult to state precisely how many unique protein structures the data bank contains, but this number may be estimated at 6000.

Each set of coordinates deposited with the PDB becomes a separate entry, and is assigned an identifier, e.g. 1HHO for Human Oxyhaemoglobin. The first character is a version number. If the scientists who deposited the data set 1HHO in the PDB redetermined this structure at higher resolution, they might provide another data set 2HHO. An identifier beginning with the numeral 0 signifies that the entry is purely bibliographic and contains no coordinates.

Submitted coordinates are subjected to a set of standard stereochemical checks, and translated into a standard entry format. This format includes information about the structure determination (e.g. for crystal structures: the unit cell dimensions and symmetry, the resolution, the R-factor), references to papers describing the structure, and the atomic coordinates.

The PDB newsletter is currently published four times a year. It contains a catalog of current holdings, ordering information, and short news items; no one with a serious interest in the field should be without it. A list of other available documents is available on the PDB web site, http://www.rcsb.org. This site is under very active development.

The PDB web site and its 'mirrors' around the world permit retrieval of entries in computer-readable form. They provide methods for identifying entries that match specifications including molecule name, depositor name, experimental technique, resolution range, etc.

An alternative route to data retrieval are the sites that present classifications of protein structures (see Chapter 4), including:

SCOP http://scop.mrc-lmb.cam.ac.uk/scop/
CATH http://www.biochem.ucl.ac.uk/bsm/cath/
DALI http://www2.embl-ebi.ac.uk/dali/

The World Wide Web

One goal of this book is to steer readers to World Wide Web (WWW) sites that contain useful resources for the field of protein architecture. It is likely that most readers will already have used the Web, for reference material, for access to data bases in molecular biology, for checking out personal information about individuals—friends or colleagues or celebrities—or just for browsing. The Web provides a complete global village, containing the equivalent of library, post-office, shops, and schools. It is not only the number of entries in the Web but the density of the connections between them—their reticulation—that makes it such a powerful information retrieval tool.

Your access to the Web is mediated by a browser program. It presents the page you are viewing and its embedded links. It presents control information, allowing you to follow trails forward and back, and to initiate or interrupt a side excursion. It allows you to download information to your local computer, including text, pictures, and audio. It is possible to run browsers from terminals that do not have graphics facilities, but these will be restricted in what they can show you.

Once you have used the Web, the intersession memory facilities of the browsers allow you, the next time, to pick up cleanly where you left off. Files saved in your own local area contain a personal history of sites you have visited. When you find yourself viewing a document to which you will want to return, you can save a link to it in a file of 'bookmarks.' In subsequent sessions you can view your list of bookmarks and select any of them to return to a site directly, without having to follow the trail that brought you there the first time.

Every location on the Web is identified by a *URL—Uniform Resource Locator*. The URL specifies the format of the material and where it is located. After all, every document on the Web must ultimately be a file on some computer somewhere. An example of a URL is: http://www.oucs.ox.ac.uk/web/wwwfaq/index.html. This is the URL of a document of frequently asked questions about the Web itself. The prefix http:// stands for hypertext transfer protocol. This informs your browser of the format in which to expect the document. The next section www.oucs.ox.ac.uk is the name of a computer, in this case at Oxford within the United Kingdom academic community. The rest of the URL is the name of the file on the computer at Oxford which your browser will read.

Perhaps more important than any fixed indices or listings are information-retrieval programs that accept combinations of keywords, or 'search engines'. You can enter one or more indexing terms, such as 'haemoglobin', 'allosteric change', and 'crystal structure'. The search program will return a list of links to documents on the Web that contain these terms. You will thereby be able to identify a set of documents relevant to the topic of your interest. We are in an era of transition to paper-free publishing. As the Web grows, more and more scientific publications are appearing on it. A journal may post its table of contents, a table of contents together with abstracts of articles, or even full articles. It is already a good idea to include, in your own printed articles, your e-mail address and the URL of your home page. The Web has already transformed the world. Its professional and personal implications are of revolutionary proportions, that are almost impossible to overestimate. The implications of higher bandwidth are very hard to imagine.

Throughout this book useful URLs will be noted (these are collected in 'Useful web sites' at the end of the book). The problem is that many sites are volatile, and the web is cluttered with 'dead' links. I have tried to choose those that are likely to remain active and be kept up-to-date.

Databanks for molecular biology

The Web contains many compilations of information in the field of molecular biology. Some are general and comprehensive databanks of sequence, structure, function or bibliography; others are specialized 'boutique' collections. Still others are primarily indices of other web sites. Indeed, given the crucial role of the web in the reticulation of information sources, the utility of a site depends on the quality of the links it contains as well as the data it presents. The January issues of the journal *Nucleic Acids Research* contain annual contributions from most of the important data base projects.

Archival databanks

Major primary data curation projects in molecular biology collect nucleic acid sequences, protein sequences, and macromolecular structures.

1. *Nucleic acid sequences* are treated by a collaboration of GenBank (Washington, D.C.), the EMBL Data Library (Hinxton, U.K.) and the DNA Data Bank of Japan (Mishima, Japan) (the URLs are collected at the end of this section).
2. Two organizations present annotated collections of *amino acid sequences of proteins*. One is another triple partnership: the Protein Information Resource (Georgetown, U.S.A.), the Munich Information Centre for Protein Sequences (Martinsried, Germany), and the International Protein Information Database in Japan (Noda, Japan). The other is SWISS-PROT, a collaboration between the University of Geneva and the European Bioinformatics Institute (Hinxton, U.K.)
3. *Macromolecular structures* are archived by the Protein Data Bank (for many years at Brookhaven National Laboratories, New York, U.S.A., but now managed by the Research Collaboratory for Structural Bioinformatics (RCSB), a consortium based at Rutgers University, New Jersey, the U.S. National Institute of Standards and Technology, Maryland, and the San Diego Supercomputer Center, California), by the Nucleic Acid Data Bank (a component of the RCSB) and by the BioMagRes Data Bank (Madison, Wisconsin) (NMR structures). The Cambridge (U.K.) Crystallographic Data Centre archives structures of small molecules.

Specialized databanks

Many individuals or groups select, annotate, and recombine data focused on particular topics, and include selected links affording steamlined access to information about the topic of interest. Databases about specific protein families are collected at http://msd.ebi.ac.uk/add/Links/family.html For instance, the *Protein kinase resource* is a compilation that includes sequences, structures, functional information, laboratory procedures, lists of interested scientists, tools for analysis, a bulletin board, and links (http://www.sdsc.edu/kinases/). The HIV protease database archives structures of Human Immunodeficiency Virus 1 (HIV-1) proteinases, Human Immunodeficiency Virus 2 (HIV-2) proteinases, and Simian Immunodeficiency Virus (SIV) proteinases, and their complexes, and provides tools for their analysis and links to other sites with AIDS-related information (http://www.ncifcrf.gov/HIVdb/). This database contains some crystal structures not deposited in the PDB.

Information-retrieval tools

Databanks without effective access become data graveyards. Effective access requires computer programs that make it easy for the user to identify and retrieve data according to versatile sets of useful criteria. (It is the responsibility of the database designers to organize the data with an adequate internal logical structure that makes writing such programs possible.)

Major data archive projects in molecular biology

Name of data bank (home URL)	Type of data	Base
GenBank www.ncbi.nlm.nih.gov/	Nucleic acid sequences	National Library of Medicine, Washington, D.C., U.S.A.
EMBL Data Library www.ebi.ac.uk/ebi_docs/embl_db/ebi/topembl.html	Nucleic acid sequences	European Bioinformatics Institute, Hinxton, U.K.
DNA Data Bank of Japan www.ddbj.nig.ac.jp/	Nucleic acid sequences	National Institute of Genetics, Mishima, Japan
Protein Identification Resource www-nbrf.georgetown.edu/pir/	Amino acid sequences	Georgetown University, Washington, D.C., U.S.A.
Munich Information Center for Protein Sequences (MIPS) speedy.mips.biochem.mpg.de/	Amino acid sequences	Max-Planck-Institute für Biochemie Martinsried, Germany
International Protein Information Database in Japan (JIPID)	Amino acid sequences	Science University of Tokyo Noda, Japan
SWISS–PROT www.expasy.ch/sprot/	Amino acid sequences	Geneva, Switzerland and Hinxton, U.K.
Protein Data Bank www.rcsb.org	Protein structures	Rutgers University, New Jersey, U.S.A.
Nucleic Acid Data Bank ndbserver.rutgers.edu/	Nucleic acid structures	Rutgers University, New Jersey, U.S.A
BioMagResBank www.bmrb.wisc.edu/	NMR structure determinations	Madison, Wisconsin, U.S.A.
Cambridge Structural Database www.ccdc.cam.ac.uk/	Small-molecule crystal structures	Cambridge, U.K.

In addition to scanning databanks for items of interest, facilities available on the web provide a wide range of computational tools for data analysis. They take the form of 'web servers'—sites that allow the user to select input and launch calculations. Some of these are straightforward operations, such as the calculation of the molecular weight of a protein. Some, at the other extreme, are at the cutting edge of research—examples are the servers that attempt to predict the three-dimensional structure of a protein from its amino acid sequence.

If the computation is a fast one, the results may be returned to the browser 'on the fly'; or, if the computation is slow, the program may send the results to the user by e-mail. The old model of 'install programs on your computer and download the data on which to run them' is giving way to world-wide distributed computer facilities as the mode of research. Only connect.

The general goal of bioinformatics is to map out the world of sequences, structures, and functions of biological molecules, and the relations between them. What would a minimal bioinformatics tool kit contain? The subject matter includes:

- nucleic acid and protein sequences
- macromolecular structures
- protein function
- the scientific literature.

What are the essential operations? You must be able to retrieve individual sequences and structures. Perhaps you can name exactly what you want; for instance, the gene sequence of sperm whale myoglobin. Or you may specify criteria that define a class of items; for instance, all genes for zinc-finger proteins in *E. coli.* You should be able to display the results in some intelligible form, such as a sequence coloured by physico-chemical properties of the amino acids, or a picture of a protein structure. Many web sites provide these facilities.

You will want to analyse and compare the material identified; for example, to align all the gene sequences of *E. coli* zinc-finger proteins. Then you will want to search more widely, for related items—for example, find all proteins that have the same topology as the immunoglobulin domains, and show their optimal superposition.

A very important problem is the detection of sequences related to a newly-determined gene. It is essential to be able reliably to detect similarities between very distant relatives. The most powerful current methods are PSI–BLAST, and techniques based on a mathematical approach called hidden Markov models. This problem is a focus of current research.

Some of the most complex queries involve several databanks—find all proteins from eukaryotes that have the OB fold and bind DNA. Enhancement of the links among databanks will facilitate such queries. It is not an unjustified extrapolation to suggest that someday there will be one databank, distributed around the world, and various avenues of access to it, most coordinated but, undoubtedly, a few resolutely independent.

A number of web sites provide indices of computational tools for molecular biology. The following box gives a few suggested sites, selected in the hope that they will remain stable, although it is quite likely that the URLs will change, as equipment is replaced or scientists change institutions. In that event one can consult one of the sites that index web tools for molecular biology, or even a general search engine. In some cases, sites are 'mirrored'—i.e. cloned—to appear on computers in different continents. It will greatly enhance the speed of browsing to select a site near you.

Summary

All proteins have the common chemical feature of the polypeptide backbone, but each protein has a different sequence of pendant sidechains.

A toolkit of web sites

Task and web site	Program name
Retrieve one sequence http://srs.ebi.ac.uk/	Sequence Retrieval System (SRS)
http://www.ncbi.nlm.nih.gov/Entrez	Entrez
Retrieve one structure www.rcsb.org	Protein Data Bank (PDB)
Match one sequence to one sequence http://www-hto.usc.edu/software/seqaln/seqaln-query.html or http://vega.igh.cnrs.fr/bin/align-guess.cgi	ALIGN
Multiple sequence alignment http://www.ebi.ac.uk/clustalw	ClustalW
Probe sequence databank with sequence http://www.ncbi.nlm.nih.gov/blast/psiblast.cgi http://www.sanger.ac.uk/Software/Pfam/search.shtml http://www.cse.ucsc.edu/research/compbio/HMM-apps/HMM-applications.html	 PSI-BLAST Hidden Markov models
Probe structure databank with structure http://www2.embl-ebi.ac.uk/dali/	DALI
Probe structure databank with sequence (fold recognition, threaders) http://www.embl-heidelberg.de/predictprotein/predictprotein.html http://www.doe-mbi.ucla.edu/people/frsvr/frsvr.html	
Bibliographical search http://www.ncbi.nlm.nih.gov/PubMed/	U.S. National Library of Medicine (PubMed)

Because the backbone is flexible, it can adopt an infinite number of spatial conformations, just as a piece of string thrown repeatedly on the floor will form random and unrelated patterns. In natural proteins, interactions among the sidechains select one three-dimensional structure that, under physiological conditions of solvent and temperature, is distinguished by much greater thermodynamic stability than any other. Each protein thereby takes up a unique conformation. Such definite structures are essential for proteins to perform their biological roles.

We now have an answer—at least in general terms—to one crucial question: how is a one-dimensional genetic message translated into a three-dimensional protein structure? First, the genetic message is translated into the linear sequence of amino acids in a protein molecule. Then the three-dimensional structure of a protein is dictated automatically by the sequence of sidechains along its polymer backbone.

A native protein consists of one or more compact assemblies of residues. Any region in the protein is characterized by several features consistent with low free energy: dense packing of residues in the interior, satisfaction of hydrogen-bonding potential of polar groups, and burying of hydrophobic surface. For natural amino acid sequences, the optimal solution of this thermodynamic jigsaw puzzle is a special conformation in which the backbone describes a curve traversing the space occupied by the molecule. The shape of this curve usually includes standard elements of secondary structure—helices and sheets—and makes use of a common repertoire of ways in which helices and sheets pack against one another.

Protein architecture is the study of how protein folding patterns may be classified, how to understand the relationships between the local interactions and the overall folding pattern of the chain, and how the entire structure is determined by the amino acid sequence.

Bioinformatics is the systematization, organization, and interlinking of data on sequences, structure and function of biological molecules. The World Wide Web is the avenue for access to the data and to tools for their analysis.

Glossary

Native state

The unique stable, active structure of a protein, taken up spontaneously, in most cases, under appropriate conditions of solvent and temperature.

Structure determination

The identification of the actual atomic positions of all atoms in a protein, usually excluding hydrogens, carried out in almost all cases by X-ray crystallography or NMR spectroscopy.

Folding pattern

The spatial course of the main chain of a native structure of a protein.

Sasisekharan–Ramakrishnan–Ramachandran diagram

A graph showing the conformation of each individual residue in a protein, with boundaries delineating regions of sterically favourable conformation.

Hydrophobic residues

Residues with uncharged and non-polar sidechains, for example phenylalanine. Like molecules of oil, these have thermodynamically unfavourable interactions with water, and tend to be sequestered together in the interior of a protein structure.

Secondary structure

The formation of standard conformations—helices and sheets—by hydrogen bonds between main chain atoms. The fact that these energetically-favourable structures can be formed with little restriction on amino acid sequence accounts for their appearance in most protein structures.

Supersecondary structure

A standard conformation of a region of polypeptide chain formed by the interaction of two or more successive regions of secondary structure; for example, the β–α–β unit formed by two parallel strands of β-sheet connected by a α-helix.

Common core

The main central elements of secondary structure—including the active site—that retain their folding pattern during the evolution of a family of protein structures.

Useful web sites

Primer on molecular genetics:
http://www.ornl.gov/TechResources/Human_Genome/publicat/primer/prim1.html

Human genome project information:
http://www.ornl.gov/hgmis/

Genome sequencing project information:
http://www.ebi.ac.uk/genomes/info.html
http://www-biol.univ-mrs.fr/english/genome.html
http://megasun.bch.umontreal.ca/ogmpproj.html (organelles)

The following four sites are general ones, and each would be a good starting point for 'surfing':

Home page of National Center for Biotechnology Information:
http://www.ncbi.nlm.nih.gov/

Home page of the European Bioinformatics Institute, an outstation of the European Molecular Biology Laboratory:
http://www.ebi.ac.uk

Home page of the Expasy molecular biology site of the Swiss Institute of Bioinformatics:
http://www.expasy.ch/

Home page of the GenomeNew WWW server, based at the Institute for Chemical Research, Kyoto University, and the Human Genome Center, Institute of Medical Science, University of Tokyo, Japan. In addition to access to standard data bases, and graphs of their growth (http://www.genome.ad.jp/dbget/db_growth.gif), there are links to specialized data bases and information-retrieval tools developed in Japan:
http://www.genome.ad.jp/

Sequence retrieval:
http://srs.ebi.ac.uk/
http://www.ncbi.nlm.nih.gov/Entrez/

Definition of polypeptide conformation:
http://www.chem.qmw.ac.uk/iupac/misc/biop.html

Home page of Protein Data Bank:
http://www.rcsb.org

Home page of ReLiBase, a system for analysing receptor–ligand complexes in the Protein Data Bank:
http://rcsb.rutgers.edu:8081/home.html

Promise: Prosthetic groups and metal ions in protein active sites:
http://bmbsgi11.leeds.ac.uk/promise/MAIN.html

Collection of on-line analysis tools, including database searches:
http://www-biol.univ-mrs.fr/english/logligne.html

Index to web sites in molecular biology, including specialized databases:
http://www.cbs.dtu.dk/biolink.html

BCM search launcher—various database searches and associated tools:
http://kiwi.imgen.bcm.tmc.edu:8088/search-launcher/launcher.html

The sequence and structure searching site:
http://sss.berkeley.edu/sss/

Collection of protein analysis tools:
http://www.graylab.ac.uk/cancerweb/research/protanal.html

Site about electronic scholarly publishing, with emphasis on genetics:
http://www.esp.org/

Sidechain rotamer libraries:
http://duc.urbb.jussieu.fr/rotamer.html
http://www.fccc.edu/research/labs/dunbrack/sidechain.html

Recommended reading and references

World Wide Web, information retrieval

Bishop, M.J. (ed.) (1998). *Human genome computing*. Academic Press, New York and London.

Bishop, M.J. (1999). *Genetic databases*. Academic Press, New York and London.

Durbin, R., Eddy, R., Krogh, A., Mitchison, G. and Eddy, S. (1998). *Biological sequence analysis : probabilistic models of proteins and nucleic acids*. Cambridge University Press, Cambridge.

Frishman, D., Heumann, K., Lesk, A.M. and Mewes, H.-W. (1998). Comprehensive, comprehensible, distributed and intelligent databases: current status. *Bioinformatics* **14**, 551–61.

Kreil, D.P. (1999). DATABANKS—a catalogue data base of molecular biology data banks. *Trends Biochem*. Sci. **24**, 155–7.

Lesk, M. (1997). *Practical digital libraries: books, bytes and bucks*. Morgan Kaufmann, San Francisco.

Trends Guide to Bioinformatics. (1998). Elsevier Trends Journals, Amsterdam.

Trends Guide to Pharma Informatics. (1999). Elsevier Trends Journals, Amsterdam.

Structure determination

Baker, T.S. and Johnson, J.E. (1996). Low resolution meets high: towards a resolution continuum from cells to atoms. *Curr. Opin. Struc. Biol.* **6**, 585–94.

Cruickshank, D.W.J. (1999). Remarks about protein structure precision. *Acta cryst.* **D55**, 583–601.

Dauter, Z., Lamzin, V. and Wilson, K.S. (1997). The benefits of atomic resolution. *Curr. Opin. Struc. Biol.* 7, 681–7.

De Rosier, D.J. (1997). Electron cryomicroscopy. Who needs crystals anyway? *Nature* **386**, 26–7.

Drenth, J. (1994). *Principles of protein X-ray crystallography*. Springer-Verlag, New York & London.

Kay, L.E. (1997). NMR methods for the study of protein structure and dynamics. *Biochem. Cell. Biol.* **75**, 1–15.

Longhi, S., Czjzek, M. and Cambillau, C. (1998). Messages from ultrahigh resolution crystal structures. *Curr. Opin. Struc. Biol.* **8**, 730–7.

Wüthrich, K. (1995). NMR—this other method for protein and nucleic acid structure determination. *Acta Cryst.* **D51**, 249–70.

Sequence retrieval and similarity searching

Altschul, S.F. and Koonin, E.V. (1998). Iterated profile searches with PSI-BLAST— a tool for discovery in protein databases. *Trends Biochem. Sci.* **23**, 444–7.

Bateman, A., Birney, E., Durbin, R., Eddy, S.R., Howe, K.L. and Sonnhammer, E.L. (2000). The Pfam protein families database. *Nucl. Acid Res.* **28**, 263–6.

Protein structure and folding, sequence-structure relationships

Baldwin, R.L. and Rose, G.D. (1999). Is protein folding hierarchic? I. Local structure and peptide folding. II. Folding intermediates and transition states. *Trends Biochem. Sci.* **24**, 26–33, 77–83.

Chothia, C. (1984). Principles that determine the structures of proteins. *Ann. Revs. Biochem.* 53, 537–72.

Dobson, C.M., Šali, A. and Karplus, M. (1998). Protein folding: a perspective from theory and experiment. *Angewandte Chemie Int. Ed.* **37**, 868–93.

Dunbrack, R.L., Jr. and Karplus, M. (1994). Conformational analysis of the backbone-dependent rotamer preferences of protein sidechains. *Nature Struct. Biol.* **1**, 334–40.

Gerstein, M. (1998). Patterns of protein-fold usage in eight microbial genomes: a comprehensive structural census. *Proteins: Struct. Funct. Genet.* **33**, 518–34.

Lesk, A.M. (1991). *Introduction to physical chemistry*. Prentice-Hall, Englewood Cliffs, New Jersey.

Ponder, J.W. and Richards, F.M. (1987). Tertiary templates for proteins. Use of packing criteria in the enumeration of allowed sequences for different structural classes. *J. Mol. Biol.* **193**, 775–91.

Richards, F.M. (1991). The protein folding problem. *Sci. Amer.* **264**(1), 54–7, 60–3.

Richards, F.M. and Lim, W.A. (1993). An analysis of packing in the protein folding problem. *Quart. Revs. Biophys.* **26**, 423–98.

Rose, G.D. and Wolfenden, R. (1993). Hydrogen bonding, hydrophobicity, packing, and protein folding. *Ann. Revs. Biophys. Biomol. Struct.* **22**, 381–415.

Samudrala, R. and Moult, J. (1998). Determinants of sidechain conformational preferences in protein structures. *Prot. Eng.* **11**, 991–97.

Structural genomics

Šali, A. (1998). 100,000 protein structures for the biologist. *Nature Struc. Biol.* **5**, 1029–32.

Exercises, problems and weblems

Exercises

2.1. Draw the chemical structures of tripeptides (a) Ala–Leu–Phe, (b) Ser–Pro–Asn (assuming that the peptide preceding the Pro is in the *cis* conformation).

2.2. What is the sequence of the peptide shown in Figure 2.13?

2.3. Which tripeptide would you expect to be more water soluble: (a) Ala–Thr–Ser or (b) Phe–Ile–Trp?

2.4. Figure 2.14 shows the active site of *E. coli* isocitrate dehydrogenase, binding its substrate [5ICD]. Cells regulate the activity of this enzyme by phosphorylating Ser113. (a) Write the structural formula for the substrate isocitrate. (b) Explain why phosphorylating Ser113 inactivates the enzyme. (c) What would you expect to be the effect on the activity of the unphosphorylated enzyme of mutating Ser113 to Asp? Explain your answer. (d) From an inspection of Figure 2.14a, would you expect the main chain of Ser113 to be (relatively) mobile or rigid, in the absence of substrate? Explain your answer.

2.5. From an inspection of the pictures of myoglobin and cytochrome c in

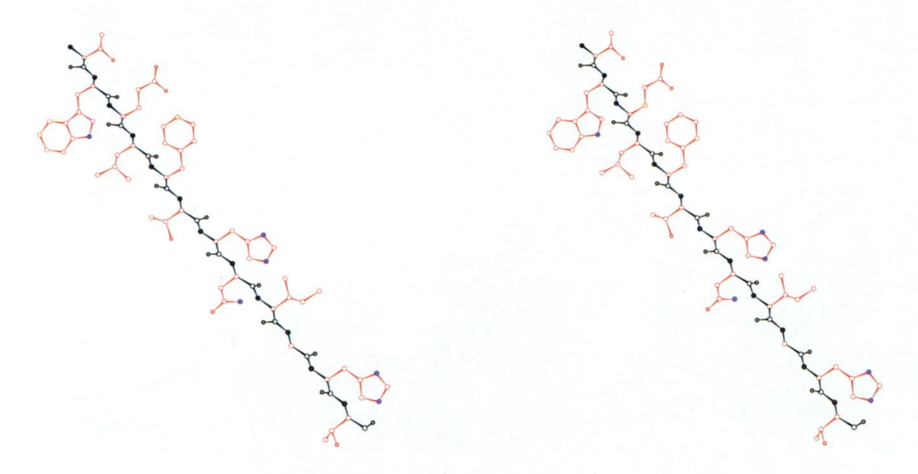

Fig. 2.13 A polypeptide in the extended chain conformation. What is the amino acid sequence? Remember that the amino acid sequence is always stated in the direction from N-terminus to C-terminus.

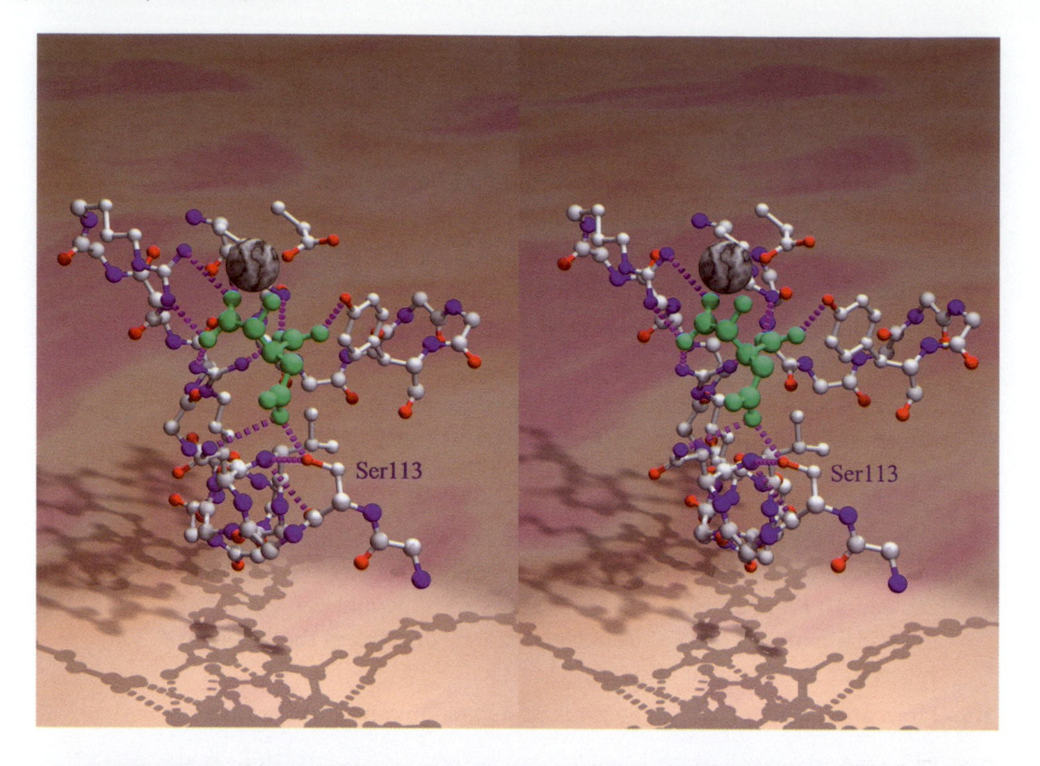

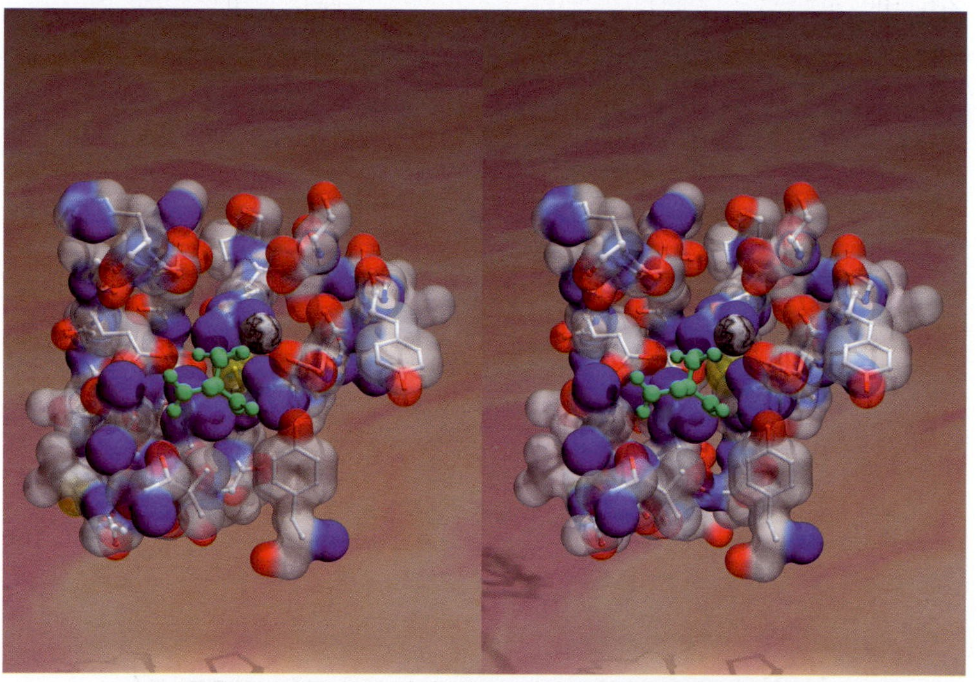

Fig. 2.14 The active site of *E. coli* isocitrate dehydrogenase [5ICD]. (a) Ball-and-stick representation. The enzyme backbone is shown in black, the enzyme sidechains in red (oxygen atoms are blue) and the substrate in green. The mottled grey sphere is a magnesium ion. Broken lines show enzyme–substrate hydrogen bonds. (b) Smoothed surface showing the charge distribution of the active site.

Figure 2.6, describe qualitatively the differences in the way the haem environment is constructed in these two molecules.

2.6. On a photocopy of the NAD-binding domain of horse liver alcohol dehydrogenase, (Figure 2.7a) label the helices αA, αB, . . . according to their order of appearance along the chain, and label the strands of sheet βA, βB, . . . according to their order of appearance along the chain.

2.7. In protein engineering, it is often convenient to synthesize two independent parts of a protein in a single polypeptide chain, with a linker polypeptide between them. Why is the sequence $(gly_3ser)_n$ often used as a linker?

2.8. A Sasisekharan–Ramakrishnan–Ramachandran plot specialized to particular sidechains can be constructed as follows: for any amino acid, e.g. Val, create in a computer a model of the tripeptide Ala–Val–Ala. For each combination of ϕ–ψ values of the central residue, find the conformation with lowest energy, and plot that energy as a function of ϕ–ψ. Regions with unavoidable steric collision will appear as high energy, and can be marked as disallowed regions. Suppose such plots are constructed for Ala and Val. Which would contain the larger disallowed area?

Problems

2.1. The table of the standard genetic code shows 64 codons each of three bases, three of which are STOP signals (Figure 2.1). Using the classification of amino acids on page 21, what is the probability that a single base change in a gene (a) leaves the amino acid unchanged (a silent mutation), or (b) replaces an amino acid by one in the same class?

2.2. The standard free energy difference between *cis*- and *trans*-proline conformations is about 5 $kJ \cdot mol^{-1}$ (1.2 $kcal \cdot mol^{-1}$). For an isolated proline residue in a denatured protein at 300 K, estimate the fraction that are in the *cis* conformation. If two states differ in a standard free energy by Δ, at equilibrium their populations will be in the ratio $\exp(-\Delta/RT)$, where $R = 8.314472\ J \cdot mol^{-1} \cdot K^{-1}$ and T is the absolute temperature. For a denatured protein containing two isolated prolines, estimate the fraction of molecules for which both prolines are in the *trans* conformation. (Proline isomerization is one of the factors governing the kinetics of refolding. Prolyl isomerases catalyse the conversion of prolines between *cis* and *trans* isomers and speed up the folding of proteins containing proline residues.)

2.3. Suppose the genetic code were extended to code for a set of 40 amino acids, the natural ones and their D isomers. (a) Would a stretch of residues with values of the mainchain conformational angles in the α_R region be able to form a helix if the sequence contained a mixture of L and D isomers? (b) Would a stretch of residues with values of the mainchain conformational angles in the sheet region be able to form a sheet if the sequence contained a mixture of L and D isomers? (c) How could proteins form structures containing helices and sheets?

2.4. The crystal structure of pepsin (the subject of J.D. Bernal's original protein X-ray diffraction photograph) was solved in the late 1980's. The amino acid sequence, and the residues on the surface (s) and buried in the interior (b) are:

```
        10        20        30        40        50        60
IGDEPLENYLDTEYFGTIGIGTPAQDFTVIFDTGSSNLWVPSVYCSSLACSDHNQFNPDD
sbsbsbsbsssssbsbsbbbbsssssbsbbbbbsbbbbbbbbssbsssbbsssssbsbss

        70        80        90       100       110       120
SSTFEATSQELSITYGTGSMTGILGYDTVQVGGISDTNQIFGLSETEPGSFLYYAPFDGI
ssssssssssbssssssssbsbsbbsbsbsbssssbssbbbbbbssbsssssssbsbbbs

       130       140       150       160       170       180
LGLAYPSISASGATPVFDNLWDQGLVSQDLFSVYLSSNDDSGSVVLLGGIDSSYYTGSLN
bbbbbsssbsssbsbbbssbsssssbsssbbbbsbbssssssbsbbbbbsssssssssss

       190       200       210       220       230       240
WVPVSVEGYWQITLDSITMDGETIACSGGCQAIVDTGTSLLTGPTSAIANIQSDIGASEN
sbssssssbsbbbsbssbsssssssbssssssbbbbbsssbbbssssbssbbssbsbsss

       250       260       270       280       290       300
SDGEMVISCSSIDSLPDIVFTIDGVQYPLSPSAYILQDDDSCTSGFEGMDVPTSSGELWI
ssssssbssssssssssbsbsbssssbsbsbbbssssssssbsbbbsssssssssssbbb

LGDVFIRQYYTVFDRANNKVGLAPVA
bbbbbbssbbbbbbssssbbbbsss
```

(a) What percent of the residues is buried? (b) For each of the following physico-chemical classes: (1) non-polar (GASTCVILPFYMW); (2) polar (NQH); and (3) charged (DEKR), what percent of the amino acids in each of these classes is buried and what percent is on the surface?

2.5. Suppose that a structure contains only three atoms, which are at the vertices of an equilateral triangle of edge 1. You measure the interpoint distances approximately, as $d_{12} = 1.0 \pm 0.2$, $d_{13} = 1.0 \pm 0.2$, $d_{23} = 1.0 \pm 0.2$. Estimate the maximum possible value of the average difference in atomic position (after optimal superposition) between the correct structure and a model structure that fits the measurements to within the stated errors. (Hint: consider models that have the form of equilateral triangles.)

2.6. The accessible surface area (A.S.A.) of small monomeric proteins varies with molecular weight M according to the relationship: A.S.A. $= 11.1\, M^{\frac{2}{3}}$. If the amino acid sequence of these proteins is placed on a polypeptide in the extended chain conformation, the A.S.A. is of course higher, and is given by: A.S.A.$_{\text{extended chain}} = 1.45\, M$. (a) Explain why the accessible surface area of a native protein varies as the 2/3 power of M, and that of the extended chain varies linearly with M. (b) What is the formula for the buried surface area—relative to the extended chain—of proteins as a function of M? (c) What is the expected buried surface area per residue for a monomeric protein of 100 residues? (Assume that the average M of a residue is 110.) (d) For this example, what is the approximate contribution to the free energy of stabilization of the native state from the hydrophobic effect?

Weblems

2.1. Retrieve the sequence of myosin light chain kinase from SWISS-PROT.
http://srs.ebi.ac.uk/

2.2. Find human proteins containing the pentapeptide Glu–Leu–Val–Ile–Ser.
http://www-nbrf.georgetown.edu/pirwww/search/patmatch.html

2.3. Find five PDB entries containing immunoglobulin fragments determined at 2.1 Å or better. (Remember that a lower number specifies a better, i.e. higher, resolution.)
http://www.rcsb.org

2.4. In which year did the number of bases in the GenBank nucleic acid sequence database first exceed 10^9 bases?
http://www.genome.ad.jp/dbget/db_growth.html

2.5. Approximately 10 years ago, scientific journals adopted the policy that publication of a protein structure required deposition of the coordinates in the Protein Data Bank. From the discontinuity in the rate of deposition of entries, estimate the year of adoption of this policy.
http://www.genome.ad.jp/dbget/db_growth.gif

2.6. (a) What are the conformational angles that specify the common rotamers of tyrosine, independent of secondary structure? (b) Which is the most commonly observed rotamer?
ftp://fccc.edu/dunbrack/pub/uncompressed_files/bbind98.Feb.lib

(c) What is the most common tyrosine rotamer observed for a tyrosine residue in an α-helix?
ftp://fccc.edu/dunbrack/pub/compressed_files/bbdep98.Feb.lib.

Is this the same as the answer to (b), the most commonly observed tyrosine rotamer independent of secondary structure?

2.7. What is the accessible surface area of pepsin?
http://www.bork.embl-heidelberg.de/ASC/scr1-form.html

2.8. Calculate the hydrophobicity profile of baboon α-lactalbumin, SWISS–PROT ID LCA_PAPCY.

```
KQFTKCELSQNLYDIDGYGRIALPELICTM
FHTSGYDTQAIVENDESTEYGLFQISNALW
CKSSQSPQSRNICDITCDKFLDDDITDDIM
CAKKILDIKGIDYWIAHKALCTEKLEQWLC
EK
```

(Suggested web server: http://bmbsgi11.leeds.ac.uk/bmb5dp/profiles.html; use Kyte and Doolittle option, window size 10.) Predict where turns between elements of secondary structure might occur by determining the local minima in the hydrophobicity. Compare these with the starting and ending residues of the helices and sheets in the X-ray crystal structure, listed in the PDB entry 1ALC.

2.9. Submit the amino acid sequence of the highly immunogenic outer capsid protein (Hoc) of bacteriophage T4 (SWISS-PROT ID HOC_BPT4) to at least three sequence database similarity searches.

```
MTFTVDITPKTPTGVIDETKQFTATPSGQTGGGTITYAWSVDNVPQDGAEATFSYVLKG
AGQKTIKVVATNTLSEGGPETAEATTTITVKNKTQTTTLAVTPASPAAGVIGTPVQFTA
LASQPDGASATYQWYVDDSQVGGETNSTFSYTPTTSGVKRIKCVAQVTATDYDALSVTS
EVSLTVNKKTMNPQVTLTPPSINVQQDASATFTANVTGAPEEAQITYSWKKDSSPVEGS
NVYTVDTSSVGSQTIEVTATVTAADYNPVTVTKTGNVTVTAKVAPEPEGELPYVHPLPH
SSAYIWCGWWVMDEIQKMTEEGKDWKTDDPDSKYYLHRYTLQKMMKDYPEVDVQESRNG
IIHKTALETGIIYTYP
```

Compare the results returned. Can you propose a structural affinity with any known protein family?

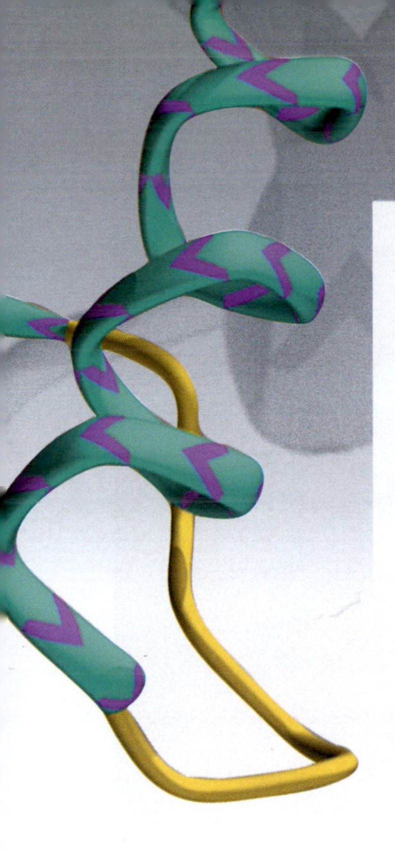

CHAPTER 3

Pattern and form in protein structure

Helices and sheets

Consider the following three goals that protein structures must achieve:

1. low-energy conformations of individual residues
2. hydrogen-bonding by polar groups, including buried ones
3. formation of compact, well-packed structures.

The Sasisekharan–Ramakrishnan–Ramachandran plot shows how to solve the first problem, at least for the main chain: just keep the conformations of the residues in the α_R or β region. For the second, consecutive residues, all in the α_R region, produce a hydrogen-bonded helical conformation; and consecutive residues, all in the β region, produce a nearly-extended chain, which can interact by lateral hydrogen bonding to form a sheet. For fibrous proteins, this is enough. Hair and wool are based on separate molecules that form bundles of long helices, wound around each other like the strands of a rope. Silk, in contrast, contains sheets. The proteins of wool and silk do not form compact, water-soluble structures (but, hey, two out of three ain't bad).

To form a compact structure, globular proteins contain regions of helix and strand, linked by regions in which the chain alters direction, called turns or loops. Usually, the helices and strands traverse the structure, with the loops on the surface. The study of protein architecture is largely a description of the spatial assembly of helices and strands of sheet within the structure, and its relation to their distribution along the amino acid sequence. To focus attention on the helices and sheets, we often draw them in simplified 'cartoon' representations. Figure 3.1 shows conventional depictions of helices and strands of sheet in proteins, and also of nucleic acids.

Hydrogen-bonding patterns of helices and sheets

The most common type of helix in proteins is the α-helix, in which the NH group of residue i forms a hydrogen bond to the O = C group of residue $i + 4$

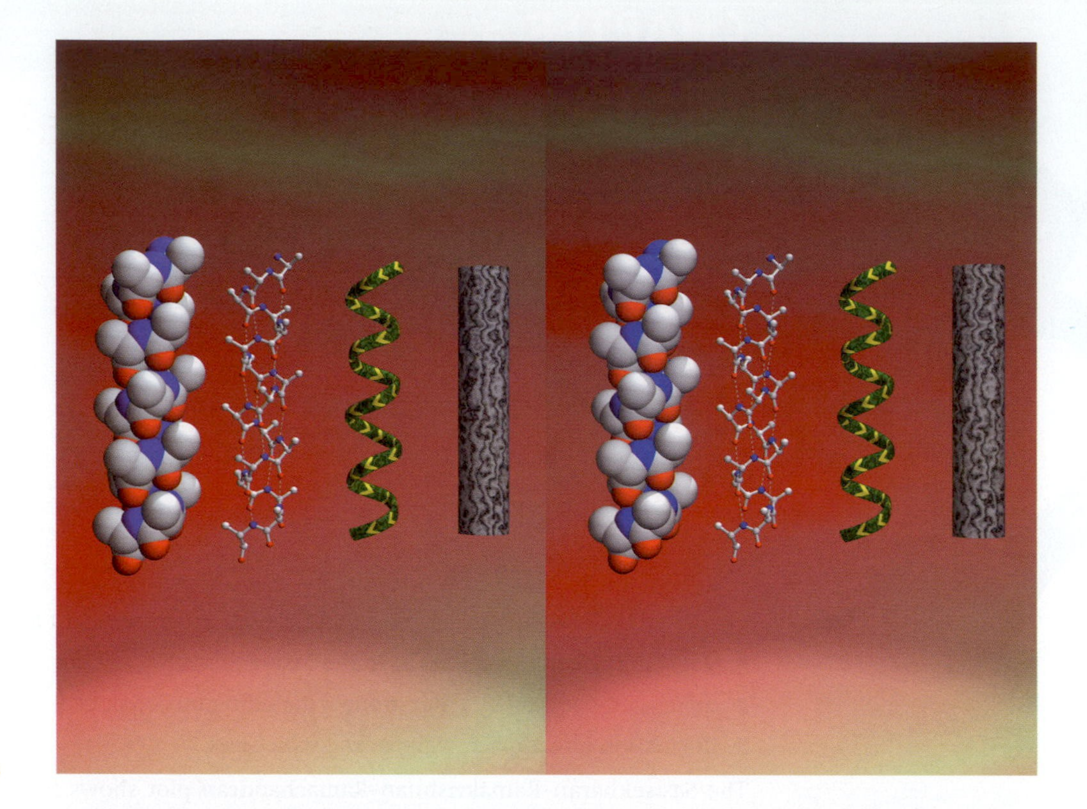

Fig. 3.1a

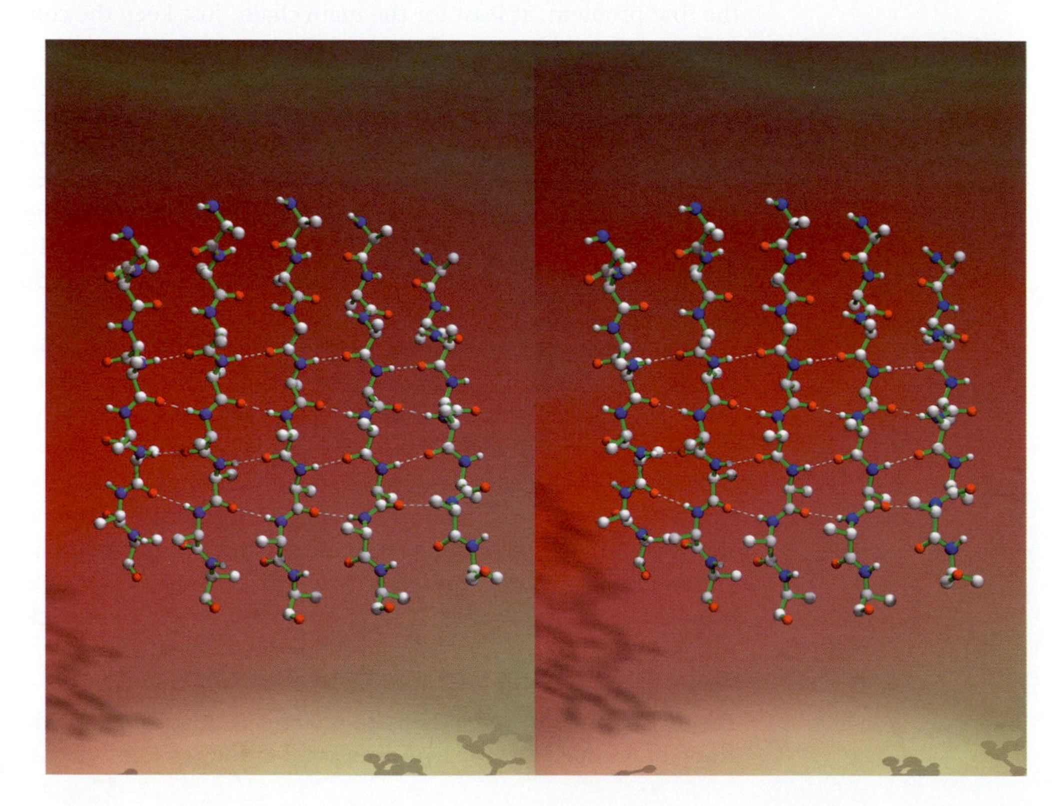

Fig. 3.1b

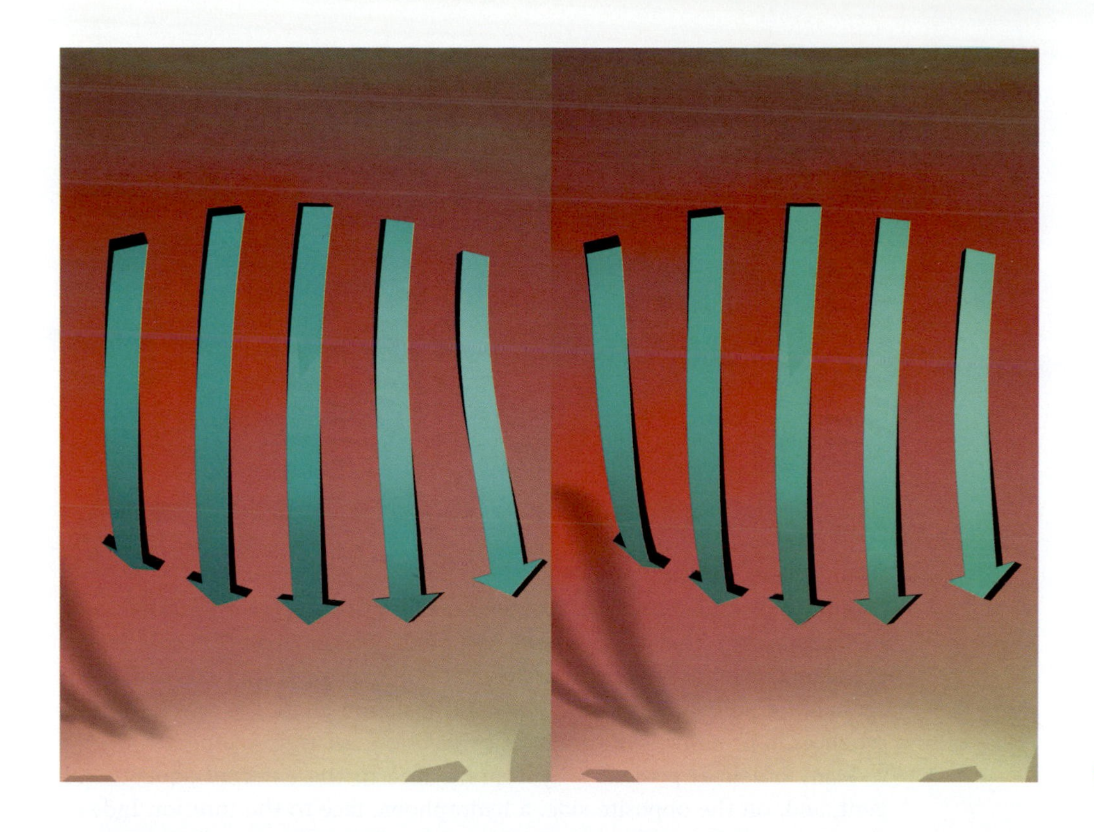

Fig. 3.1c

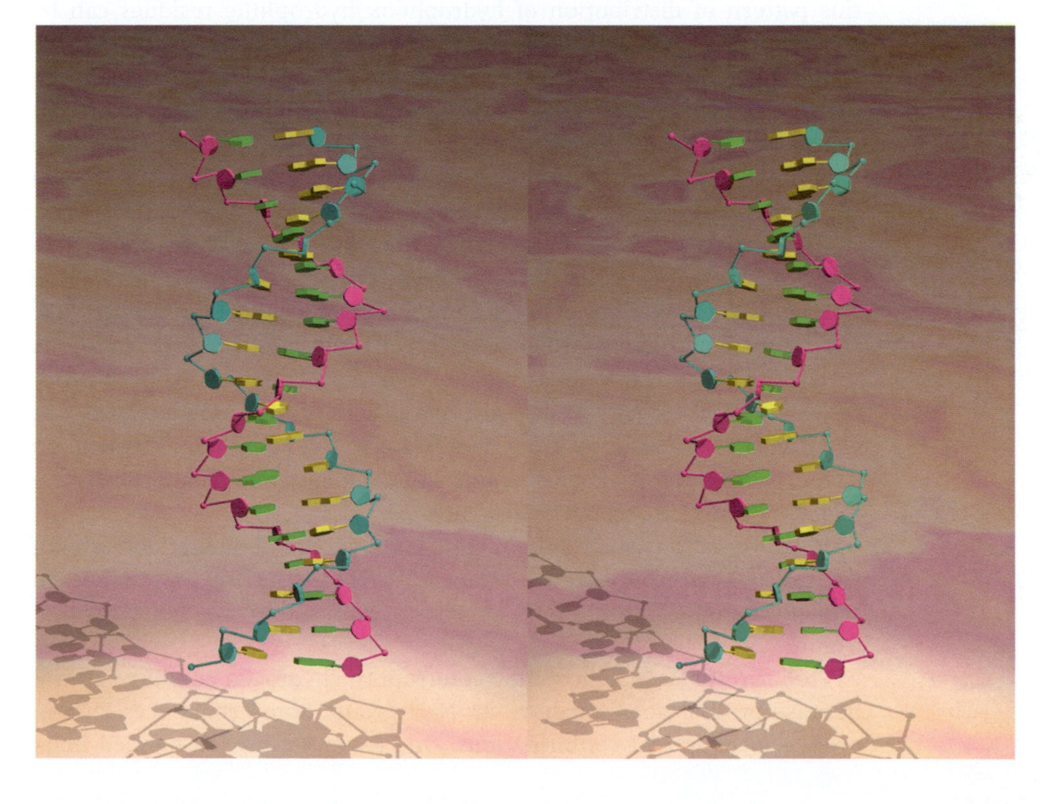

Fig. 3.1d

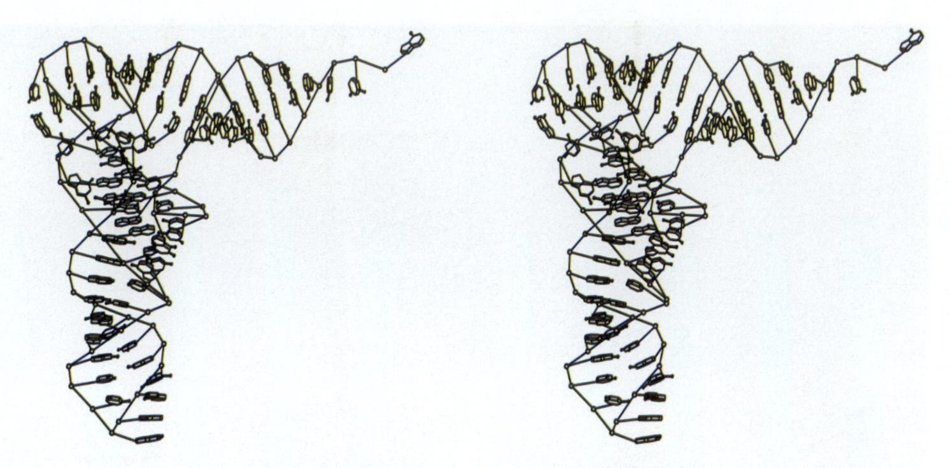

Fig. 3.1e

Fig. 3.1 Representations of (a) helix, (b, c) sheet, (d) DNA, and (e) transfer RNA [4TNA].

(Figures 3.2a and 3.2b). The geometric relationship between residues in a helix is another kind of symmetry. In an α–helix, the operation that relates the main chain of one residue to the next is a rotation of approximately 100° around the helix axis, together with a translation along the axis by 1.50 Å. α-helices in proteins, almost without exception, are right-handed.

Many α-helices present a hydrophilic face to the external aqueous solvent, and, on the opposite side, a hydrophobic face to the interior. Indeed, this pattern of distribution of hydrophobic/hydrophilic residues can be spotted, in the sequence of a region, as suggestive of a helical segment. We should expect to observe this pattern in the amino acid sequences of the helical segments of the cytochrome and heavy subunits of the reaction centre, but not for the membrane-traversing segments. A diagram called the 'helical wheel' (a projection of the sequence down the axis of an α-helix) shows the nature of the faces of the helix (Figure 3.3).

If the chain winds up more tightly than in an α-helix, an alternative hydrogen-bonded structure can form, called the 3_{10} helix. In a 3_{10} helix, the N–H of residue i is hydrogen bonded to the O=C of residue $i + 3$. (Figures 3.2c and 3.2d). If the chain winds up less tightly than in the α-helix, it can form a π-helix, in which the N–H of residue i is hydrogen bonded to the O=C of residue $i + 5$. Extended π-helices are very rare in proteins. Figures 3.2e, 3.2f and 3.2g show an example from photoreactive yellow protein. The box on page 67 shows the backbone conformational angles and geometrical parameters of common secondary structures.

Helix formation satisfies the main chain hydrogen bonding potential of the residues within the helix, except for those at the ends. Often special sequences and conformations appear as caps stabilizing helix termini (Figure 3.3h).

Proline residues in helices must interrupt the hydrogen bonding pattern, because proline does not have an N–H group. Some sequences rich in proline form special conformations. One example is a left-handed helix

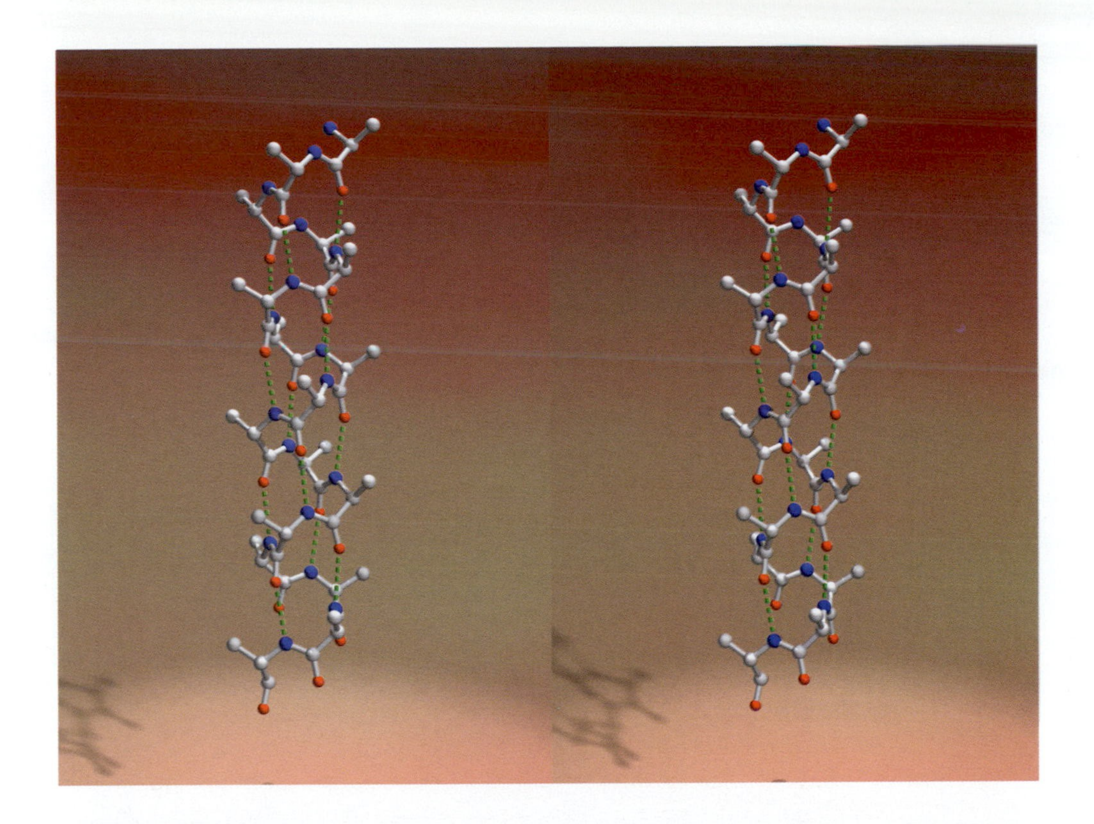

Fig. 3.2a

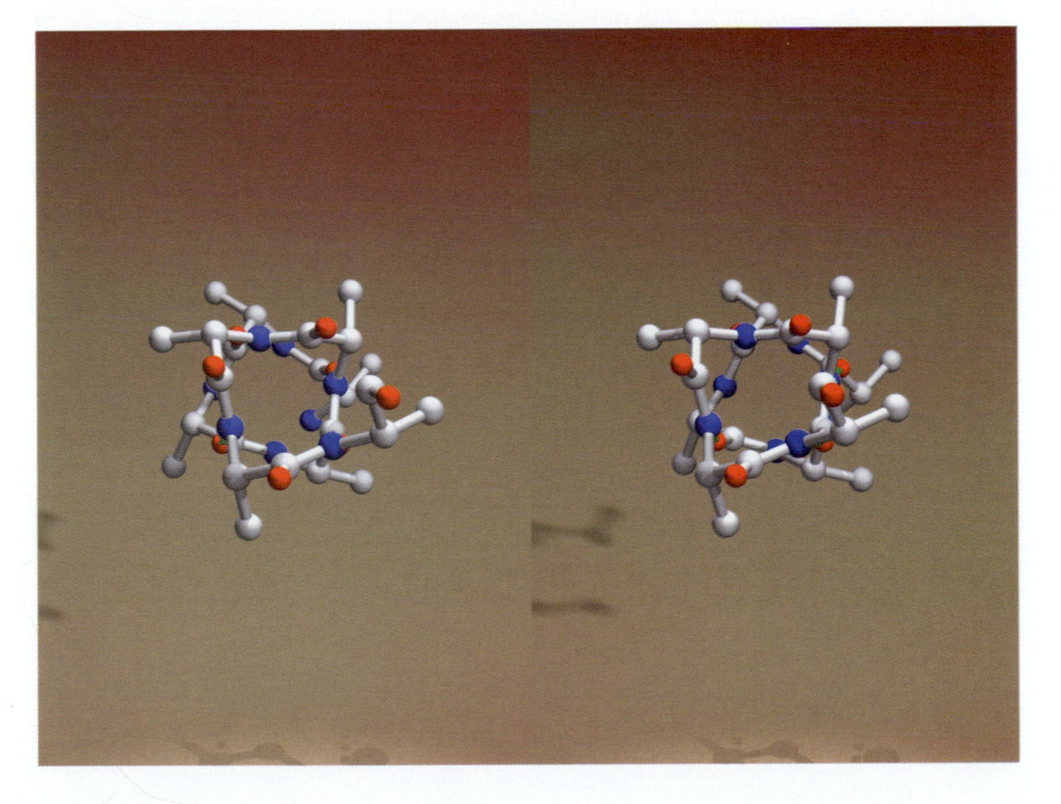

Fig. 3.2b

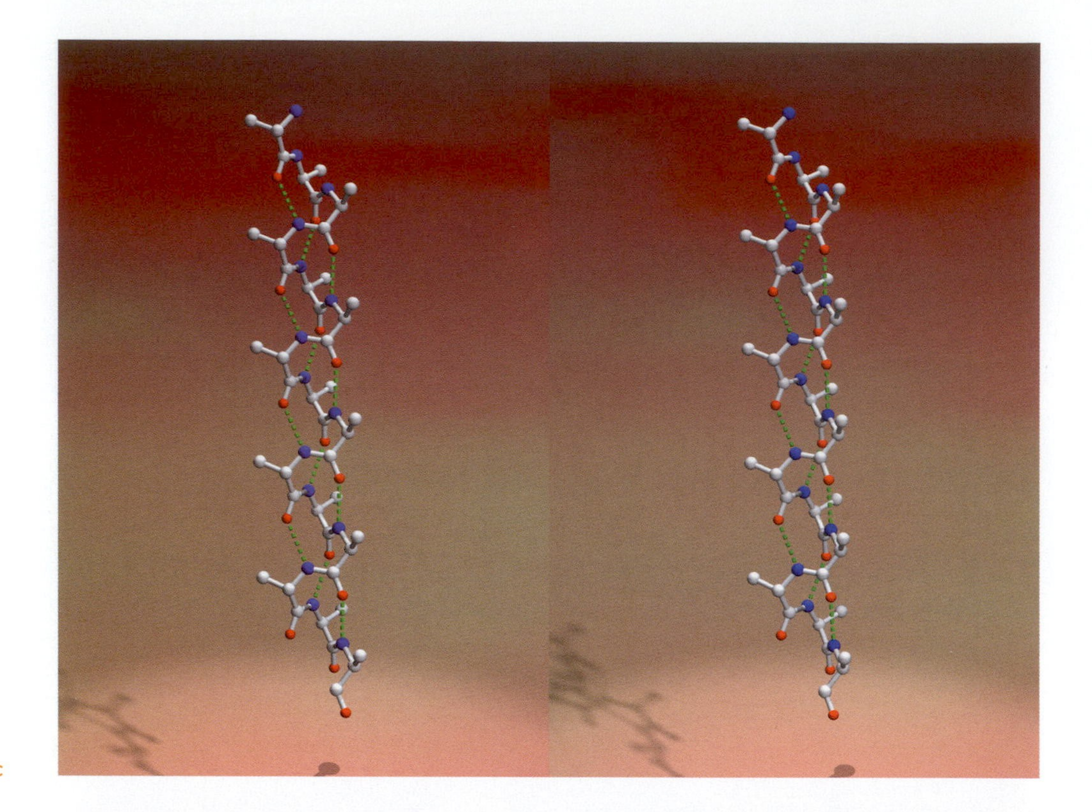

Fig. 3.2c

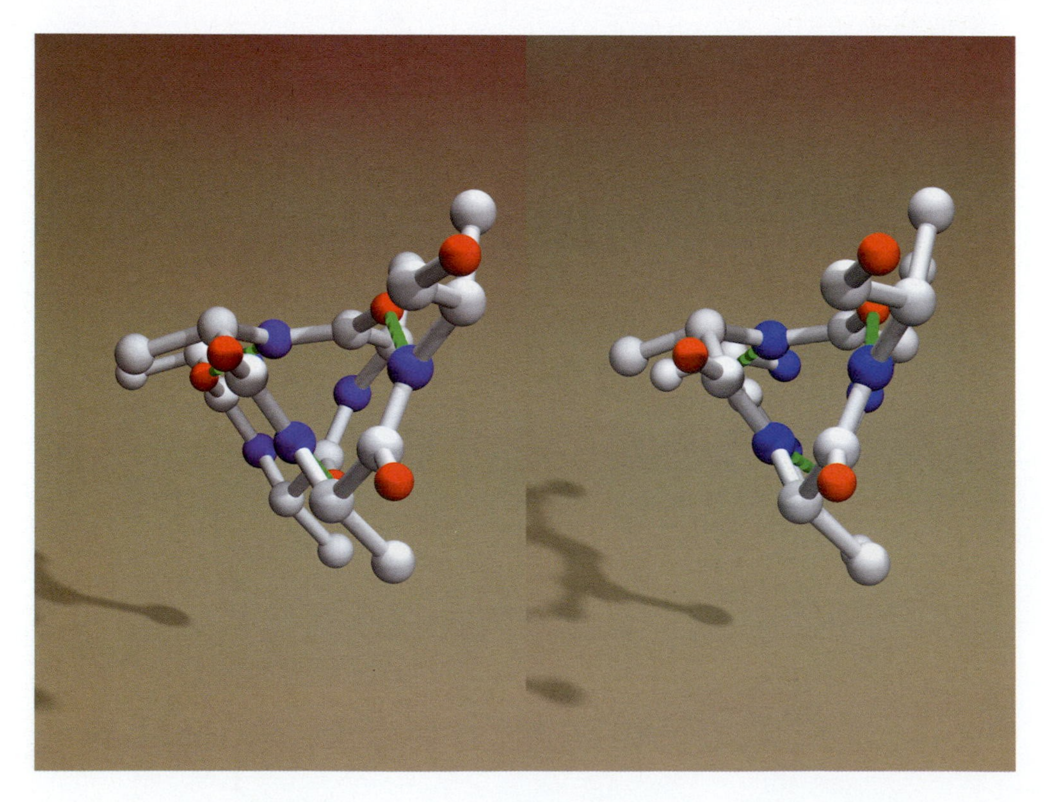

Fig. 3.2d

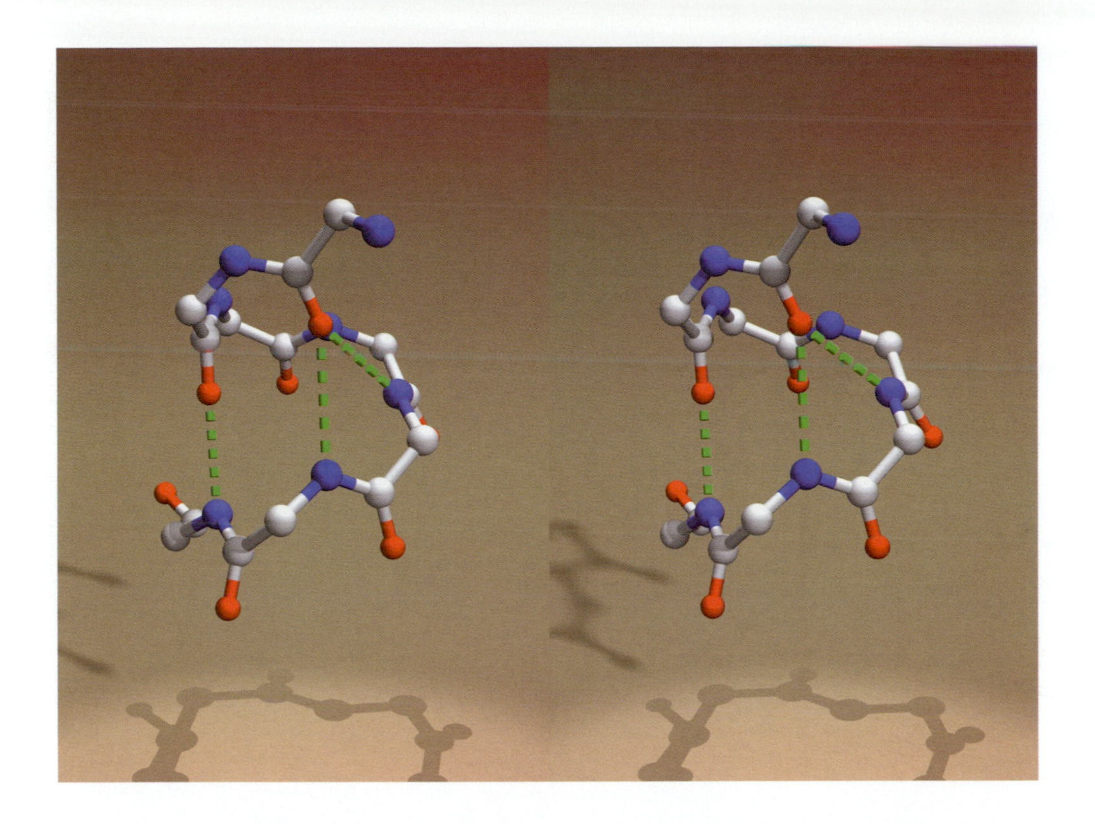

Fig. 3.2e

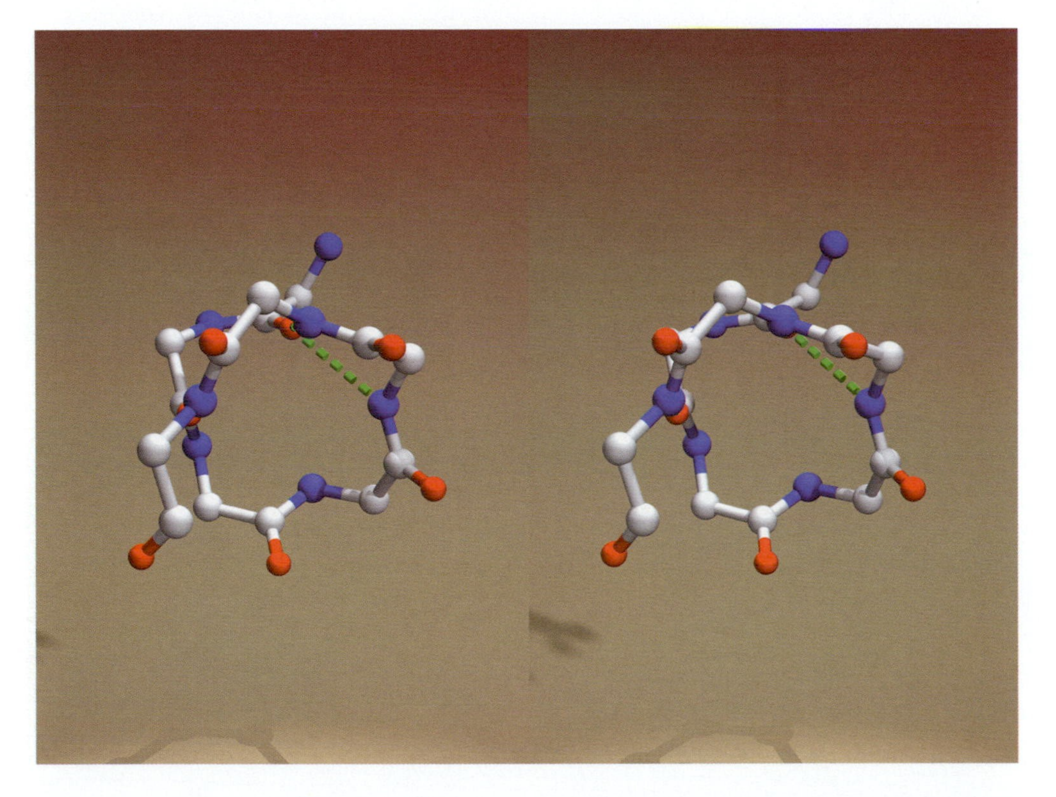

Fig. 3.2f

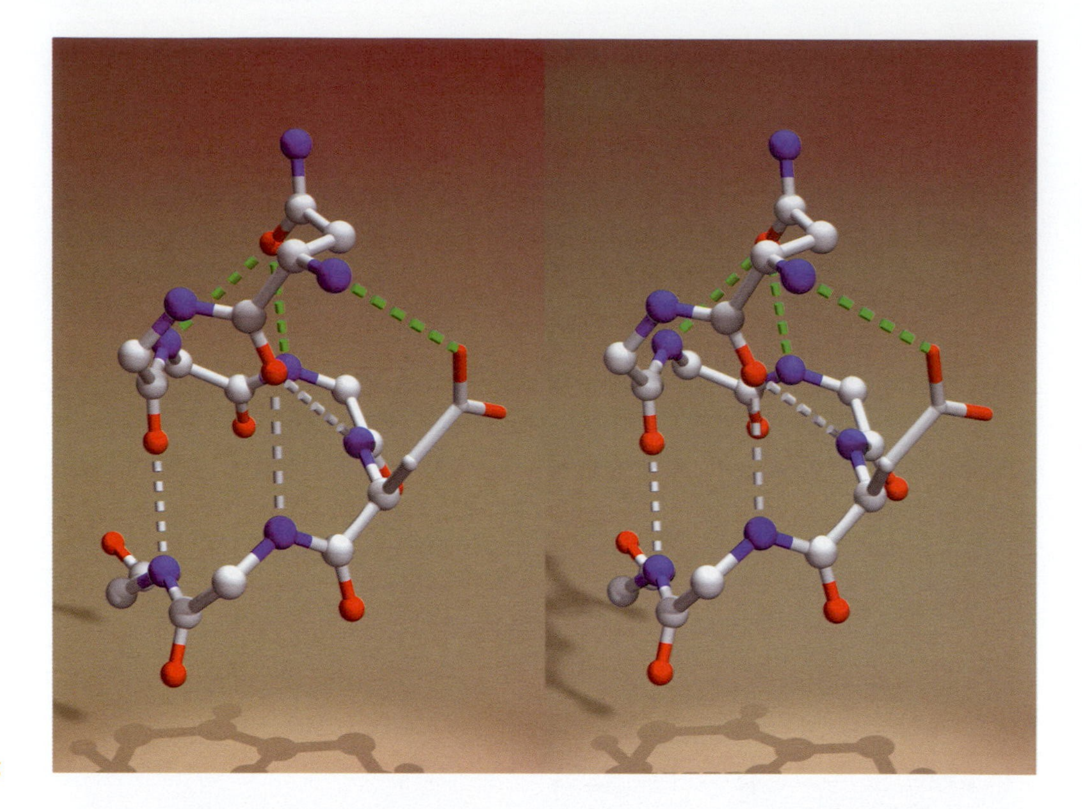

Fig. 3.2g

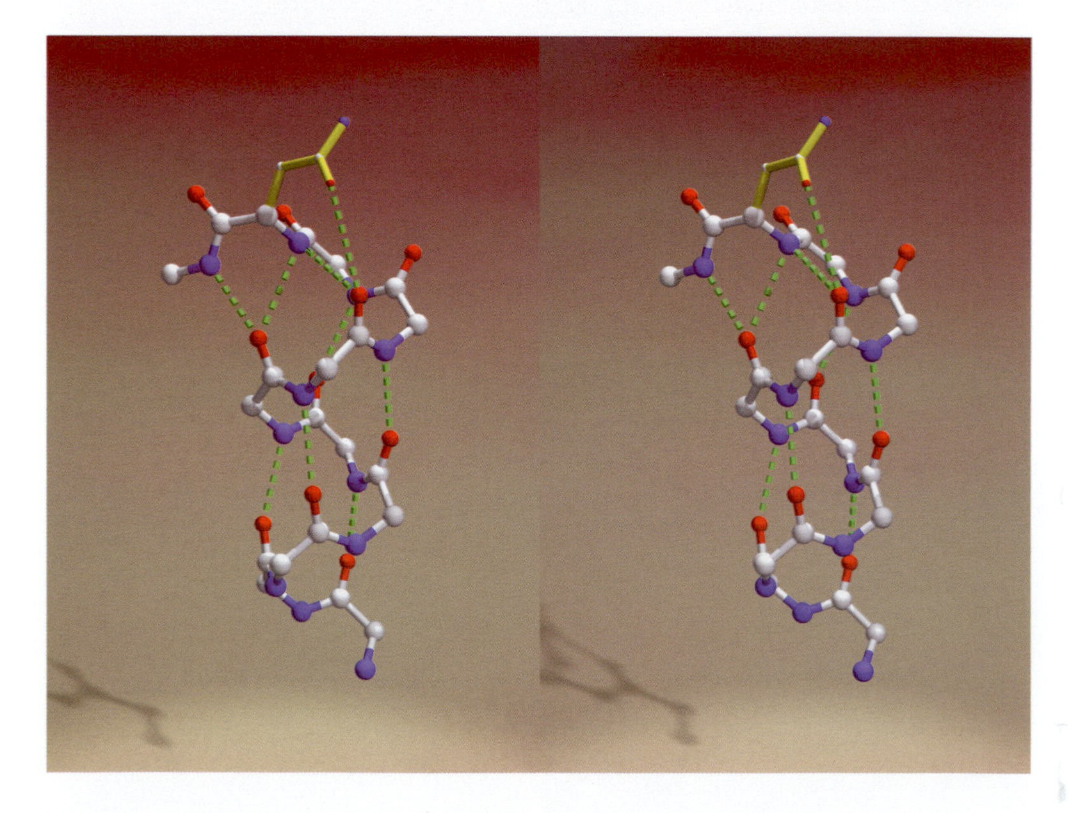

Fig. 3.2h

Fig. 3.2 Comparison of (a,b) ideal α-helix, (c,d) ideal 3_{10} helix, (e,f) very unusual example of π-helix, from photoactive yellow protein [2PHY]. (g) How does photoactive yellow protein stabilize this unusual structure? Very carefully—it buttresses the π-helix with sidechain hydrogen bonds. (h) A common helix-capping motif: residues from the last turn of this helix, from cytochrome c_6 [1CTJ], form extra hydrogen bonds to main chain and sidechain atoms. The sequence Gly–Asn at the end of the helix provides both conformational freedom and a hydrogen bonding partner. (The Asn, coloured yellow, is shown forming a hydrogen bond between its sidechain and the main chain of the residue three before it in the sequence. It is in the α_L conformation, and the Gly that precedes it has an unusual backbone conformation also.)

Structural parameters for protein secondary structures

Structure	ϕ	ψ	n	d	p
α-helix	−57	−47	3.6	1.5	5.5
3_{10} helix	−49	−26	3.0	2.0	6.0
β-helix	−57	−70	4.4	1.1	5.0
Polyproline II helix	−79	+149	3.0	3.1	9.4
Parallel β strand	−119	+113	2.0	3.2	6.4
Antiparallel β strand	−139	+135	2.0	3.4	6.8

ϕ and ψ are the conformational angles of the mainchain, with $\omega \sim 180°$ (the trans conformation)

n = the number of residues per turn.

d = the displacement between sucessive residues along the helix axis.

p = the pitch of the helix, the distance along the helix axis of a complete turn. Note that $p = n \times d$. (The equation is exact; the values of p, n and d in the table have been rounded to two significant figures.)

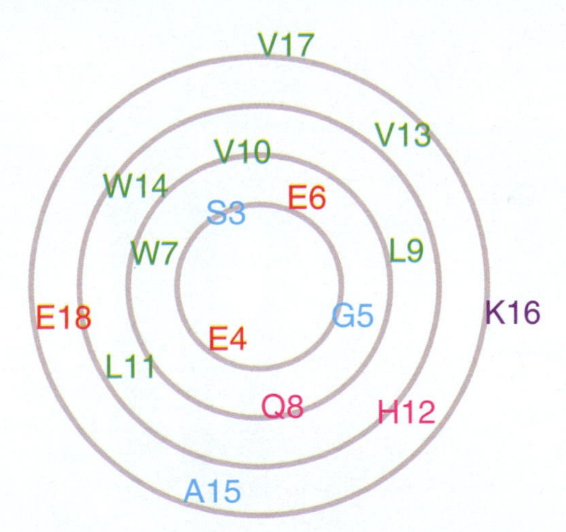

Fig. 3.3 A 'helical wheel' drawn from the sequence of the A helix of sperm whale myoglobin. Note the that lower face is charged, and the upper face is hydrophobic. The total angle of rotation between a residue and the residue seven down the chain is 7 × 100° = 700° or approximately two full turns. Colour code: small, cyan; medium–large hydrophobic, green; polar, magenta; positively charged, blue; negatively charged, red.

called a polyproline II helix. Ligands bound to SH3 domains, molecules active in signal transduction from the cell surface to the nucleus, form the polyproline II conformation.

Hydrogen-bonding patterns in sheets

A β-sheet is formed from separate strands, which may arise from regions distant in the sequence. (In contrast, a helix contains a single consecutive stretch of residues.) A pair of adjacent strands in a β-sheet may interact in two possible orientations: parallel or antiparallel. β-sheets may have all strands parallel (called a parallel sheet for short), all adjacent strands antiparallel (called an antiparallel sheet), or mixed (Figure 3.4). The relationship between the positions of the strands in a sheet in space, and their positions in the sequence, is quite variable. However, two special cases are common. (1) It is possible to form an antiparallel sheet in which adjacent strands appear successively in the sequence, separated by turns called β-hairpins. (2) In contrast, two adjacent parallel strands require a bridging segment, often an α-helix. Many proteins contain a succession of β–α–β units; for example the NAD-binding domains of dehydrogenases (see Figure 2.7).

Note the very different hydrogen-bonding patterns in parallel and antiparallel sheets (Figure 3.5). The principle is that the mainchain hydrogen-bonding groups (N–H and O=C) are in the plane of the sheet, with N–H and O=C groups from successive residues pointing in opposite directions.

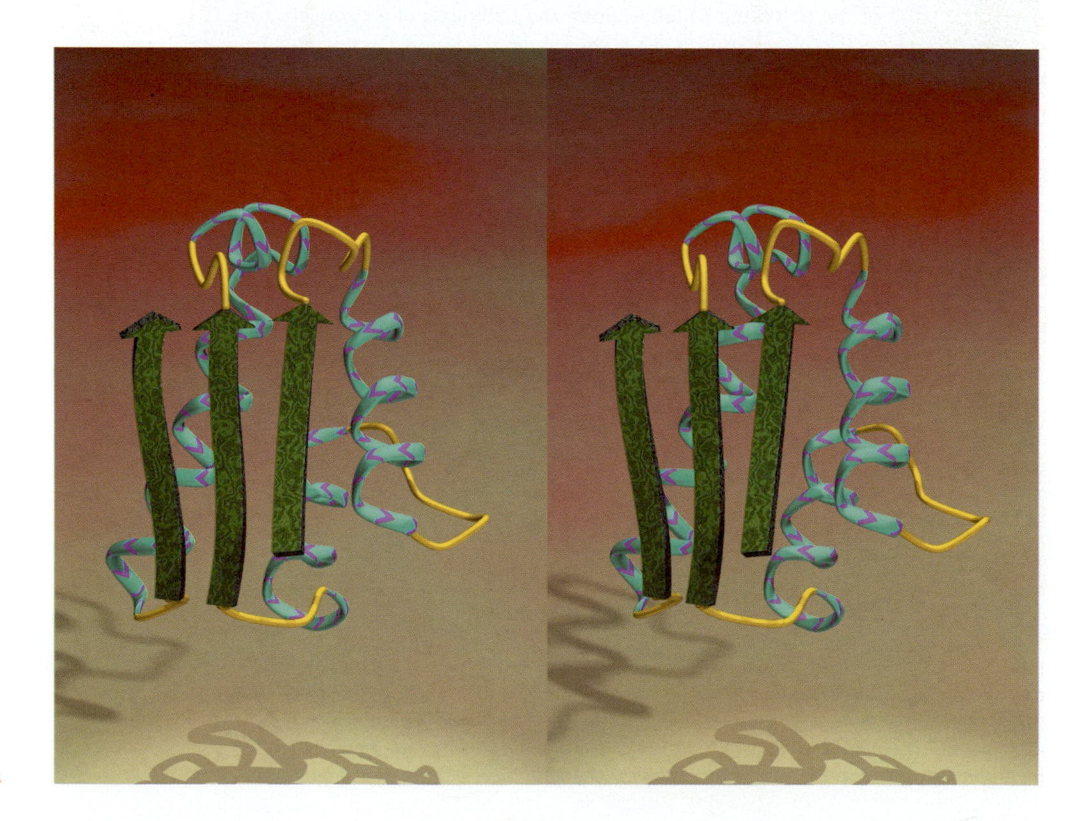

Fig. 3.4a

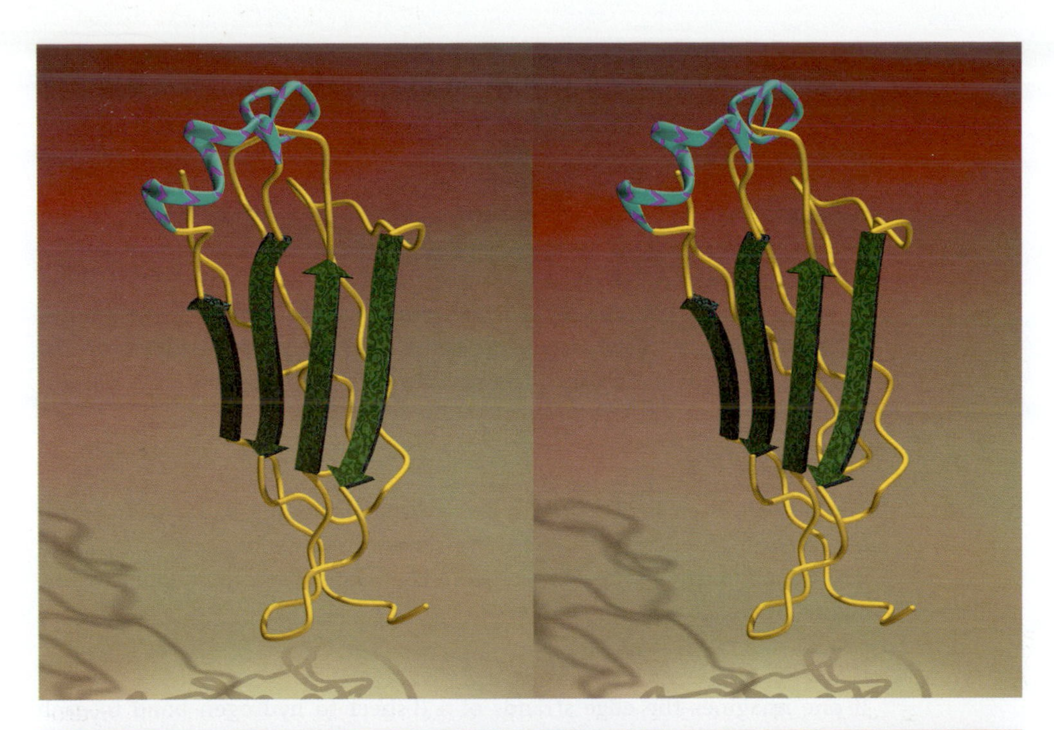

Fig. 3.4b

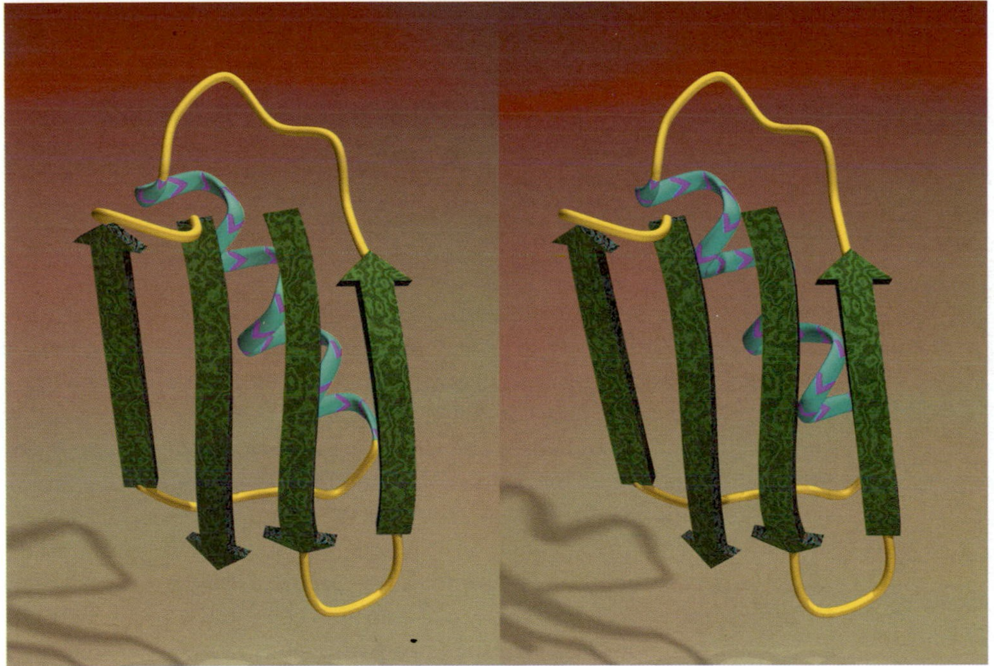

Fig. 3.4c

Fig. 3.4 Examples of small protein domains containing (a) a parallel β-sheet: Barstar (an inhibitor of Barnase = BActerial RiboNucleASE), with a mutant sequence C40→A, C82→A [1BRS]. (b) an antiparallel β-sheet: the CL domain from immunoglobulin TE33 [1TET]. This is in fact a 'β-sheet sandwich': there is a second sheet at the back of the picture. (c) a 'mixed' β-sheet: streptococcal protein G B1 immunoglobulin binding domain [1PGA].

Therefore if the N–H group of one residue forms a hydrogen bond to the C=O of a residue from the strand on its *left*, the N–H of the *next* residue will form a hydrogen bond to the C=O of a residue from the strand on the *right*. Successive residues continue this alternation (see Figure 3.5). In antiparallel strands, each residue forms two hydrogen bonds with a single residue on the adjacent strand. In parallel strands, each residue forms hydrogen bonds to two residues, separated by two in the sequence, on the adjacent strand.

The hydrogen-bonding pattern of a sheet can be summarized in a diagram in which the residues are projected onto the plane of the sheet and reduced to a circle. Broken lines indicate the hydrogen-bonding pattern (Figure 3.5).

The β bulge

The β bulge is an irregularity in the hydrogen-bonding pattern of a sheet observed in an edge strand. Some residues—usually only one or two, but occasionally many more—deviate from the relatively straight course of the chain (Figure 3.6).

β-barrels

If one imagines the edge strands of a β-sheet to hydrogen bond to each other, a closed structure is created, called a β-barrel. Think of the sheet as wrapped around a cylinder (Figure 3.7). The β-sheet in glycolate oxidase is an eight-stranded, all-parallel sheet (Figure 3.8). The eight strands are linked by eight helices packed around the outside of the sheet. This type of β-barrel was originally discovered in triose phosphate isomerase, and is often called a 'TIM barrel.'

The hierarchical nature of protein architecture

K.U. Linderstrøm-Lang described the following levels of protein structure: The sequence—the set of primary chemical bonds—is called the *primary structure*. The assignment of helices and sheets—the hydrogen-bonding pattern of the main chain—is called the *secondary structure*. The assembly and interactions of the helices and sheets is called the *tertiary structure*. For proteins composed of more than one subunit, for instance the photosynthetic reaction centre, the assembly of the monomers is the *quaternary structure*. In some cases, evolution can merge proteins—changing tertiary to quaternary structure. For example, five separate enzymes in the bacterium *E. coli*, that catalyse successive steps in the pathway of biosynthesis of aromatic amino acids, correspond to five regions of a single protein in the fungus *Aspergillus nidulans*. Sometime homologous monomers form oligomers in different ways; for instance, globins form tetramers in mammalian haemoglobins, and dimers—using a different interface—in the ark clam *Scapharca inaequivalvis*.

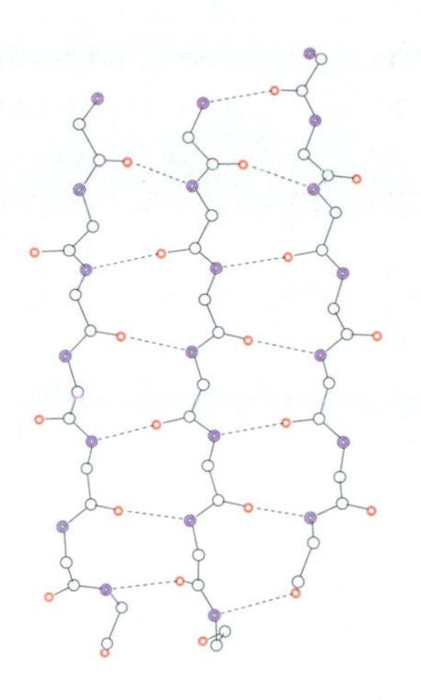

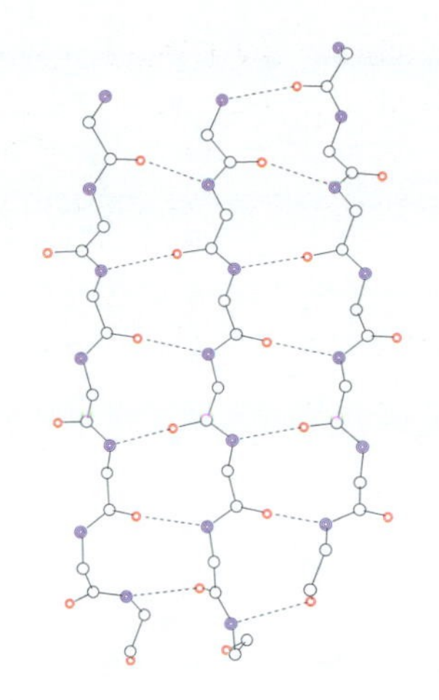

Fig. 3.5a

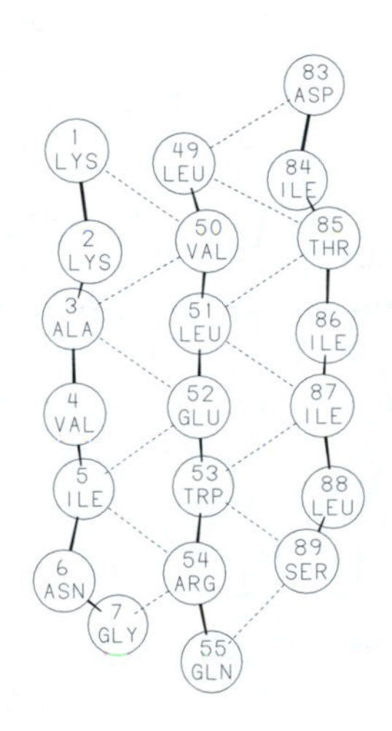

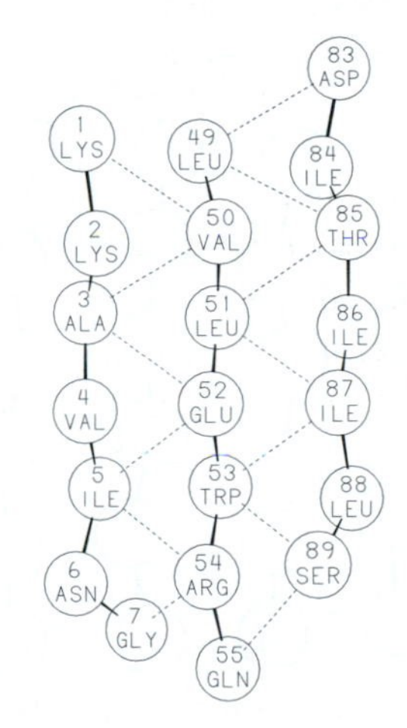

Fig. 3.5b

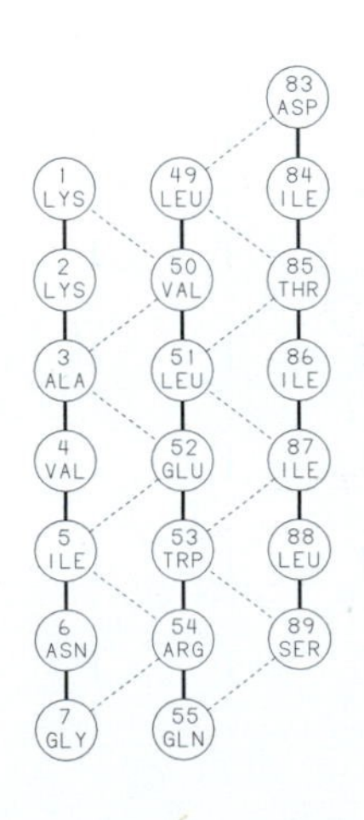

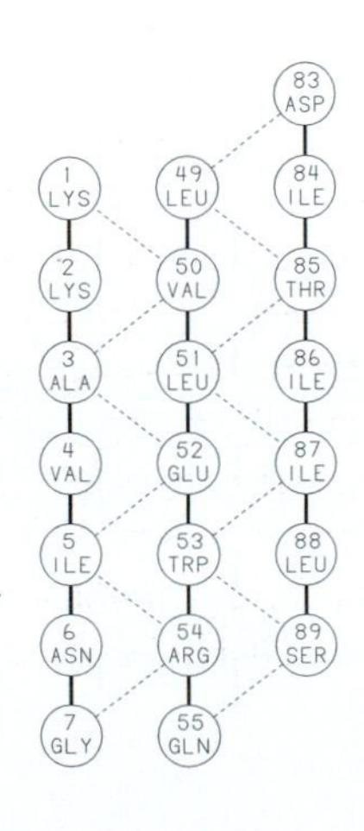

Fig. 3.5c

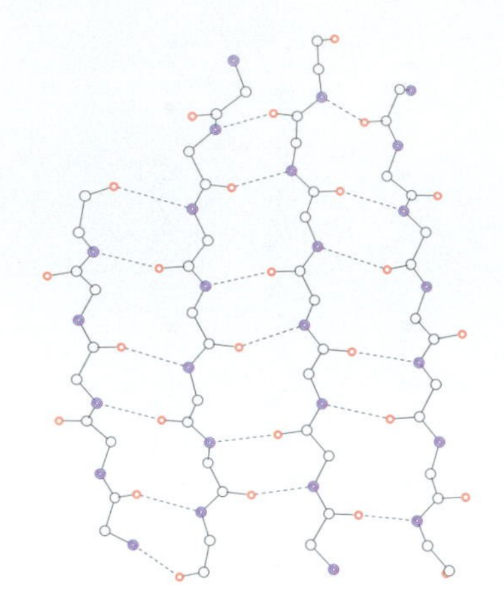

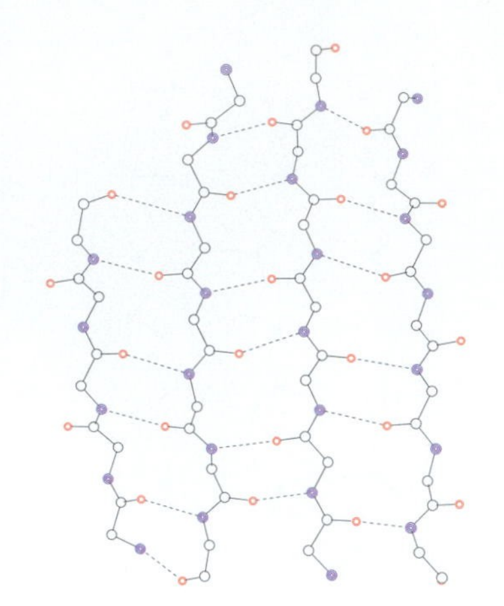

Fig. 3.5d

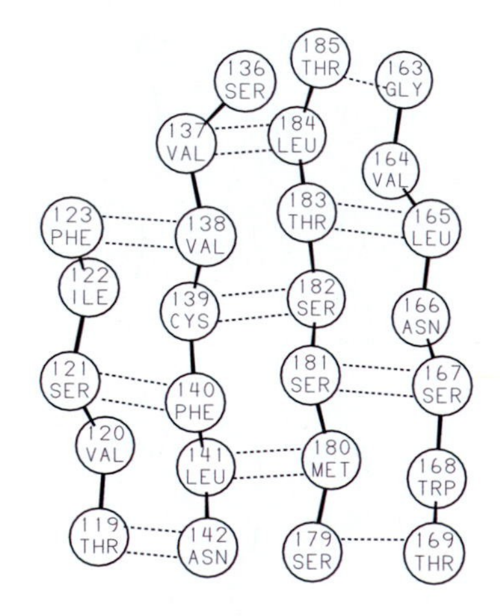

Fig. 3.5e

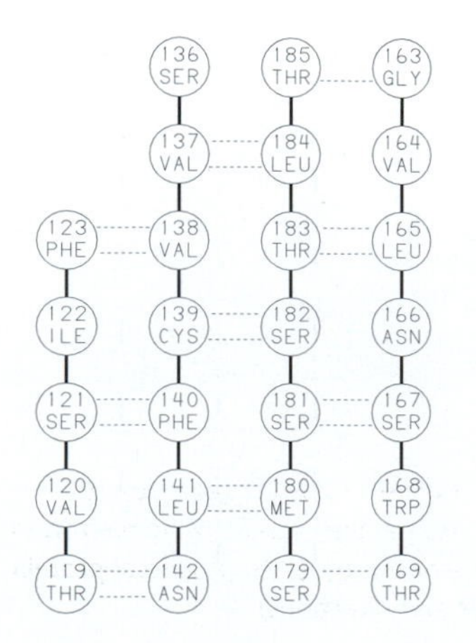

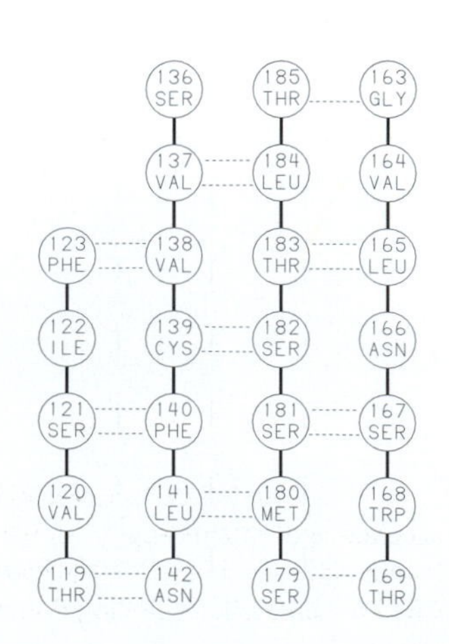

Fig. 3.5f

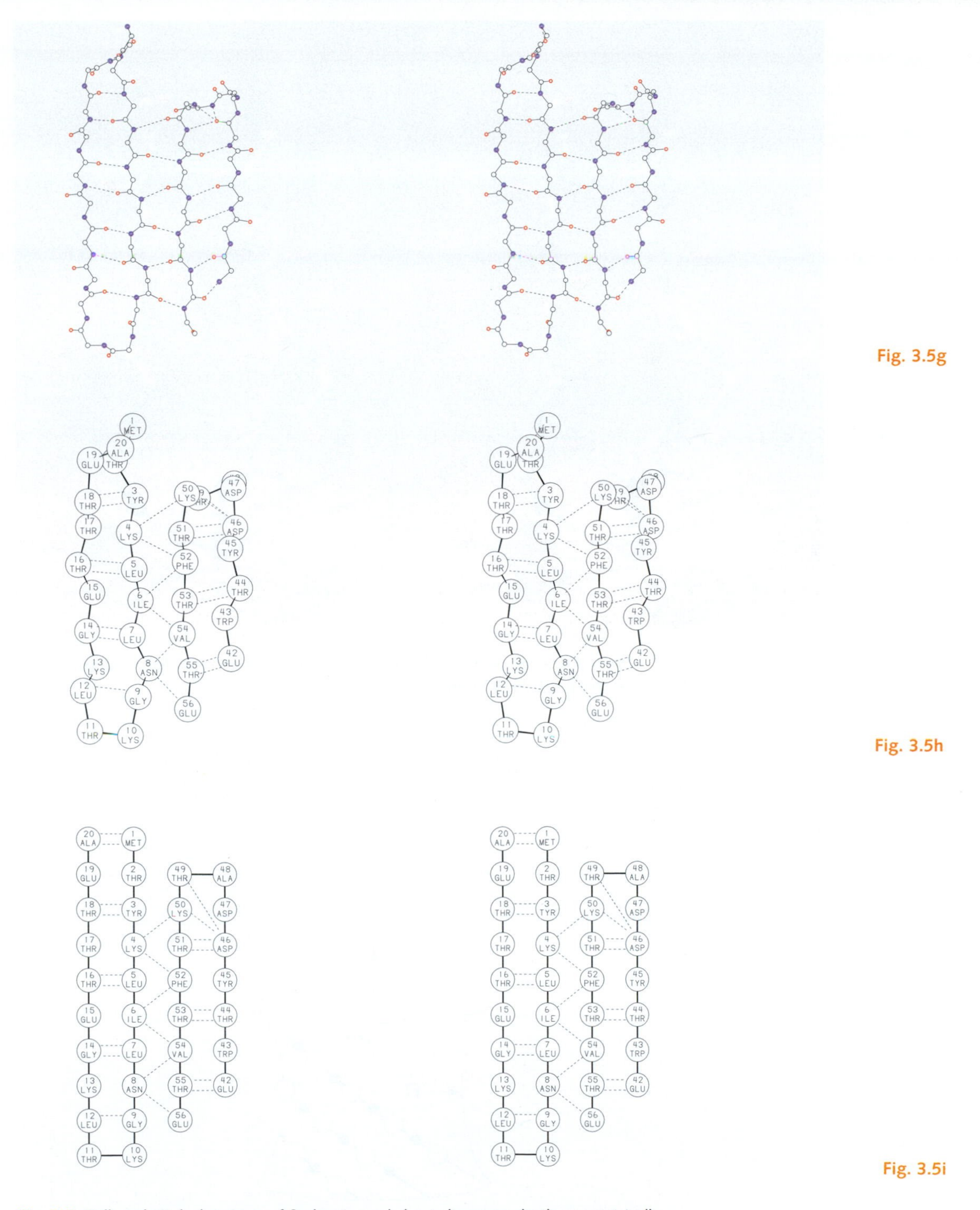

Fig. 3.5 Ball-and-stick drawings of β-sheets, and sheet diagrams, both geometrically accurate and 'squared-up.'(a–c) Barstar [1BRS] (all strands parallel). (d–f) CL domain from immunoglobulin TE33 [1TET] (all adjacent strands antiparallel). (g–i) Streptococcal protein GB1 immunoglobulin binding domain [1PGA] (mixed strand directions).

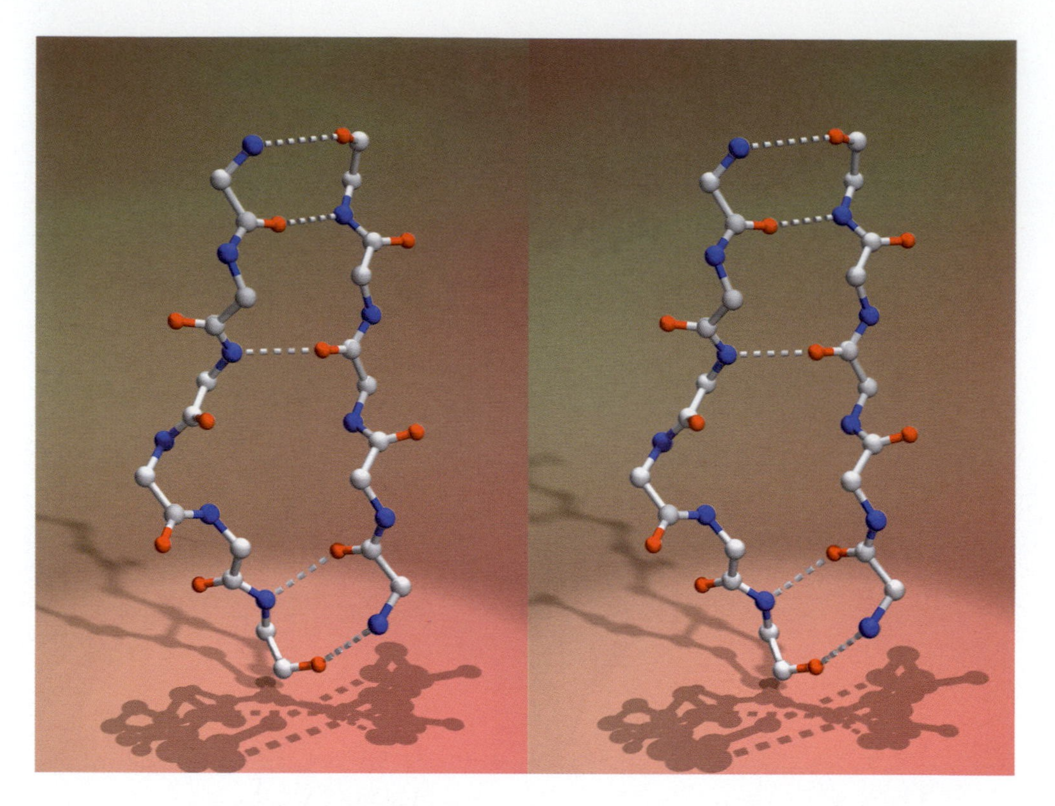

Fig. 3.6 The β bulge, an insertion of one or more non-hydrogen-bonded residues into a sheet (from Fab J539 [2FBJ].)

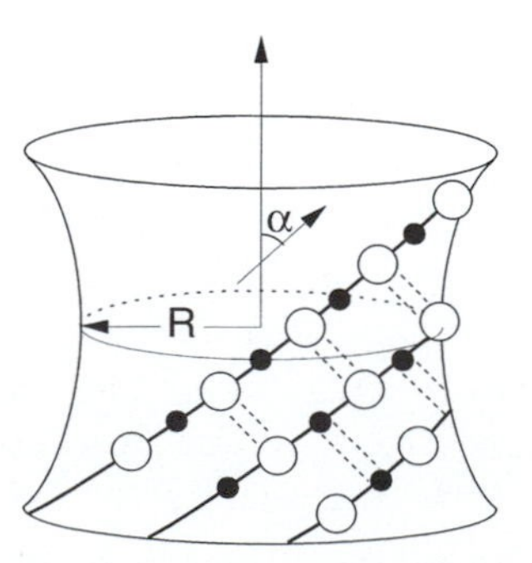

Fig. 3.7 Generalized and schematic idea of a β-barrel. The topology of a β-barrel determines its geometric properties such as its radius, R, and the angle between the strands and the barrel axis, α.

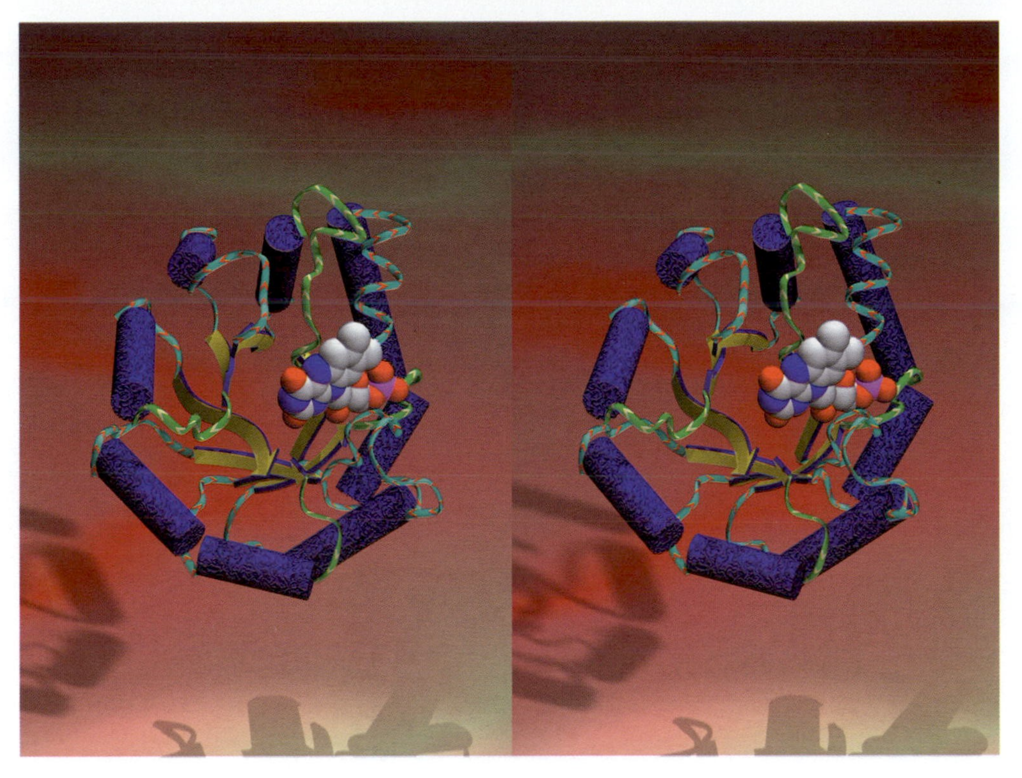

Fig. 3.8 Spinach glycolate oxidase, a β-barrel, binding FMN [1GOX].

It has proved useful to add additional levels to Linderstrøm-Lang's classification:

1. **Supersecondary structures.** Proteins show recurrent patterns of interaction between helices and sheets close together in the sequence. These supersecondary structures include the α-helix hairpin, the β-hairpin, and the β–α–β unit (Figure 3.9). In fact there are examples of each of these in structures previously illustrated.
2. **Domains.** Many proteins contain compact units within the folding pattern of a single chain, that look as if they should have independent stability. These are called domains. Phosphoglycerate kinase is an enzyme with its active site in a cleft between two domains (Figure 3.10). In the hierarchy, domains fall between supersecondary structures and the tertiary structure of a complete monomer. It is difficult to define the idea of a domain precisely, and most people would concur with Mr Justice Stewart in saying that 'perhaps I could never succeed in intelligibly doing so. But I know it when I see it.' Indeed, the web sites that present systematic classifications of protein structures differ, in many cases, in the way that they divide structures into domains.
3. **Modular proteins.** Modular proteins are multidomain proteins which often contain many copies of closely related domains. The domains can appear in many proteins in different structural contexts; that is, different modular proteins can 'mix and match' sets of domains. For example,

Fig. 3.9a

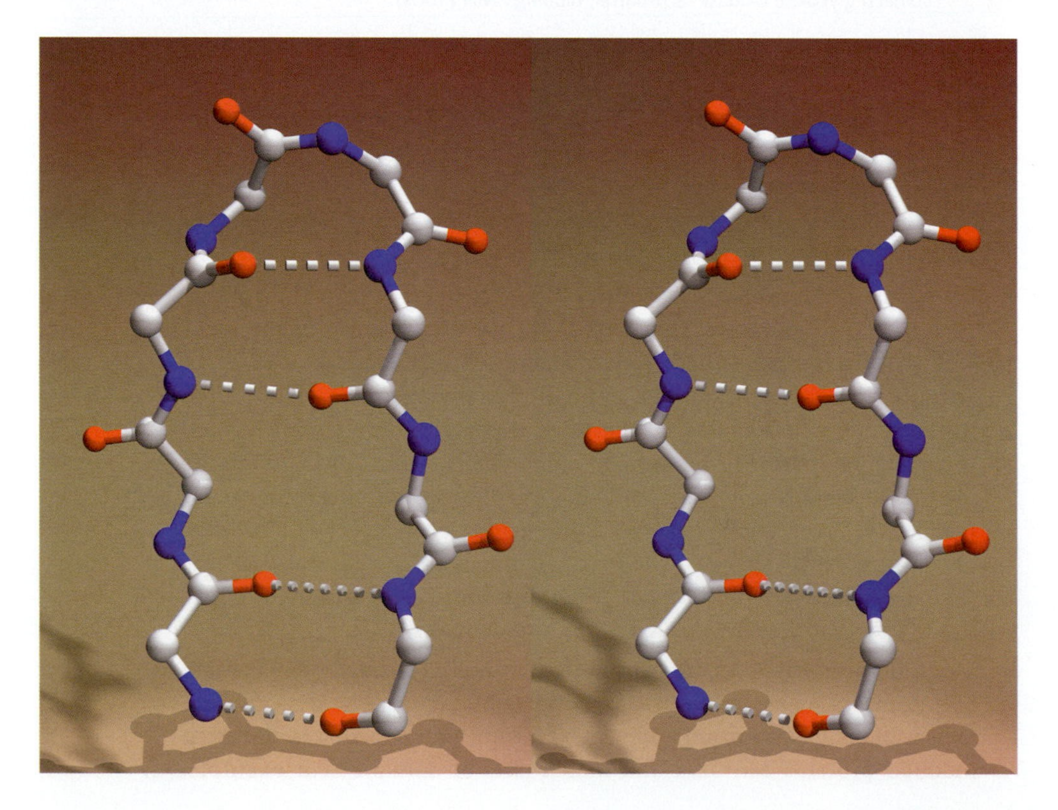

Fig. 3.9b

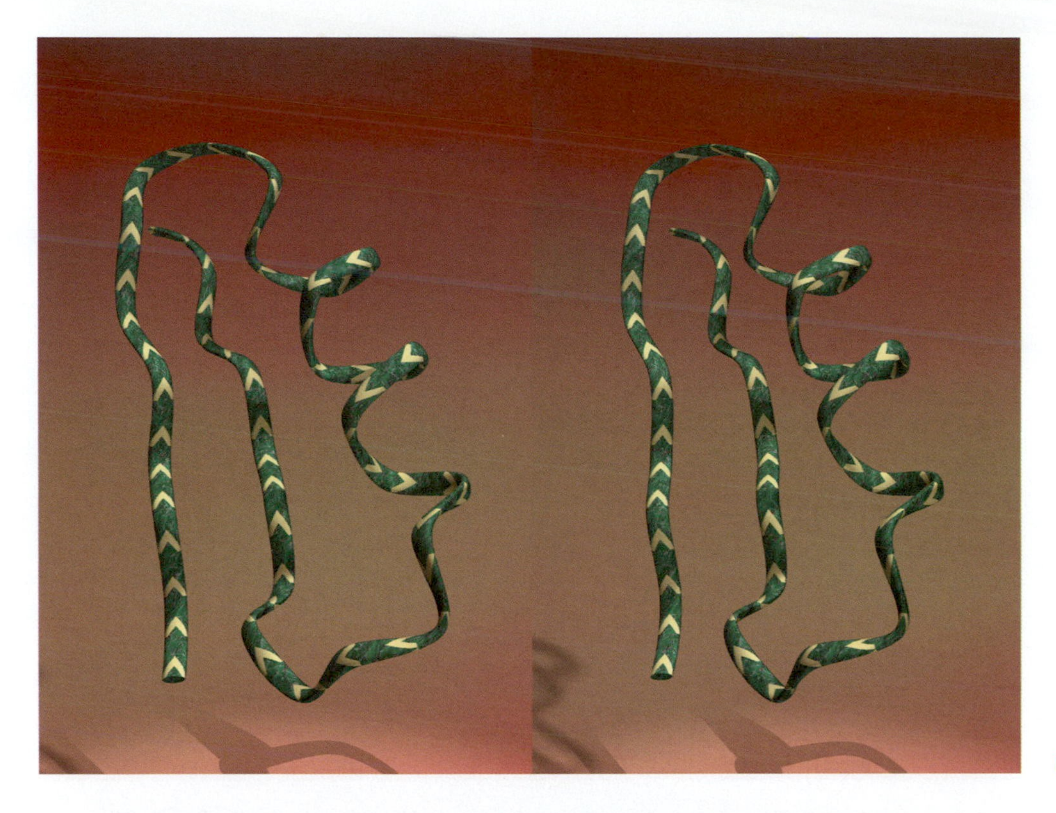

Fig. 3.9c

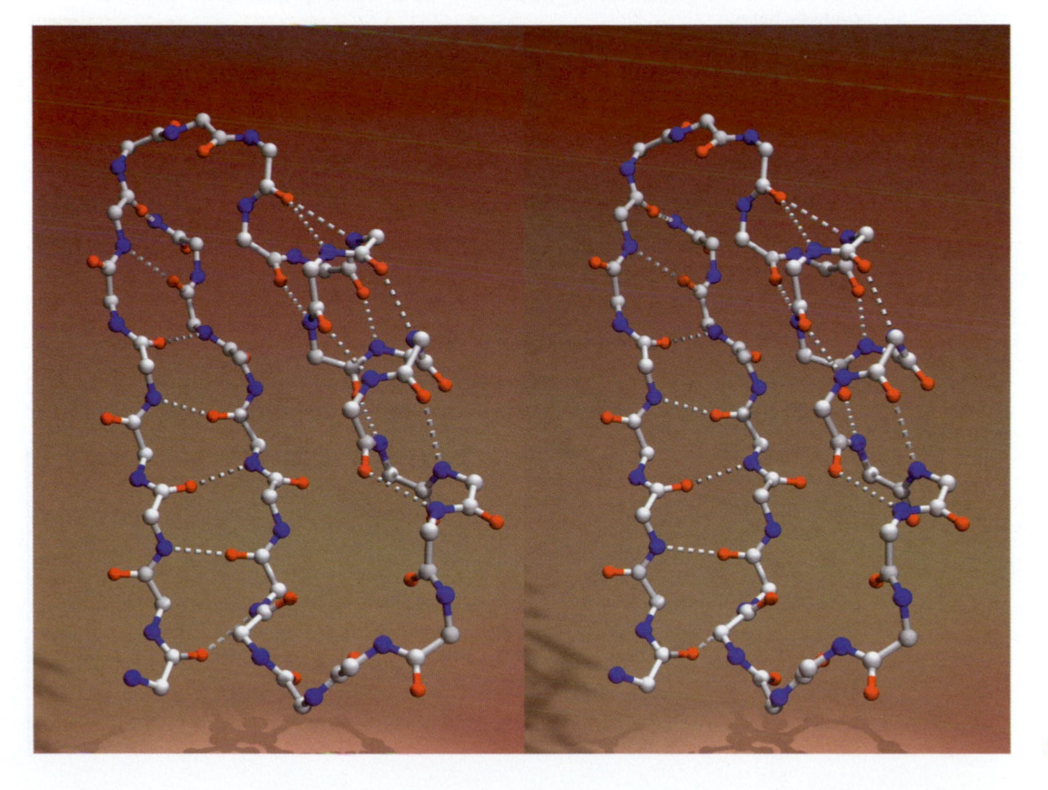

Fig. 3.9d

Fig. 3.9e

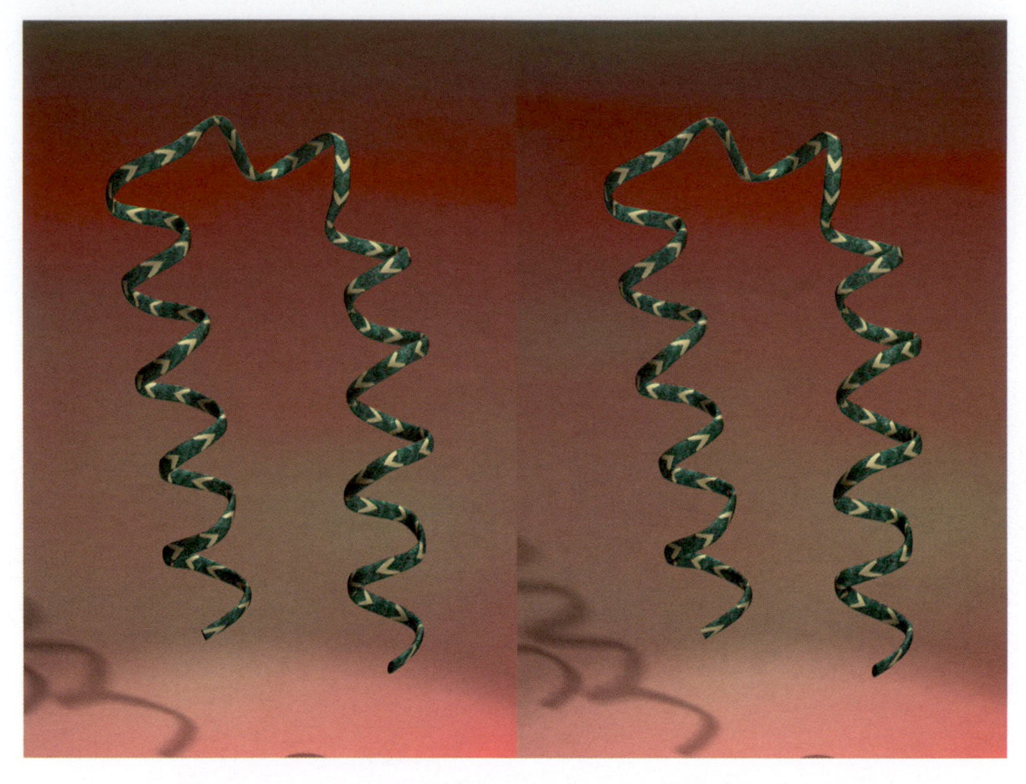

Fig. 3.9 Common supersecondary structures: (a, b) a β hairpin, (c, d) a β–α–β unit (e) helix hairpin.

fibronectin, a large extracellular protein involved in cell adhesion and migration, contains 29 domains including multiple tandem repeats of three types of domains called F1, F2, and F3. It is a linear array of the form: $(F1)_6\ (F2)_2\ (F1)_3\ (F3)_{15}\ (F1)_3$. Fibronectin domains also appear in other modular proteins.

(See http://www.bork.embl-heidelberg.de/Modules/ for pictures and nomenclature.)

Assignment of helices and sheets

A first step in the analysis of a protein structure is the assignment of the secondary structure: where do the helices and strands of sheet begin and end?

There is a rough analogy between the analysis of protein structures at different levels, and the analysis of text.

The amino acids correspond to letters, the secondary structures to words, supersecondary structures to phrases (or even to clichés), elements of tertiary structure to sentences—this is the level at which true individuality makes its appearance—domains to paragraphs, the structure of a full polypeptide chain to a chapter, and the quaternary structure to the assembly of chapters into a complete book.

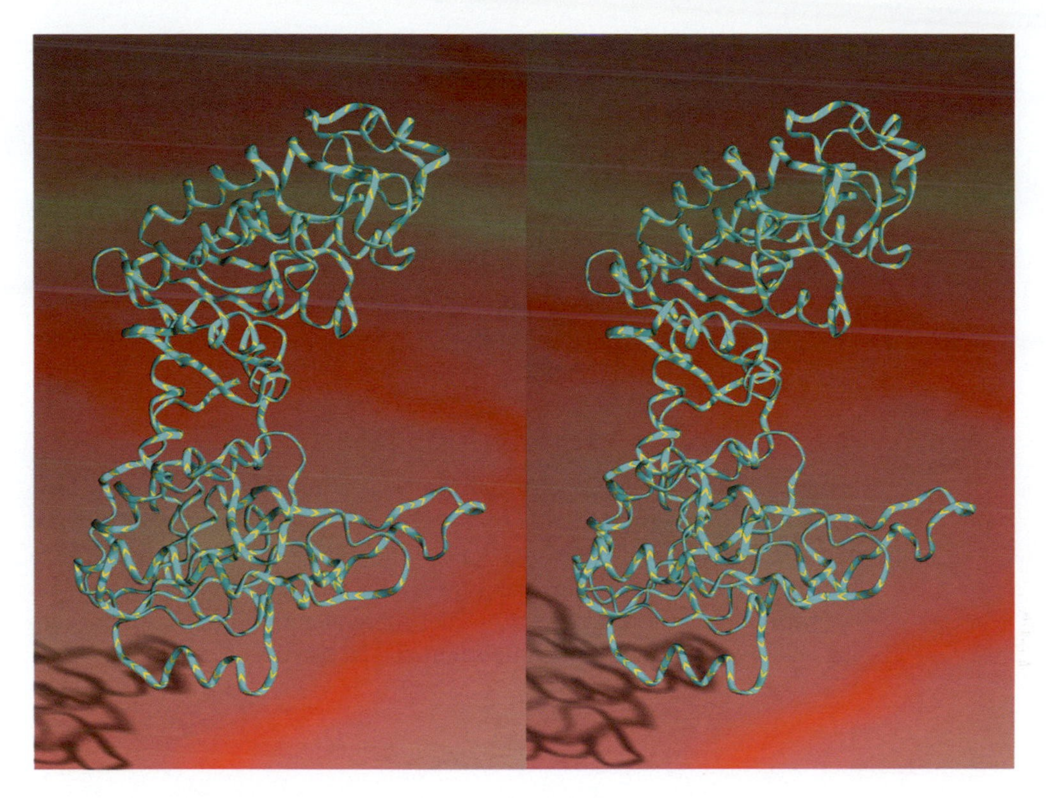

Fig. 3.10 Phosphoglycerate kinase, a protein composed of two domains [3PGK]. The substrate binds between the domains.

It is a fact of life that helices and sheets in globular proteins can be irregular, especially at their ends. Helices frequently either unravel or tighten up in their last turn (see Figure 3.3h), and can bend in the middle, to the extent of losing a hydrogen bond or two (often but not exclusively at a proline). β-sheets may be interrupted by β bulges, and can also show irregularities at ends of strands.

Of course these effects make any definition of the limits of the regions of secondary structures somewhat imprecise. Indeed, for different purposes different definitions may be necessary and appropriate. For example, if one wants to determine all the pairs of residues in contact at a helix–helix interface, the most generous definition of the helical regions should be used. But to measure the interaxial distance and angle between two helices, the definitions should be restricted to those residues that form a regular helix, frayed ends omitted.

Protein Data Bank entries contain helix and sheet assignments reported by the individual crystallographers and NMR spectroscopists who solved the structure. However, scientists do not always agree on the criteria for deciding where regions of secondary structure begin and end. In consequence, there have been attempts to write computer programs to identify regions of secondary structures consistently and objectively. The first of these was by M. Levitt and J. Greer, and there are recent ones by W. Kabsch and C. Sander, and by D. Frishman and P. Argos. Indeed, many crystallographers and NMR spectroscopists now use such programs to define the secondary structures of proteins that they solve. These programs do have the merit that they apply the same criteria, in a consistent way, to all proteins. (Of course it is essential to have an objective way to assign secondary structure, for tests of secondary-structure prediction methods.) What is unfortunate is that people use these secondary structure assignments unquestioningly; perhaps the greatest damage the programs do is to create an impression (for which Levitt, Greer, *et al.* cannot be blamed) that there is A RIGHT ANSWER. Provided that the danger is recognized, such programs can be useful.

An album of small structures

Figure 3.11 contains a dozen small protein structures or domains. These illustrate the structural themes that we have discussed. The reader is urged to trace the chains visually, picking out the helices and sheets, disulphide bridges (are the residues they link close or distant in the sequence?) Can you see supersecondary structures? Consider these pictures as exercises in training your eye to recognize the important spatial patterns.

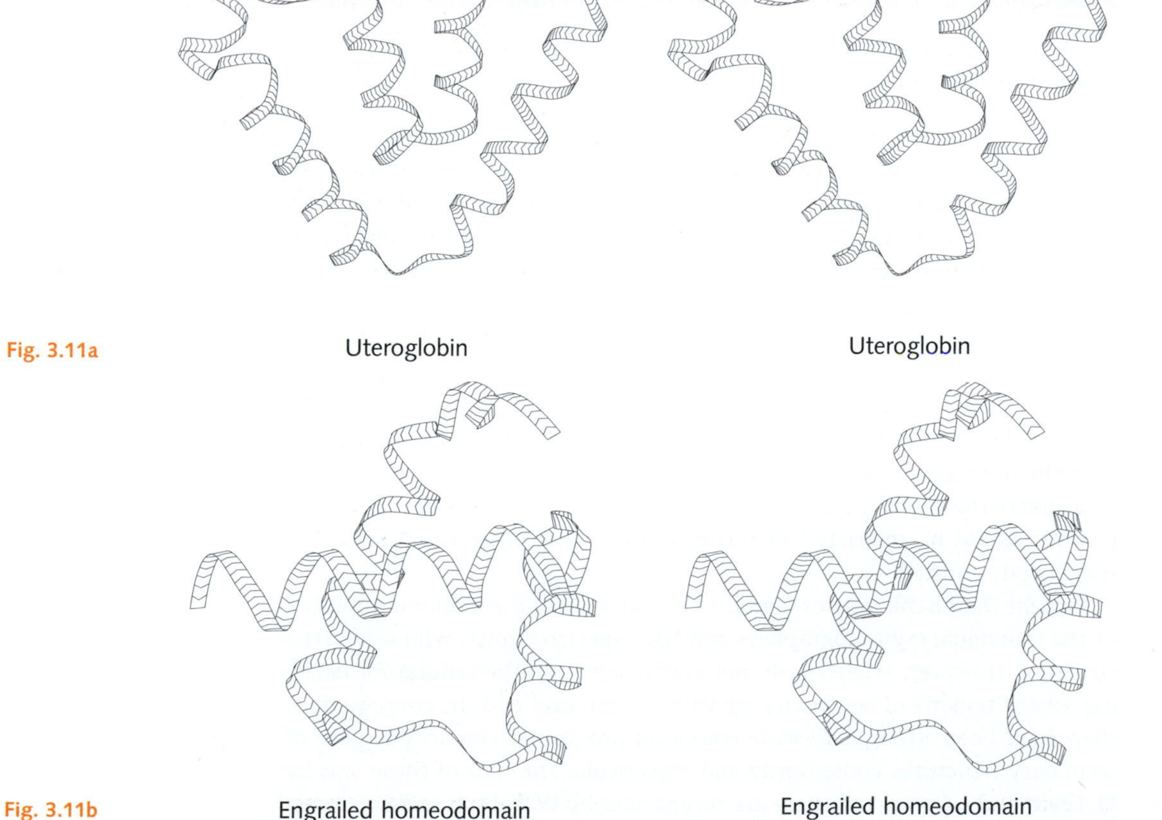

Fig. 3.11a Uteroglobin Uteroglobin

Fig. 3.11b Engrailed homeodomain Engrailed homeodomain

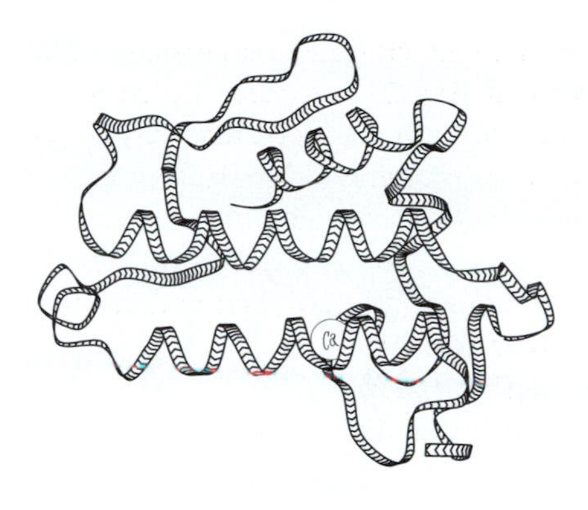

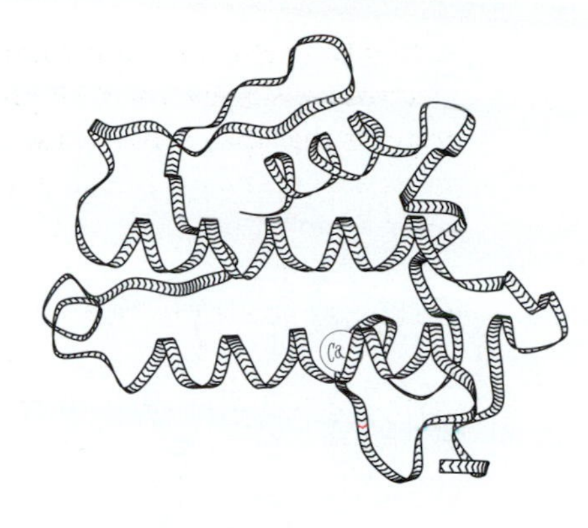

Phospholipase A_2 Phospholipase A_2

Fig. 3.11c

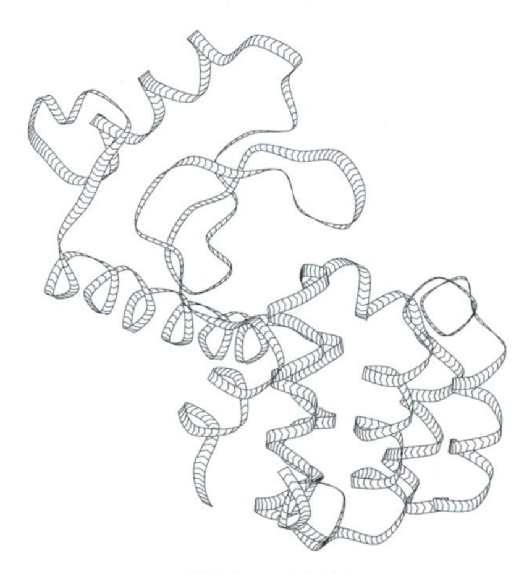

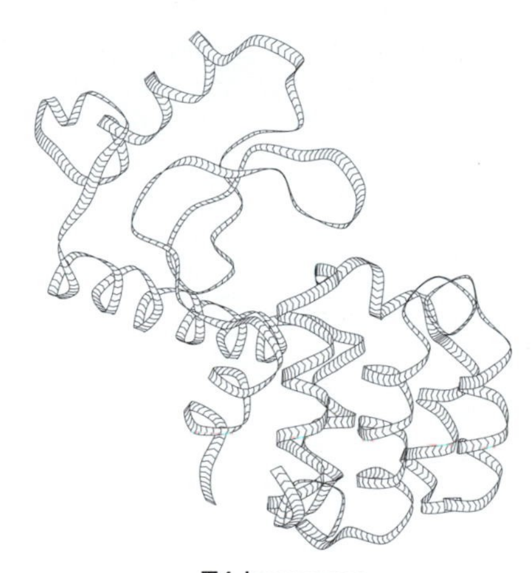

T4 Lysozyme T4 Lysozyme

Fig. 3.11d

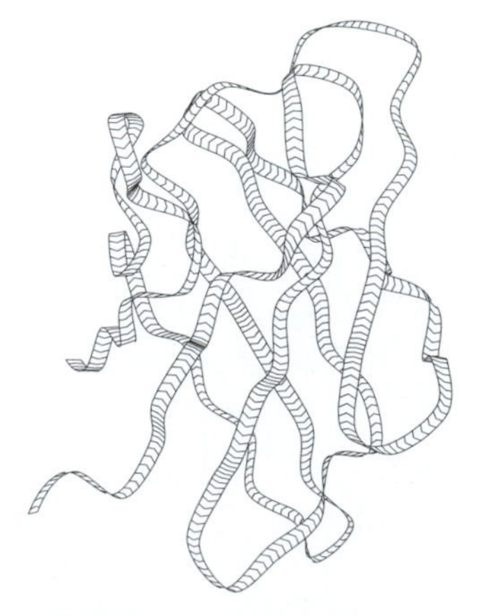

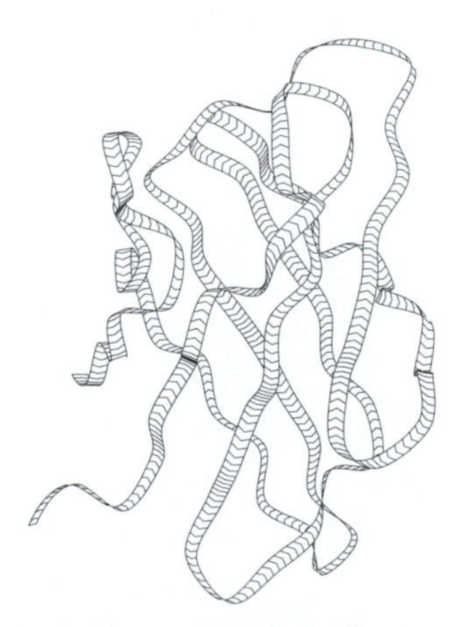

Timothy grass pollen allergen Timothy grass pollen allergen

Fig. 3.11e

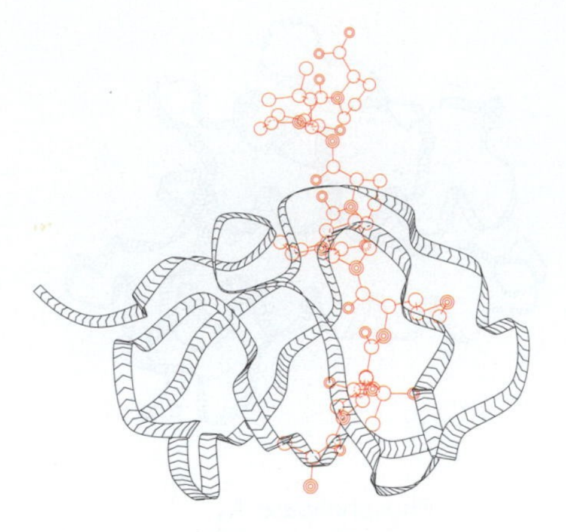

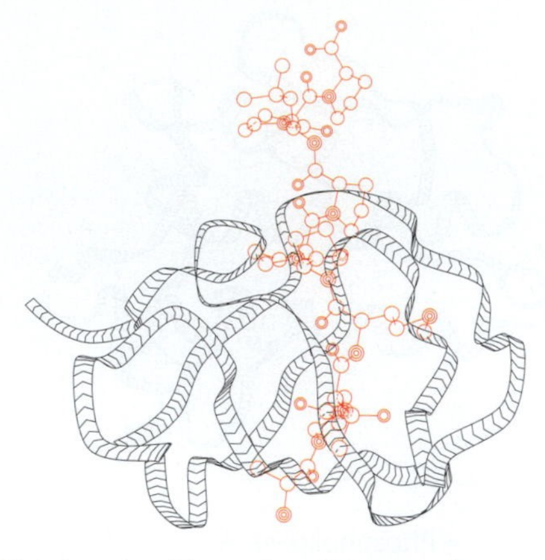

Fig. 3.11f Ab1 tryosine kinase/peptide complex

Ab1 tryosine kinase/peptide complex

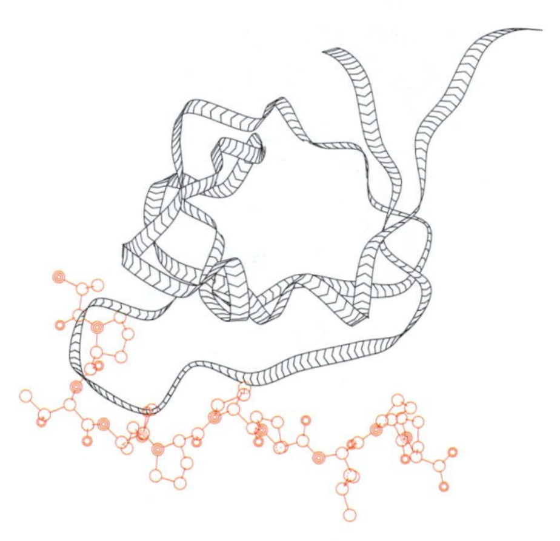

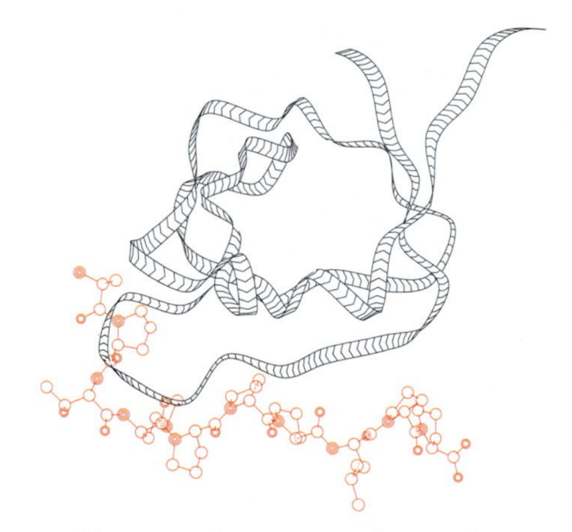

Fig. 3.11g Abl tryosine kinase/peptide complex

Abl tryosine kinase/peptide complex

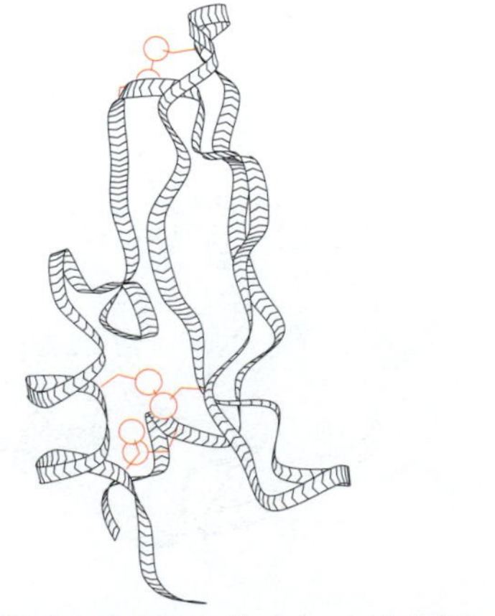

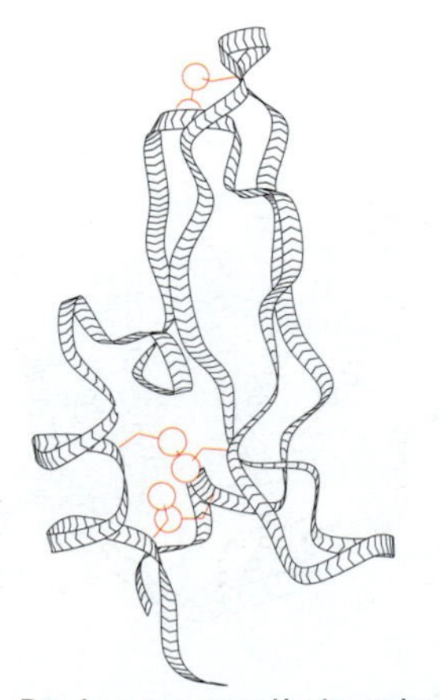

Fig. 3.11h Bovine pancreatic trypsin inhibitor

Bovine pancreatic trypsin inhibitor

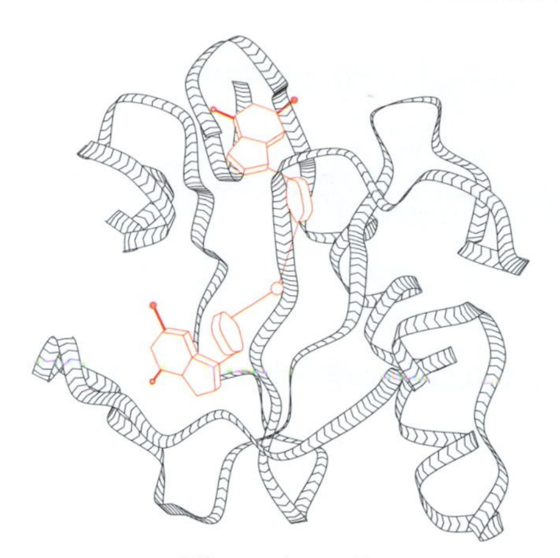

Ribonuclease T1

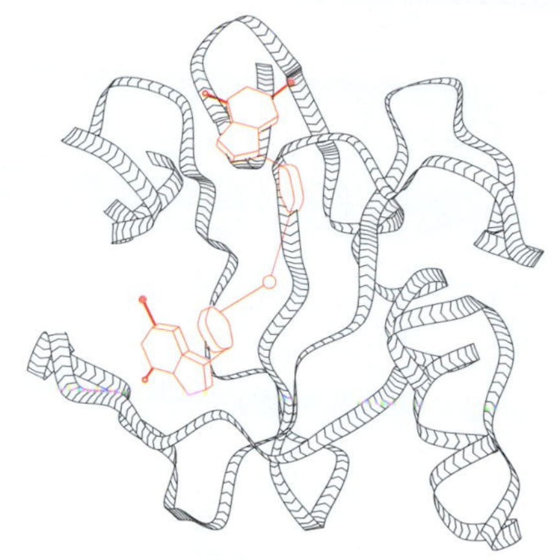

Ribonuclease T1

Fig. 3.11i

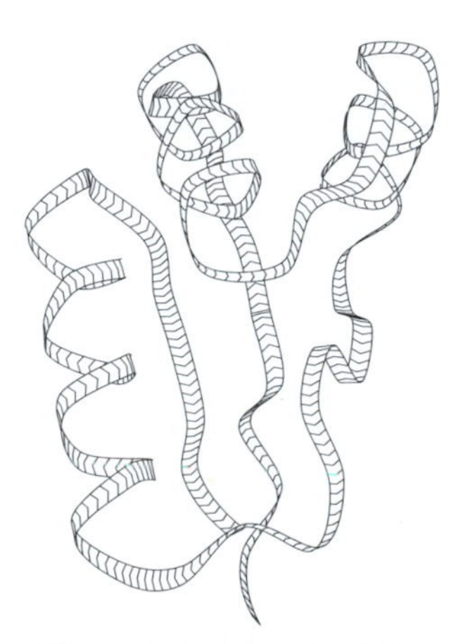

Ribosomal protein L7/L12

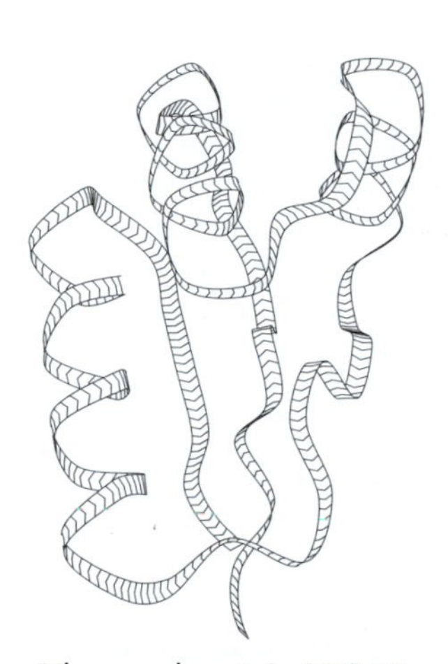

Ribosomal protein L7/L12

Fig. 3.11j

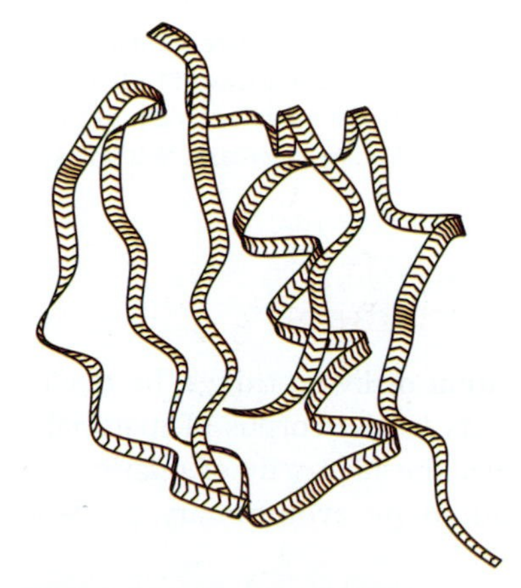

Barley chymotrypsin inhibitor

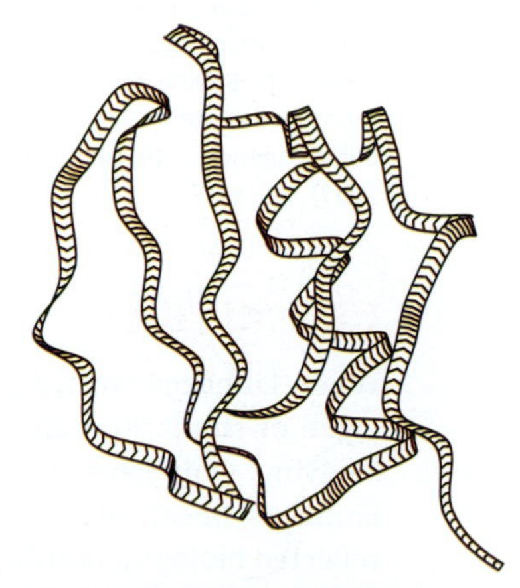

Barley chymotrypsin inhibitor

Fig. 3.11k

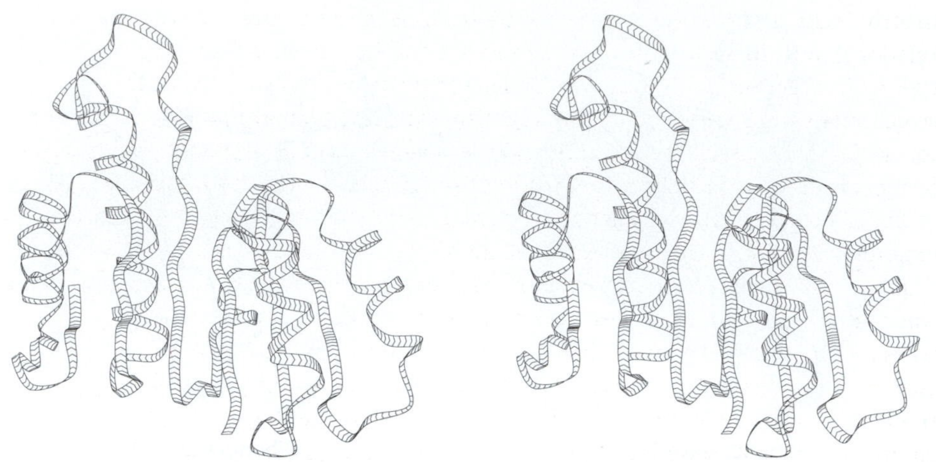

Fig. 3.11l Malate dehydrogenase, NAD-binding domain Malate dehydrogenase, NAD-binding domain

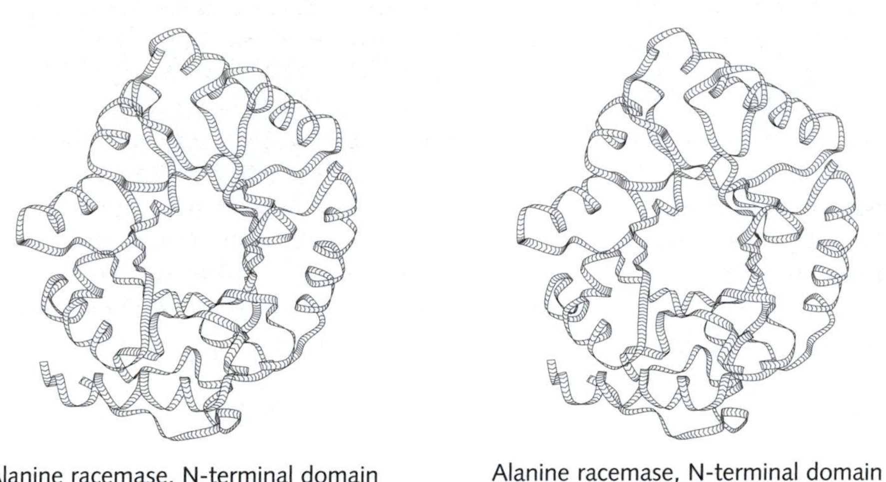

Fig. 3.11m Alanine racemase, N-terminal domain Alanine racemase, N-terminal domain

Fig. 3.11 An album of small protein domains, which will repay study. 'To see is itself a creative operation, requiring an effort.' (H. Matisse) (a) Uteroglobin [1UTG]. (b) Engrailed homeodomain [1ENH]. (c) Phospholipase A_2 [1BP2]. (d) T4 Lysozyme [3LZM]. (e) Timothy grass pollen allergen [1WHO]. (f and g) Abl tyrosine kinase / peptide complex (two views) [1ABO]. (h) Bovine pancreatic trypsin inhibitor [5PTI]. (i) Ribonuclease T1 [2RNT].(j) Ribosomal protein L7/L12 [1CTF]. (k) Barley chymotrypsin inhibitor [2CI2]. (l) Malate dehydrogenase, NAD-binding domain [1MLD]. (m) Alanine racemase, N-terminal domain [1SFT].

Classification of protein structures

When Linnaeus created his classification of living things, he made a catalogue of similarities among the objects in the corpus of material he was studying. Only later did it emerge that the hierarchy of relationships that Linnaeus observed was indeed induced by evolutionary processes and reflected biological kinship.

With protein structures, we can observe similarities in topology or even in structural details. But only in some cases can we infer genuine biological relationship in the sense of descent from a common ancestor. The alternative is 'convergent evolution'; the independent generation of favourable structural features. Given two protein structures with apparently similar conformations, but dissimilar function and no clear amino acid sequence similarity, it is usually very difficult to distinguish homology from convergence.

But not always. There are many well-characterized protein 'families'. For example, we know the structures of globins from 30 species, and the sequences of several hundred. Analysis of the sequences and structures of this and other protein families shows that one can construct evolutionary trees on the basis of molecular data from related molecules in different species. These are in most cases equivalent in their branching structure to the evolutionary trees constructed by the classical taxonomic methods of comparative anatomy, embryology, and the fossil record.

It may be instructive to pursue the analogy with biological classification a bit further. Comparing human haemoglobin with dog haemoglobin would be like comparing the skeleton of a human hand with the skeleton of a dog's forepaw; it is well-known that the skeletons contain a bone-for-bone correspondence, with a topologically similar spatial relationship, but that quantitatively the sizes, shapes, and exact arrangement of the corresponding bones differ. But comparing human haemoglobin with human (or dog) chymotrypsin is like comparing a lung with a stomach. Only an understanding of the evolutionary history can assess the nature of a relationship. Superficial similarities in structure and function may not reflect a true kinship—for example the detailed form of the eye of a human and the eye of an insect arose independently—and divergent evolution may conceal a homology—for example, bones of the human ear are related to bones in the jaw of a fish. Similar problems lurk among the similarities between protein structures.

Comparisons of protein sequences and structures

In devising measures of similarity and difference between two proteins, it is sometimes clearer to see how to proceed in comparing sequences, than in comparing structures. If the amino acid sequences of two proteins can be aligned, then we can either count the number of identical residues, or use a more subtle measure based on an index of similarity between amino acids. The similarity between two sequences is then the sum of values of the indices of similarity for each pair of aligned amino acids, plus a correction to account for insertions or deletions of amino acids. Indeed, it is by the maximization of such a score that the optimal alignment of two sequences is conventionally calculated.

We can also compare proteins in three dimensions. Given the structures of two related proteins we can superpose their structures using computer graphics (see Figure 3.12). This makes it easy to see the similarities and differences between the structures. A commonly-used measure of the difference between two structures is the root-mean-square deviation in atomic position after optimal superposition, the *r.m.s. deviation* for short (see

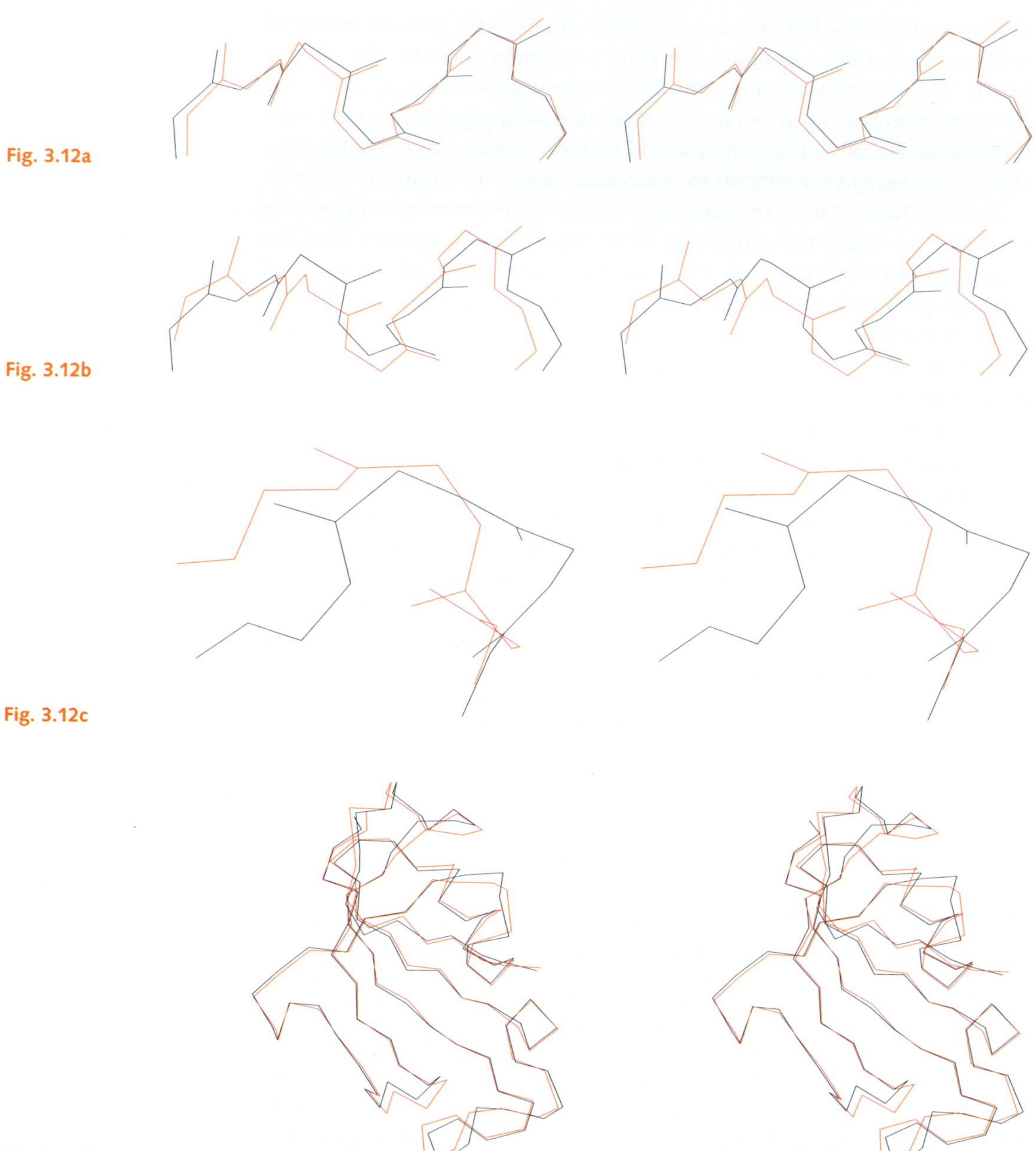

Fig. 3.12 Superposition of structures is one of the most important tools for comparing protein structures. Superpositions indicate clearly the overall quality of the fit, and distinguish the parts of the structure that fit relatively well and relatively poorly. The root-mean-square (r.m.s.) deviation is a measure of the average distance between corresponding atoms in two structures, after they have been optimally superposed. To give the reader some feeling for their values, consider these examples of superpositions of short loop regions, and of entire proteins. (a) Superposition of corresponding loops from immunoglobulins J539 [2FBJ] (black) and TE33 [1TET] (red). The r.m.s. deviation of all mainchain atoms (N, Cα, C, O) is 0.40 Å. These structures are the same to within

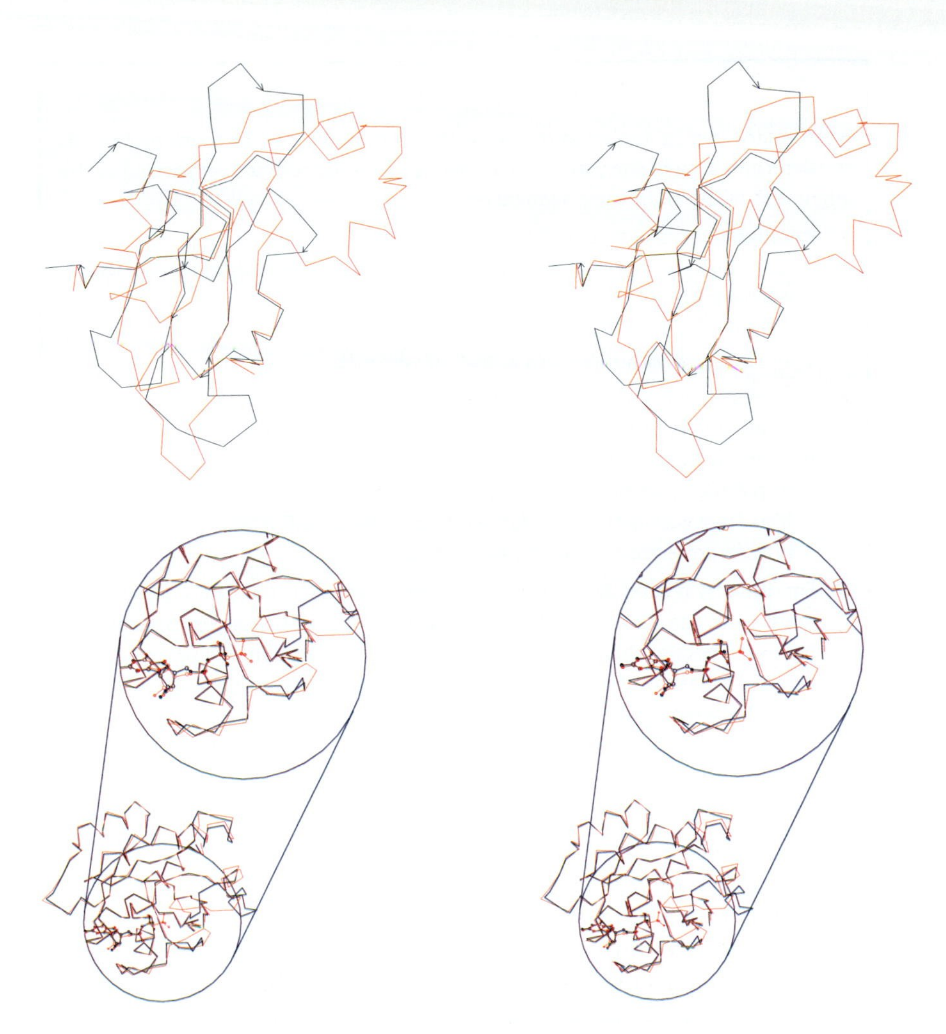

Fig. 3.12e

Fig. 3.12f

experimental error. (b) Superposition of the same region from J539 (black) with the corresponding region from immunoglobulin D1.3 [1VFB] (red). The r.m.s. deviation of all main chain atoms (N, Cα, C, O) is 0.94 Å. These conformations are similar topologically, but distorted. (c) Superposition of the same region of J539 (black) with a non-corresponding region from the same molecule (red). The r.m.s. deviation of all main chain atoms (N, Cα, C, O) is 1.02 Å. The conformations of these regions are quite different. (d) Superposition of human and yeast FK506–binding proteins [2FKE] (black) and [1YAT] (red). The r.m.s. deviation of all main chain atoms (N, Cα, C, O) is 0.78 Å. These very closely-related proteins have very similar structures. (e) Superposition of two distantly-related structures *Streptomyces aureofaciens* ribonuclease [1RGE] (black) and barnase (*Bacillus amyloliquefaciens*) [1BRN] (red). Because there are insertions and deletions, as well as large distortions, the superposition must be restricted to selected residues. In this case, the r.m.s. deviation of the main chain atoms of 64 residues is 1.4 Å. (*S. aureofaciens* RNAse contains 96 residues and barnase contains 107.) (f) Superposition of two forms of the catalytic domain of c-Ha-ras encoded p21, binding GDP (black) [4Q21] and a modified GTP (red) [6Q21]. Although most of the structure fits very well, specific regions have very different conformations depending on whether the ligand is GDP or GTP. The r.m.s. deviation of 149 Cα atoms, selected by omitting the mobile regions, is 0.5 Å. The r.m.s. deviation of all 168 Cα atoms is 1.7 Å.

Alignment

A necessary step in analysing the relationship between two or more proteins is the determination of the correspondences between the residues, expressed as the alignment of the amino acid sequences.

1. An alignment may be based on sequence similarity alone. Here are two character strings that are sufficiently similar that the correct alignment is clear:

   a b c d e f g
   a b z d – f g

   There is one substitution (c ↔ z) and one deletion.
2. A *multiple sequence alignment* is more informative than the pairwise sequence alignment in (1) because it can better reveal patterns. Several web sites collect multiple sequence alignments:

   http://ebi.ac.uk/dali/fssp/fssp.html
   http://www.sanger.ac.uk/Pfam *or* http://pfam.wustl.edu/
   http://coot.embl-heidelberg.de/SMART/
3. For distantly related proteins, it may be impossible to deduce the correct alignment from the sequences. A *structural alignment* establishes correspondences between residues that have the same spatial disposition in two or more structures. Structural alignment can identify affinities between more distantly-related proteins than sequence-based methods can.

Figure 3.12). Protein structure tends to change more conservatively than amino acid sequence, and it is not uncommon to be able to recognize a relationship between proteins from their structure, even if no evidence of homology appears in the sequences.

Indeed, it is crucial to understand the nature of the changes that occur during molecular evolution, because they reflect the rules that govern the dynamic development of living species (see following box).

Stasis and change in evolution of protein structures

1. Related structures retain most of the elements of secondary structure, the helices and strands of sheet.
2. A *core* of the structure—the assembly of the central helices and/or sheets—retains its topology or folding pattern. For closely-related proteins the core comprises almost the entire structure. For distantly-related proteins, the core may amount to only half the residues or even less.
3. Peripheral regions, outside the core, may change their folding pattern entirely.
4. The relative geometry of the secondary structures, even in the core, is variable. As a result of mutations, helices and sheets can shift and rotate with respect to one another.
5. For evolution with retention of function, the structural changes are subject to constraints that conserve function, for example, to maintain the integrity of the active site. For evolution with change in function, these constraints are replaced by other constraints, required by the altered function, producing greater structural change.

Even when we compare the structures of proteins from different families, we see a recurrence of structural themes. We can compare and classify the conformations of even apparently unrelated proteins on the basis of secondary and tertiary structures, and on their topology or 'fold'. In this way we can achieve a classification encompassing all known protein structures, a useful thing to have. However, within the hierarchy of such a classification, only the relationships among classes of proteins within the same family reflect evolutionary divergence. At higher levels of the hierarchy, the classification of sets of unrelated proteins is based purely on architectural similarity, independent of provable evolutionary history and relationship.

Classification of protein topologies

The most general classification of families of protein structures—a classification general enough to encompass even unrelated proteins—was first proposed by M. Levitt and C. Chothia. Their classification is based on the secondary and tertiary structures of domains. Because the individual domains in multidomain or modular proteins often have different structures, each must be classified individually. (See Box.)

Within these broad categories, protein structures sort themselves into varieties with recognizably different folding patterns. For instance, acylphosphatase, thioredoxin, and signal-transduction protein CheY are all of the α/β type, but have different topologies (Figure 3.13).

Within the sets of proteins of similar topology, there are families that share enough features of structure and/or sequence, to suggest evolutionary relationship. Hierarchical classifications of the entire set of known protein structures appear on the World Wide Web. The first of these, SCOP

Classification of proteins according to secondary and tertiary structure

Class	Characteristic	Examples
α-helical	Secondary structure exclusively or almost exclusively α-helical	Myoglobin, cytochrome c, citrate synthase
β-sheet	Secondary structure exclusively or almost exclusively β-sheet	Chymotrypsin, immunoglobulin domain
α + β	α-helices and β-sheets separated in different parts of molecule. Absence of β–α–β supersecondary structure	Papain, staphylococcal nuclease
α/β	Helices and sheet assembled from β–α–β units	
α/β linear	Line through centres of strands of sheet roughly linear	Alcohol dehydrogenase, flavodoxin
α/β-barrels	Line through centres of strands of sheet roughly circular	Triose phosphate isomerase, glycolate oxidase
Little or no secondary structure		Wheat germ agglutinin, ferredoxin

(Structural Classification of Proteins) is based in Cambridge, England. CATH (Class, Architecture, Topology, Homology) is based in London. We shall discuss these in detail in the next chapter.

Protein structure prediction

Nature has an algorithm which specifies the three-dimensional structures of proteins from their amino acid sequences alone. We should be able to discover this algorithm, and be able to predict the structures of proteins from their sequences. This would unlock the secrets inherent in the bacterial, viral and eukaryotic genome sequences now emerging; and in the Human genome.

The only adequate methods for judging techniques for predicting protein structures are blind tests. To this end, J. Moult initiated biennial CASP (Critical Assessment of Structure Prediction) programmes. Crystallographers and NMR spectroscopists who are in the process of determining the structure of a protein are invited to:

1. publish the amino acid sequence several months before the expected date of completion of their experiment, and
2. commit themselves to keeping the results secret until an agreed date.

Predictors submit models, which are held under seal until the deadline for release of the structure. Then the predictions and experiments are compared—to the delight of a few and the chagrin of most.

Prediction methods fall into two broad classes: the inductive and the deductive. Inductive methods make direct use of databanks of sequences and structures. Deductive methods are true *ab initio* approaches—the 'desert island' case—attempting to predict protein structure from general principles of physics, chemistry and biology, without explicit reference to known sequences and structures. Of course the development of *ab initio* methods depends on what we have learned from the study of known sequences and structures. The distinction is that the understanding achieved has been distilled into general principles that can be applied without looking up specific information in databases.

Prediction methods that use databanks include

1. methods for *homology modelling*—prediction of the target structure from a closely-related protein of known structure, and
2. methods for *fold recognition*—assessing the compatibility of the amino acid sequence with the library of known protein folding patterns.

These methods are growing more powerful, partly but not entirely because of the growth in the databanks. The more sequences and structures that are known, the more likely that a new protein will be similar to one that is already known. In contrast, *ab initio* methods are improving more sluggishly. A comment about them after a recent CASP competition was that at least 'failure can no longer be guaranteed.' A pessimist might predict that the growth of databanks will mean that inductive methods will provide pragmatic solutions to such a large majority of questions, that interest in and

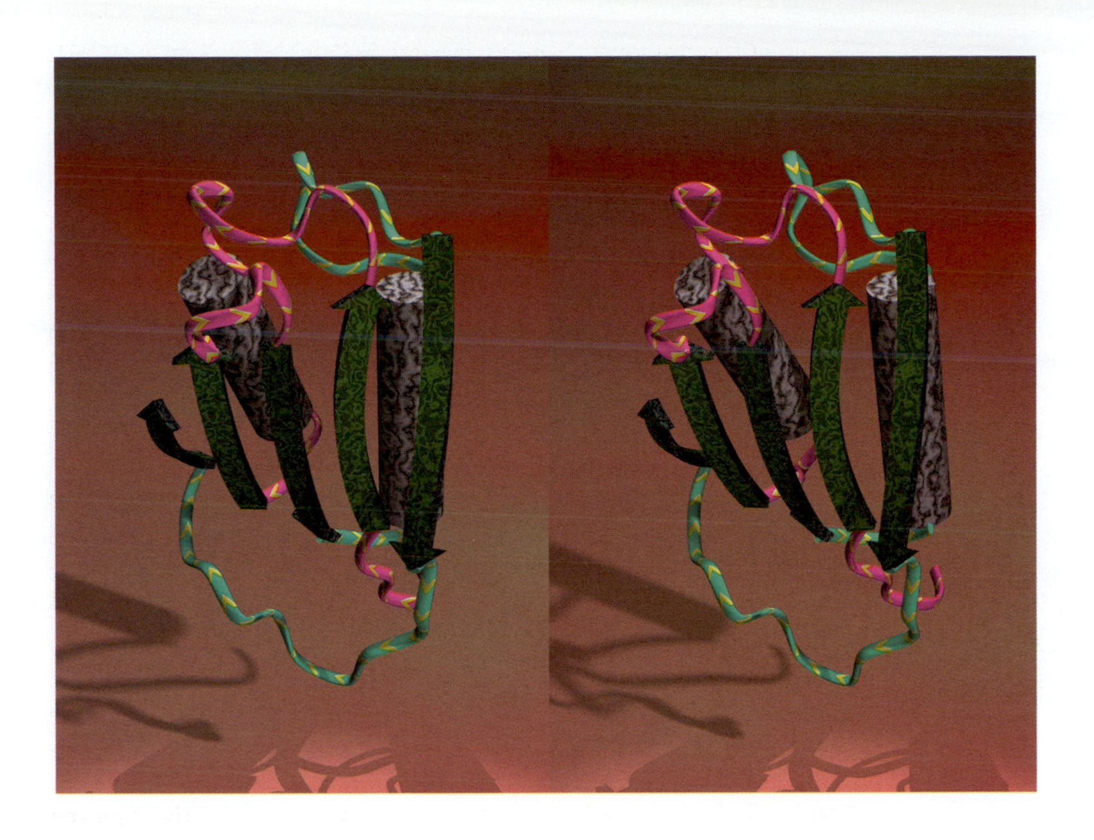

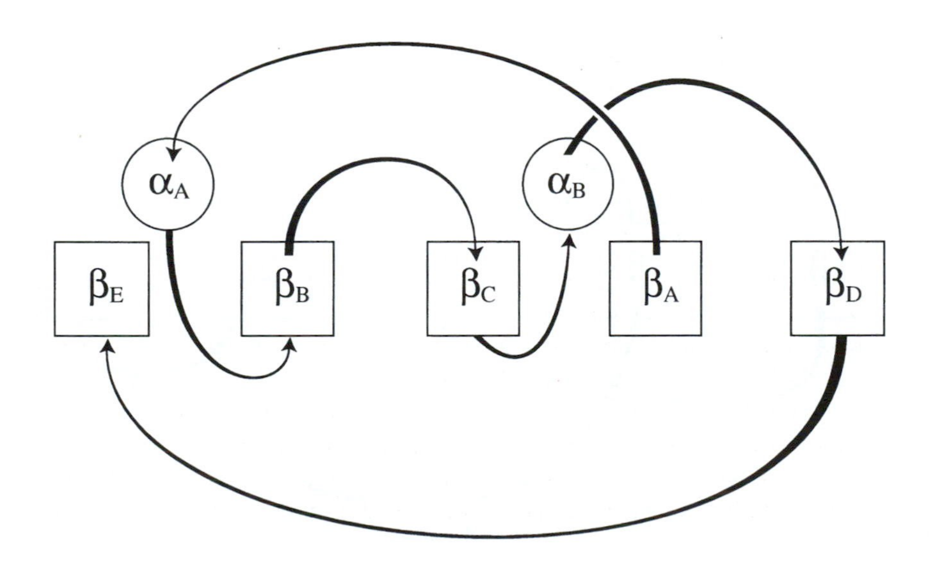

Fig. 3.13a

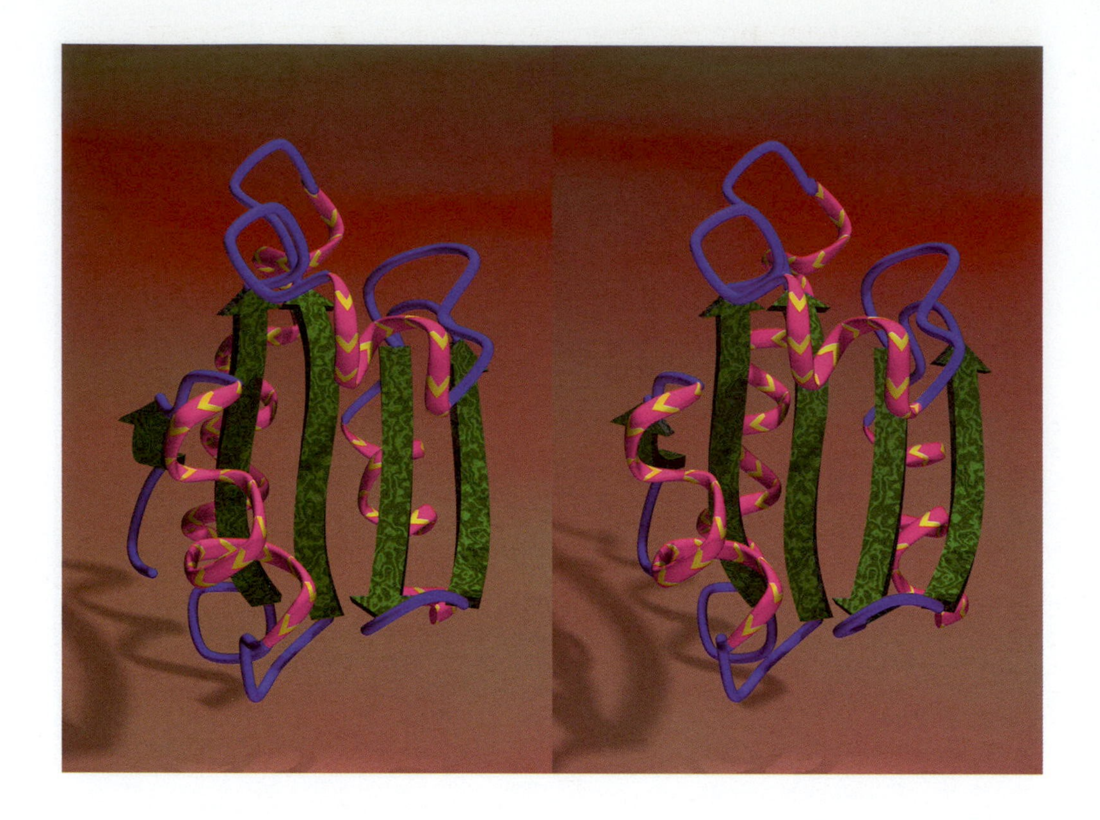

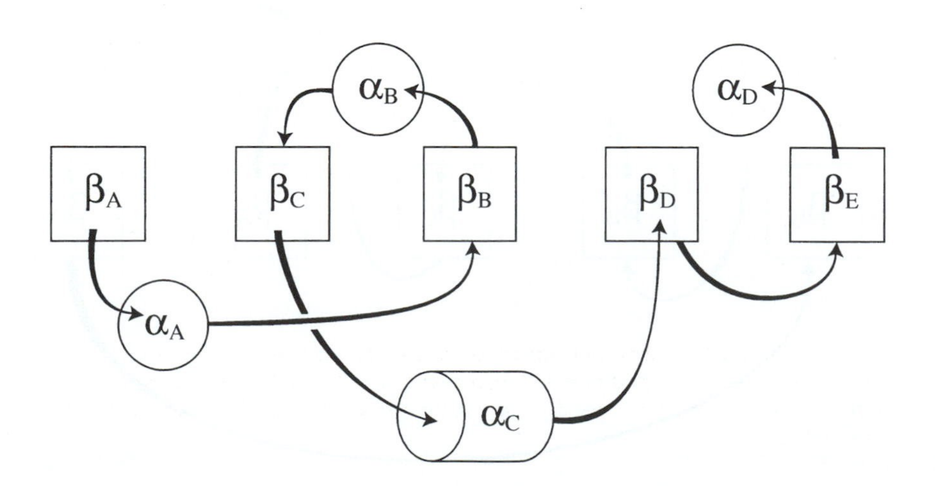

Fig. 3.13b

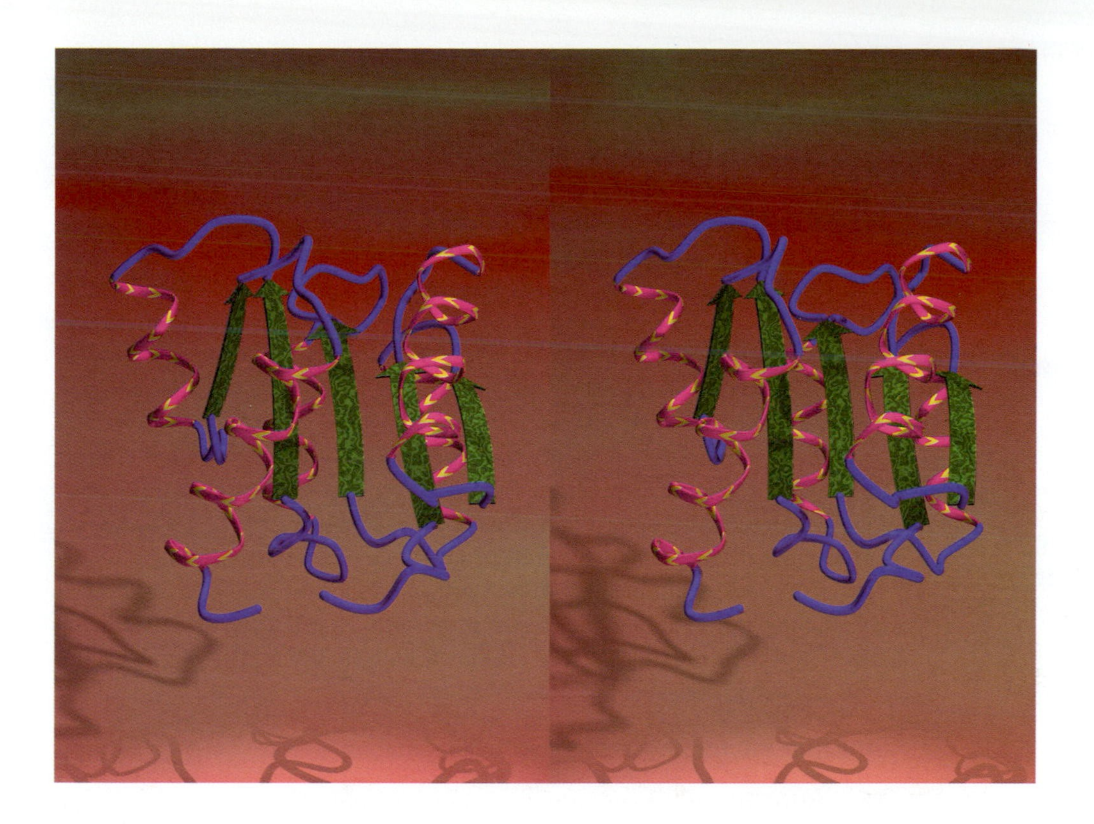

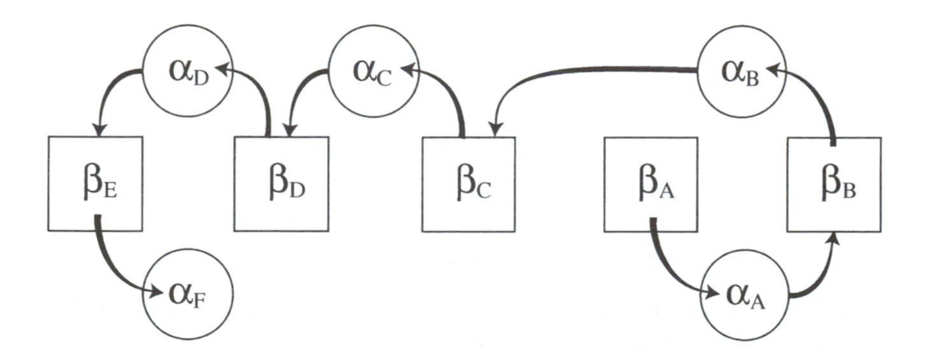

Fig. 3.13c

Fig 3.13 Three α/β proteins with different topologies. (a) Acyl phosphatase [2ACY]. (b) Thioredoxin [1THX]. (c) Signal-transduction protein CheY [3CHY]. (When loops cross in the picture, it can be difficult to trace the chain; in these cases the loops have been drawn in different colours.)

support for the development of *ab initio* methods will wane. (A reader of this chapter in draft suggested changing 'pessimist' to 'optimist' in the previous sentence. That is also a legitimate point of view.)

One use of a hierarchical classification of proteins would be in assessing the quality of our ability to predict protein structures from amino acid sequences. Ultimately, we should like a fully detailed prediction of the atomic structure. This we cannot do now. What can we do?

It should be easier to predict the general structural class of a protein (α, β, α + β, α/β) from its amino acid sequence, than to predict its tertiary structure. *The most detailed level of the hierarchical classification of protein structure, at which our predictions can correctly assign proteins from their amino acid sequences, is a measure of the state of the art of structure prediction.*

Programs to predict secondary structures can now distinguish helix, sheet and turn residues with approximately 70% accuracy. This can do a fairly good job of distinguishing the basic classes of protein families: α, β, α + β and α/β. Other types of programs seek to distinguish one protein family, e.g. the globins, from other families, based on patterns of conservation of residues in the aligned amino acid sequences of proteins of that family, or by assessing the compatibility of a sequence with a folding pattern. These approaches have made considerable progress in recent years.

Structural interpretation of genome information

The entire genome sequences for about 50 prokaryotes, plus yeast, the nematode worm *Caenorhabditis elegans* and the fruit fly *Drosophila melanogaster* are known, and completion of the human genome is expected in 2001. These have ushered in a new era in molecular biology. We now know everything about a yeast cell that a yeast cell knows! We have a complete set of blueprints, specifying the amino acid sequence of every protein a yeast cell will ever synthesize.

The first step in harvesting this information is to analyse the structures of these proteins. In many cases we do not know the structure of a protein identified from a gene sequence, but we may know the structure of one or more proteins with similar amino acid sequences from other organisms. How can we apply this knowledge?

There is a fairly well worked-out procedure for responding to the question: 'here is an amino acid sequence; what can you say about the structure or function of this protein?'

First, one does a 'screen' of the new sequence against a database of known sequences. This answers the following questions:

1. Is there any protein of known structure that has sufficient similarity to the sequence of the unknown protein to suggest a familial relationship?
2. If not, what sequence of any known protein is most similar to the sequence of the unknown protein?

In the most favourable case, the new protein will be recognizable as a member of a family of proteins of known structure. If, in an optimal align-

Some motifs in protein structures

Structure or function identified	Template
Nucleotide-binding site	G*G**G
N-glycosylation site	N*S or N*T
Nuclear protein transit sequence	KKKRKV
Factor IX proteinase cleavage site	IEGR
Serine proteinase active site	GDSGG
Acid proteinase active site	FDTGS
Fibronectin cell adhesion sequence	RGDS
Copper binding site	H***H ... H or H****H ... H

* = any single amino acid

*** = indefinite number of amino acids

ment, over about 25% of the residues are identical, it is fairly safe to conclude that the two proteins have the same fold. For more distantly-related proteins, sequence analysis will not always provide a correct answer: the sequences of two related proteins may have less than 20% residue identity in an optimal alignment, and the sequences of definitely unrelated proteins may have as much as 20% residue identity in an optimal alignment. Borderline cases remain in what R.F. Doolittle called 'the twilight zone'.

In favourable cases, when one can identify a relationship to a protein of *known structure*, it is possible to assign the general fold, or the topology, of the unknown protein, and to suggest that it shares a structure with its relative. It is possible to go on to say more: to give a quantitative assessment of the structural similarity, based on how far the amino acid sequences have diverged. This will be discussed in Chapter 5.

Suppose next that the new protein sequence is related to a known protein of unknown structure. If the known protein has a known function, the unknown protein will be expected to have a similar or at least a related activity.

Even if database screening methods based on overall sequence similarity do not pick up a relative, there is one more string to our bow. In a number of cases, the active site of a protein can be recognized by a specific 'fingerprint' or 'template': a constellation of even a fairly small set of residues that is unique to a family of proteins, usually reflecting the conserved residues in a binding site. An example is the sequence G*G**G (where G = glycine and * = any amino acid) which seems to define a binding site for nucleotides. This is quite a powerful technique, although we do not yet have templates for all protein families. The database PROSITE, available on the Web, is a compilation of sites and patterns found in protein sequences (see http://www.expasy.ch/ sprot/prosite.html).

Loops

Loops are the sections of the polypeptide chain that connect regions of secondary structure. A typical globular protein contains approximately

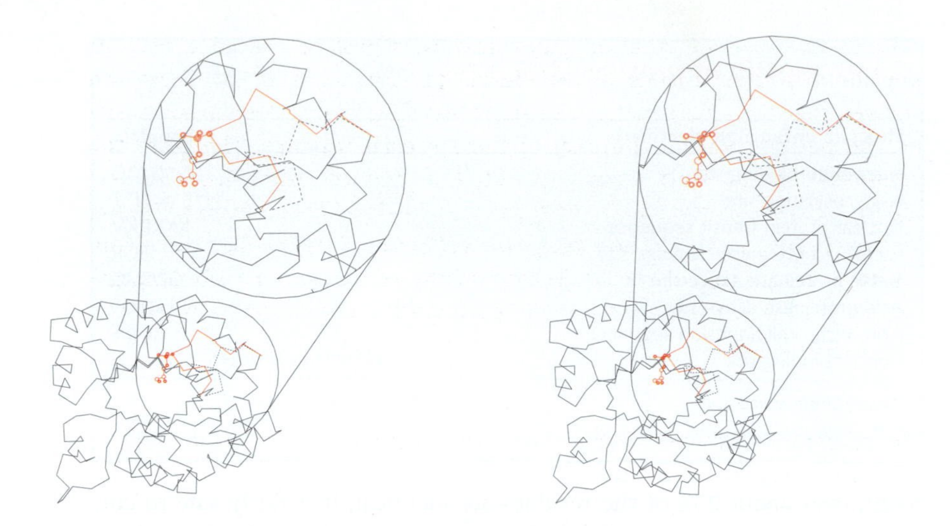

Fig. 3.14 Conformational change in the loop of triose phosphate isomerase [6TIM]. The view is down the axis of the barrel, and the region of structural change is 'blown up'. As is typical of TIM-barrel enzymes, the substrate binding site is at the mouth of the barrel at the C-termini of the strands. In the unliganded structure (black) a mobile loop (black, broken line) is in the 'open' form. With the substrate analog glycerol-3-phosphate bound (red) this loop has closed over the ligand.

two-thirds of its residues in helices and sheets, and one-third in loops. Some loops connect successive helices or strands of sheet that interact with each other to form supersecondary structures.

In many enzymes, loops contain functional residues. Because loops tend to be more flexible in conformational changes than helices and sheets, they are often used when a protein needs to respond in relatively simple ways to changes in the state of ligation, as in triose phosphate isomerase (Figure 3.14). In contrast, allosteric changes involve larger-scale changes in relative geometry and packing of entire subunits (see Chapter 8).

Other loops that have purely structural roles, and that appear on protein surfaces, change fairly rapidly in evolution. They are frequently subject to insertions and deletions as well as to amino acid substitutions. From a set of aligned sequences of related proteins, by a combination of hydrophobicity profiles and patterns of sequence change, it is possible to infer the positions of loops in a protein of unknown structure fairly well.

Sequence–structure relationships in short β hairpins

β hairpins are loops that connect successive antiparallel strands of a β-sheet; that is, hairpins connect strands of sheet that are hydrogen bonded to each other. The conformations of short β hairpins fall into distinct classes, classified first by C.L. Venkatachalam.

For a short region of polypeptide chain to reverse direction, and fold back on itself to form a hairpin, a residue in a conformation in the non-allowed region of the Sasisekharan–Ramakrishnan–Ramachandran diagram is required. This residue is usually a glycine (or sometimes Asn or Asp) that can take up the α_L conformation. Alternatively, a proline preceded by a *cis* peptide can effect a turn of the chain. B.L. Sibanda and J.M. Thornton showed that the conformations of short β–hairpin loops depend primarily on the position within the loop of these special residues. Therefore the conformation of a short hairpin can often be deduced from the sequence. Figure 3.15a shows the hairpin of sequence Glu–Gly–Gly–Val from actinidin; compare Figure 3.15b, showing the sequence Ser–Gly–Ser–Ser from elastase. A diagram of the 'trajectory' of the residues in a Sasisekharan–Ramakrishnan–Ramachandran plot gives a quick snapshot of the conformation (Figure 3.16).

Sequence–structure relationships in two-residue β hairpins

The common backbone structural pattern has residues 1 and 4 making two hydrogen bonds to each other

```
2——3
|  |
1===4
```

Sibanda and Thornton, and others, have created dictionaries of sequence-structure correlations in loops. Here is the result for two-residue β hairpins. The sequence pattern is specified as follows: Letters representing amino acids indicate conserved residues; X indicates an unconstrained residue. The residue conformations are specified according nomenclature of A.V. Efimov (Figure 3.17). The nomenclature of Types is from Venkatachalam.

Sequence	Conformation	Type
XGXX	$\beta - \varepsilon - \gamma - \beta$	II′
XXGX	$\beta - \alpha_L - \alpha_L - \beta$	I′
XXXG	$\beta - \alpha - \alpha - \varepsilon$	I
XXXX	$\beta - \alpha_L - \gamma - \beta$	III′
XGGX	$\beta - \alpha_L - \alpha_L - \beta$	I′
or	$\beta - \alpha_L - \gamma - \beta$	III′

Such dictionaries give useful predictions for short β hairpins up to about 6 residues in length.

Structural determinants of medium-sized loops

Not all loops in proteins are short β hairpins, however. Proteins often contain loops 6–10 residues in length; and some are much longer. The determinants of the conformations of such loops are not intrinsic to the amino

Fig. 3.15 (next page) (a) A Type I′ hairpin loop from actinidin [2ACT], with the sequence

```
Gly—Gly
 |   |
Glu==Val
```

(where == indicates the two β–sheet hydrogen bonds between the Glu and Val residues). (b) A Type II′ hairpin loop from elastase [3EST], with the sequence

```
Gly—Ser
 |   |
Ser==Ser
```

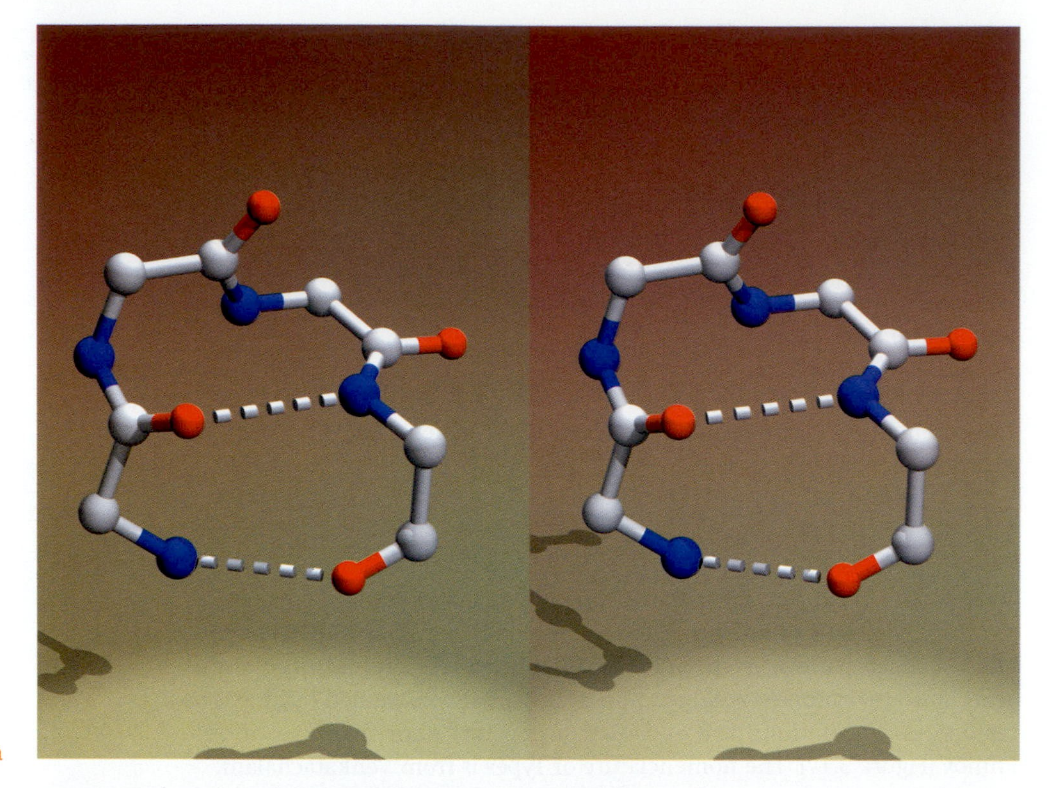

Fig. 3.15a

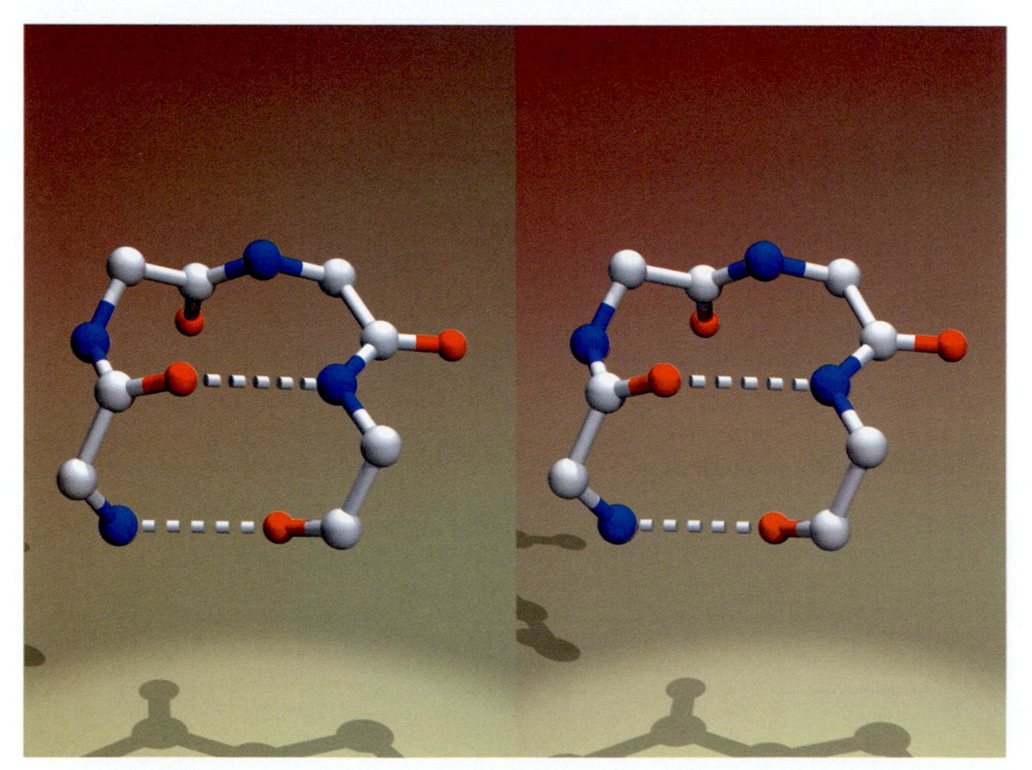

Fig. 3.15b

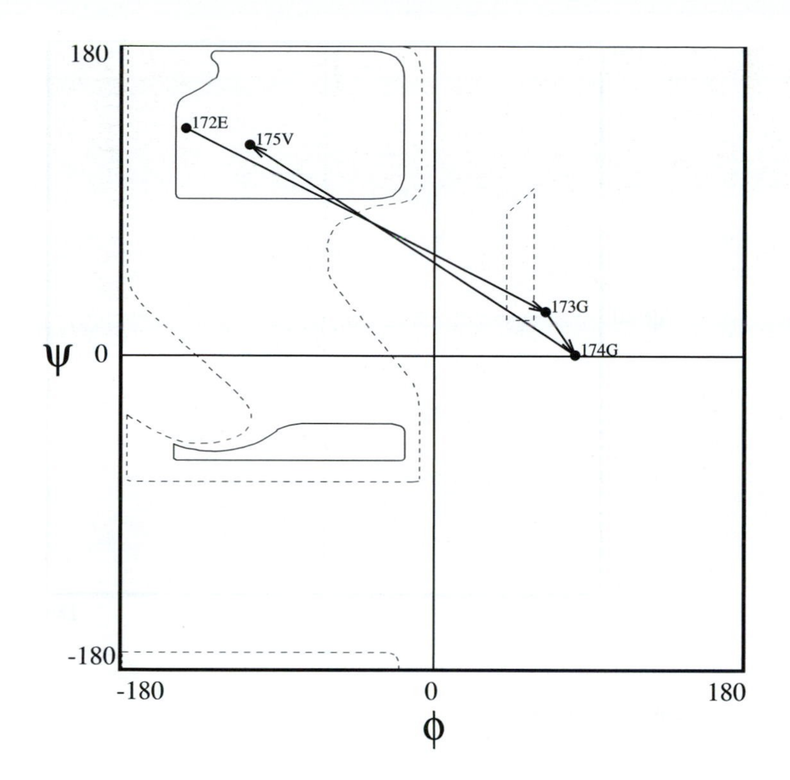

Fig. 3.16 'Trajectories' in Sasisekharan–Ramakrishnan–Ramachandran diagrams of the hairpin loops from (a) actinidin [2ACT]. (b) elastase [2EST].

Fig. 3.16a

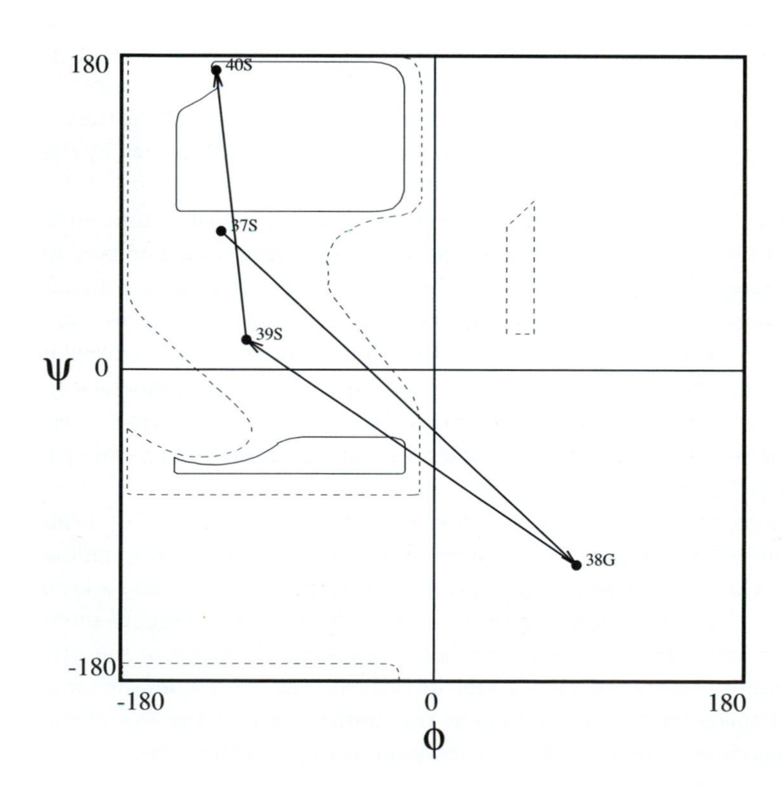

Fig. 3.16b

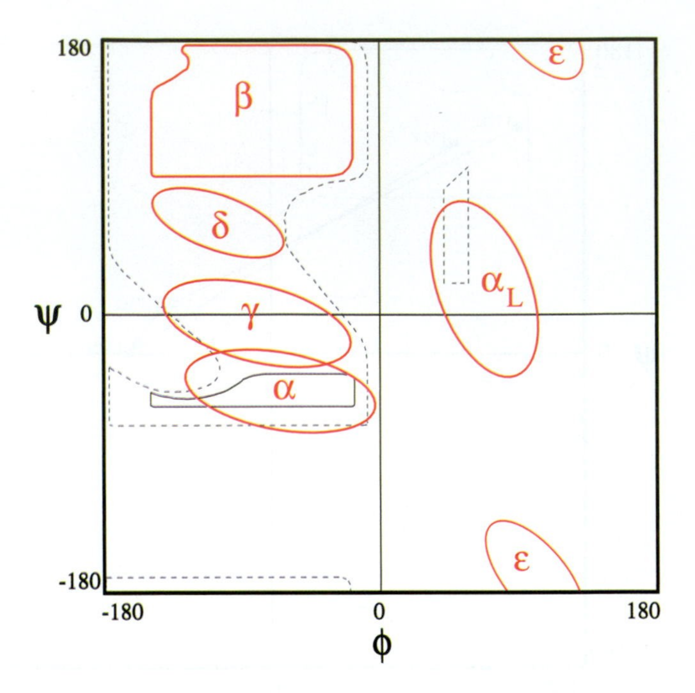

Fig. 3.17 Definitions of residue conformational states by A.V. Efimov.

acid sequence of the loop itself, and involve hydrogen bonding and packing interactions with the rest of the structure.

For medium-sized loops that form compact structures, the major conformational determinants are hydrogen bonds to the inward-pointing mainchain polar atoms of the loop. The conformation of the loop shown in Figure 3.18a, from the antigen-binding site of an immunoglobulin domain, is determined mainly by the *cis*-peptide bond before the Pro at the right, and hydrogen bonds formed by the sidechain of the Asn residue N-terminal to the loop.

A loop of very similar conformation occurs in tomato bushy stunt virus (Figures 3.18b and 3.18c). These loops have different structural contexts: In the immunoglobulin it is a hairpin; in the virus it connects strands from different sheets (Figure 3.19). Nevertheless, the structural determinants are very similar. There is a *cis*-proline at the equivalent position. Hydrogen bonds, similar to those made by the Asn sidechain in the immunoglobulin domain, are made in the virus by the main chain nitrogen of an alanine, distant in the sequence, but which occupies the same position in space relative to the loop (Figure 3.18b).

A related conformation occurs in a hairpin in cytochrome c_3 from *Desulfovibrio vulgaris* (Figure 3.20a). There is no *cis* proline: the corresponding residue is Gln, and the peptide is *trans*. However, this produces only a local distortion of the loop. Again there is a similarity in the stabilizing interactions. In cytochrome c_3, hydrogen bonds are formed by one of the propionyl groups of a haem (Figure 3.20b). Its carboxyl group occupies the same region of space relative to the loop as the amide group of the Asn of the immunoglobulin domain, and the mainchain nitrogen of the virus.

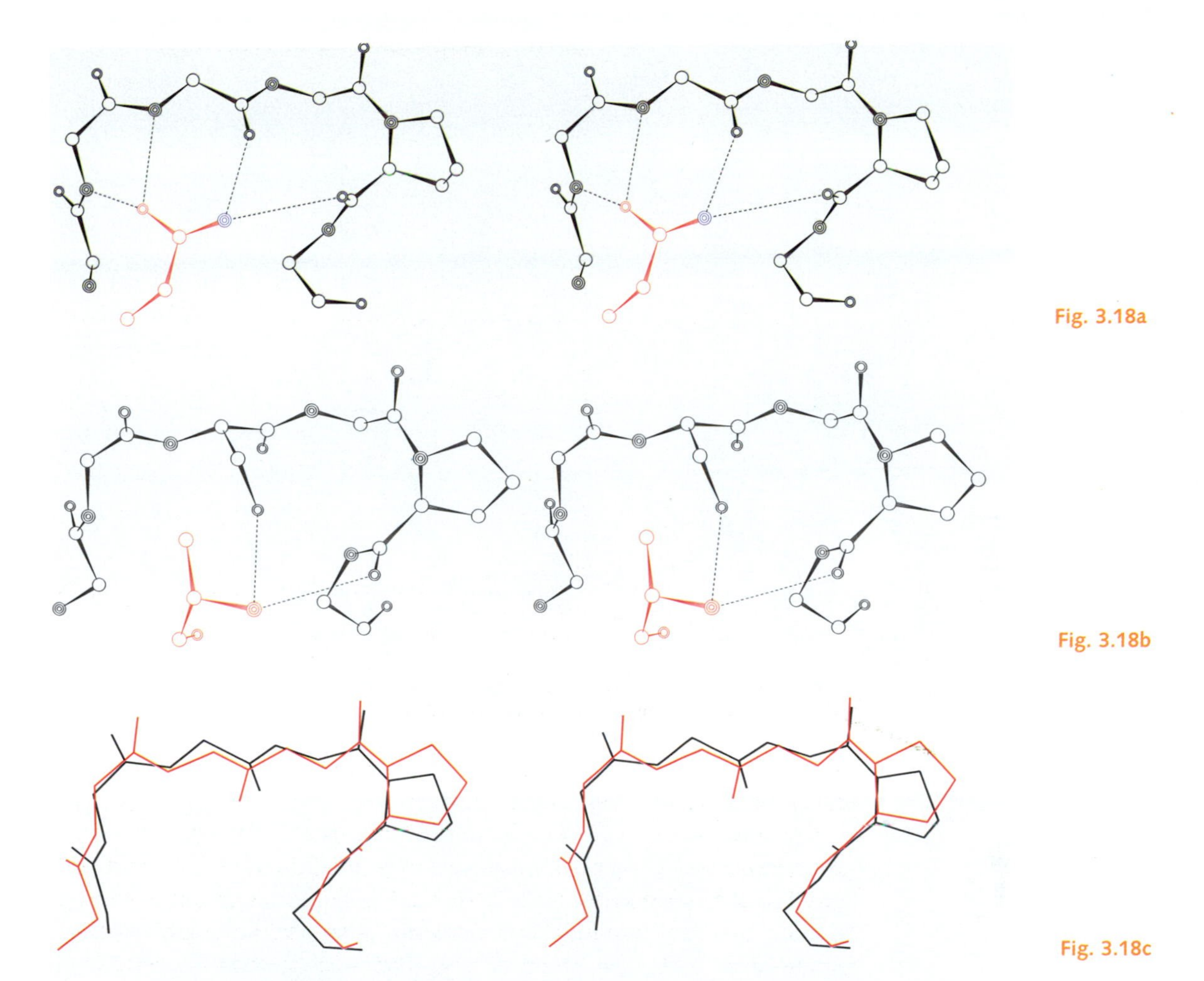

Fig. 3.18 (a) Hairpin loop from the immunoglobulin McPC603 [1MCP]. (b) Loop of similar structure from tomato bushy stunt virus [2TBV]. (c) Superposition of loops from McPC603 and tomato bushy stunt virus.

For loops not compact in themselves, the major determinant of the observed conformation is the packing of residues against the rest of the protein. Another loop, not a hairpin, from the antigen-binding site of immunoglobins, connects strands in two different β-sheets (Figure 3.21a). The loop contains a distorted helix, and a hydrophobic side chain—usually a Phe—is buried in a cavity between the sheets.

A region from the insect globin, *Chironomus* erythrocruorin, has the same conformation (Figure 3.21b). In the globin the loop links two α-helices in a 'helix hairpin'—one of the standard supersecondary structures (Figure 3.21c). As in the immunoglobulin loop, the fourth residue is a buried Phe. These two proteins form cavities to pack a Phe sidechain out of very different structural elements: β-sheets in the immunoglobulins and α-helices in erythrocruorin.

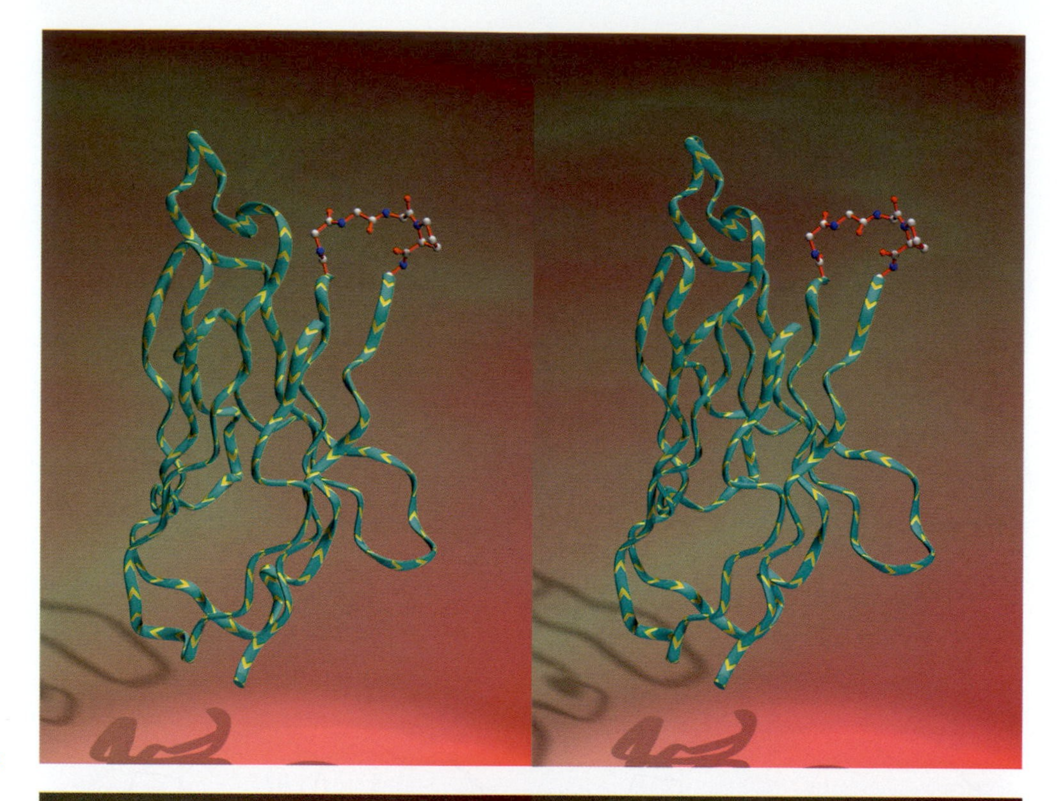

Fig. 3.19a

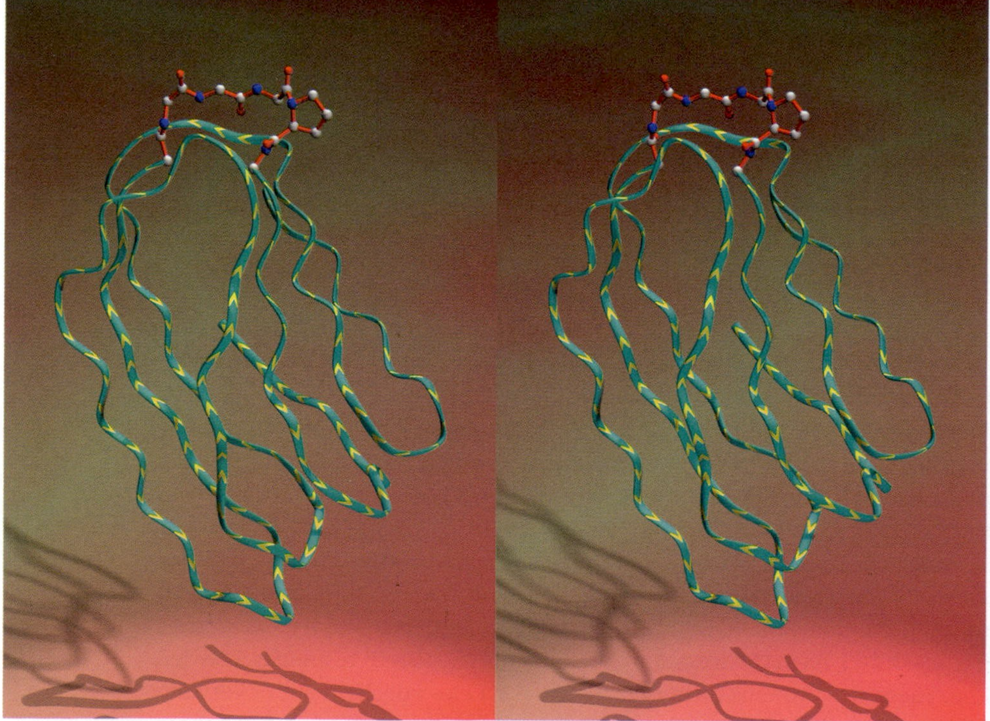

Fig. 3.19b

Fig. 3.19 Structural contexts of loops from (a) McPC603 and (b) tomato bushy stunt virus.

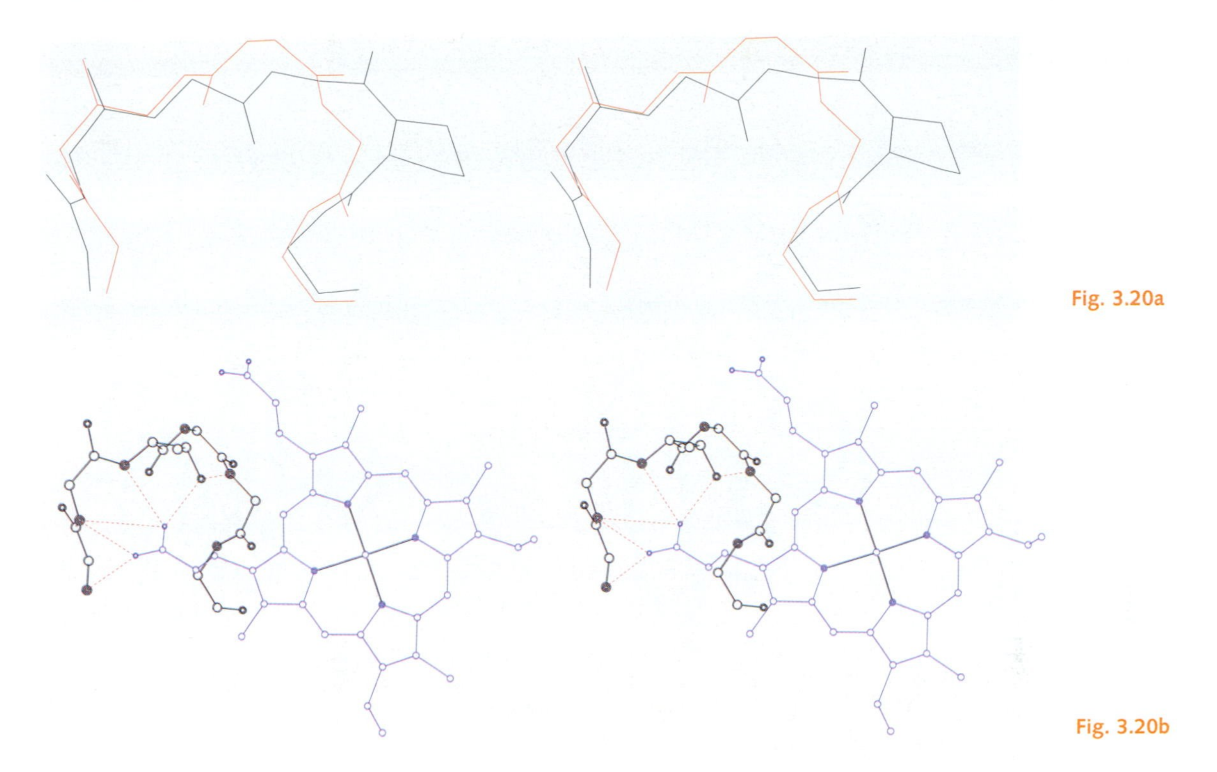

Fig. 3.20a

Fig. 3.20b

Fig. 3.20 (a) Loop from *Desulfovibrio vulgaris* cytochrome c_3 [2CTH] (red) with structure similar to the loop from McPC603 (black). (b) Interaction of loop from *Desulfovibrio vulgaris* cytochrome c_3 with haem group.

In both these examples, a loop makes demands on the protein for specific interactions. Different proteins meet these demands in different ways.

Protein–ligand interactions

Many protein structures contain inorganic ions, water, or small inorganic or organic molecules. These may be intrinsic parts of the structures, as in the case of the zinc ions of Zn-fingers or the haem groups of globins and cytochromes c; or they may be substrates, inhibitors or effectors, or antigens bound to antibodies, which interact with a structure that is fully native even in their absence. Some ligands may themselves be proteins; for example, bovine pancreatic trypsin inhibitor in its complexes with serine proteinases, or hen egg white lysozyme in its complexes with immunoglobulins. Others may be nucleic acids, such as the oligonucleotides bound to repressors, tRNA bound to amino acyl-tRNA synthetases, or the nucleic acids of viruses enfolded within coat protein aggregates.

Usually, ligands participate directly in the function of a protein, and are therefore among the most interesting components of a structure to study.

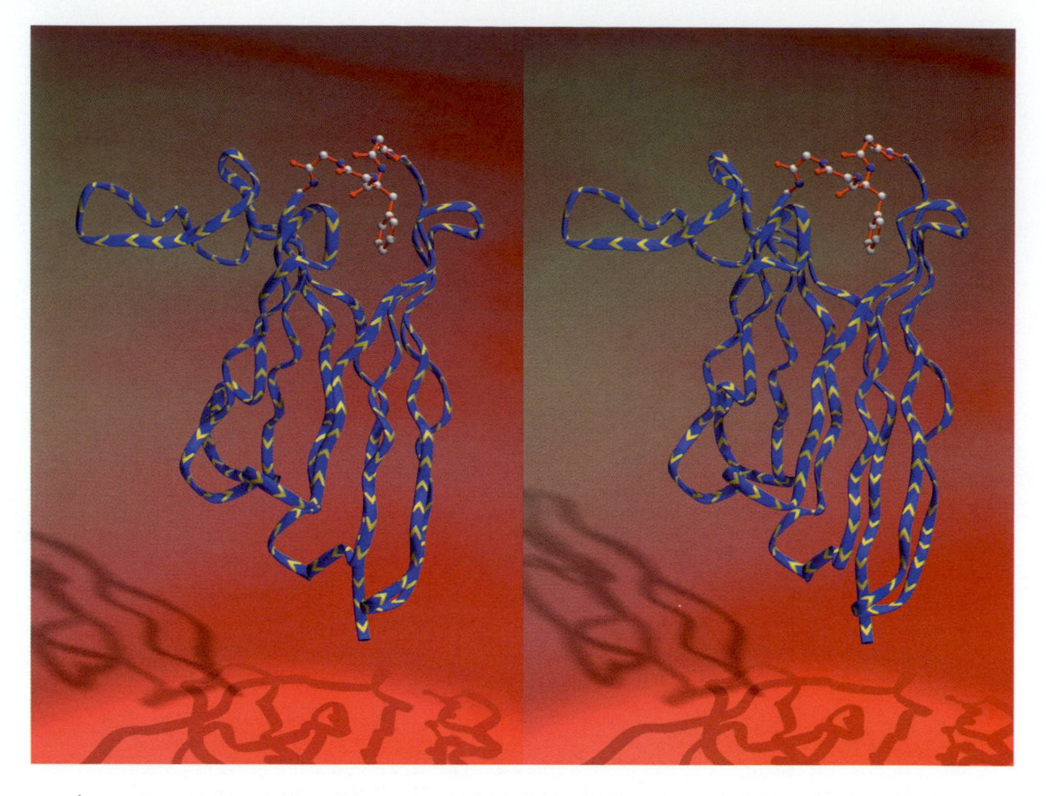

Fig. 3.21a

Fig. 3.21b

The environment within the protein that surrounds and interacts with the ligands is in many cases equivalent to the active site. Determination of the structure of a protein with bound substrate/inhibitor/cofactor is often necessary to elucidate the mechanism of function.

In some cases, changes in the state of ligation produce little conformational change in the protein. In others, ligation produces either simple or profound changes in protein structures. Some of these are allosteric changes, for instance in haemoglobin, aspartate carbamoyltransferase and phosphofructokinase. Allosteric changes alter the activity of the protein to permit fine control over function. For example, it is the allosteric change in

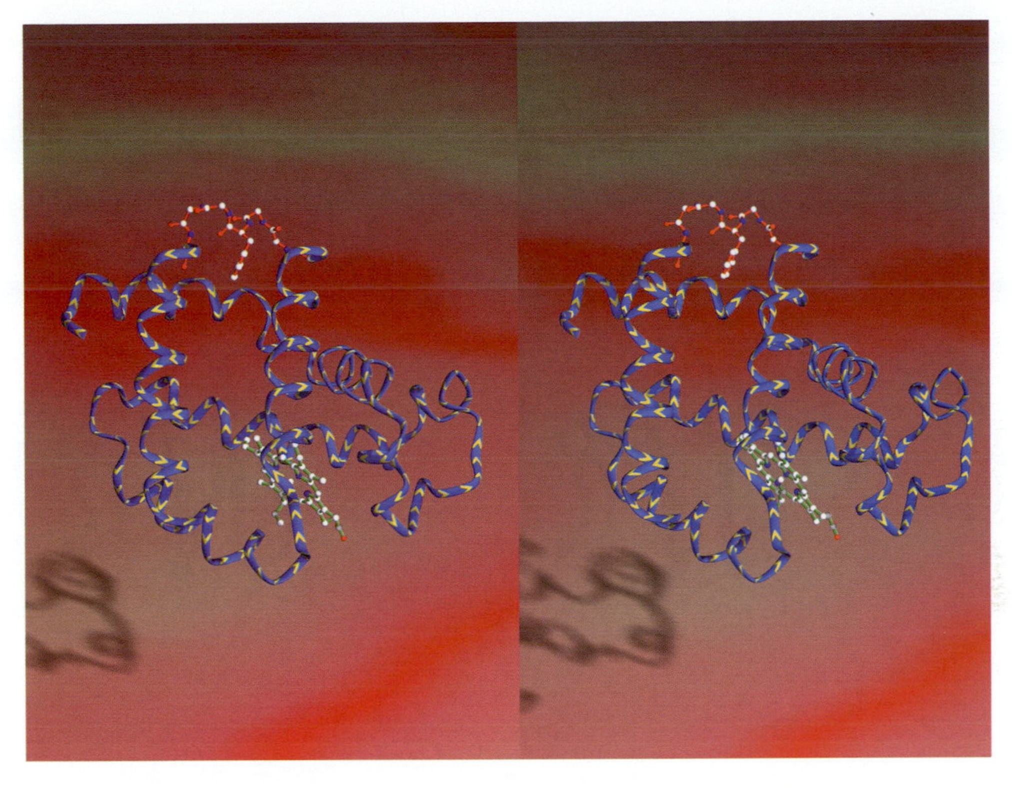

Fig. 3.21c

Fig. 3.21 (a) Another loop from an antigen-binding site, of immunoglobulin KOL [2FB4]. (b) Superposition of loop from KOL (black) and loop of similar structure from *Chironomus* erythrocruorin [1ECD] (red). (c) Structural context of the loop from *Chironomus* erythrocruorin: it links two α–helices.

haemoglobin that makes oxygen binding cooperative. Conformational change is the subject of Chapter 8.

Figure 3.22 shows typical binding sites for:

1. **Inorganic ions:** (a) the copper-binding site of poplar leaf plastocyanin (the shadows of the sidechains ligating the copper are pictorially effective), and (b) sulphate-binding site of a transport protein of *S. typhimurium*.
2. **Small organic molecules:** including but not limited to substrates, inhibitors and effectors (c) hen egg white lysozyme binding a trisaccharide inhibitor, (d) retinol-binding protein enclosing the retinol molecule, (e) a dimer fragment of the tetrameric allosteric enzyme phosphofructokinase from *B. stearothermophilus*, containing the substrate, fructose-6-phosphate, the cofactor ATP and the effector, ADP.
3. **Other proteins:** (f) the complex between the serine proteinase trypsin and bovine pancreatic trypsin inhibitor.
4. **Nucleic acids:** (g) binding of DNA to λ cro, (h–j) the interaction of the RNA and coat protein in tobacco mosaic virus.

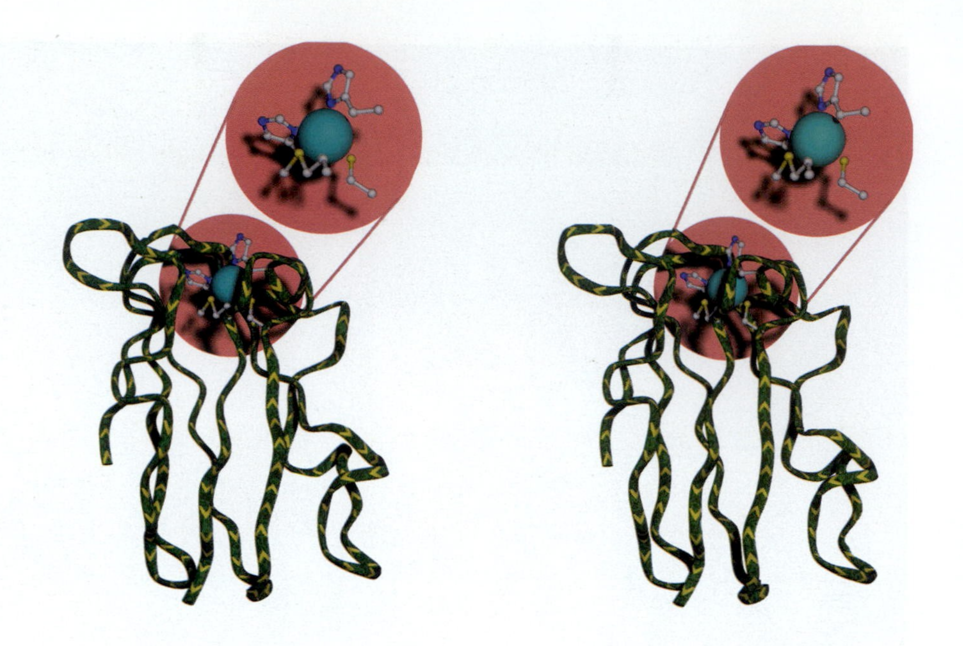

Fig. 3.22a

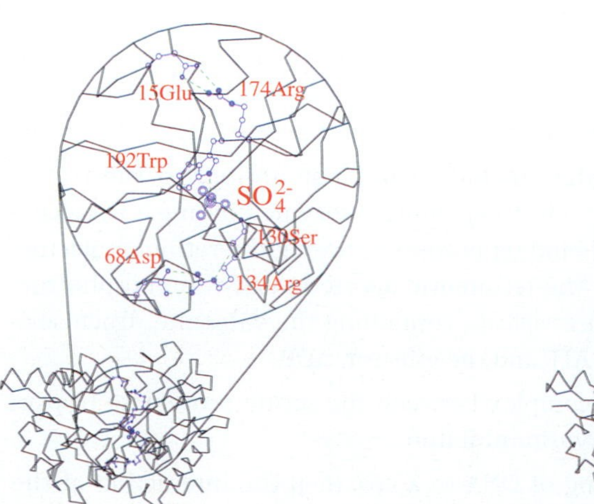

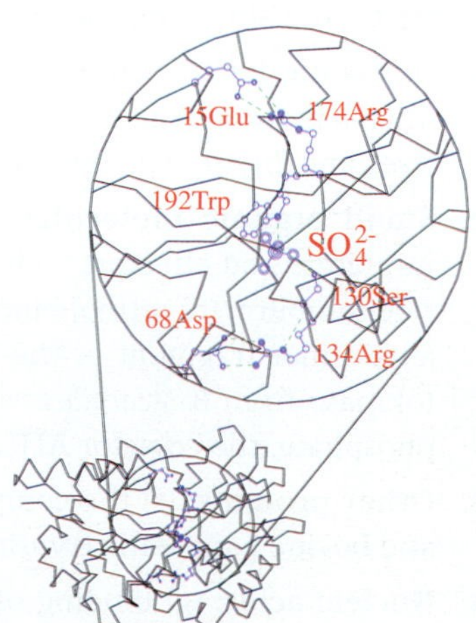

Fig. 3.22b

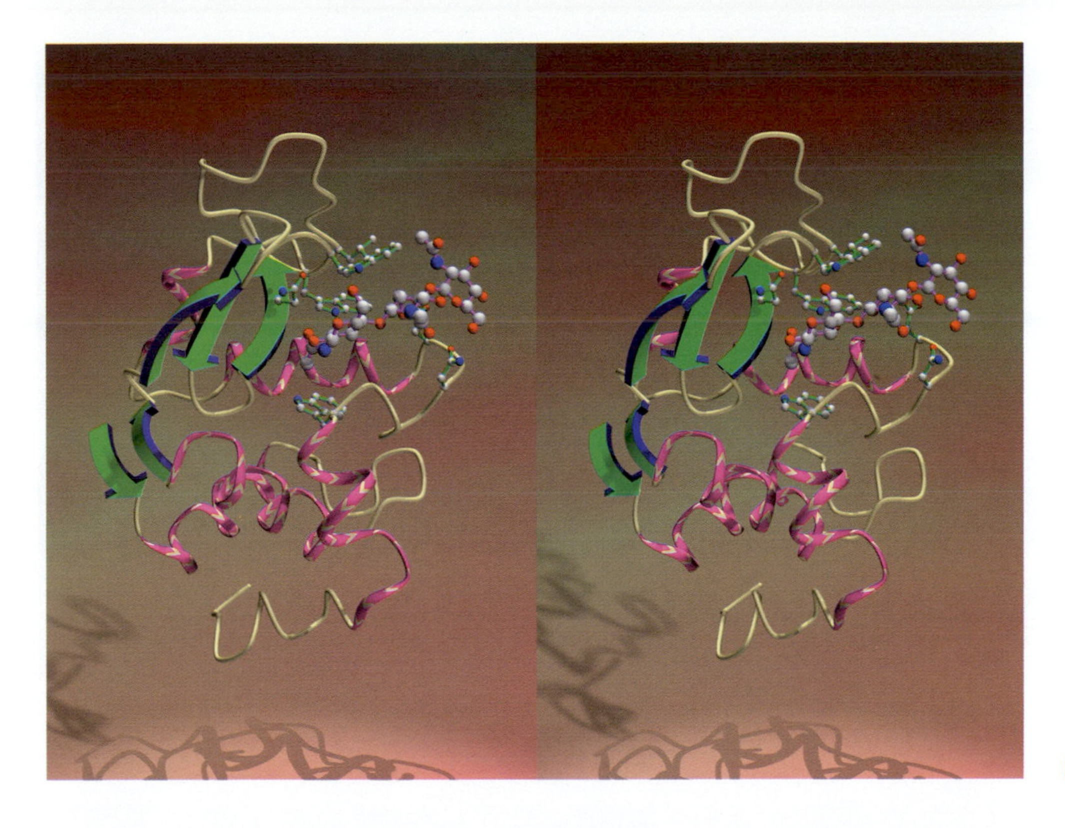

Fig. 3.22c

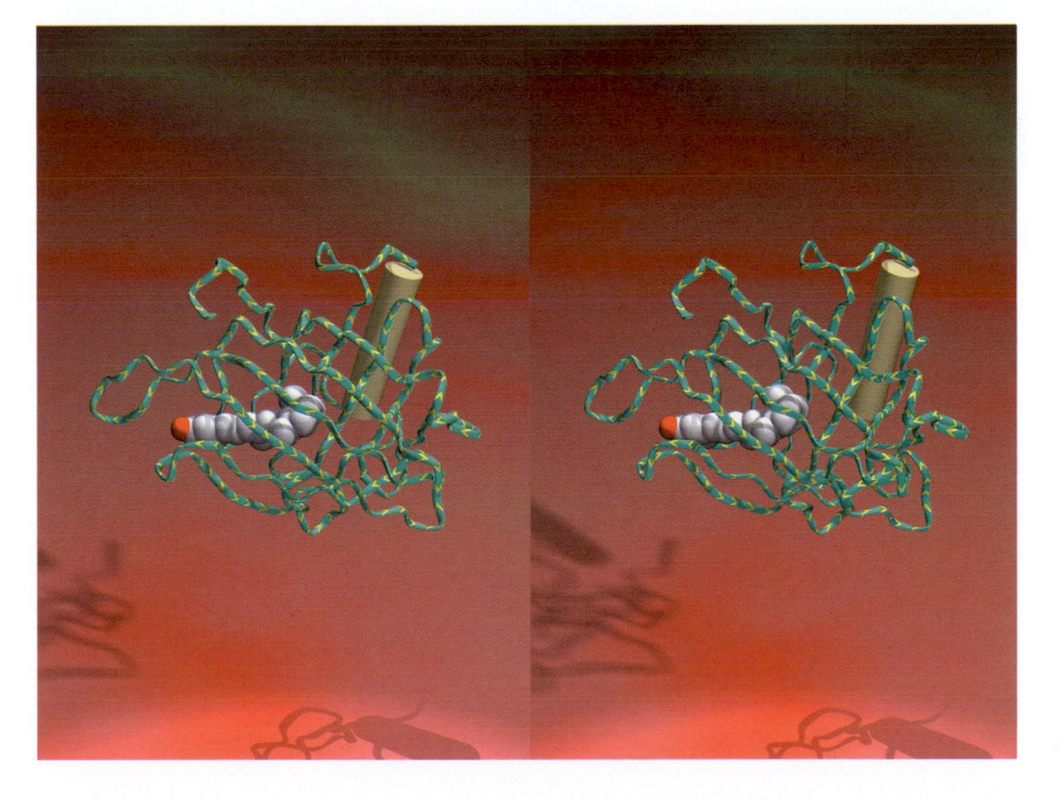

Fig. 3.22d

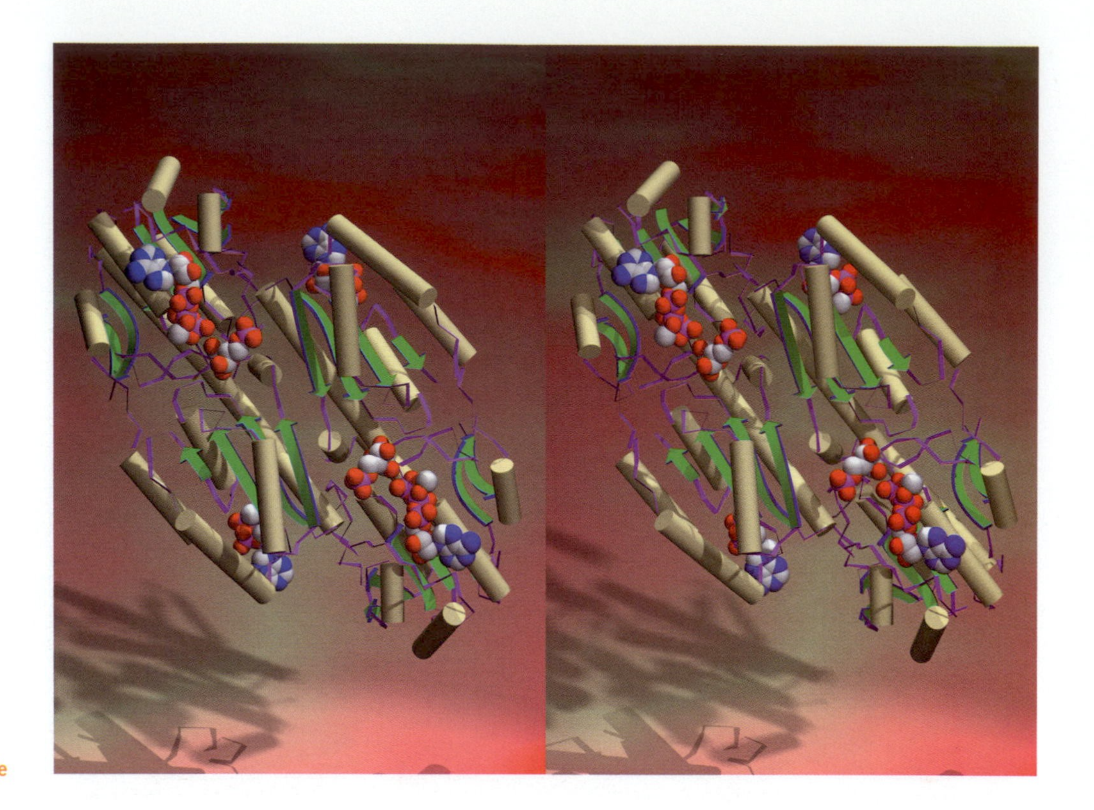

Fig. 3.22e

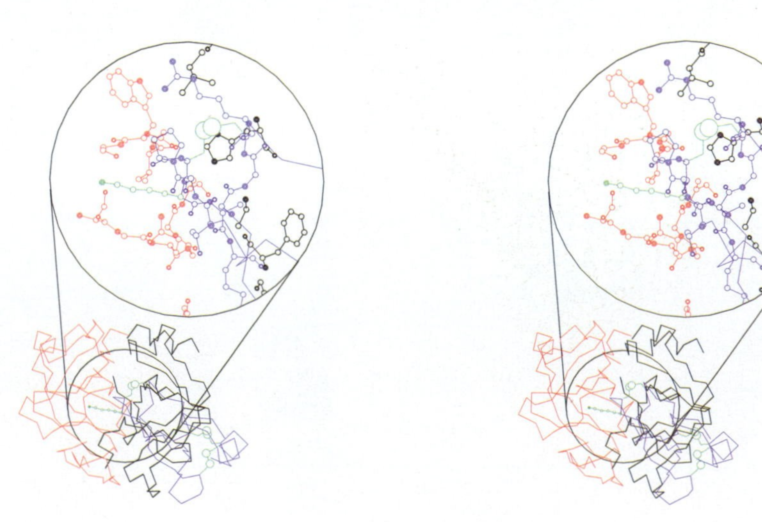

Fig. 3.22f

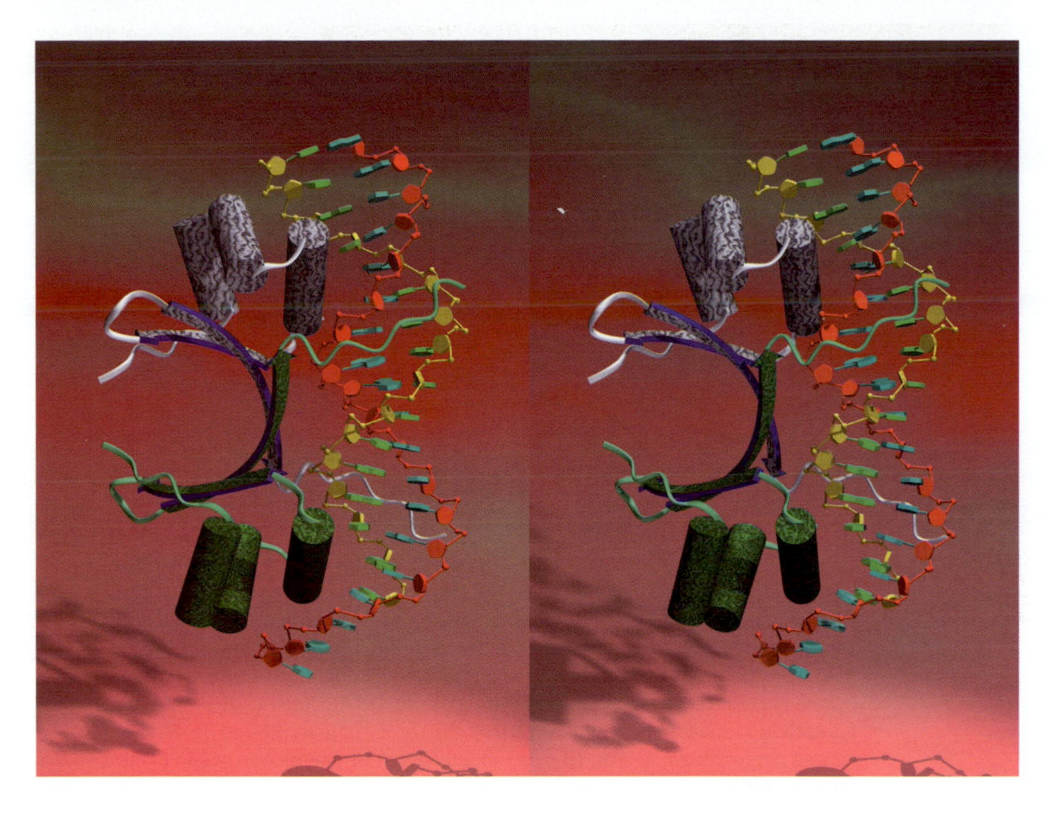

Fig. 3.22g

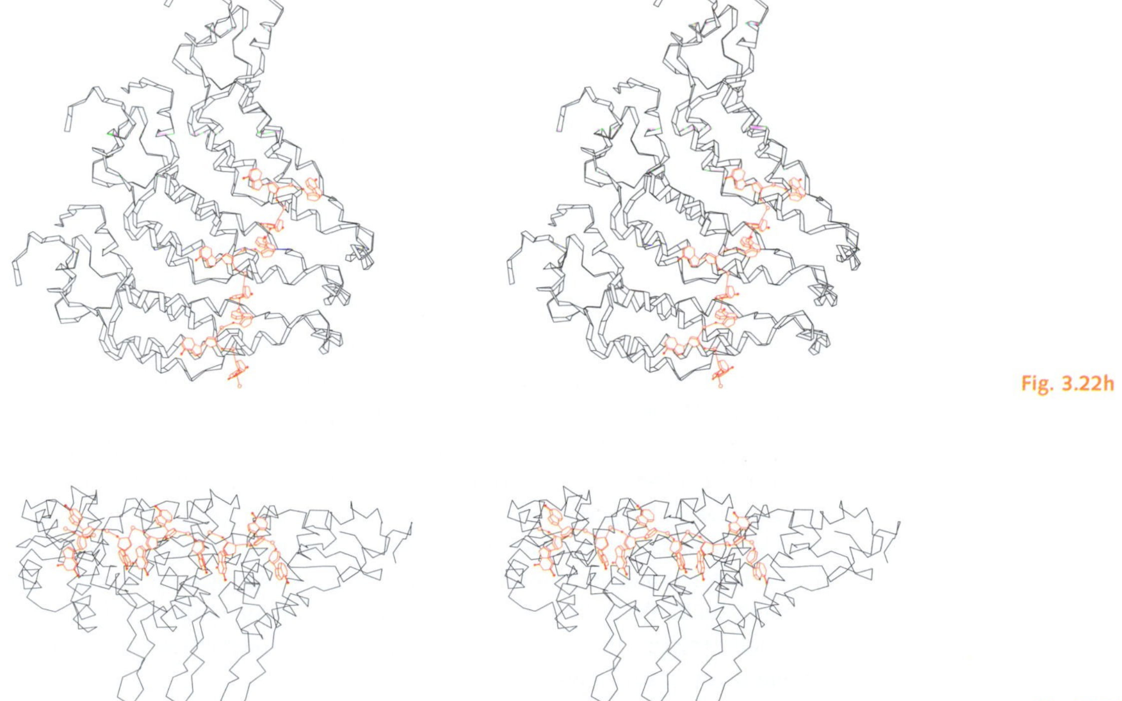

Fig. 3.22h

Fig. 3.22i

Fig. 3.22j

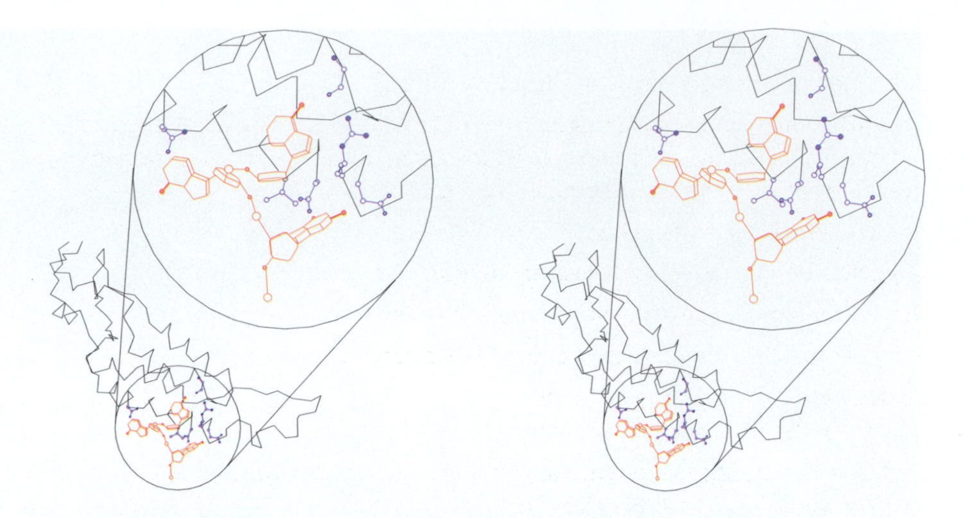

Fig. 3.22 (a) The copper binding site of poplar leaf plastocyanin [5PCY]. (b) Sulphate-binding site of transport protein [1SBP]. (c) Hen egg white lysozyme binding a trisaccharide inhibitor [1HEW]. (d) Retinol-binding protein, enclosing the retinol [1RBP]. (e) Dimer of phosphofructokinase from *Bacillus stearothermophilus* showing substrate, fructose-6-phosphate, cofactor, ATP and effector ADP [4PFK]. (f) Complex of trypsin with bovine pancreatic trypsin inhibitor [1TPA]. (g) λ cro–DNA complex. (h–j) RNA–protein interaction in tobacco mosaic virus [2TMV]. (h) Three subunits viewed down axis, (i) viewed perpendicular to axis, (j) detail of interaction of one protein subunit with a trinucleotide.

Water molecules

In the X-ray crystal structure analysis of proteins at typical resolutions, it is not possible to see hydrogen atoms in electron-density maps. Water molecules, provided their sites are well-ordered and fully occupied, present themselves as single peaks at the positions of the oxygen atoms. It requires a high-resolution structure determination, and nice judgement, to distinguish water molecules from noise. Many crystallographers accept a peak in the electron density as a water molecule only if it is in a position at which a water molecule could make reasonable hydrogen bonds.

Unambiguous identification of water molecules is possible using neutron diffraction. The relative scattering power of hydrogen for neutrons is higher than for X-rays; and it is the nuclei that are responsible for the scattering, so that the withdrawal of electrons from the hydrogen atoms to the more electronegative oxygen atom does not alter the effective shape of the molecule. As a result, water molecules can be seen as bent triatomic molecules, and their position and orientation identified unambiguously. (Currently, the Protein Data Bank contains neutron structures of myoglobin, trypsin, bovine pancreatic trypsin inhibitor, and ribonuclease A, and bibliographic entries for a neutron study of lysozyme.)

Useful web sites

Definition of secondary structure from coordinates

DSSP (Dictionary of Secondary Structures of Proteins): database of secondary structure assignments of protein in PDB available via SRS or ftp; program available but no web server for new coordinate sets:

http://www.sander.ebi.ac.uk/dssp/

STRIDE: http://www.embl-heidelberg.de/cgi/stride_serv

Protein loop classification: http://bonsai.lif.icnet.uk/bmm/loop/

Protein modules: http://www.bork.embl-heidelberg.de/Modules/

Prosite data base of sequence motifs:
http://www.expasy.ch/sprot/prosite.html

Collections of multiple sequence alignments:
http://www.ebi.ac.uk/dali/fssp/fssp.html
http://www.sanger.ac.uk/Pfam *or* http://pfam.wustl.edu/
http://coot.embl-heidelberg.de/SMART

FSSP (Fold classification based on Structure–Structure alignment of Proteins): http://www.embl-ebi.ac.uk/dali/fssp/

Analysis of genome sequences

PEDANT: http://pedant.mips.biochem.mpg.de/

Protein structure prediction centre: http://predictioncenter.llnl.gov/

Recommended reading and references

Campbell, I.D. and Downing, A.K. (1998). NMR of modular proteins. *Nature Structural Biol.* **5**, Supplement, 496–99.

Doolittle, R.F. and Bork, P. (1993). Evolutionarily mobile modules. *Sci. Amer.* **269**(**4**), 50–6.

Efimov, A.V. (1993). Standard structures in proteins. *Prog. Biophys. Mol. Biol.* **60**, 201–39.

Efimov, A.V. (1997). Structural trees for protein superfamilies. Proteins **28**, 241–60.

Lesk, A.M. (1994). Computational molecular biology. In *Encyclopedia of computer science and technology*, Vol. 31 (A. Kent and J.G. Williams, ed.). Marcel Dekker, Inc., New York. pp. 101–65.

Murzin, A.G., Brenner, S.E., Hubbard, T. and Chothia, C. (1995). SCOP: a structural classification of proteins database for the investigation of sequences and structures. *J. Mol. Biol.* 247, 536–40.

Oliva, B., Bates, P.A., Querol, E., Aviles, F.X. and Sternberg, M.J.E. (1997). An automated classification of the structure of protein loops. *J. Mol. Biol.* **266**, 814–30.

Patthy, L. (1994). Introns and exons. *Curr. Opin. Struc. Biol.* **4**, 404–12.

Rost, B. and Schneider, R. (2001). Pedestrian guide to analysing sequence databases. In: *Core techniques in biochemistry* (K. Ashman, ed.). Springer-Verlag, Heidelberg.

Sibanda, B.L and Thornton, J.M. (1985). β hairpin families in globular proteins. *Nature (London)* **316**, 170–4.

Sibanda, B.L., Blundell, T.L. and Thornton, J.M. (1989). Conformation of β hairpins in protein structures. A systematic classification with applications to modelling by homology, electron density fitting and protein engineering. *J. Mol. Biol.* **206**, 759–77.

Supplement 3 (1999) of the journal *Proteins: Structure, Function and Genetics* contains reports of the latest round of CASP (Critical Assessment of Structure Prediction).

Exercises, problems and weblems

Exercises

3.1. Carboxypeptidase is a ~300-residue protein that is approximately spherical with a radius of 25 Å. Estimate: (a) How many residues could there be in the longest possible α-helix in carboxypeptidase? (b) How many residues could there be in the longest possible β–strand in carboxypeptidase?

3.2. Describe the symmetry of the λ cro repressor–DNA complex (see Figure 3.22g).

3.3. In the α-helix shown in Figure 3.2a, is the direction of the chain (N → C) going up the page or down the page?

3.4. In Figure 1.5a, the helix at the left is an α–helix. Is the helix at the right also an α-helix? If not, what kind is it?

3.5. For the amino acid sequence DVAGHGQDIL, between which main-chain atoms of which residues would hydrogen bonds be formed if the region had the conformation of (a) an α-helix, (b) a 3_{10} helix, (c) a π-helix?

3.6. Draw a helical wheel for the following amino acid sequence, which corresponds to the B helix in sperm whale myoglobin: DVAGHGQDILIRLFKSH. Which face would you expect to point towards the interior of the protein?

3.7. (a) How many strands does the β-sheet shown at the front of of the CL domain from immunoglobulin TE33 contain (Figure 3.4b)? (b) How many hairpin loops does this β-sheet contain?

3.8. (a) How many strands does the β-sheet of *Streptococcal* protein G (B1 immunoglobulin binding domain) contain (Figure 3.4c)? (b) How many hairpin loops does this β-sheet contain? (c) How many pairs of adjacent strands in this β-sheet are parallel and how many are antiparallel?

3.9. Classify the proteins appearing in Figure 3.11 into folding classes: α, β, α + β and α/β.

3.10. For each of the structures appearing in Figure 3.11, (a) describe the secondary structure verbally. (b) State the order of appearance in the sequence of helices and strands of β-sheet.

3.11. On a photocopy of Figure 3.11, indicate and identify the types of supersecondary structures.

3.12. Which, if any, of the structures appearing in Figure 3.11 contains (a) a β-bulge? (b) a β-barrel?

3.13. Figure 3.23 shows one of the structures shown in Figure 3.11, but in a different orientation. Which one?

3.14. Figure 3.24 shows the structure of the catalytic part of DNA polymerase β from rat. (a) How many domains does this protein have? (b) On a photocopy, circle the individual domains, and indicate the points at which the chain passes between domains.

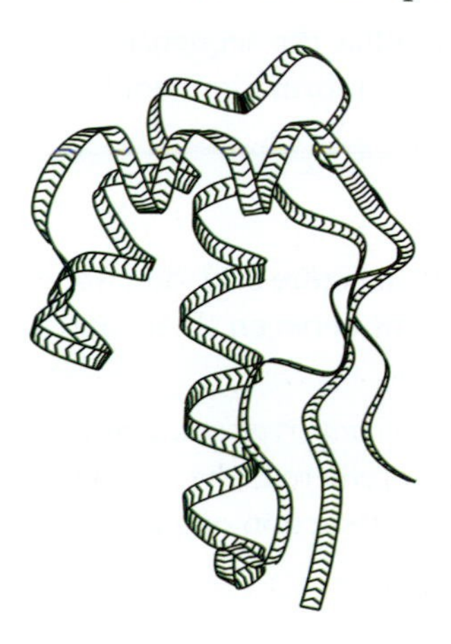

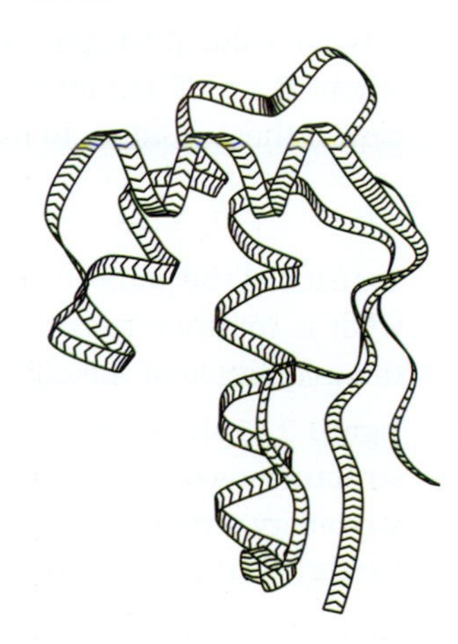

Fig. 3.23 One of the structures illustrated in Figure 3.11, shown in a different orientation. Which one?

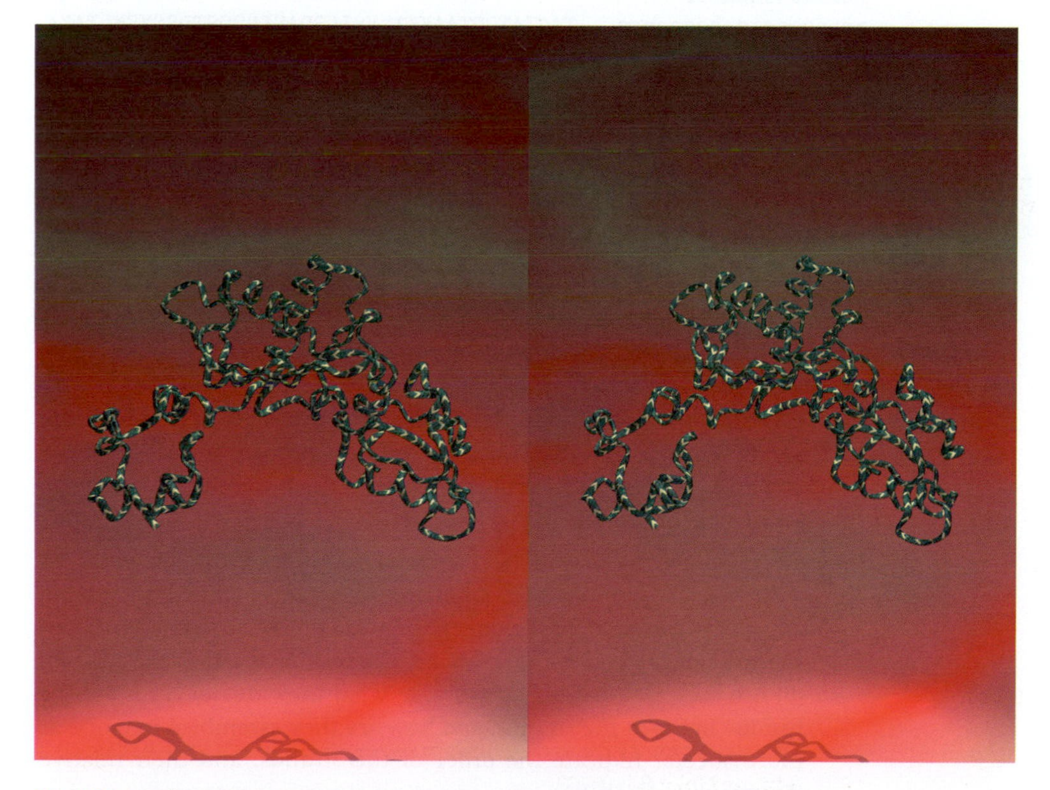

Fig. 3.24 Catalytic domains of DNA polymerase β from rat [1RPL].

3.15. Draw the different possible topologies of a three-stranded β-sheet, ignoring the connectivities of the strands. Represent each strand simply by an arrow pointing up or down. Note that ↑↑↑ and ↓↓↓ are the same topology, with the structure merely reoriented.

3.16. Why would it be impossible to form a seven-stranded β-barrel with every pair of adjacent strands antiparallel?

3.17. A two-residue β hairpin in α-lytic proteinase has the sequence RGAT, where R and T are in the β strands. What conformation would you expect these residues to have?

Problems

3.1. For each of the pairs of structures appearing in Figure 3.12, describe (a) what is common to both, (b) what is different between them, and (c) the magnitude of the differences.

3.2. Figure 3.25 shows β-cryptogein from *Phytophthora cryptogea*. Before its structure was determined, a program predicted the following secondary structure (where H and E signify prediction of a Helix and Extended strand of β-sheet):

```
residue number                  10        20        30        40
amino acid sequence    TACTATQQTAAYKTLVSILSDASFNQCSTDSGYSMLTAKA
prediction                     HHHHHHHEEEEE            EEEEEE

residue number                  50        60        70
amino acid sequence    LPTTAQYKLMCASTACNTMIKKIVTLNPPNCDLTVPTS
prediction                 HHHHHHHHHHHHHHHHHHHHH      EEE

residue number         80        90
amino acid sequence    GLVLNVYSYANGFSNKCSSL
prediction             EEEEEEEEE
```

On a photocopy of the figure, indicate the regions in which the prediction was in error. Do not consider minor differences in the beginnings or ends of segments, but only the misprediction of entire segments.

3.3. There follows on pages 115–19 a list of backbone hydrogen bonds for a 108-residue protein, a thioredoxin. Note that each hydrogen bond appears twice, once for the donor and once for the acceptor. Work out the secondary structure: (a) what are the residue limits of the helices? (b) Are they all α-helices? If not state what they are. (c) What are the residue limits of the strands, and how are they assembled into one or more sheets? Indicate the order of the strands, and their relative directions.

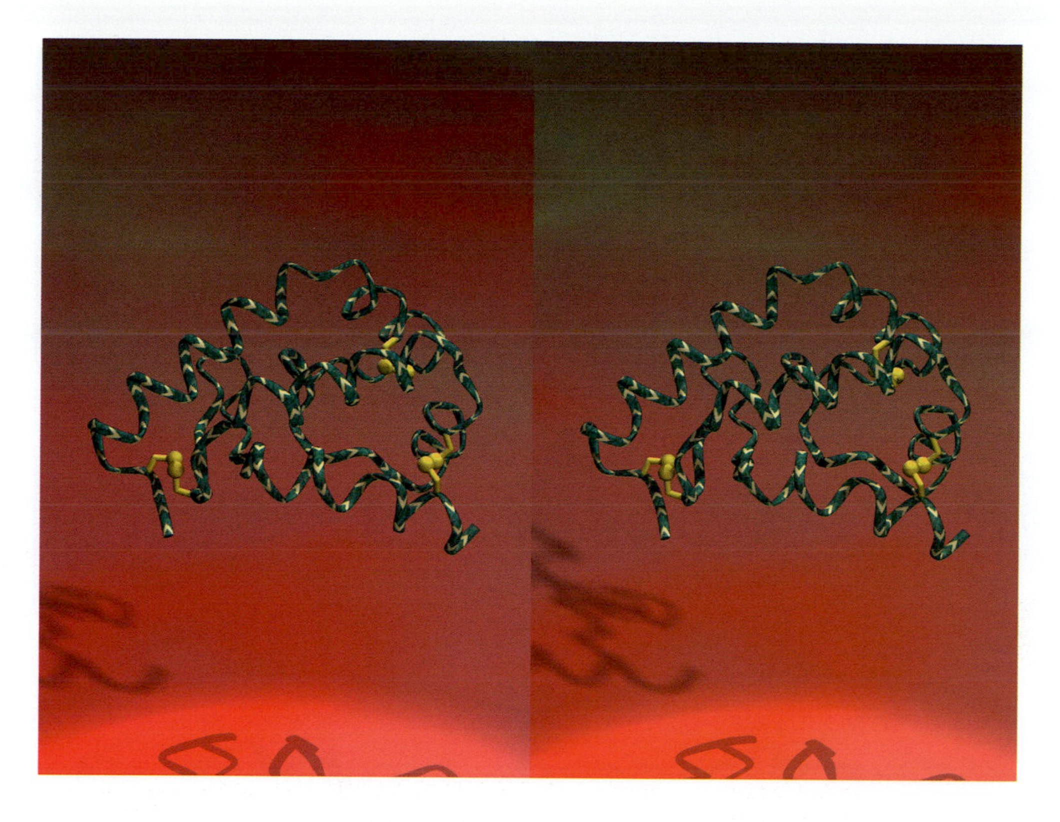

Fig. 3.25a

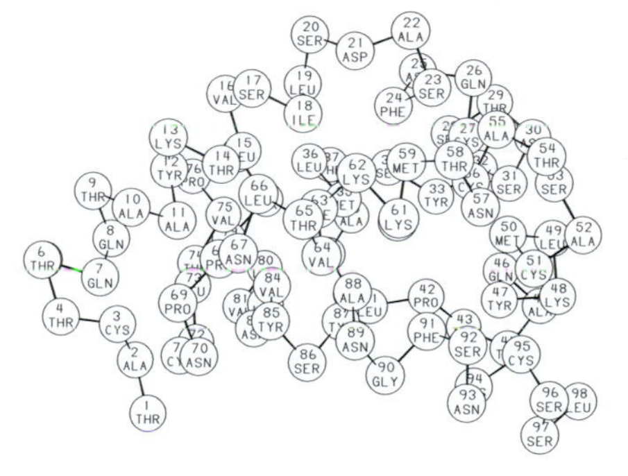

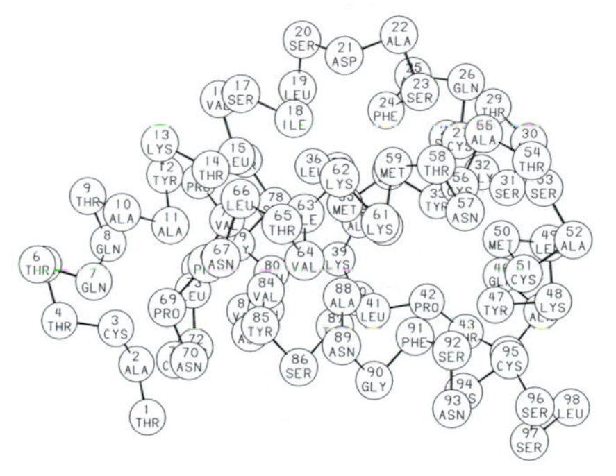

Fig. 3.25b

```
47ASN N    ... O      2LYS
                      5ILE N    ... O      55VAL
57LYS N    ... O      5ILE
                      7ILE N    ... O      57LYS
11GLU N    ... O      8THR
12PHE N    ... O      8THR
12PHE N    ... O      9ASP
                     11GLU N    ... O       8THR
14SER N    ... O     11GLU
15GLU N    ... O     11GLU
                     12PHE N    ... O       8THR
                     12PHE N    ... O       9ASP
16VAL N    ... O     12PHE
```

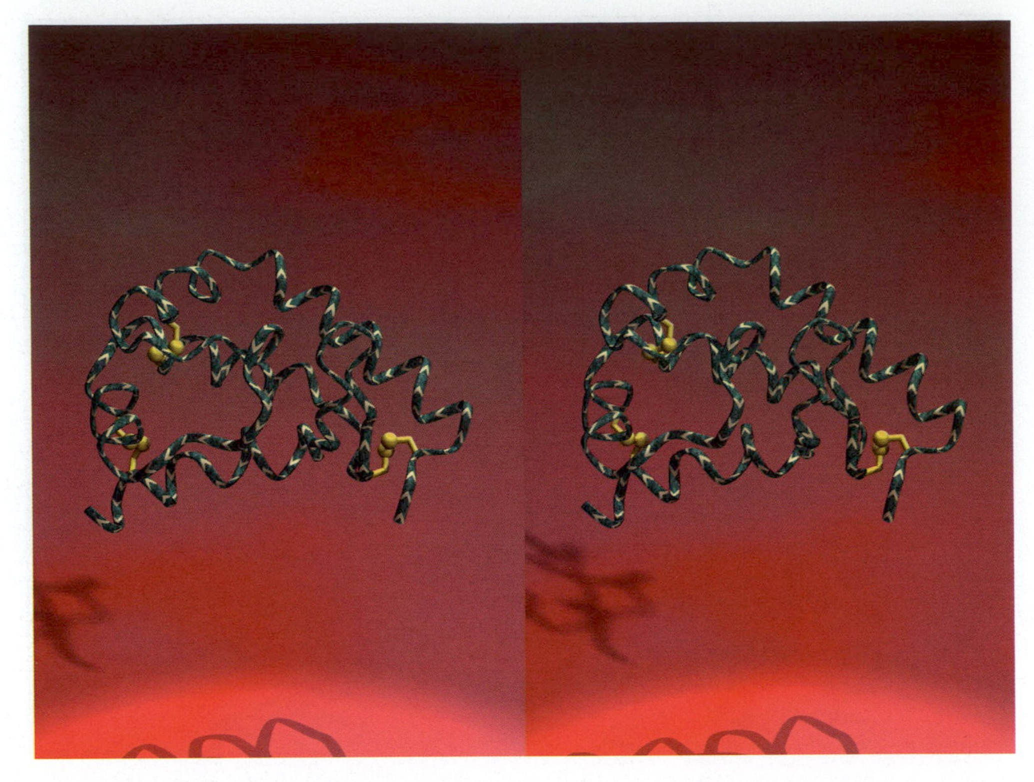

Fig. 3.25c

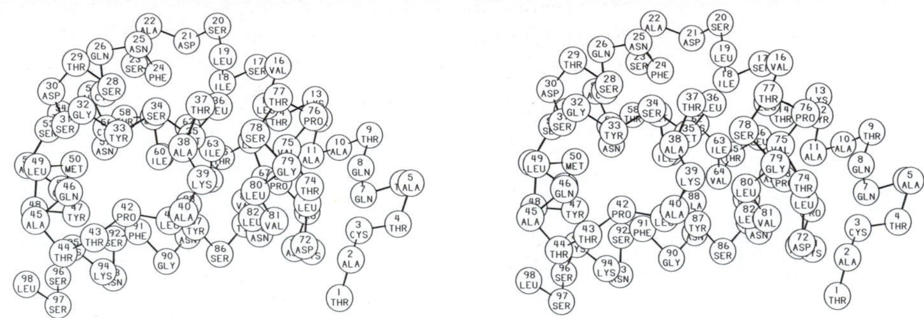

Fig. 3.25d

Fig. 3.25 β cryptogein from *Phytophthora cryptogea* [1BEO]. (a, c) Front and back views. (b, d) Same views as (a) and (c), respectively.

```
17LEU N   ... O     12PHE
17LEU N   ... O     13GLU
18LYS N   ... O     13GLU
                    14SER N   ... O     11GLU
                    15GLU N   ... O     11GLU
                    16VAL N   ... O     12PHE
19ALA N   ... O     16VAL
                    17LEU N   ... O     12PHE
                    17LEU N   ... O     13GLU
                    18LYS N   ... O     13GLU
                    19ALA N   ... O     16VAL
```

```
54LYS N    ... O     22PRO
                     23VAL N    ... O     81VAL
81VAL N    ... O     23VAL
                     24LEU N    ... O     54LYS
56VAL N    ... O     24LEU
79ARG N    ... O     25VAL
                     25VAL N    ... O     79ARG
                     26TYR N    ... O     56VAL
58LEU N    ... O     26TYR
77ALA N    ... O     27PHE
                     27PHE N    ... O     77ALA
                     28TRP N    ... O     58LEU
60ILE N    ... O     28TRP
32CYS N    ... O     29ALA
                     32CYS N    ... O     29ALA
35CYS N    ... O     32CYS
36GLN N    ... O     33GLY
37LEU N    ... O     34PRO
38MET N    ... O     34PRO
                     35CYS N    ... O     32CYS
39SER N    ... O     35CYS
                     36GLN N    ... O     33GLY
                     37LEU N    ... O     34PRO
                     38MET N    ... O     34PRO
41LEU N    ... O     38MET
42ILE N    ... O     38MET
                     39SER N    ... O     35CYS
43ASN N    ... O     39SER
44LEU N    ... O     40PRO
                     41LEU N    ... O     38MET
45ALA N    ... O     41LEU
                     42ILE N    ... O     38MET
46ALA N    ... O     42ILE
                     43ASN N    ... O     39SER
46ALA N    ... O     43ASN
47ASN N    ... O     43ASN
                     44LEU N    ... O     40PRO
48THR N    ... O     44LEU
                     45ALA N    ... O     41LEU
49TYR N    ... O     45ALA
                     46ALA N    ... O     42ILE
                     46ALA N    ... O     43ASN
50SER N    ... O     46ALA
                     47ASN N    ... O      2LYS
                     47ASN N    ... O     43ASN
                     48THR N    ... O     44LEU
                     49TYR N    ... O     45ALA
52ARG N    ... O     49TYR
53LEU N    ... O     49TYR
                     50SER N    ... O     46ALA
                     52ARG N    ... O     49TYR
                     53LEU N    ... O     49TYR
                     54LYS N    ... O     22PRO
```

```
 24LEU N    ... O     54LYS
  5ILE N    ... O     55VAL
                      56VAL N    ... O     24LEU
 26TYR N    ... O     56VAL
                      57LYS N    ... O      5ILE
  7ILE N    ... O     57LYS
                      58LEU N    ... O     26TYR
 28TRP N    ... O     58LEU
                      60ILE N    ... O     28TRP
 63ASN N    ... O     60ILE
                      63ASN N    ... O     60ILE
 67VAL N    ... O     63ASN
 68LYS N    ... O     64PRO
 69LYS N    ... O     65THR
 70TYR N    ... O     66THR
                      67VAL N    ... O     63ASN
 71LYS N    ... O     67VAL
 72VAL N    ... O     67VAL
                      68LYS N    ... O     64PRO
 71LYS N    ... O     68LYS
                      69LYS N    ... O     65THR
                      70TYR N    ... O     66THR
                      71LYS N    ... O     67VAL
                      71LYS N    ... O     68LYS
                      72VAL N    ... O     67VAL
 92GLY N    ... O     76PRO
 27PHE N    ... O     77ALA
                      77ALA N    ... O     27PHE
                      78LEU N    ... O     90THR
 90THR N    ... O     78LEU
                      79ARG N    ... O     25VAL
 25VAL N    ... O     79ARG
 87LEU N    ... O     80LEU
                      80LEU N    ... O     88ASP
 88ASP N    ... O     80LEU
                      81VAL N    ... O     23VAL
 23VAL N    ... O     81VAL
 85GLN N    ... O     82LYS
                      82LYS N    ... O     85GLN
 82LYS N    ... O     85GLN
                      85GLN N    ... O     82LYS
                      87LEU N    ... O     80LEU
                      88ASP N    ... O     80LEU
 80LEU N    ... O     88ASP
                      90THR N    ... O     78LEU
 78LEU N    ... O     90THR
                      92GLY N    ... O     76PRO
 98LYS N    ... O     95SER
 99LEU N    ... O     95SER
100LEU N    ... O     96LYS
101SER N    ... O     97ASP
                      98LYS N    ... O     95SER
102PHE N    ... O     98LYS
```

```
                              99LEU N    ... O       95SER
103LEU N    ... O             99LEU
                              100LEU N   ... O       96LYS
104ASP N    ... O             100LEU
                              101SER N   ... O       97ASP
105THR N    ... O             101SER
                              102PHE N   ... O       98LYS
                              103LEU N   ... O       99LEU
106HIS N    ... O             103LEU
107LEU N    ... O             103LEU
                              104ASP N   ... O      100LEU
108ASN N    ... O             104ASP
                              105THR N   ... O      101SER
                              106HIS N   ... O      103LEU
                              107LEU N   ... O      103LEU
                              108ASN N   ... O      104ASP
```

3.4. G. Zubay and P.M. Doty suggested in 1959 that an α-helix was the right size to fit into the major groove of DNA. This interaction has been found in the structures of many transcription factors such as Skn-1 (Figure 3.26a.) Figure 3.26b shows the a view down the major groove, showing the α-helix end-on, and also the mainchain atoms only of a π-helix seen end-on. Would it be possible for a protein containing a π-helix to act as a transcription factor by inserting a π-helix into the major groove of DNA?

3.5. (a) Identify the helices and strands of sheet in the protein shown in Figure 3.27. Give the residue numbers at which they begin and end. (b) State how the strands are arranged in the sheet. Draw a simple diagram showing the positions of the strands and their relative directions.

3.6. (a) Identify the helices and strands of sheet in the protein shown in Figure 3.28. Give the residue numbers at which they begin and end. (b) State how the strands are arranged in the sheet. Draw a simple diagram showing the positions of the strands and their relative directions. This example is somewhat more difficult than the preceding one.

Weblems

3.1. Using SCOP or CATH, find examples of proteins not illustrated in this chapter, that contain (a) an all-parallel β-sheet, (b) an all-antiparallel β-sheet, (c) a β-sheet in which some adjacent strands are parallel and others are antiparallel.

3.2. Find an example, not illustrated in this chapter, of a β-sheet containing a β bulge.

3.3. Find 10 examples of proteins containing TIM barrels.

3.4. A student in a hurry to finish his or her assignment lacks the patience to work out the secondary structure of thioredoxin (see Problem 3.2).

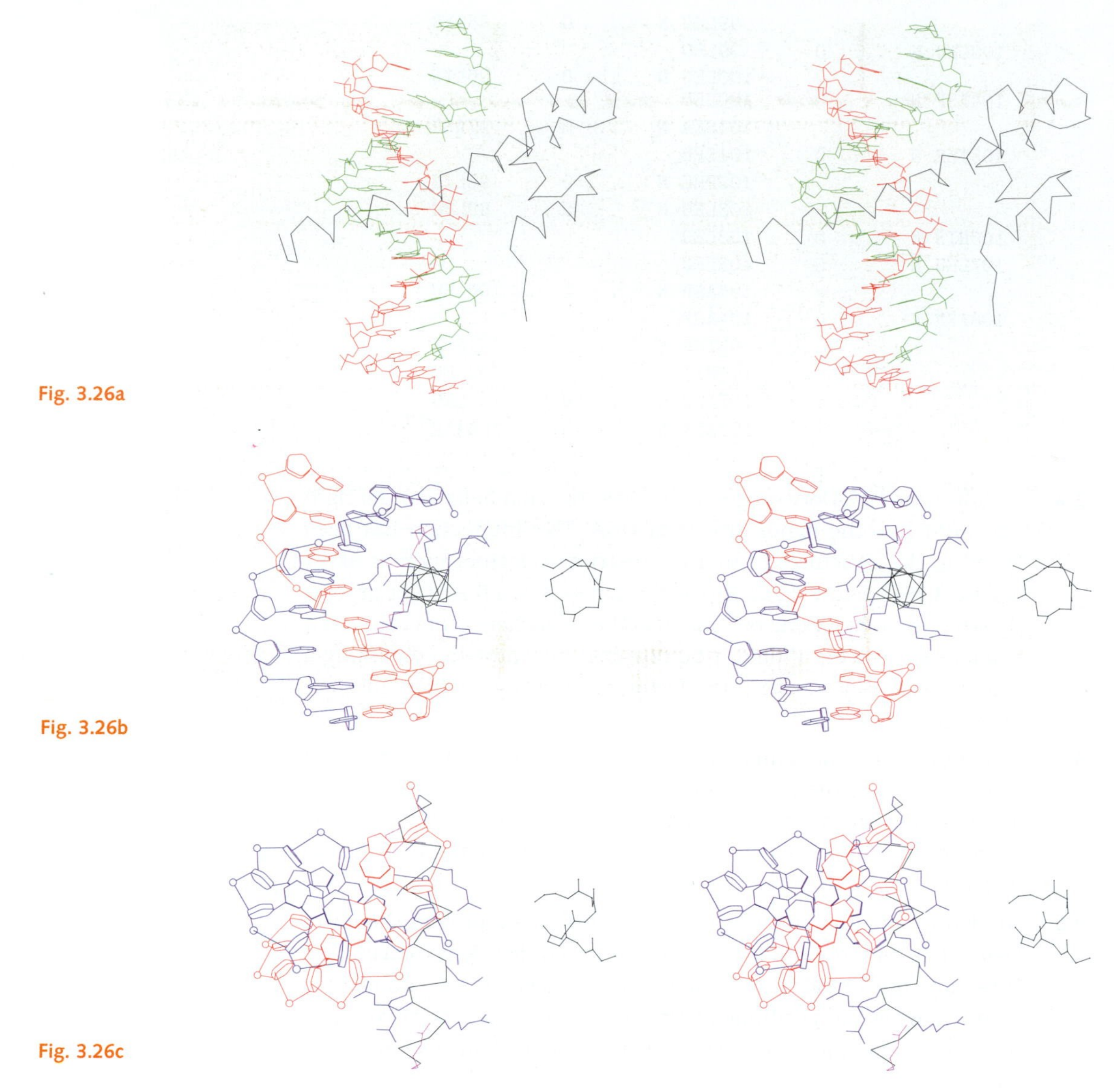

Fig. 3.26 (a) DNA complex of transcription factor Skn-1, with an α-helix binding in the major groove of the DNA [1SKN]. (b) Part of the DNA–Skn-1 complex viewed down the axis of the α-helix; with the mainchain atoms of a π-helix, also viewed down its axis. (c) View perpendicular to axes of protein helices.

How might the student find an answer (not necessarily the correct one) on the web?

3.5. A student in a hurry to finish his or her assignment lacks the patience to work out the secondary structure of the protein in Figure 3–27 (see Problem 3.5). How might the student find an answer (not necessarily the correct one) by searching the web?

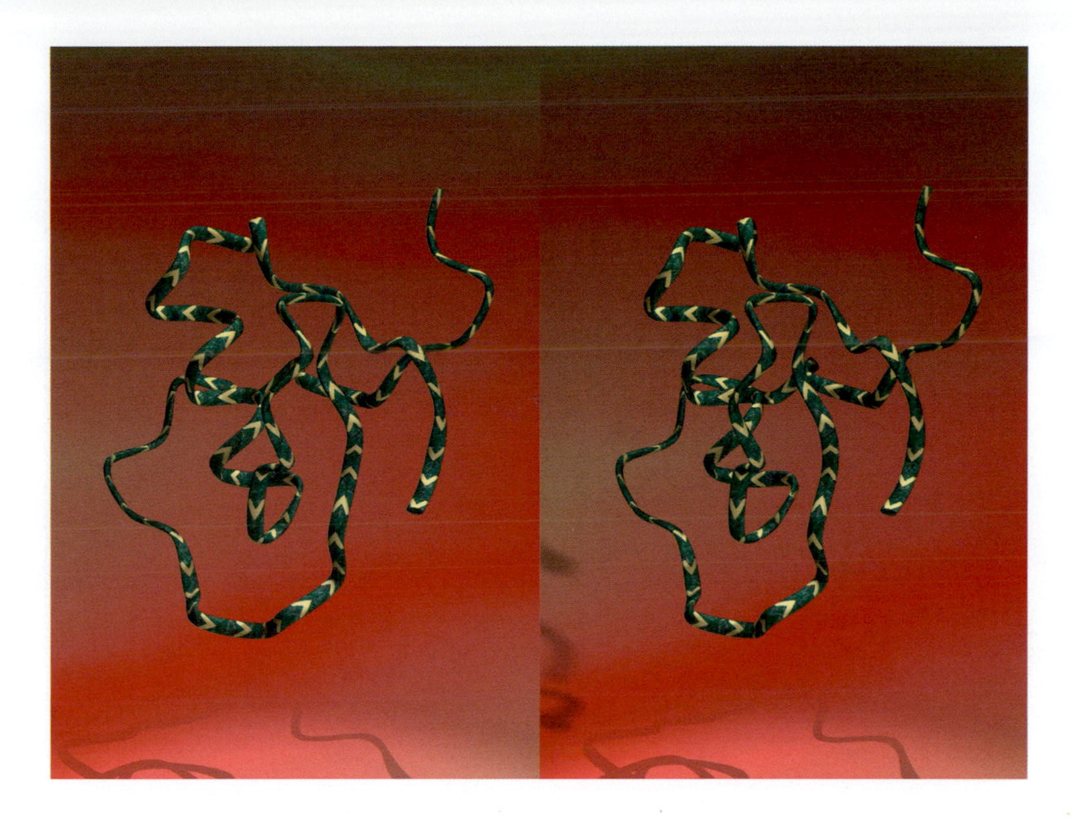

Fig. 3.27a

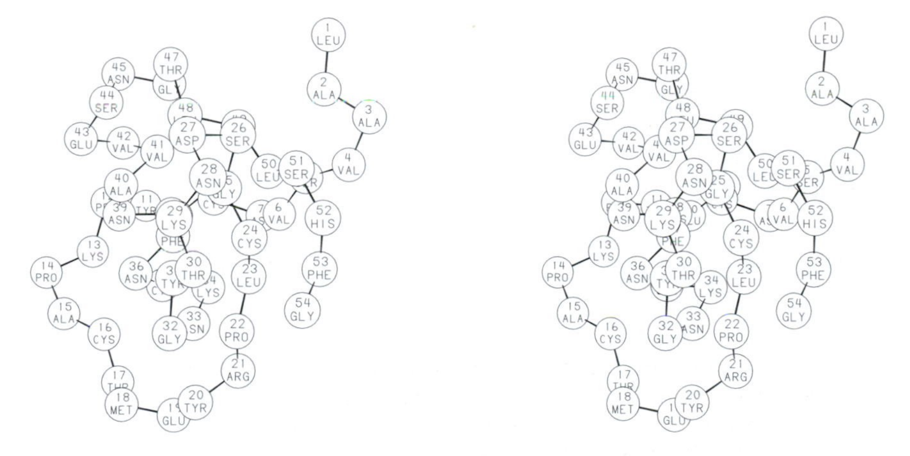

Fig. 3.27b

3.6. Retrieve the sequence of UTRO-HUMAN from SWISS-PROT. What is the function of this molecule?

3.7. From the Pfam alignment of C-type lysozymes, try to suggest where the turns between elements of secondary structure in hen egg white lysozyme might be. Give reasons.

3.8. The PROSITE pattern [LIVMA]-G-[EQ]-H-G-[DN]-[ST] specifies a region in which the first position may be any of the amino acids L, I, V, M or

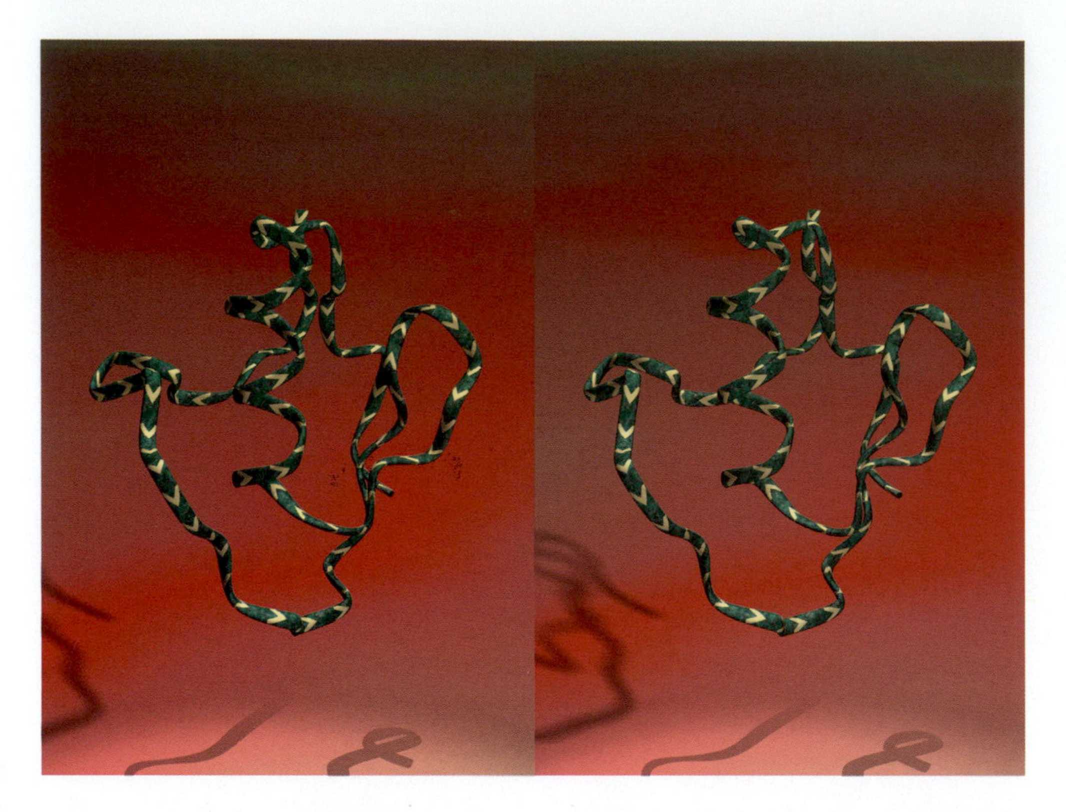

Fig. 3.27c

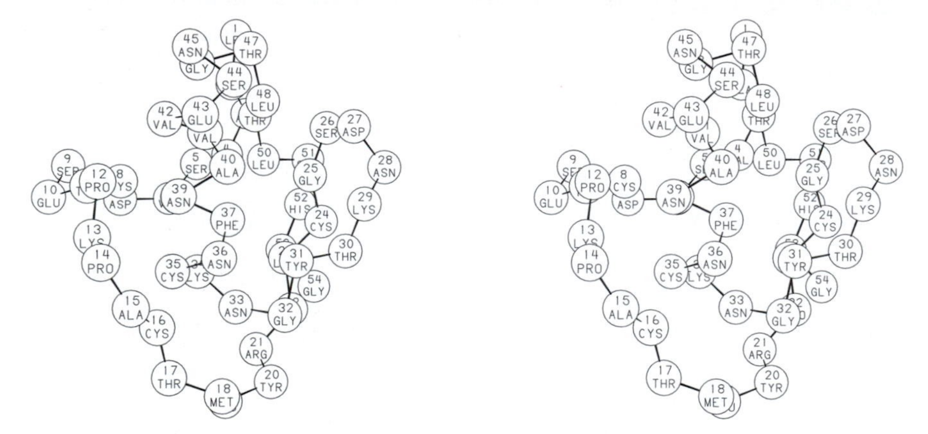

Fig. 3.27d

Fig. 3.27 Identify the helices and strands of sheet in this protein. The viewpoint in parts (a) and (b) differs from that of parts (c) and (d) by a rotation of 90°. Use parts (a) and (c) to identify the secondary structure, parts (b) and (d) to determine the residue numbers.

A; the second position must be G, etc. What proteins does this pattern characterize?

3.9. Submit the following sequence

MYKLTVFLMF IAFVIIAEAQ LTFTSSWGGK RAMTNSISCR NDEAIAAIYK AIQNEAERFI MCQKN

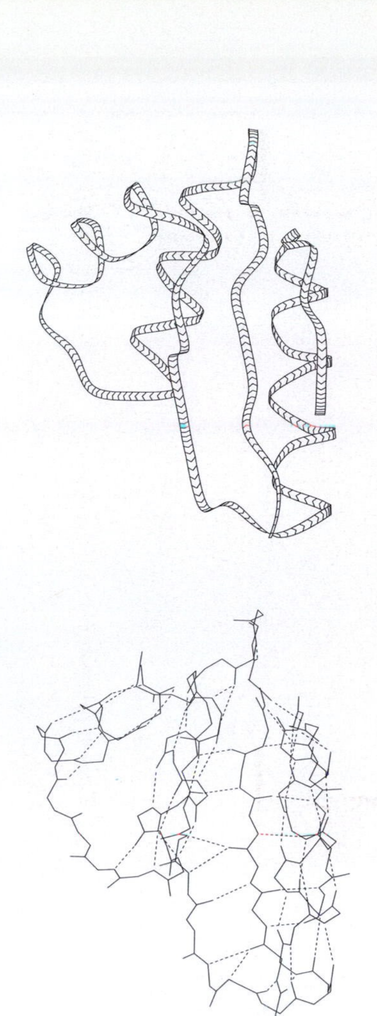

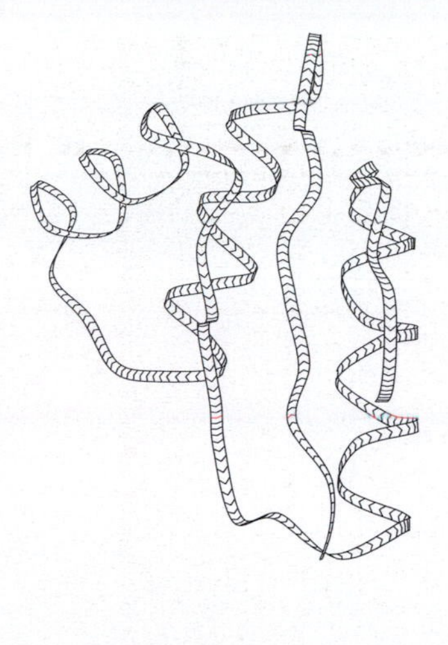

Fig. 3.28a

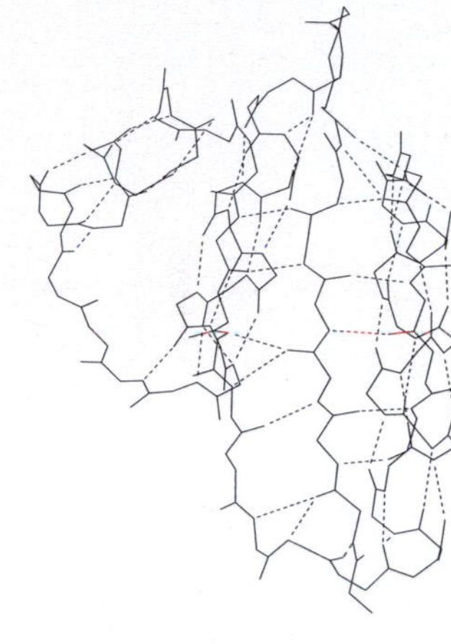

Fig. 3.28b

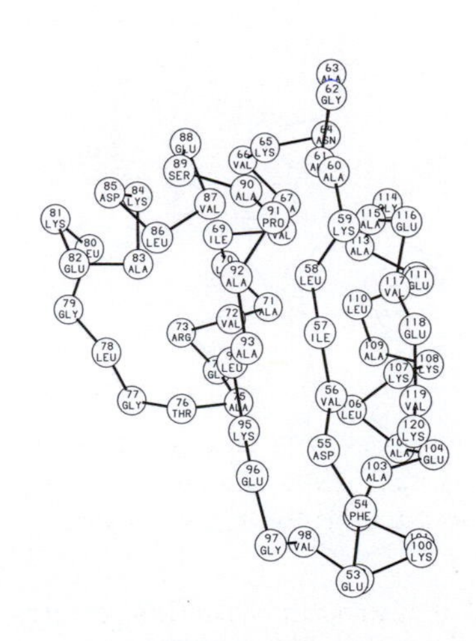

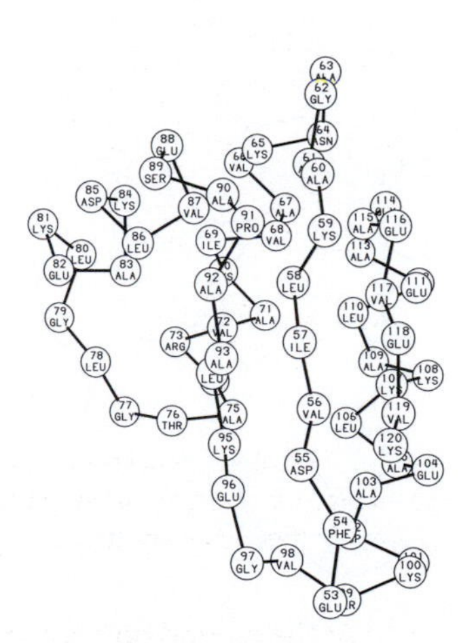

Fig. 3.28c

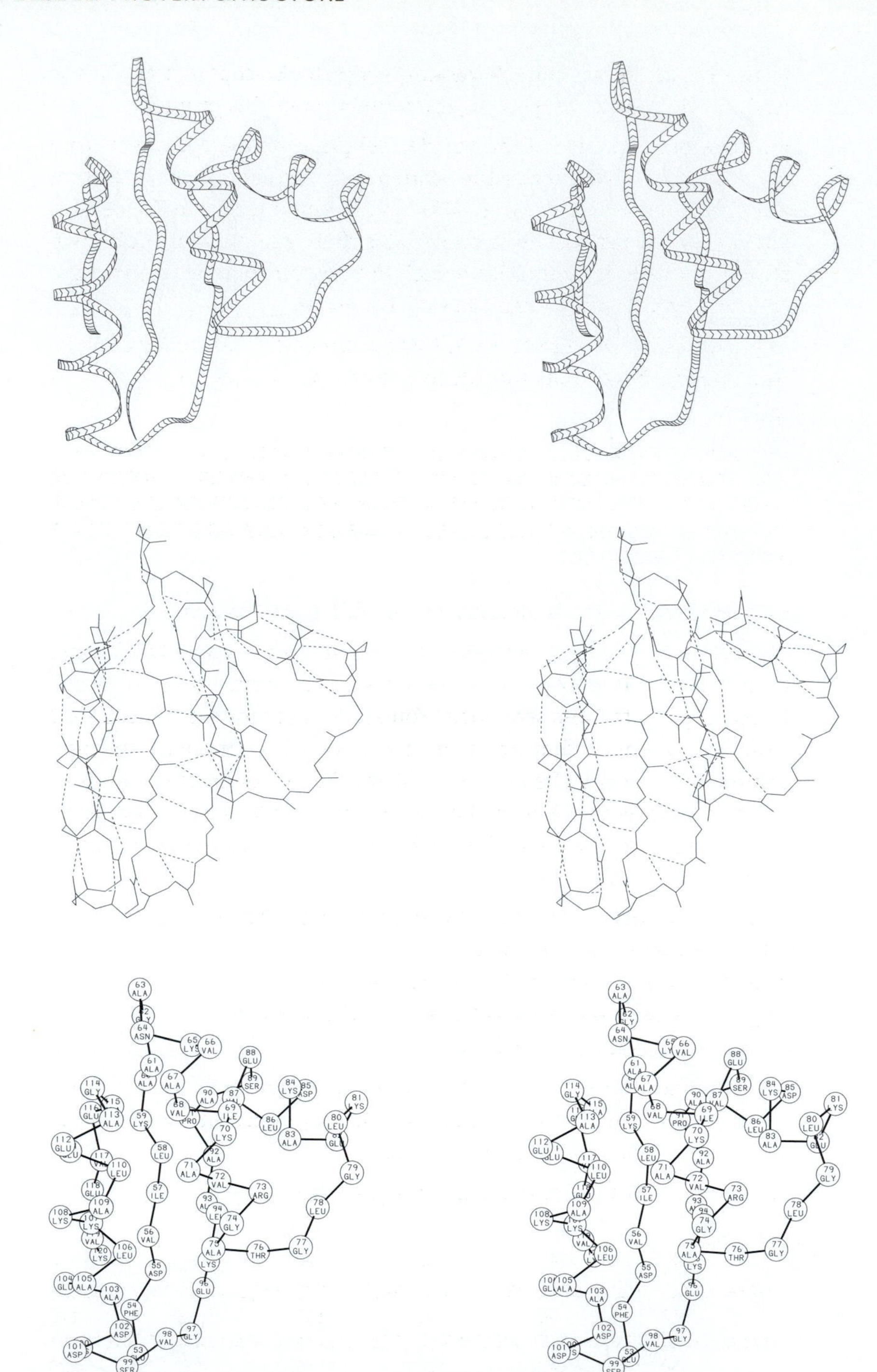

Fig. 3.28d

Fig. 3.28e

Fig. 3.28f

Fig. 3.28 Identify the helices and strands of sheet in this protein. The viewpoint of parts (a)–(c) differs from that of parts (d)–(f) by a rotation of 180° around an axis vertical in the page. Use parts (a, b, d and e) to identify the secondary structure, and parts (c) and (f) to determine the residue numbers. A somewhat more complicated case.

to the Prosite server: http://www.expasy.ch/tools/scnpsite.html On the basis of the results can you suggest a function of this protein?

3.10. Retrieve the sequence of the purple acid phosphatase precursor from *Arabidopsis thaliana*. Submit the sequence to servers on web sites that identify sequence similarities. (a) What can you infer, from the results, about the structure of this protein? (a) Is there a homologue of known structure in the Protein Data Bank? (b) Can you thereby confirm the similarity you found on the basis of sequence?

3.11. The Dickkopf-1 protein from *X. laevis* (GenBank accession AF030434) functions in head induction during development. Its sequence is as follows:

```
MGSNMFPVPL IVFWGFILDG ALGFVMMTNS NSIKNVPAAP AGQPIGYYPV SVSPDSLYDI
ANKYQPLDAY PLYSCTEDDD CALDEFCHSS RNGNSLVCLA CRKRRKRCLR DAMCCTGNYC
SNGICVPVEQ DQERFQHQGY LEETILENYN NADHATMDTH SKLTTSPSGM QPFKGRDGDV
CLRSTDCAPG LCCARHFWSK ICKPVLDEGQ VCTKHRRKGS HGLEIFQRCH CGAGLSCRLQ
KGEFTTVPKT SRLHTCQRH
```

Can you identify a homologous protein of known structure?

3.12. The protein aquaporin-1 forms water channels in red blood cell membranes. Can you identify putative regions that form transmembrane helical segments in its sequence? You are looking for segments 17–22 residues in length that are almost exclusively non-polar, and you should not expect to be able to estimate the positions of the ends of the helices to within better than about five residues. Try submitting the sequence to the following web servers that offer predictions of transmembrane helices:

http://www.embl-heidelberg.de/predictprotein/predictprotein.html
http://www.enzim.hu/hmmtop/
http://www.cbs.dtu.dk/services/TMHMM-1.0/
http://www.isrec.isb-sib.ch/software/TMPRED_form.html
http://www.tuat.ac.jp/~mitaku/adv_sosui/

and compare the results. (It is recommended that you do a general search on the Web for 'transmembrane helix prediction' to determine an up-to-date list of available methods.)

The amino acid sequence of human aquaporin-1 is:

```
AQP1_HUMAN
         10         20         30         40         50         60
  MASEFKKKLF WRAVVAEFLA TTLFVFISIG SALGFKYPVG NNQTAVQDNV KVSLAFGLSI
         70         80         90        100        110        120
  ATLAQSVGHI SGAHLNPAVT LGLLLSCQIS IFRALMYIIA QCVGAIVATA ILSGITSSLT
        130        140        150        160        170        180
  GNSLGRNDLA DGVNSGQGLG IEIIGTLQLV LCVLATTDRR RRDLGGSAPL AIGLSVALGH
        190        200        210        220        230        240
  LLAIDYTGCG INPARSFGSA VITHNFSNHW IFWVGPFIGG ALAVLIYDFI LAPRSSDLTD
        250        260
  RVKVWTSGQV EEYDLDADDI NSRVEMKPK
```

CHAPTER 4

The varieties of protein structure

How could one hope to govern a country that has two hundred fifty-eight varieties of cheese?
—attributed to Charles de Gaulle

Catalogues of protein structures

There are now so many known protein structures that we need directories and catalogues to keep track of them. 'White pages', in which the structures are listed by name, are useful provided we know what we are looking for. More generally helpful are 'Yellow pages', which organize the entries in some reasonable classification.

SCOP (Structural Classification of Proteins), CATH (Class, Architecture, Topology, Homologous superfamily), and DALI/FSSP/DDD (Fold classification based on Structure–Structure alignment of Proteins/Dali Domain Dictionary) are databases built around classifications of protein structures. Accessible over the web, they have many useful features as information-retrieval engines, including search by keyword or sequence, presentation of structure pictures (this requires a suitable terminal), and links to other related sites including bibliographical databases.

SCOP

SCOP organizes protein structures in a hierarchy according to evolutionary origin and structural similarity. It is based on protein domains rather than full protein structures. At the lowest level of the SCOP hierarchy, then, are the individual *domains*, extracted from the Protein Data Bank entries. Sets of domains are grouped into *families* of homologues. These comprise domains for which the similarities in structure, function and sequence imply a common evolutionary origin. Families that share common structure and function, but lack adequate sequence similarity, so that the evidence for evolutionary relationship is suggestive but not compelling, are grouped into *superfamilies*. Superfamilies that share a common folding topology, for at least a large central portion of the structure, are grouped as *folds*. Finally, each fold group falls into one of the general *classes*. The major classes in SCOP are α, β, α + β, α/β, and 'small proteins', which often have little secondary structure and are held together by disulphide bridges or ligands (for instance, wheat-germ agglutinin).

Figure 4.1 shows the SCOP classification of flavodoxin from *Clostridium beijerinckii*.

SCOP classification of flavodoxin

Class	α and β (α/β)
Fold	Flavodoxin-like
core contains α-helices packed onto both sides of a sheet to form a three-layered α–β–α structure; β-sheet contains five parallel strands, with order of appearance in the sequence: 21345	
Superfamily	Flavoproteins
Family	Flavodoxin-related
Protein	Flavodoxin
Species	*Clostridium beijerinckii*
Domain	PDB code 5nll

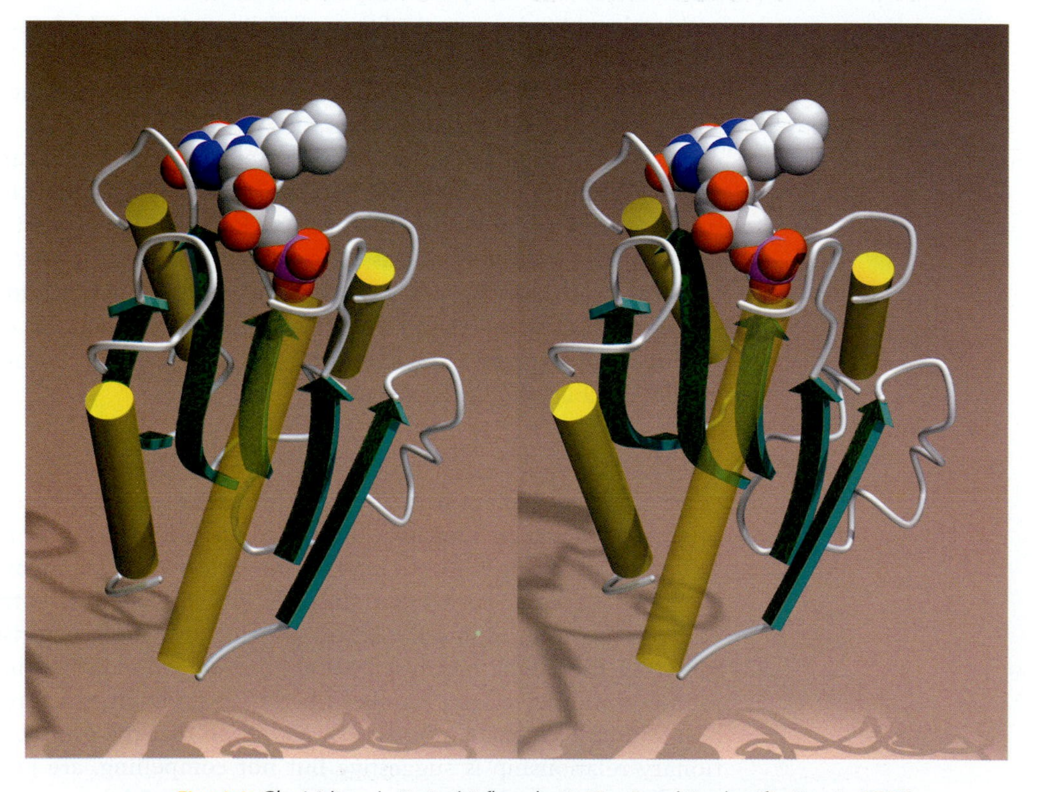

Fig. 4.1 *Clostridium beijerinckii* flavodoxin [5NLL] and its classification in SCOP.

The SCOP hierarchy (on January 1, 2000)

Level	Number of cases
Classes	7
Folds	520
Superfamilies	771
Families	1212
Domains	21529

CATH

CATH presents a classification scheme similar to that of SCOP. The letters in its name stand for the levels of its hierarchy: Class, Architecture, Topology, Homologous superfamily. In the CATH classification, proteins with very similar structures, sequences and functions are grouped into *sequence families*. A *homologous superfamily* contains proteins for which there is evidence of common ancestry, based on similarity of sequence and structure. A *topology* or *fold family* comprises sets of homologous superfamilies that share the spatial arrangement *and* connectivity of helices and strands of sheet. In CATH, *architectures* are groups of proteins with similar arrangements of helices and sheets, but with different connectivity. For instance, different four α-helix bundles with different connectivities would share the same architecture but not the same topology in CATH (Figure 4.2). Finally, the general *classes* of architectures in CATH are: α, β, α – β (subsuming α + β and α / β), and domains of low secondary structure content.

The correspondence between the levels of the hierarchy in SCOP and CATH is rough but not exact (see following box).

Correspondence between SCOP and CATH hierarchies

SCOP	CATH
Class	Class
	Architecture
Fold	Topology
	Homologous superfamily
Superfamily	
Family	Sequence family
Domain	Domain

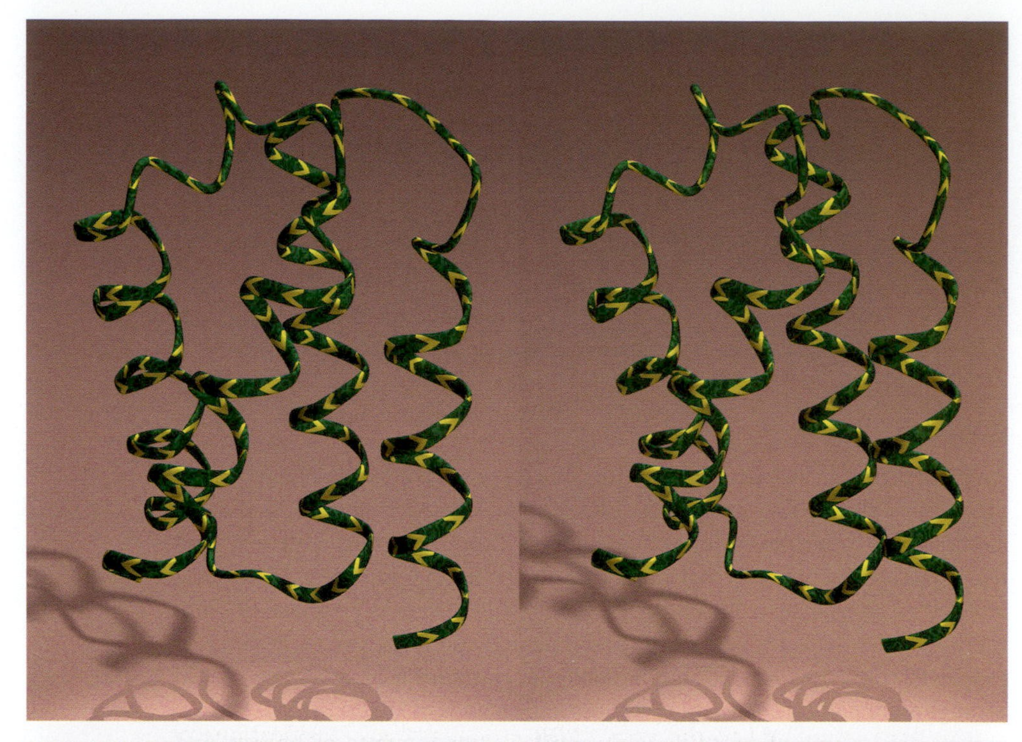

Fig. 4.2a

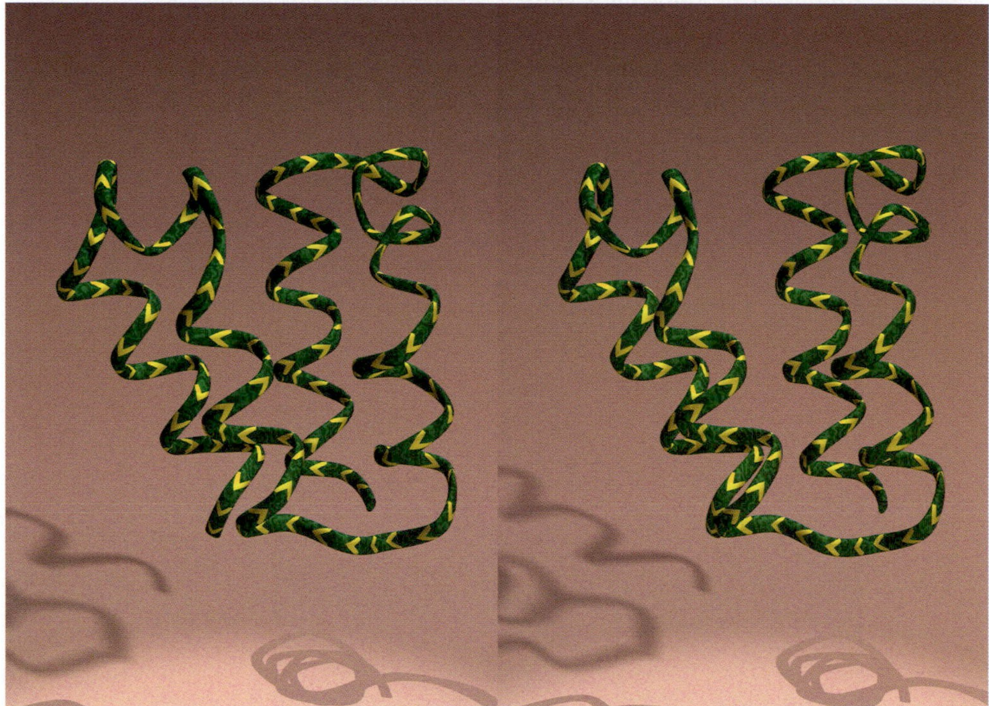

Fig. 4.2b

Fig. 4.2 Two four-α-helix bundles with similar arrangements of helices but different connectivities. (a) Domain from signal sequence recognition protein FFH [1FFH]. (b) Region from T4 lysozyme [1LYD].

Both SCOP and CATH emphasize the *presentation* of the results of the classification. The methods for comparing proteins that underlie the classification remain internal to the project. A third classification of protein structures—FSSP and the DALI Domain Dictionary—is built around a method for structure comparison. Application of the method to the entire PDB induces a complete classification of known protein structures.

FSSP, and the DALI domain dictionary

The program DALI, by L. Holm and C. Sander, provides a general and sensitive method for comparing protein structures. Given two sets of coordinates, it determines the maximal common substructure and provides an alignment of the common residues. DALI is the best of the many programs that address this problem, because of its ability to recognize distant relationships, and its speed of execution. DALI is fast enough to scan the entire PDB routinely for proteins similar to a probe structure.

A web server allows users to submit a protein structure for scanning by DALI against the PDB. Crystallographers and NMR spectroscopists who solve a new structure routinely submit the coordinates to the DALI server, to detect similarities to known structures. If the function of the new protein is unknown, a successful hit will often but not always permit a good guess, or at least suggest experiments.

Application of DALI to the entire PDB produces two classifications of structures:

1. FSSP (Fold classification based on Structure–Structure alignment of Proteins) presents the results of applying DALI to all *chains* from proteins in the PDB. For all protein chains longer than 30 residues in the known structures, structural redundancy is removed by selecting a single representative for clusters of structures which are more than 25% identical in sequence. The DALI program then detects the similarities among these representatives, inducing a classification presented in the FSSP data bank. The FSSP entry for each chain includes its alignment with proteins of similar structure, and reports the structurally-equivalent residues. From the web site the user can display multiple sequence alignments and superimposed structures.
2. The *DALI domain dictionary* is a corresponding classification of recurrent *domains* automatically extracted from known proteins.

Numerous other web sites offering classifications of protein structures are indexed at the following sites:
http://www.bioscience.org/urllists/protdb.htm and
http://msd.ebi.ac.uk/add/Links/fold.html

Do the different classification schemes agree? Recognize that to classify protein structures (or any other set of objects) you need to be able to measure the similarities among them. The measure of similarity induces a tree-like representation of the relationships. CATH, SCOP, DALI and the others, agree, for the most part, on what is similar, and the tree structures of their classifications are therefore also similar. However, what an objective measure of similarity does *not* specify is how to define the categories that you call the different levels of the hierarchy. These are interpretative decisions, and any apparent differences in the names and distinctions between the levels only disguise the underlying general agreement about what is similar and what is different.

To give some idea of the nature of the similarities expressed by the different levels of the hierarchies, Figure 4.3 shows protein domains that are classified, in SCOP, into the same superfamily, fold and class. Flavodoxin from *Clostridium beijerinckii* [5NLL] and NADPH-cytochrome P450 reductase [1AMO] are in the same superfamily, but different families. Flavodoxin and the signal transduction protein CHEY [1CHN] are in the same fold category, but different superfamilies. Flavodoxin and spinach ferredoxin reductase [1FNB] are in the same class (α / β) but have different folds.

The known structures

In succeeding sections we shall survey the known structures to illustrate the variety of their folding patterns. It will be difficult to escape the feeling that there is rather a bewildering variety of patterns, for classification itself does not provide any general principles that explain why certain folds are observed and others are not. What one would really like is a *periodic table* for folding patterns. When Mendeleev wrote down the periodic table, it did more than classify the known elements, it provided a framework for deciding what elements, not yet observed, *would* exist, and even predicted some of their properties.

Only for a few subsets of protein folding patterns can we achieve something similar. Consider the four α-helix bundles, examples of which are illustrated in Figure 4.2. Suppose we insist that each pair of successive helices be antiparallel. Then there are only a small number of ways to connect the helices, to produce bundles of different topology. We can enumerate all the possibilities, and look for them within the known structures. If one or more possibilities were not present in any known protein structure, we could predict that they would eventually be found, as Mendeleev did for three missing chemical elements.

For the totality of folds, it would be very difficult to circumscribe the possibilities. C. Chothia estimated that there are of the order of 1000 protein folds in nature. But why do *these*—however many there turn out to be—exist, but not others equally consistent with the laws of physics, chemistry and biology? Just lucky, I guess! Perhaps.

α-helical proteins

In the class of α-helical proteins, SCOP contains a total of 70 folds.

The first protein structures to be determined, myoglobin and haemoglobin, were rich in helices. In each of these molecules the monomer binds a

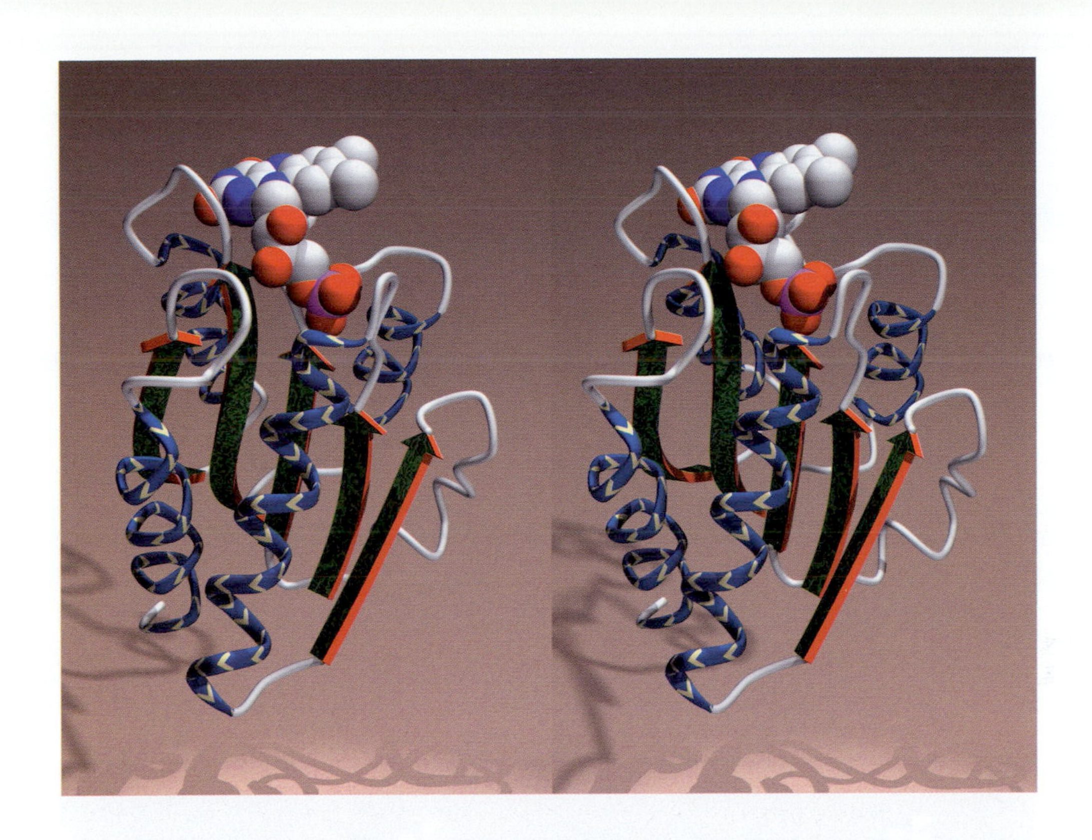

Fig. 4.3a

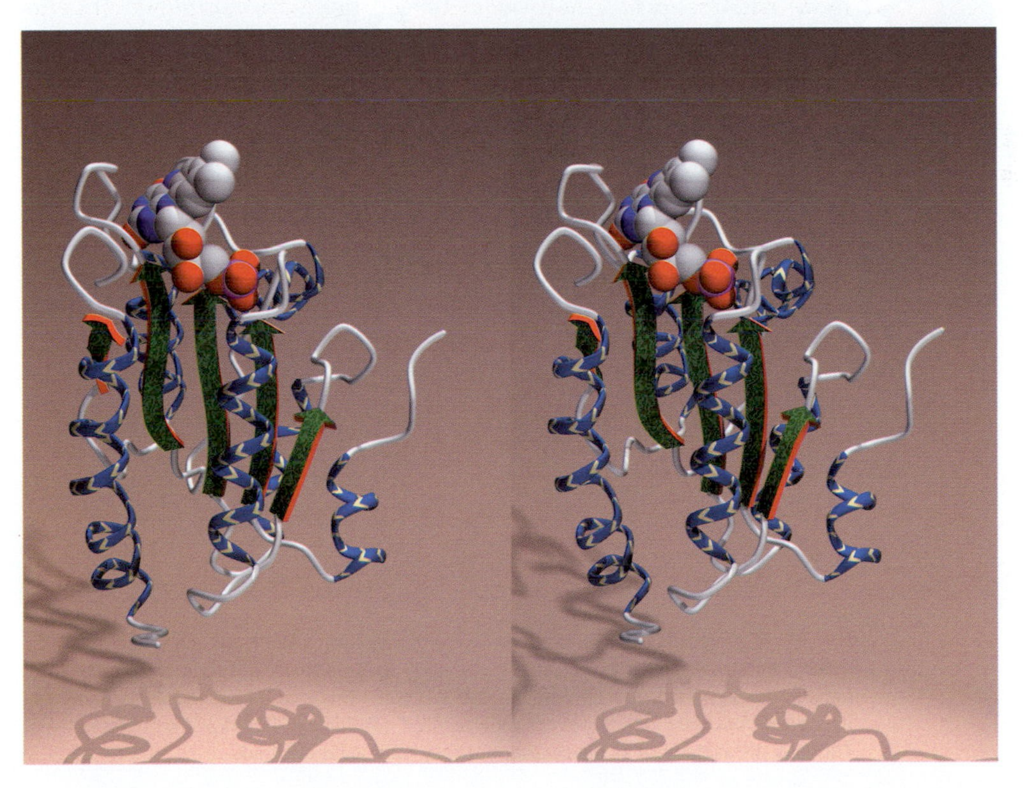

Fig. 4.3b

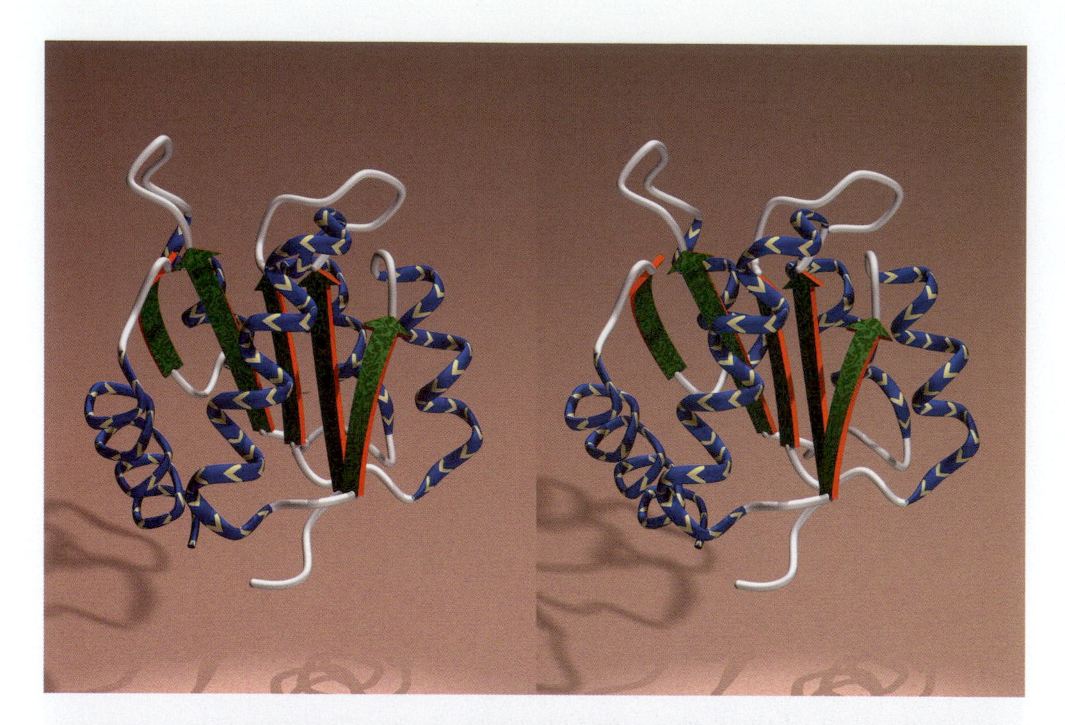

Fig. 4.3c

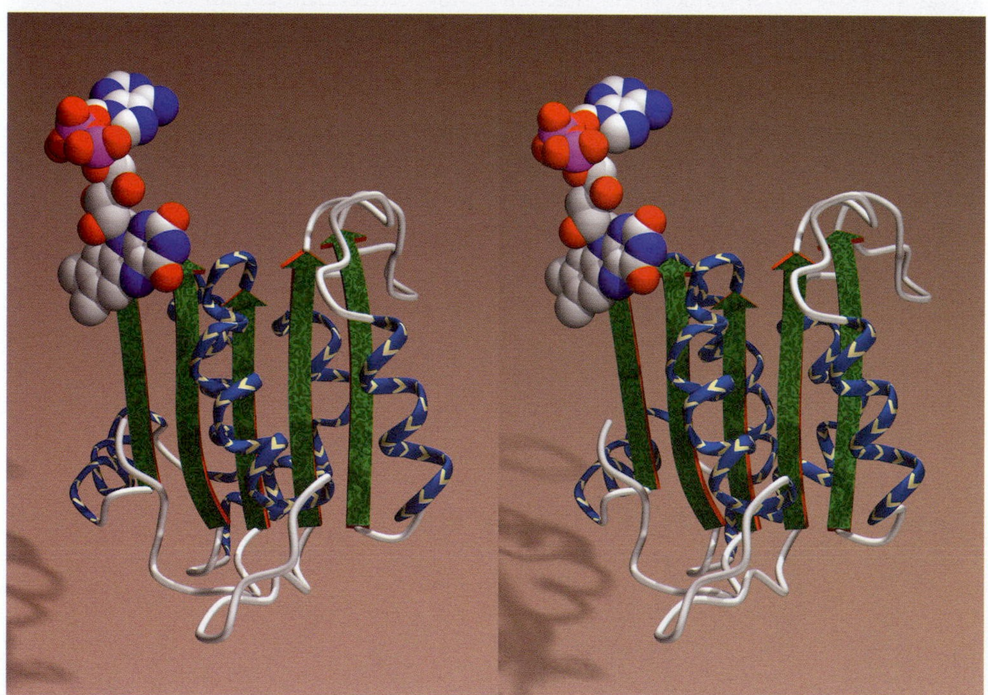

Fig. 4.3d

Fig. 4.3 (a) *Clostridium beijerinckii* flavodoxin [5NLL]. (b) NADPH-cytochrome P450 reductase [1AMO]. This is in the same superfamily as flavodoxin, but they are in different families. (c) Signal transduction protein CHEY [1CHN]. This is in the same fold category as flavodoxin, but they are in different superfamilies. (d) Spinach ferredoxin reductase [1FNB]. This is in the same class as flavodoxin, but they have different folds.

haem group flanked in most cases by two histidine residues. The globins include monomeric proteins in many species including animals, plants and bacteria, dimers in some invertebrates, tetramers such as mammalian haemoglobins (Figure 4.4) and higher aggregates. The quaternary structure of haemoglobin is necessary for its allosteric properties (see Chapter 8). Crystallographic studies of globins continues to be an active field; the PDB contains over 200 globin structures, from 30 species.

The cytochromes c are another family of haem proteins, the secondary structures of which are exclusively helical (Figure 4.6). Although both globins and cytochromes c are α-helical proteins that bind haem groups, they are very different in folding topology, and in the manner in which they bind the haem. It would be a mistake to think that they are related, based on the superficial similarities.

Even very large proteins can have a purely helical secondary structure, the ras-GTPase activating domain of P120gap being an impressive example (Figure 4.5). Citrate synthase (Chapter 8) is another.

Some globular proteins contain α-helices that form supercoils, like those in fibrous proteins. Figure 4.7 shows the 'leucine zipper,' GCN4, a transcriptional activator, binding a stretch of DNA.

Transmembrane segments of proteins tend to be helices, as we saw in the L and M segments of the *R. viridis* reaction centre. Bacteriorhodopsin contains seven transmembrane helices (Figure 4.8).

Principles of the architecture of α-helical proteins

The classifications contained in SCOP, CATH and other databases are invaluable but they are purely empirical. What governs, what limits the observed protein folding patterns? Here we consider three possible approaches, applying, respectively: enumeration, geometry, and structural chemistry.

Complete enumeration within restricted classes of folds

Myohaemerythrin contains four α-helices packed into a compact bundle (see Figure 4.2a). Successive helices have their axes antiparallel, which permits the lengths of the connections between them to be short. Within the class of protein topologies that, like myohaemerythrin, form four-α-helix bundles with each pair of successive helices antiparallel, how many topologically distinct structures could there be? How many are actually observed?

Each four-helix bundle has a cross-section that looks like this:

○ ○
○ ○

We want to avoid counting the same structure in different orientations more than once, so let us rotate the molecule so that helix 1—the first helical region in the sequence—is at the upper right corner, with its chain direction out of the page. Let ⊙ symbolize a helix for which the overall N to C direction of the chain is out of the page, and ⊗ symbolize a helix for which the overall N to C direction of the chain is into the page. Then myohaemerythrin has the topology shown at the top of page 140.

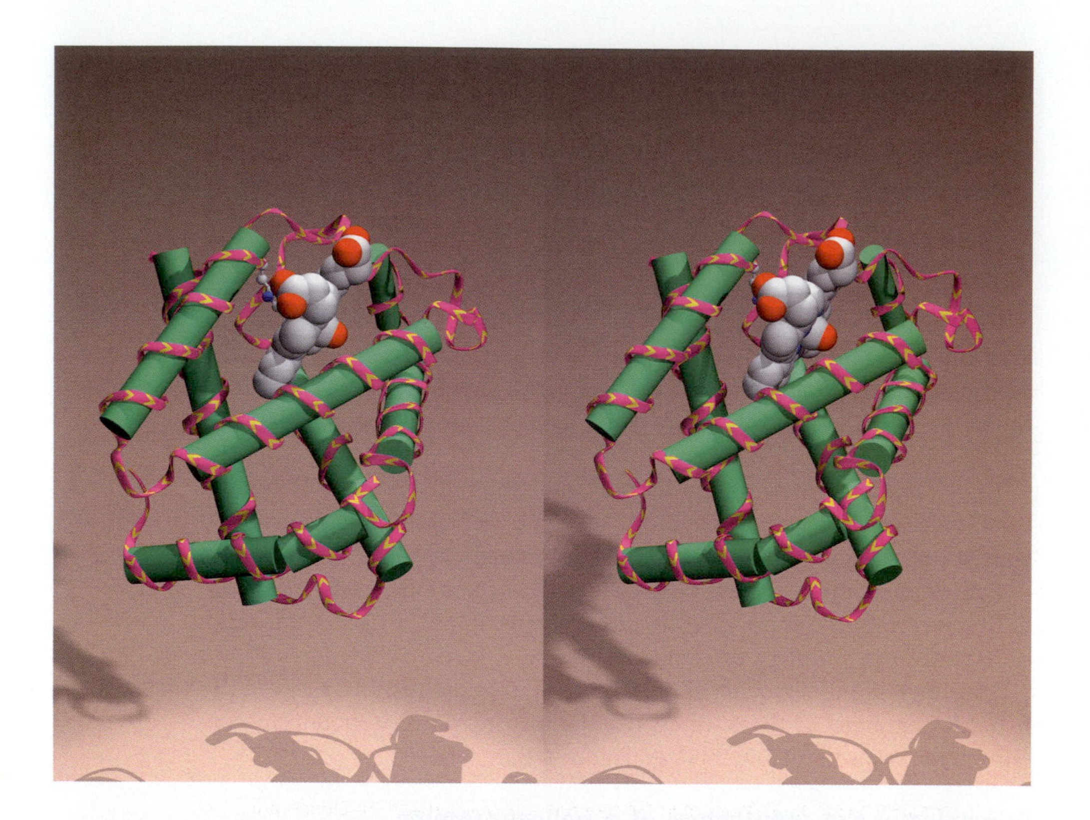

Fig. 4.4a

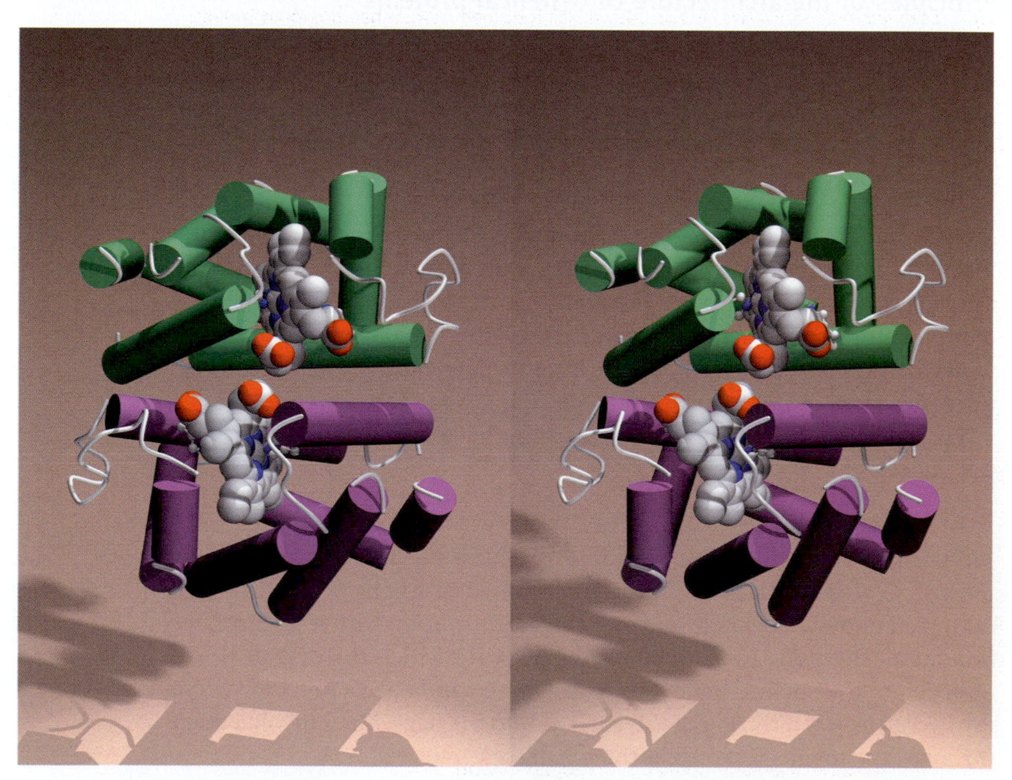

Fig. 4.4b

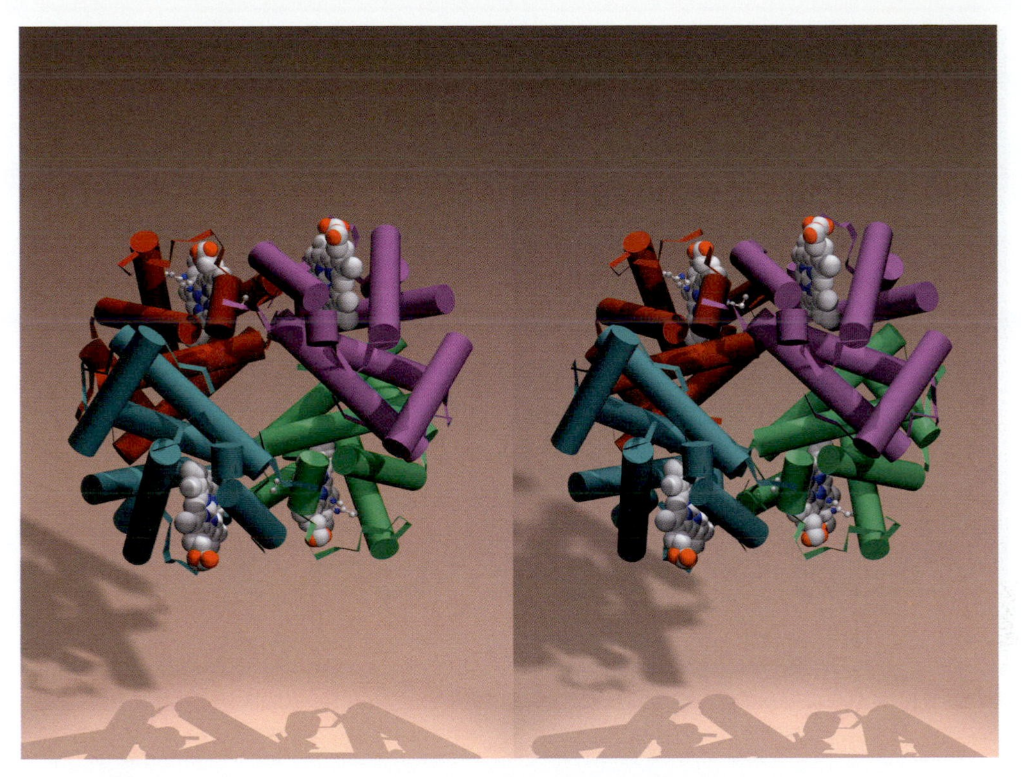

Fig. 4.4c

Fig. 4.4 Globin structures. (a) Monomer: Glycera haemoglobin [1HBG]. (b) Dimer: ark clam (*Scapharca inaequivalvis*) globin [4SDH]. (c) Tetramer: human haemoglobin [4HHB].

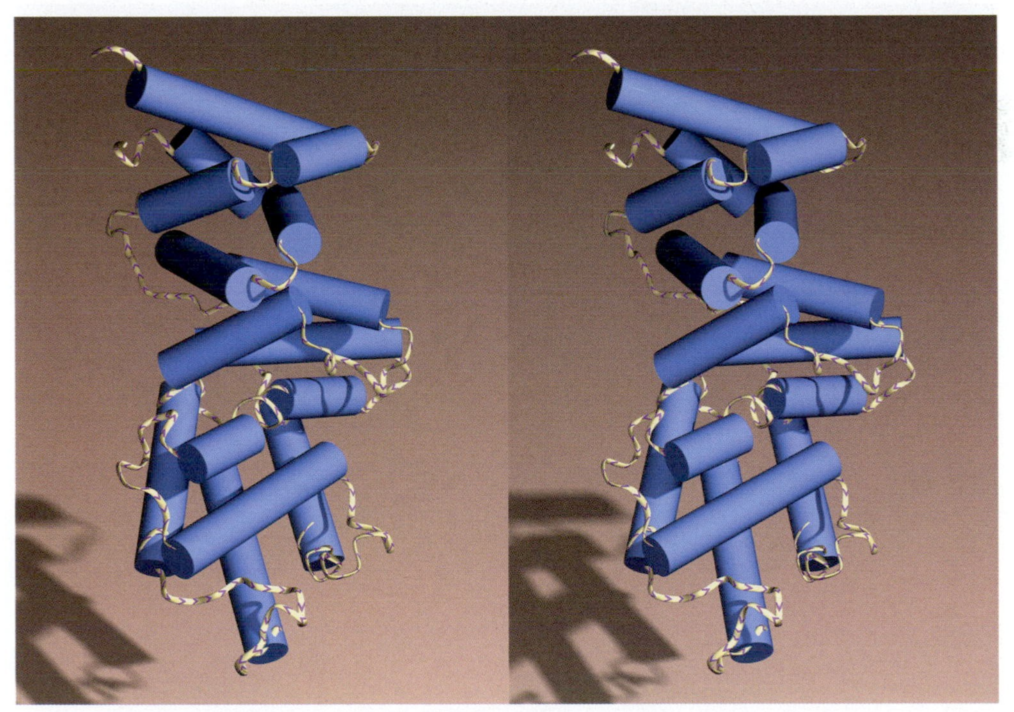

Fig. 4.5 Ras-GTPase activating domain of P120gap [1WER].

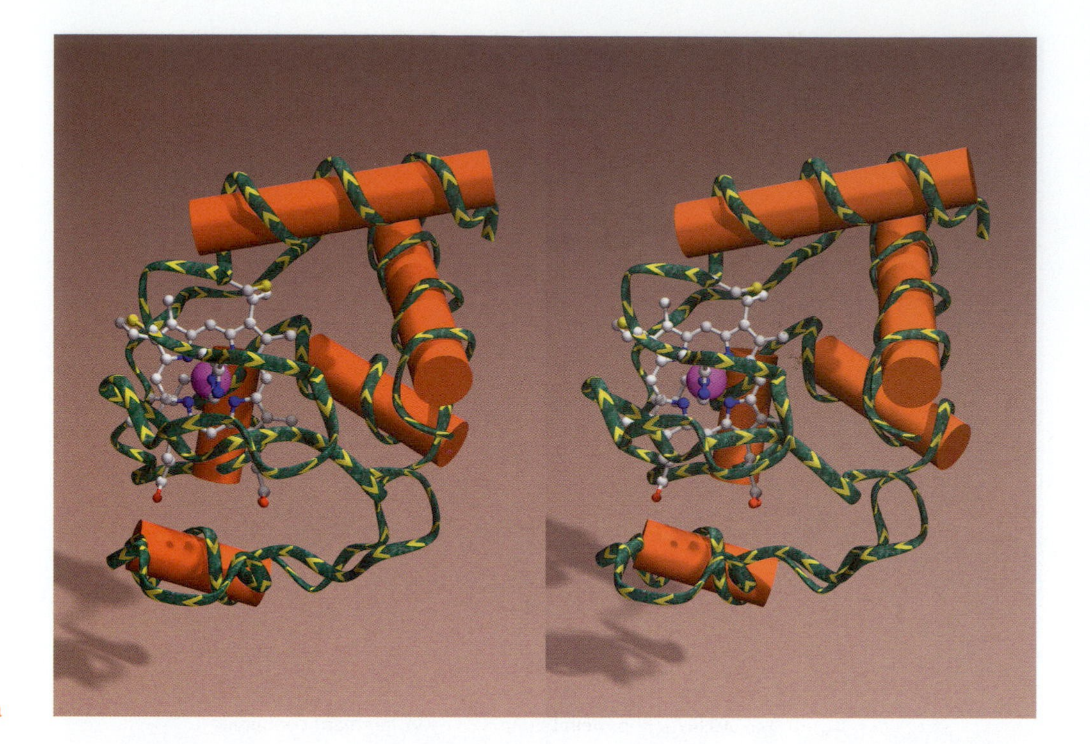

Fig. 4.6a

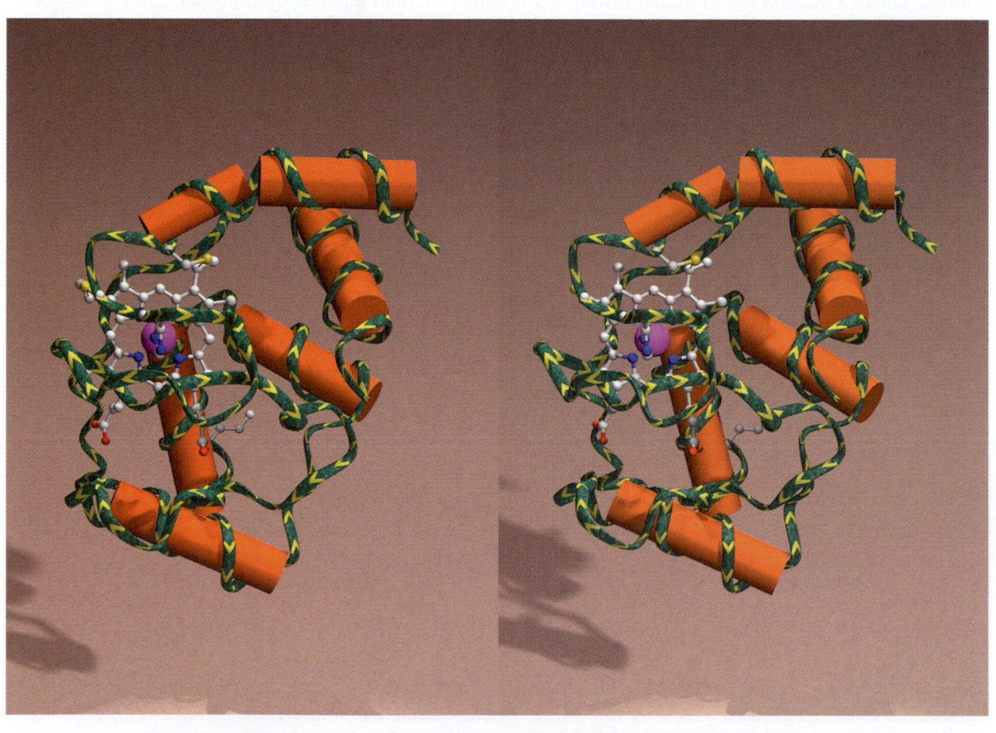

Fig. 4.6b

Fig. 4.6 Cytochromes c, from: (a) tuna [5CYT]; (b) rice [1CCR].

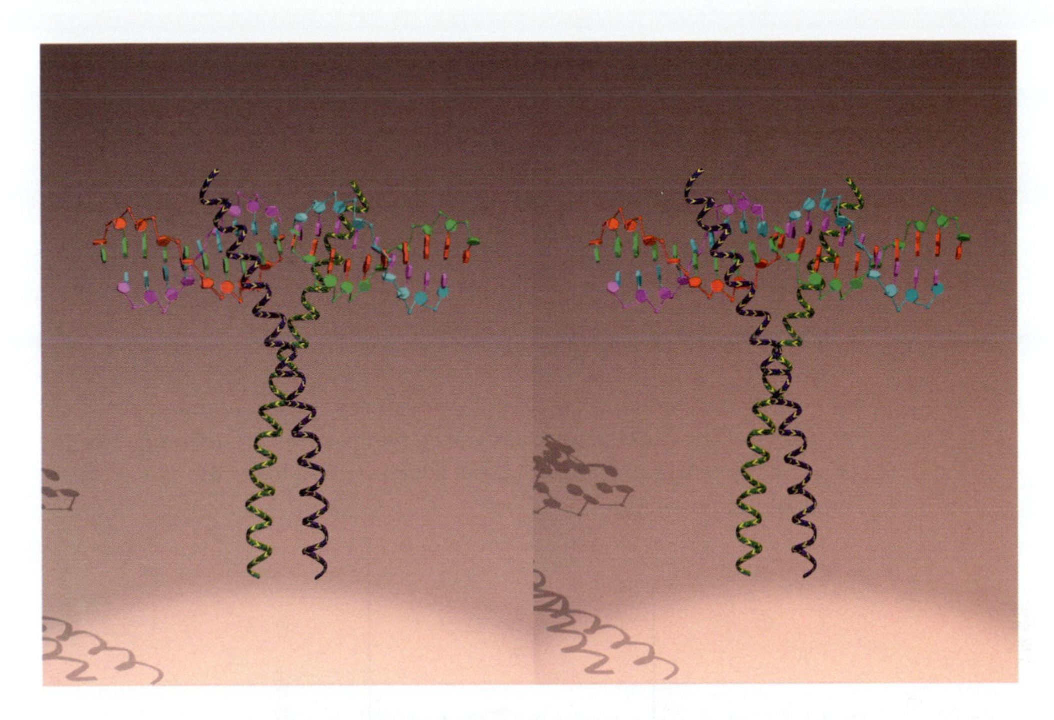

Fig. 4.7 A 'leucine zipper', the GCN4-bZIP protein bound to a DNA fragment containing the ATF/CREB recognition sequence [2DGC].

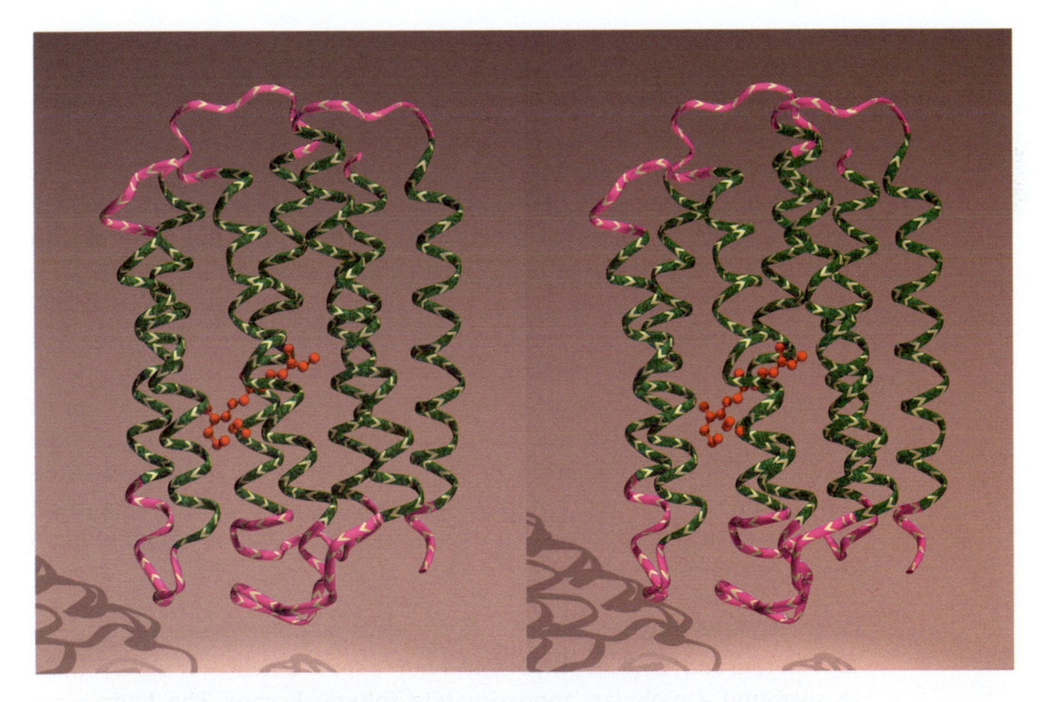

Fig. 4.8 Bacteriorhodopsin from *Halobacterium halobium* [1BRD]. As in the L and M subunits of the reaction centre (see Figure 1.4), a set of parallel helices spans a membrane. The chromophore is retinal.

⊗ 2 ⊙ 1

⊙ 3 ⊗ 4

We can indicate the connections explicitly:

⊗ 2 ⊙ 1

⊙ 3 ⊗ 4

However, this is unnecessary; the order of the helices and the directions of their axes specify the connections unambiguously.

Are there other four-helix bundle topologies of this type? Yes. For, keeping helix 1 fixed, helix 2 could occupy any of the three remaining positions, and then helix 3 could occupy any of the two positions remaining after the second helix has been assigned, after which helix 4 takes the last position. There are six possibilities:

⊗ 2	⊙ 1	⊗ 2	⊙ 1	⊙ 3	⊙ 1
⊙ 3	⊗ 4	⊗ 4	⊙ 3	⊗ 2	⊗ 4
A		B		C	
⊗ 4	⊙ 1	⊙ 3	⊙ 1	⊗ 4	⊙ 1
⊙ 3	⊗ 2	⊗ 4	⊗ 2	⊗ 2	⊙ 3
D		E		F	

We have enumerated *all* possible structures of this kind, without any consideration of chemistry or the details of protein conformation. The price we paid was to remain within a *very* restricted set of folds.

Most of the observed four-helix bundles of this type have topologies A or D. Why are these popular? Two systematic differences suggest themselves: (1) Helix–helix contacts in bundles are mainly around the periphery. In topologies A and D, all the helix–helix contacts around the periphery are antiparallel. In the others there are two parallel and two antiparallel contacts. (2) All topologies other than A and D have at least one connection that goes diagonally across the end of the barrel.

Although it is not possible to explain the observed distribution of topologies with confidence, it is noteworthy that (1) there are now enough structures known that the distribution of topologies is likely to be significant, and (2) the enumeration has led us to frame structural hypotheses. If it has not led us to the answers, it has at least led us to some questions.

Geometry: the polyhedral model of α-helical globular proteins

A.G. Murzin and A.V. Finkelstein asked how a set of α-helices could assemble to surround a globular, approximately spherical, core. The helices are assumed to present a hydrophilic face to the solvent and the opposite, hydrophobic face to the interior. They proposed a model in which the axes of α-helices lie along the edges of a convex polyhedron with triangular faces, with all edges of equal length. Up to half the edges are 'occupied' by helices, as two helices cannot intersect at any vertex. For each polyhedron, different assignments of edges to helix axes create different assemblies of helices. Varying the directions of the helices and adding different combinations of loops—that is, different patterns of connectivity—generate a set of different folding topologies. Murzin and Finkelstein found that four polygons describe the packings of three, four, five and six helices (see following table).

Polygon	Number of helices	Number of models	
Octahedron	3	2	
Dodecahedron	4	10	(Includes four-helix bundles)
Hexadecahedron	5	10	
Icosahedron	6	8	

For more than six helices, one helix will be entirely buried by other helices packing around it, and the assembly will not be described by a convex polyhedron. Murzin and Finkelstein found that their model fitted most of helical proteins to which it was applicable.

Structural chemistry: tertiary-structural interactions

The relative geometry of α-helices depends on the shapes of the helix surfaces, through the requirement—imposed by the thermodynamics of protein folding—that the interfaces between them bury hydrophobic residues in a well-packed interior. C. Chothia and coworkers studied the geometric restrictions imposed by packing requirements on interactions of elements of secondary structure. Their description of helix–helix packings helps to rationalize the folding patterns of α-helical proteins.

The structure of helix–helix packings

In many globular proteins, pairs of α-helices are in contact, burying sidechains in the interface between them. Provided that the sides rather than the ends of the helices are in contact, the interface between the helices will be formed from a patch on the surface of each helix.

The relative geometry of the helices can be described by the distance of closest approach between their axes, and the interaxial angle. Typically, interaxial distances are 6–10 Å, with approximately 1 Å interpenetration of the sidechains. In order to achieve good packing densities, the two interfaces have complementary surfaces, like the occluding surfaces of teeth.

Interaxial angles of helices in contact show preferred values. A closer examination of the nature of the packing at helix-helix interfaces reveals why this is so.

Given the basic geometry of the α-helix (3.6 residues per turn) residues separated by four in the sequence are close together on the helix surface. Sidechains at these positions are poised naturally to create ridges, which will of course have a defined angle relative to the helix axis (Figure 4.9). The ridges created by the sidechains of residue i and residues $i+4$ and $i-4$ are called $i\pm4$ ridges. Thus, in an ideal α-helix, the line joining the Cβ of residue i to the Cβ of residues $i+4$ and $i-4$ will make an angle of 26° with the helix axis. (The actual direction of the ridge may deviate moderately from this value because of the shapes and conformations of particular sidechains.) It follows that forming an interface between these ridges on the two helices will fix the interaxial angle at a value around −52°. (The negative sign follows the convention that a clockwise rotation is negative.) This is indeed near the average for a commonly-observed class of interaxial angles.

Although the formation of ridges by residues separated by four in the sequence is the most common, ridges can also be formed by residues separated by three in the sequence or by residues separated by one in the sequence. In the last case the ridge runs nearly around the helix (circumferentially) rather than nearly along it (longitudinally). These are called $i\pm3$ ridges and $i\pm1$ ridges, respectively.

Three preferred classes of helix–helix packings correspond to the interaction of different ridges and grooves:

Interacting ridges	$i\pm4/i\pm4$	$i\pm3/i\pm4$	$i\pm1/i\pm4$
Interaxial angle	−52°	+23°	−105°

Interaxial angles of pairs of packed helices show a distribution peaked around these three values, with a good correlation between interaxial angle and the ridge–groove structure of the interface.

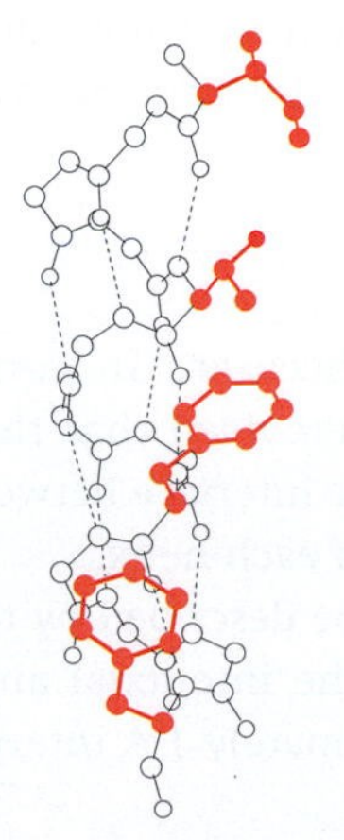
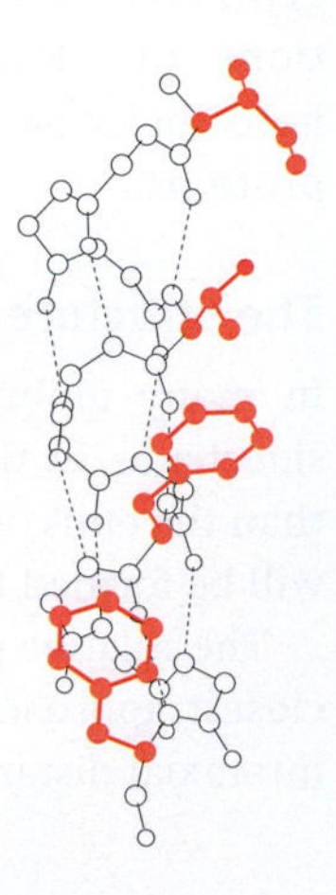

Fig. 4.9 A helix from carboxypeptidase showing a prominent ridge on its surface created by sidechains [5CPA].

Figure 4.10 shows the interface between the B and G helices of sperm whale myoglobin. The residues from the G helix form a ridge (created by the sidechains of residues 106, 110, 114 and 118) that nestles in a groove between two ridges on the surface of the B helix (created, respectively, by residues 24, 28 and 31, 35). In each case the residues are separated by four in the sequence.

Although most helix interfaces fit the 'ridges-into-grooves' model, there are exceptions. For example, in the B–E helix packing in globins, ridges from either helix cross each other, at a notch formed at a pair of glycine residues (Figure 4.11).

Native protein structures as solved jigsaw puzzles

The demonstration that tertiary structural interactions tend to be subject to geometric constraints imposes form on the original 'oil drop' model of protein interiors, that was based exclusively on the idea of burial of hydrophobic surface. A native protein structure depends on a mutually-consistent set of good packings between the elements of secondary structure that interact.

The model that emerges for a native protein is that of a three-dimensional jigsaw puzzle. However, there is this proviso: the pieces of the puzzle have no rigidity except by virtue of their interactions. The events that take place during the folding process remain obscure. As in the crystallization of a short peptide, the elements must somehow explore different individual shapes, and different relative positions and orientations, to find the proper shapes and the proper fit.

Thus, the native state of a globular protein is similar to a finished jigsaw puzzle. But in the unfolded state the pieces are not only taken apart, but their shapes are changed and disguised.

β-sheet proteins

It is rare for a protein structure to contain a β-sheet with both faces exposed. Two kinds of interactions are common: two β-sheets packed against each other, discussed here; and a β-sheet with α-helices packed against it, discussed in the section on α/β proteins.

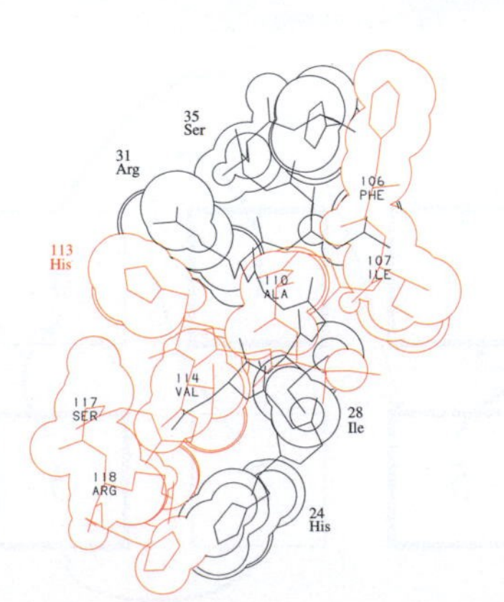

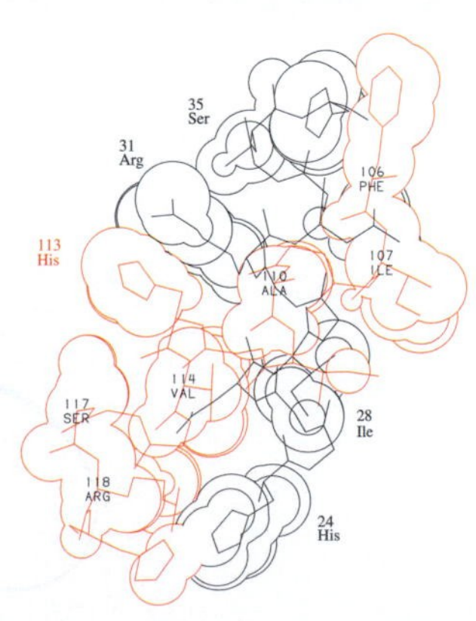

Fig. 4.10 Interface between B (in black) and G (in red) helices of sperm whale myoglobin. This interface, formed by $i \pm 4$ ridges from both helices, is the most common interface structure.

Fig. 4.11 Interface between B (in black) and E (in red) helices of sperm whale myoglobin. This interface contains an unusual 'crossed-ridge' structure, created because the relative sizes of the residues create notches in the ridges.

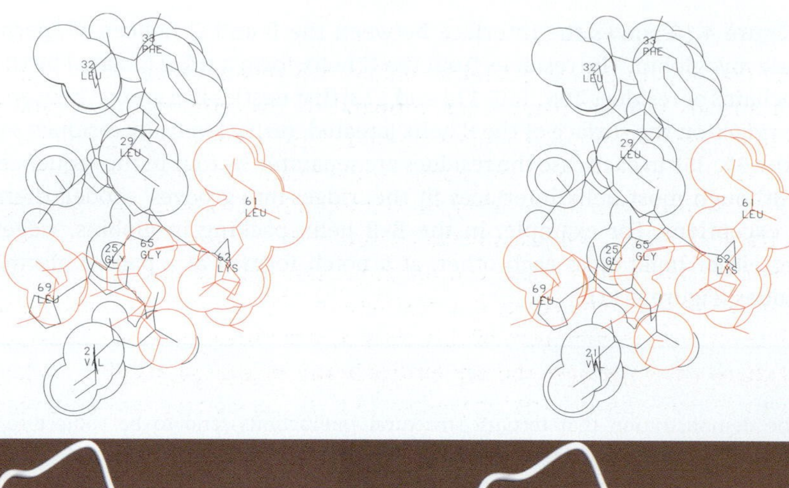

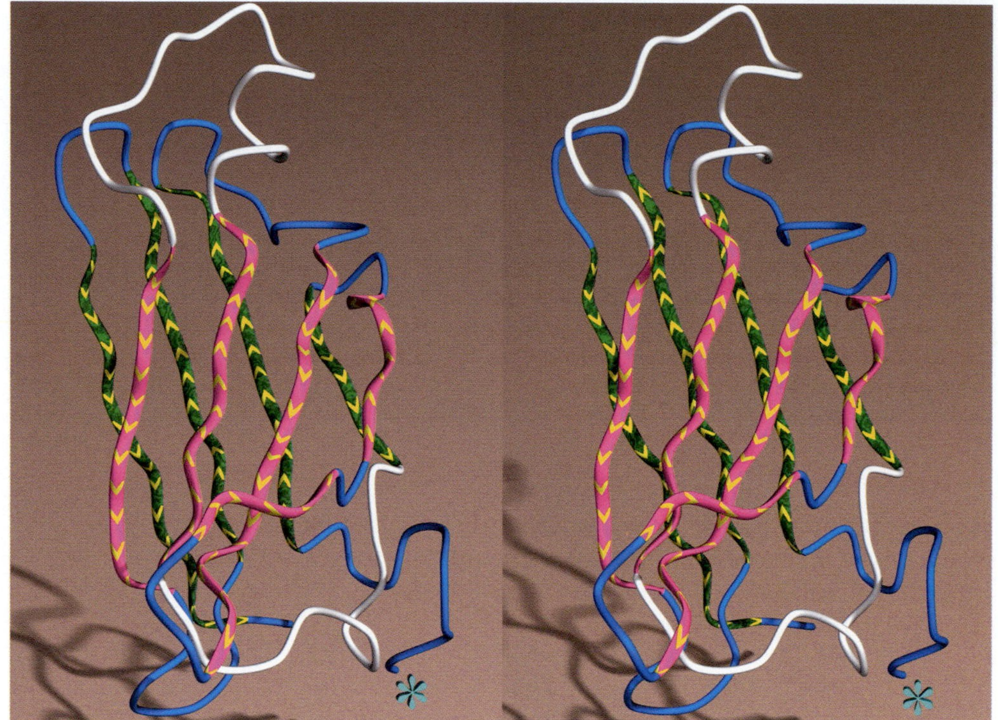

Fig. 4.12a

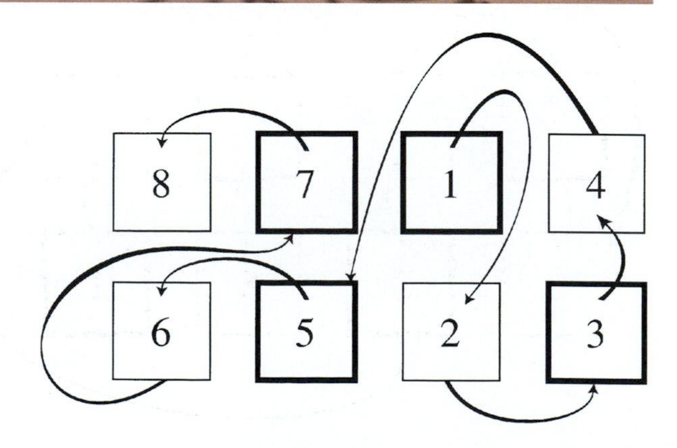

Fig. 4.12b

β-sheet 'sandwiches'

Many domains in β-sheet proteins contain two sheets packed face-to-face, with strands roughly parallel (or antiparallel). Transthyretin is a typical example (Figure 4.12a). Each sheet contains four strands. Why is this? It is likely that a larger number of strands would make it difficult to accommodate the natural twist of the sheet. In concanavalin A, for example, the large six-stranded β-sheet is unusually flat (Figure 4.13).

Figure 4.12b shows a reduced diagram showing the directions of the strands and their connectivity in transthyretin. If we compare transthyretin with several other members of this class of structures, such as glycosylasparaginase or γ-crystallin, the similarities in the arrangements of secondary structural elements are striking (Figure 4.12c–f). How do we know whether

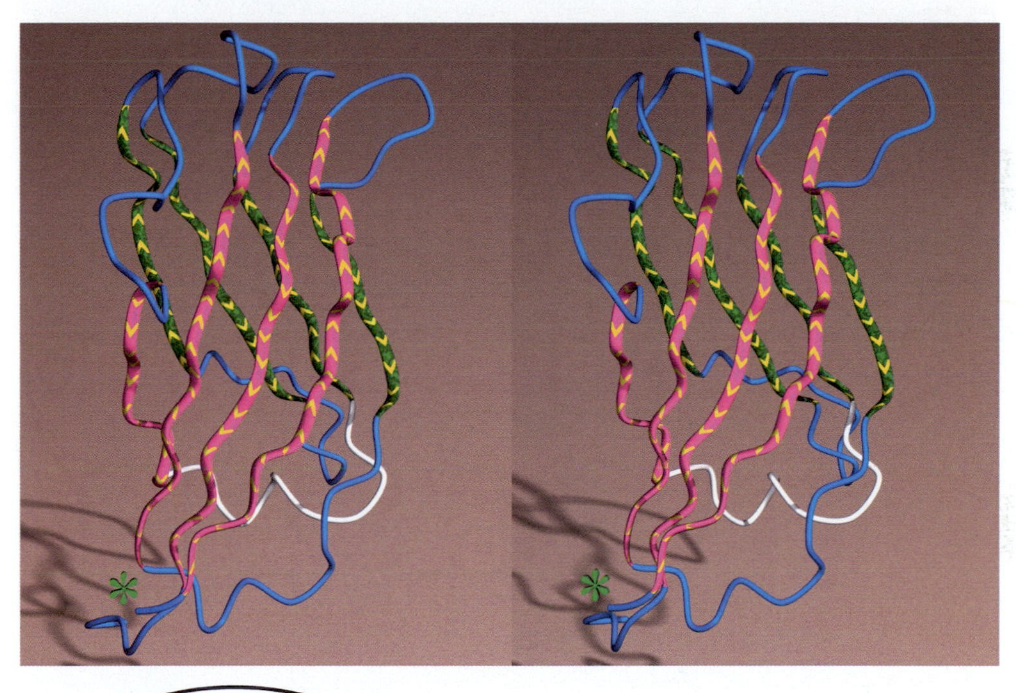

Fig. 4.12c

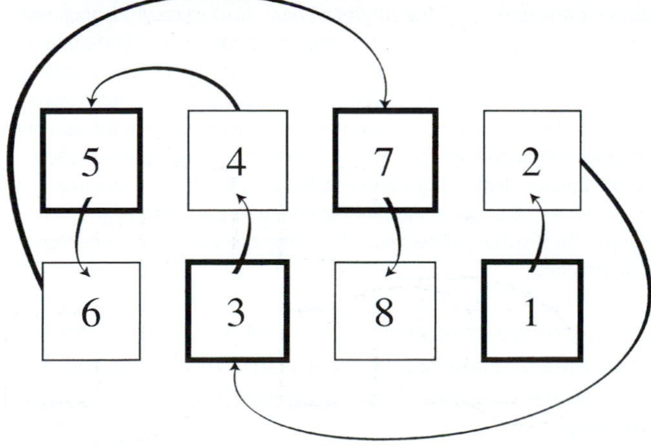

Fig. 4.12d

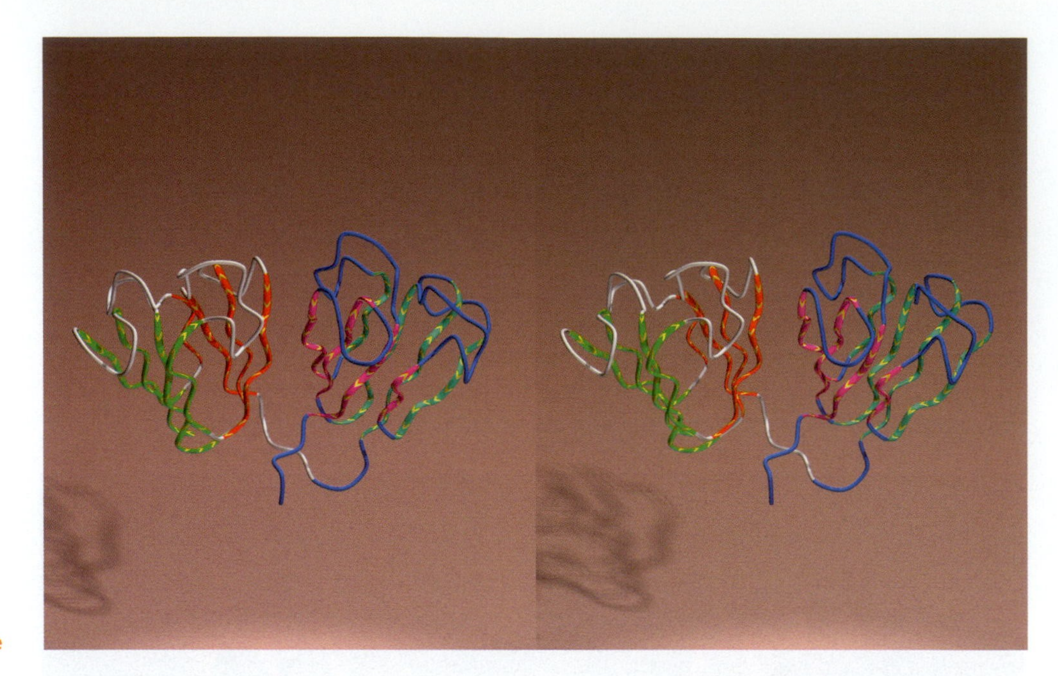

Fig. 4.12e

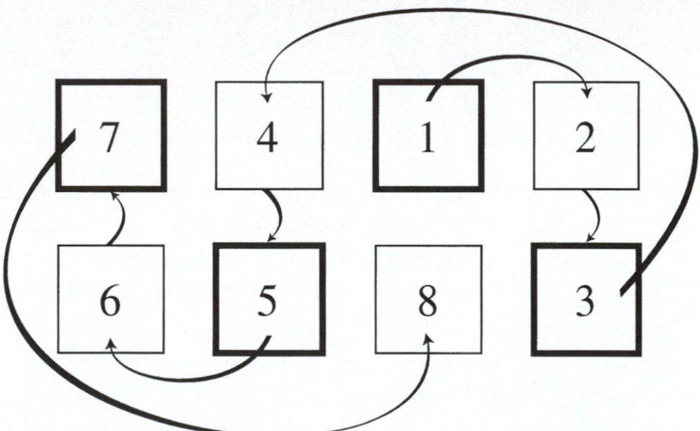

Fig. 4.12f

Fig. 4.12 β-sheet sandwich structures: (a,b) transthyretin [1ETA]. (c,d) glycosyl-asparaginase domain [1PGS]; (e,f) γ-crystallin domain, containing two equivalent sandwiches [1AMM]. In (a) and (c) the * indicates the N-terminus. In the schematic diagrams (b, d and f), the idea is that one is looking down on the sheets in a direction roughly along the strands (different viewpoint from a, c, e). Each square signifies a strand, numbered in order of appearance in the sequence. Bold squares identify strands orientated with their N to C direction pointing approximately towards the viewer; lighter squares identify strands orientated with their N to C direction pointing approximately away from the viewer. Arrows that penetrate the squares signify connections across the top of the structure; arrows cut off by the squares signify connections across the bottom of the structure.

these proteins are showing a common solution of a structural problem rather than an evolutionary relationship? Notice first of all that the directions of the strands are not all equivalent. Moreover, the connectivities of the strands are different.

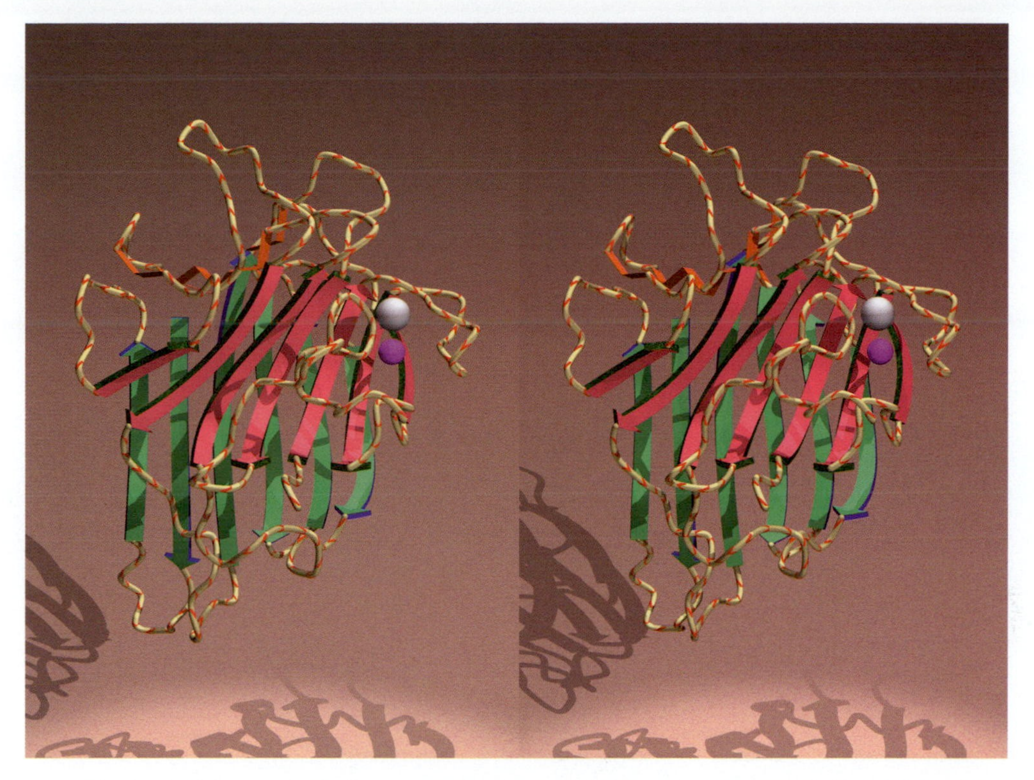

Fig. 4.13 Jackbean Concanavalin A [3CNA].

Let us examine this in more detail. If we trace the chain along the sequence, we can ask for the topological relationship between successive strands. For example, if we come up along one strand we might then find a 'hairpin' loop between two hydrogen-bonded strands on the same sheet. Alternatively, we might find that the next strand encountered, as we proceed along the sequence, is on the opposite sheet. *It is because it is difficult to envisage a continuous pathway along which evolution could proceed to interconvert the topology of these connections, that it appears that these molecules are not related by evolution.* They are merely showing a common solution to a general structural problem.

It is possible to enumerate the possible topologies of four stranded double-β-sheet sandwiches (this has been done by A.V. Finkelstein), but there are quite a large number!

Other β-sheet proteins

Ascorbate oxidase is a large β-sheet protein that contains parallel β-sheet domains (Figure 4.14). Influenza haemagglutinin contains an unusual β-sheet 'propellor' (Figure 4.15).

α+β proteins

Many proteins are known which contain both α-helices and β-sheets, but which do not have the special structures created by alternating β–α–β patterns. These are considered in the next section.

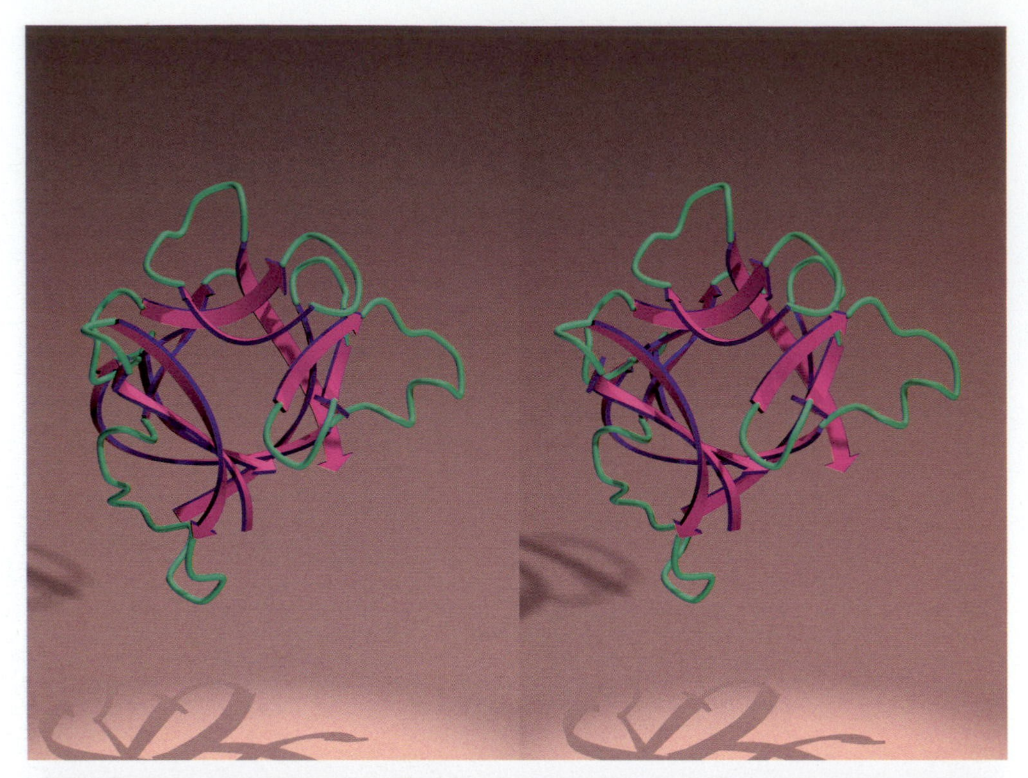

Fig. 4.14a

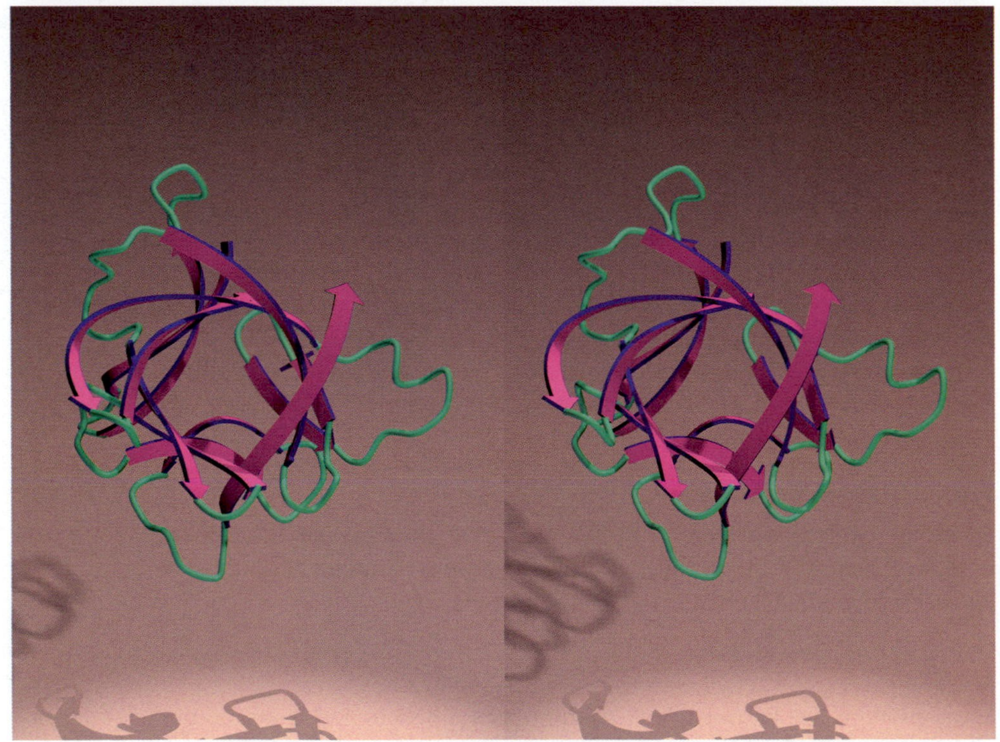

Fig. 4.14b

Fig. 4.14 Ascorbate oxidase [1ASA]. (a) Looking at the domain with the β-trefoil fold. (b) Looking at the β-barrel domain. These viewpoints are 180° apart.

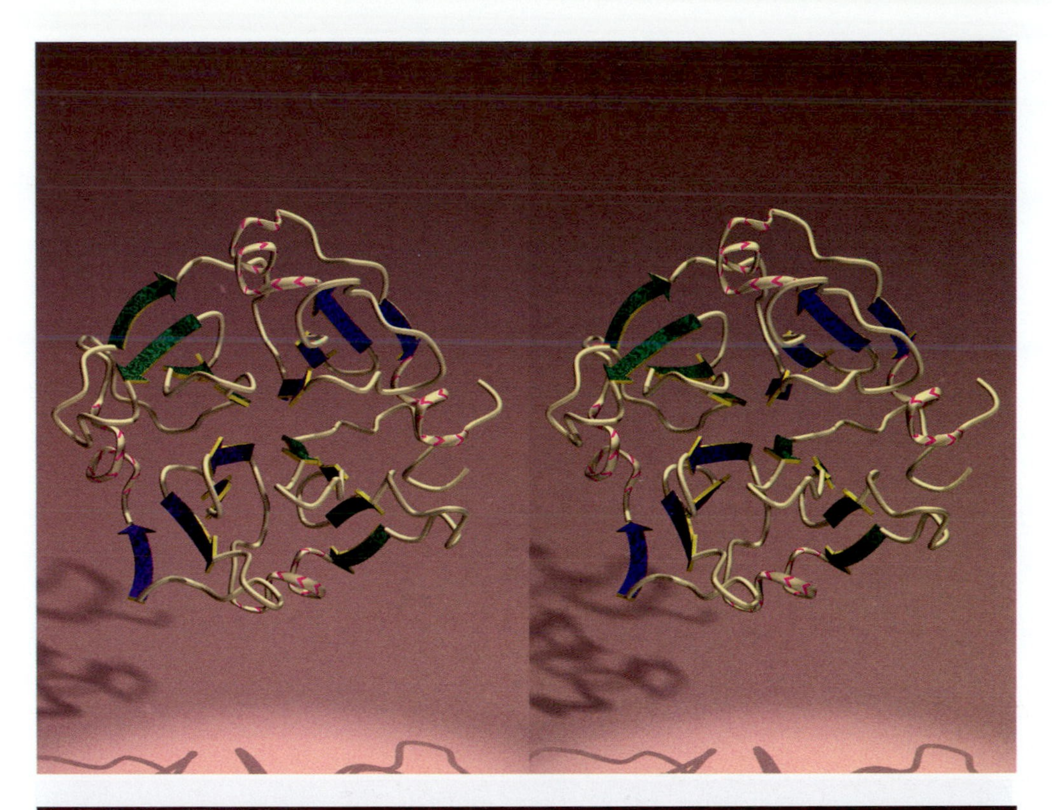

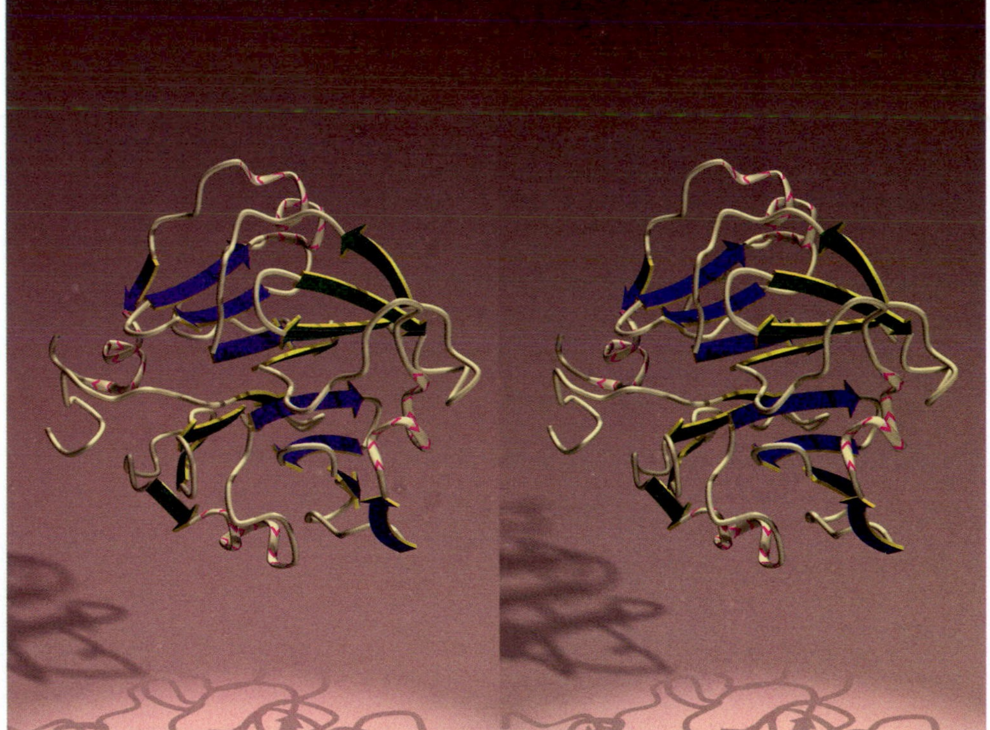

Fig. 4.15 Influenza haemagglutinin [3HMG]. Two viewpoints.

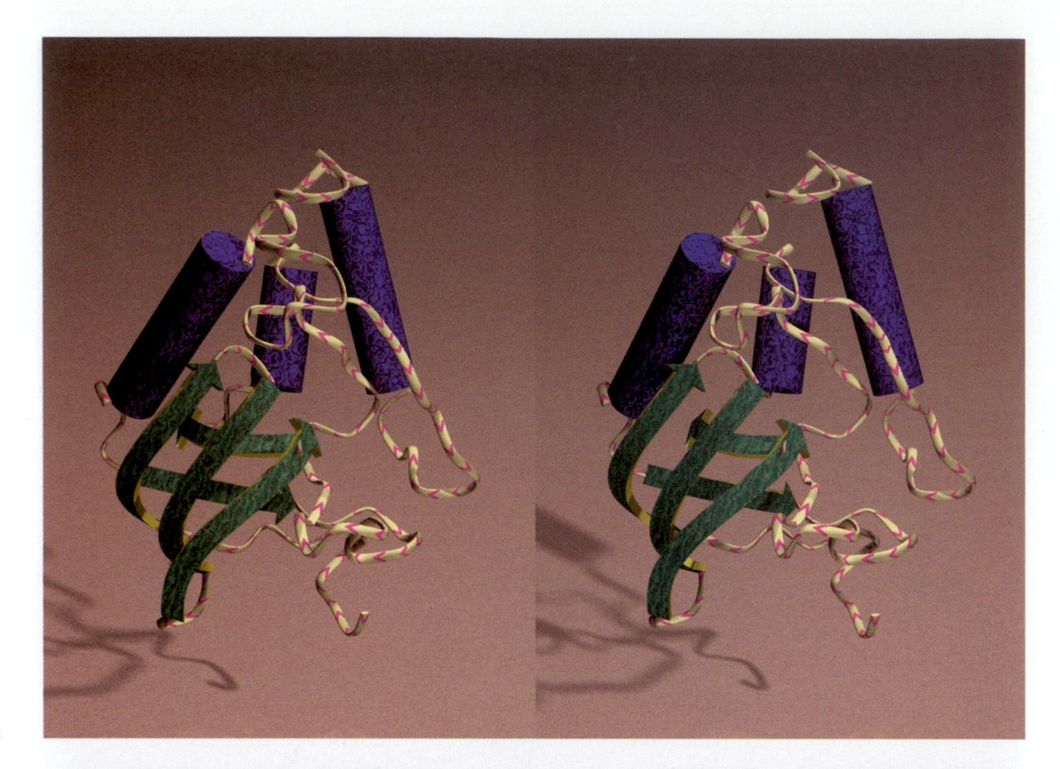

Fig. 4.16a

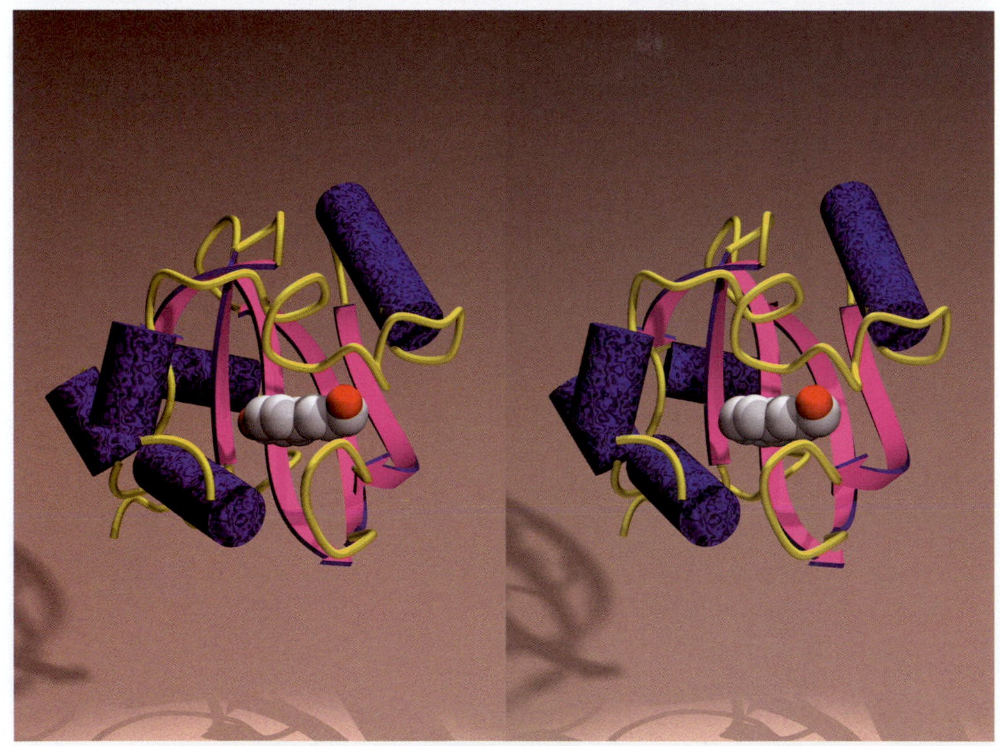

Fig. 4.16b

Fig. 4.16 α/β proteins: (a) staphylococcal nuclease [2SNS]; (b) murine/human ubiquitin-conjugating enzyme ubc9 [1U9B].

In staphylococcal nuclease, and human ubiquitin-conjugating enzyme ubc9, the β-sheets and α-helices tend to be segregated in different regions of space (Figure 4.16).

α/β proteins

The β–α–β unit

The very important supersecondary structure consisting of an β-α-β unit, with the β-strands parallel and hydrogen bonded, forms the basis for many enzymes, especially those that bind nucleotides or related molecules (Figure 4.17). Often, several units combine to form a parallel β-sheet. In some cases, there is a linear β-α-β-α-β. . . arrangement, but in other cases the β-sheet closes on itself. In triose phosphate isomerase, the eighth strand is hydrogen bonded to the first, to form the very common 'TIM-barrel' structure.

Linear or open β–α–β proteins

Many proteins that bind nucleotides contain a domain made up of six β-α units, with a special topology (see Figure 4.18). C.-I. Brändén showed that this folding pattern naturally creates a cavity adjacent to a 'crossover'—a link joining non-adjacent strands of the sheet. Here the long loop between

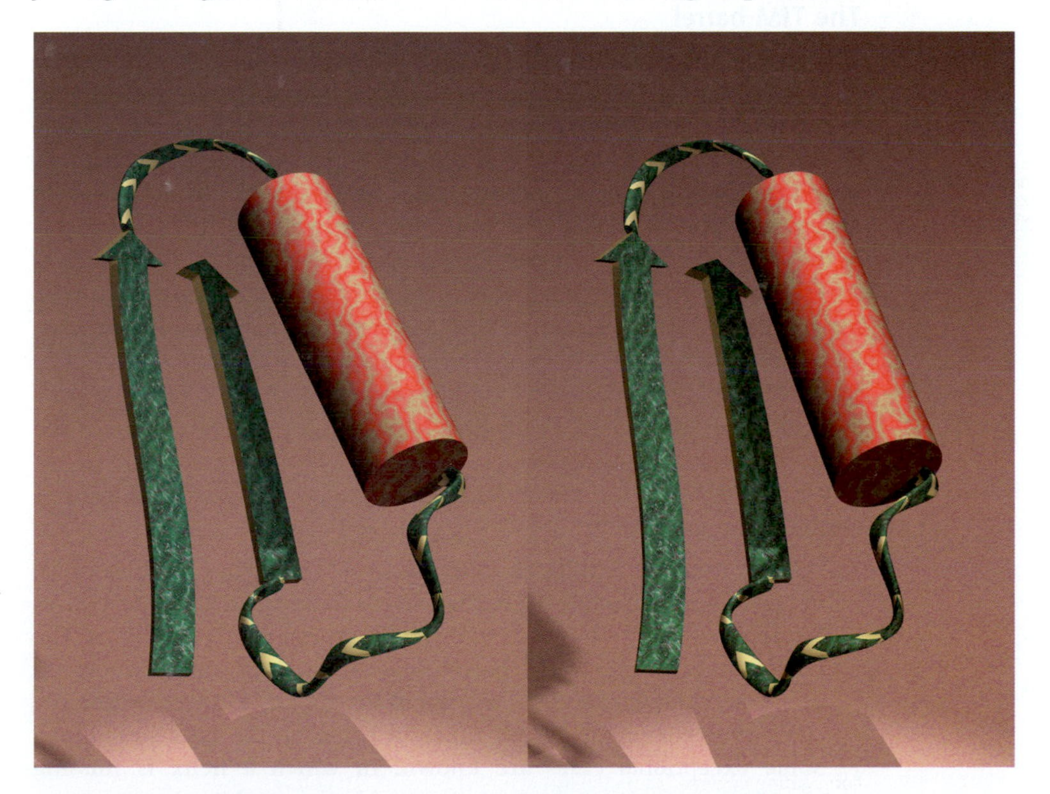

Fig. 4.17 β–α–β unit, a supersecondary structure common to many enzymes.

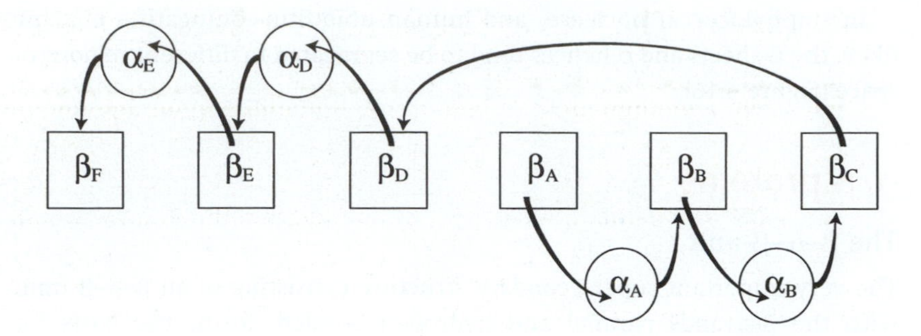

Fig. 4.18 Schematic diagram of the topology of the NAD-binding domain of horse liver ADH and many other enzymes.

strands βC and βD tends to create a natural pocket for a NAD or NADP ligand. The N–H groups in the first turn of helix A are well-positioned to form hydrogen bonds to phosphate oxygens (see Figure 2.7).

The NAD-binding domain of horse liver alcohol dehydrogenase is typical; other dehydrogenases have very similar domains (see Chapter 6). Flavodoxin and adenylate kinase contain a variation on the theme, with five strands instead of six. Dihydrofolate reductase has eight strands.

Closed β–α–β barrel structures

The TIM barrel

The enzyme glycolate oxidase is typical of a large number of structures that contain eight β–α units in which the strands form a sheet wrapped around into a closed structure, cylindrical in topology. The helices are on the outside of the sheet (Figure 4.19). Chicken triose phosphate isomerase (TIM), first solved in 1975, was for a long time the only example, but now over 40 enzymes containing TIM-like barrels are known. In all cases the active site is at the end of the barrel that corresponds to the C-termini of the strands of sheet, just as in the open β–α–β structures. Although they show similar folding patterns, TIM-barrel enzymes catalyse a variety of different reactions, and the amino acid sequences of enzymes that share this fold but differ in function are very dissimilar: we cannot prove that all are diverged from a common ancestor; indeed, many people believe that at least some of the structures arose by convergence.

With many structures available, it is possible to adduce general features of this folding pattern:

1. Looking from the outside, the sheet is formed by eight parallel strands, tipped by approximately 36° to the barrel axis (Figure 4.19b). The helices are approximately parallel to the strands, typical of α/β structures. The chain proceeds around the barrel in a consistently counterclockwise direction, viewed from the C-termini of the strands of sheet: locally the chain proceeds 'up' a strand, 'down' a helix, up the next strand, etc. Some exceptional cases are known, in which a helix is missing (muconate lactonizing enzyme) or a strand is inverted (enolase).

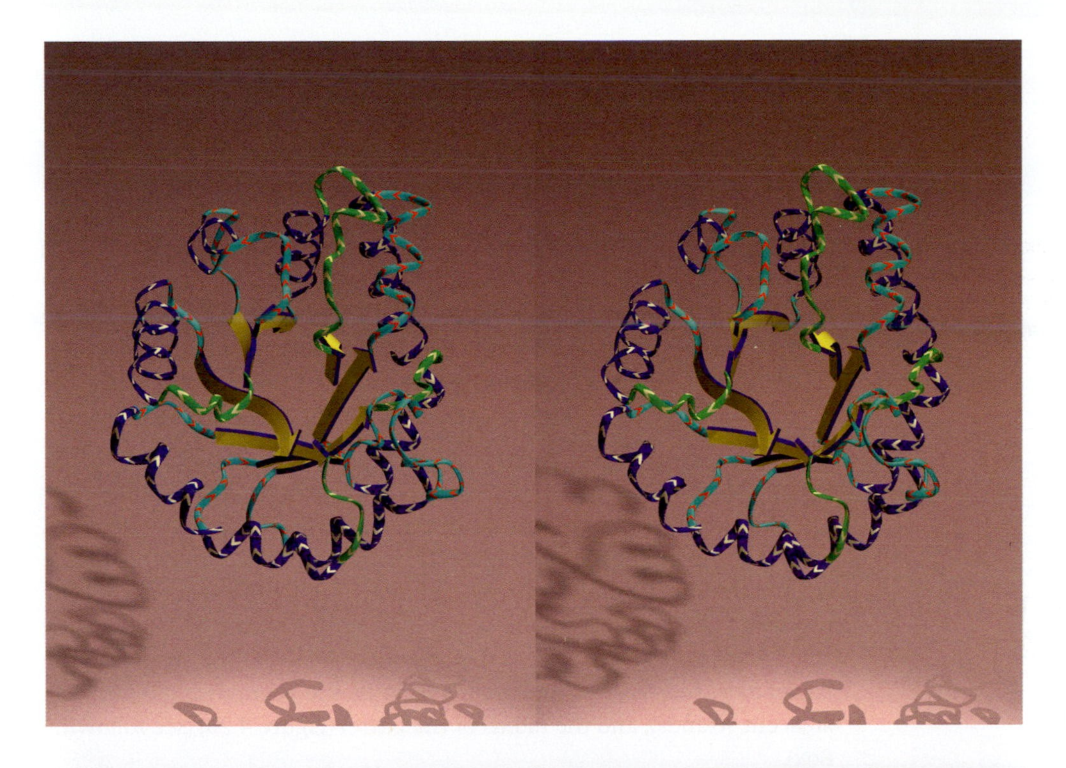

Fig. 4.19a

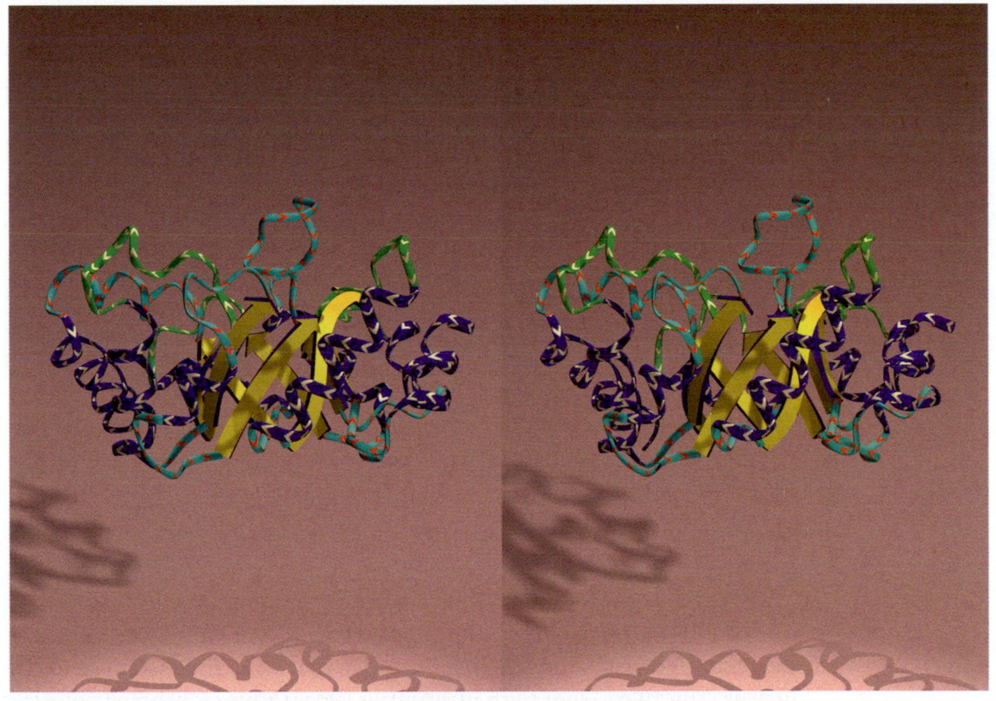

Fig. 4.19b

Fig. 4.19 The β-barrel from spinach glycolate oxidase [1GOX]. First discovered in Triosephosphate IsoMerase, β-barrels with this topology (8 strands, shear number 6) are known as 'TIM' barrels. (a) View along the barrel axis. (b) View perpendicular to the barrel axis.

2. A perspicuous view of the inside of the structure of the sheet is afforded by 'rolling out the barrel' (Figure 4.20). The leftmost (the N-terminal) strand is repeated at the right. To recover the three-dimensional eight-stranded barrel, this diagram must be folded over, and the two images of first strand glued over each other: superposing A onto A′ and B onto B′ in Figure 4.20.

   The tipping of the strands to the barrel axis (vertical in Figure 4.20) produces a layered structure. Note that the sidechains of each strand point alternately into and out of the barrel; this is an important feature of the packing inside the barrel. The strands vary in length; however, three residues from each strand appear at the same height, along the barrel axis, forming a continuous hydrogen-bonded net girdling the barrel. Sidechains from these levels pack in three layers in the interior of the barrel.

3. The different barrels are similar in topology. A.D. McLachlan classified ideal β-barrel topologies. Two integral quantities: the number of strands (n), and the shear number (S = the difference, along the sequence, of the residues forced to correspond when closing the barrel by superposing the two copies of the first strand) determine the tilt of the strands to the barrel axis, the twist of the strands (that is, the average angle between adjacent strands), and the radius of the barrel (Figure 4.20) (see following Box).

Geometric parameters of β-barrels

a = Cα–Cα distance along strand = 3.3 Å
b = perpendicular distance between strands = 4.4 Å

n = number of strands
S = shear number

R = radius of barrel = $[\,(Sa)^2 + (nb)^2\,]^{1/2} \,/\, [2n \sin(\pi/n)\,]$
α = angle between strands and barrel axis; $\tan \alpha = Sa \,/\, (nb)$

All TIM barrels have eight strands, $S = 8$, a tilt of the strands to the barrel axis of approximately 36°, and radii of 6.5–7.5 Å, depending on the eccentricity of the cross-section.

4. The packing of residues inside the sheet shows a common structural pattern.

   Figure 4.21a shows the hydrogen-bonding net obtained by 'rolling out the barrel' of glycolate oxidase. On each strand of sheet, alternate sidechains point towards the region inside the sheet and out towards the helices. The 12 residues with identifying letters have sidechains pointing inwards. The packing inside the barrel is formed by the interactions of these 12 residues. The first, third, fifth and seventh strands each contribute two sidechains, and the second, fourth, sixth and eighth strands

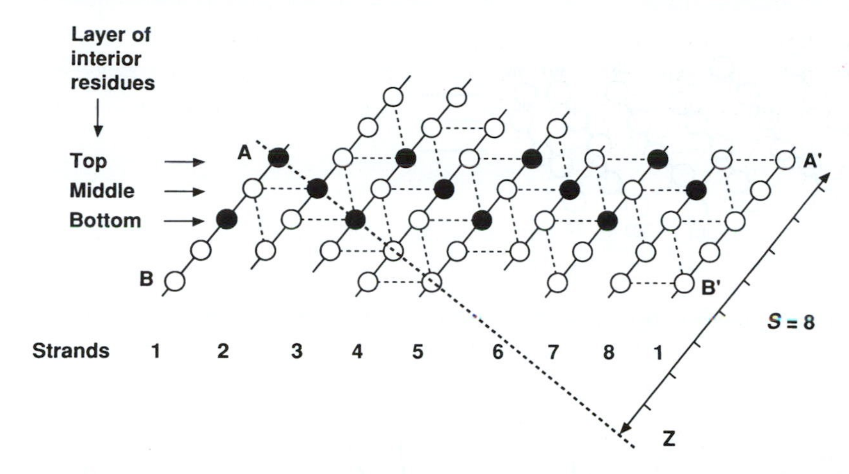

Fig. 4.20 Schematic diagram of the hydrogen-bonding nets obtained by 'rolling out the TIM barrel' — a closed β-sheet with eight strands and shear number 6, which appears in triose phosphate isomerase, glycolate oxidase and many other proteins. Each circle represents a residue; broken lines represent hydrogen bonds. Filled circles represent residues in three layers that point into the barrel interior; note their pattern of alternation.

This diagram contains nine strands, because the first strand is duplicated and appears at left and right edges. To recover the three-dimensional eight-stranded barrel, the leftmost strand must be superposed on the rightmost by folding the paper into a cylinder and glueing residue A onto residue A′ and residue B onto residue B′. This produces the barrel as it occurs in the proteins, with the strands tipped by 36° to the barrel axis.

To form an eight-stranded barrel with strands *parallel* to the axis, residue A would have to be glued onto point Z. The shear, *S*, was defined by McLachlan as a measure of the stagger of the strands. It is the number of residues by which A′ (the residue on which A is actually superposed) is displaced from Z (the residue on which A would be superposed in a barrel with strands parallel to the axis).

At the centre of the barrel, 12 inward-pointing sidechains pack together. These correspond to the filled circles. Note their arrangement in three parallel layers, in planes lying at the same height relative to the barrel axis. The symmetry of this pattern could accommodate the change of a strand from parallel to antiparallel, as observed in enolase.

each contribute one sidechain. Note that residues at the same height along the axis of the sheet (vertical in Figures 4.20 and 4.21a) are not nearest neighbours on adjacent strands, because of the tilt of the strands with respect to the barrel axis.

The packing of these sidechains in the barrel interior is shown in Figures 4.21b to 4.21e. Part (b) is a side view of the sheet of GAO, pruned to three residues per strand. The sidechains occupy three tiers or layers with almost perfect segregation.

The packing of these residues is seen in Figure 4.21c to 4.21e, which show serial sections through the three layers of GAO. Atoms from odd-numbered strands are outlined in red, and atoms from even-numbered strands in blue. In the first and third layers, the four packed sidechains are shown in black. These belong to odd-numbered strands. In the central layer, the four packed sidechains are shown in red. These belong to even-numbered strands.

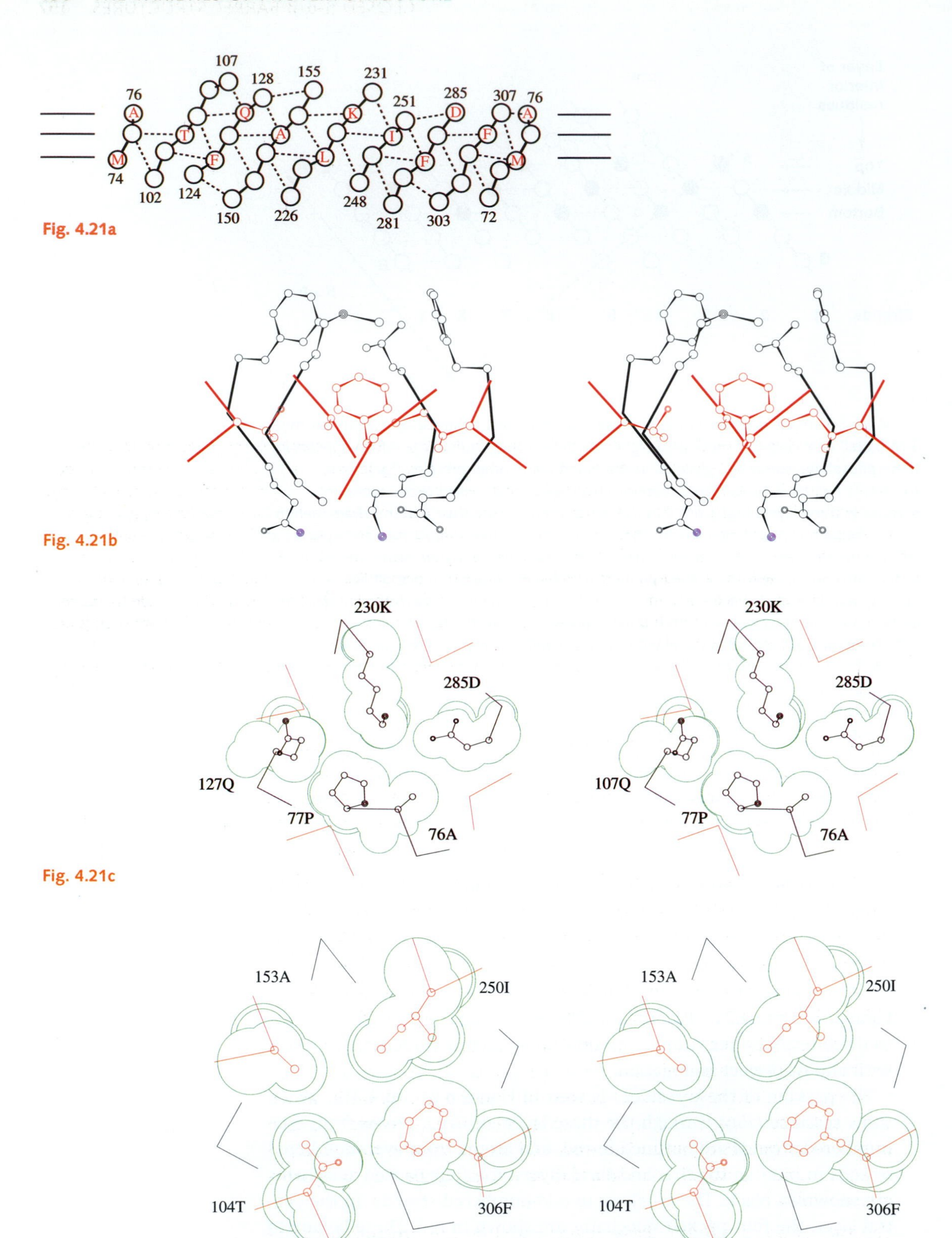

Fig. 4.21a

Fig. 4.21b

Fig. 4.21c

Fig. 4.21d

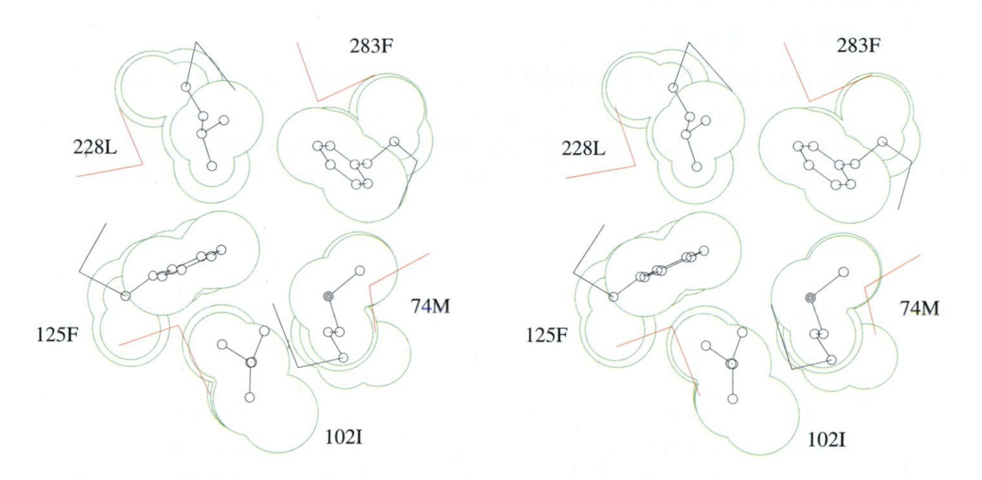

Fig. 4.21e

Fig. 4.21 The packing of residues inside the barrel of spinach glycolate oxidase [1GOX].) (a) The β-sheet of glycolate oxidase, unrolled. Each circle represents a residue; one-letter codes identify residues the sidechains of which pack inside the barrel. Broken lines represent hydrogen bonds. Numerals indicate residue numbers. Nine strands are shown: the edge strand (residues 72–76) is duplicated. This picture is a cylindrical projection drawn from atomic coordinates, and it gives an accurate picture of the positions of the residues. In contrast, Figure 4.20 is idealized. (b) The inward-pointing sidechains form three layers inside the barrel. This drawing shows the eight strands of the β-barrel, pruned to the three residues per strand, and the inward-pointing sidechains. The view is perpendicular to the barrel axis. (c, d, e) Serial sections cut through a space-filling model (van der Waals slices) of the three layers of residues packing inside the barrel of spinach glycolate oxidase. In each drawing three slices separated by 1 Å are shown. Atoms from alternate strands are shown in black and red. In (c, d, e) the view is parallel to the barrel axis.

The reader is urged to follow each of the twelve residues from part (a) to part (b) to part (c), (d) or (e) of this figure.

These results suggest a simple description of the packing of residues inside the sheet of glycolate oxidase and other TIM barrels. There is a three-tiered arrangement, involving a double alternation—a three-dimensional chessboard pattern, in which alternate strands contribute sidechains to alternate layers:

1. The tilt of the strands relative to the axis of the sheet, and the twist of the sheet, place the inward-pointing sidechains in layers. Each layer contains four sidechains from alternate strands. The sidechains that point 'in' are on odd-numbered levels on odd-numbered strands, and on even-numbered levels on even-numbered strands (or vice versa). The central region of the barrel is filled by 12 sidechains from three layers.
2. Qualitatively, the packing is a layered A–B–A type structure. (That is, A indicates the layout of units on the first layer. On the second level, the units have a different layout, B. But the units on the third layer lie above those of the first layer, with layout A again. Such structures are common in simple inorganic crystals.) Successive layers are related by a rotation by 45° around the barrel axis and a translation along the axis by approximately 3 Å.
3. The formation of a fourth layer is prevented by the protrusion of the sidechains from the top and bottom layers (Figure 4.21b).

Other β-barrel structures

The TIM fold is by far the most common β-barrel structure, but many different β-barrel folds are known. The serine proteinase domain will be discussed in Chapter 6; others appear in problems at the end of this chapter.

McLachlan's classification of β-barrel topologies by *discrete* indices—the number of strands and the shear number—permits writing down a complete 'periodic table' of possible β-barrel folds. Quantitative predictions of β-barrel geometries agree very well with experiment.

Irregular structures

A classification of proteins based on secondary structure must eventually face the structures that contain very few of their residues in helices and sheets (Figure 4.22). These tend to be stabilized by additional primary chemical bonds: in the case of wheat germ agglutinin, there are numerous disulphide bridges. In the case of ferredoxin there are iron–sulphur clusters. The 'kringle' structure, known first from a domain of prothrombin but occuring in many other proteins, contains disulphide bridges as well as several short stretches of two-stranded β-sheet.

Fig. 4.22a

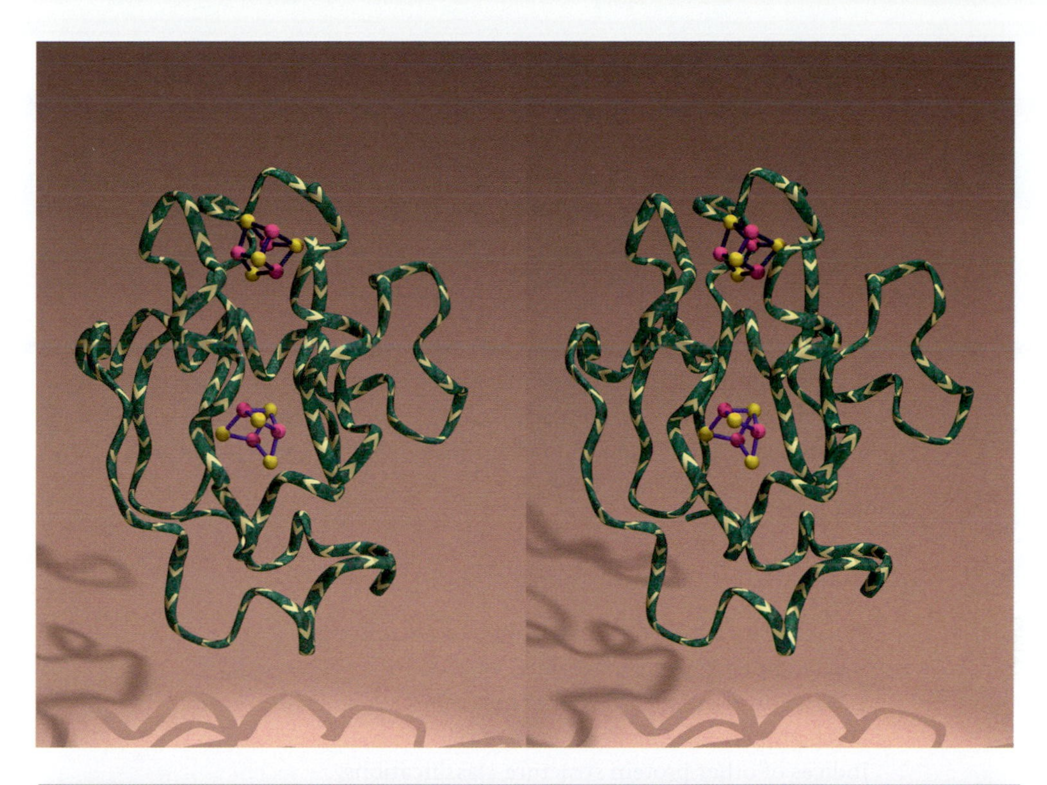

Fig. 4.22b

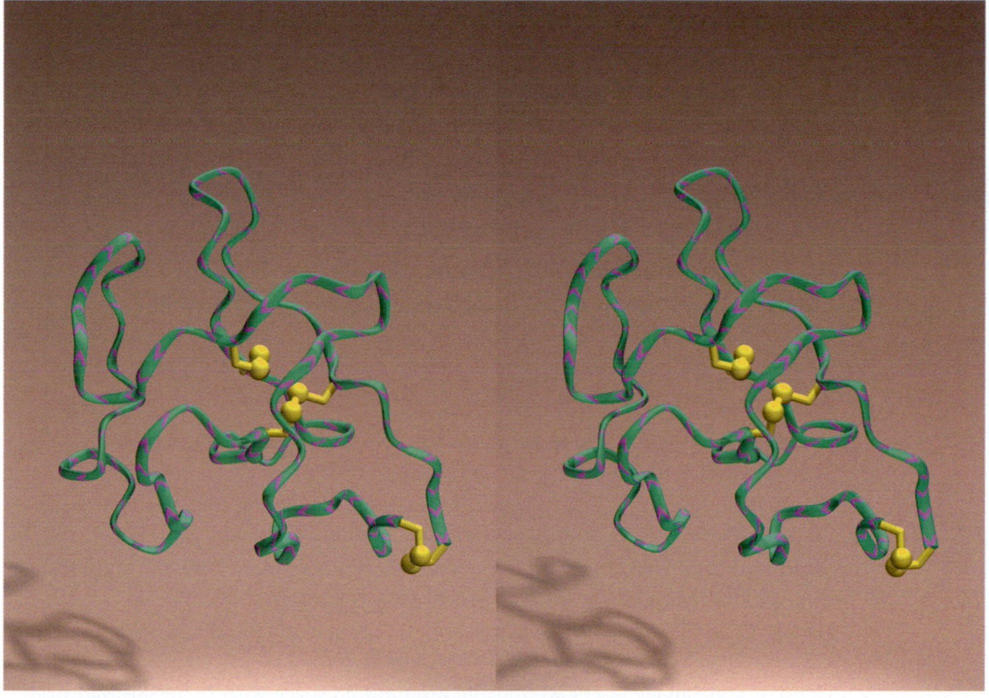

Fig. 4.22c

Fig. 4.22 Proteins with little or no secondary structure, which usually contain extra primary chemical bonds. (a) Wheat germ agglutinin [9WGA]. (b) Ferredoxin [6FD1]. (c) The 'kringle' domain of prothrombin [2PF1].

Conclusions

In this chapter we have looked at some of the variety in the spatial arrangement of secondary structural elements in globular proteins, and some of the principles governing their assembly. If one follows closely the new structures reported each year, many—not surprisingly—are similar to known structures. However, others show truly novel features. (Aficionados distinguish between, 'Oh well . . .' structures and 'Oh, wow!' structures.) In writing these pages, I wonder what it would be like to write such an essay in—what? 10 years?—when we really do know the structure of every protein of yeast. Presumably it will then be possible to give a fairly complete account of what types of proteins exist. Maybe this will be compensation for the loss of the excitement we now have, in not knowing what unexpected new structures the next few weeks will bring.

Useful web sites

Protein Data Bank home page: http://www.rcsb.org
SCOP: http://scop.mrc-lmb.cam.ac.uk/scop/
CATH: http://www.biochem.ucl.ac.uk/bsm/cath/
DALI: http://www2.embl-ebi.ac.uk/dali/
Indices of other protein structure classifications:
http://www.bioscience.org/urllists/protdb.htm
http://www2.ebi.ac.uk/msd/Links/fold.shtml.
Databases of protein sequences homologous to those of known structures:
FSSP (Fold classification based on Structure–Structure alignment of Proteins)
http://www2.embl-ebi.ac.uk/dali/fssp/
HSSP (Homology-derived secondary structure)
http://www.sander.embl-heidelberg.de/hssp/

Access to structural databases at University College London:
http://www.biochem.ucl.ac.uk/bsm/biocomp/index.html#databases

Recommended reading and references

Brändén, C.-I. (1980). Relation between structure and function of α/β proteins. *Quart. Rev. Biophys.* **13**, 317–38.

Brenner, S., Chothia, C. and Hubbard, T.J.P. (1997). Population statistics of protein structures: lessons from structural classifications. *Curr. Opin. Struc. Biol.* **7**, 369–76.

Chothia, C., Hubbard, T., Brenner, S., Barns, H. and Murzin, A. (1997). Protein folds in the all-β and all-α classes. *Ann. Revs. Biophys. Biomol. Struc.* **26**, 597–627.

Hadley, C. and Jones, D.T. (1999). A systematic comparison of protein structure classifications: SCOP, CATH and FSSP. *Structure*. **7**, 1099–112.

Holm, L. and Sander, C. (1998). Touring protein fold space with Dali/FSSP. *Nucl. Acids Res.* **26**, 316–9.

Lupas, A. (1996). Coiled coils: new structures and new functions. *Trends Biochem. Sci.* **21**, 375–82.

McLachlan, A.D. (1979). Gene duplications in the structural evolution of chymotrypsin. *J. Mol. Biol.* **128**, 49–79.

Murzin, A.G., Lesk, A.M. and Chothia, C. (1994). Principles determining the structure of β-sheet barrels in proteins: I. A theoretical analysis. II. The observed structures. *J. Mol. Biol.* **236**, 1369–81 and 1382–400.

Orengo, C.A., Michie, A.D., Jones, S., Jones, D.T., Swindells, M.B. and Thornton, J.M. (1997). CATH— hierarchic classification of protein domain structures. *Structure* **5**, 1093–108.

Przytycka, T., Aurora, R. and Rose, G.D. (1999). A protein taxonomy based on secondary structure. *Nature Structural Biology*. **6**, 672–82.

Swindells, M.B., Orengo, C.A., Jones, D.T., Hutchinson, E.G. and Thornton, J.M. (1998). Contemporary approaches to protein structure classification. *BioEssays* **20**, 884–91.

Exercises, problems and weblems

Exercises

4.1. Which topology (A, B, C, D, E, or F; see page 140) does the structure in Figure 4.2b have?

4.2. Write down all topologically possible combinations of two α-helices and two strands of β-sheet. Which of these contain standard supersecondary structures?

Problems

4.1. Figure 4.23 shows the structure of the timothy grass pollen allergen Phl p II [1WHO], a double β-sheet sandwich structure. Draw a diagram similar in style to Figure 4.12b for the tertiary structure. Does it have the same topology as transthyretin, the glycosyl-asparaginase domain, or γ-crystallin?

4.2. Draw all possible topology diagrams for four-helix bundles with one parallel connection.

4.3. Figure 4.24 shows the structure of the ribosomal protein S6. Draw a diagram similar in style to Figure 4.18 for the tertiary structure.

4.4. In the crystal structure of a fragment of Japanese quail ovomucoid, a serine proteinase inhibitor, four molecules pack together in the crystal to form a β-barrel with four strands and shear number $S = 8$. Draw a hydrogen-bonding diagram, similar to Figure 4.20, for this β-barrel topology.

4.5. Figure 4.25 contains the hydrogen-bond pattern of a β-barrel. Note that the edge strand is repeated at right and left. What are the values of n and S? What barrel radius is expected?

4.6. Calculate the inner radius of an ideal TIM barrel from the β-sheet geometry. The perpendicular distance between strands is 4.4 Å and the Cα–Cα distance along each strand is 3.3 Å.

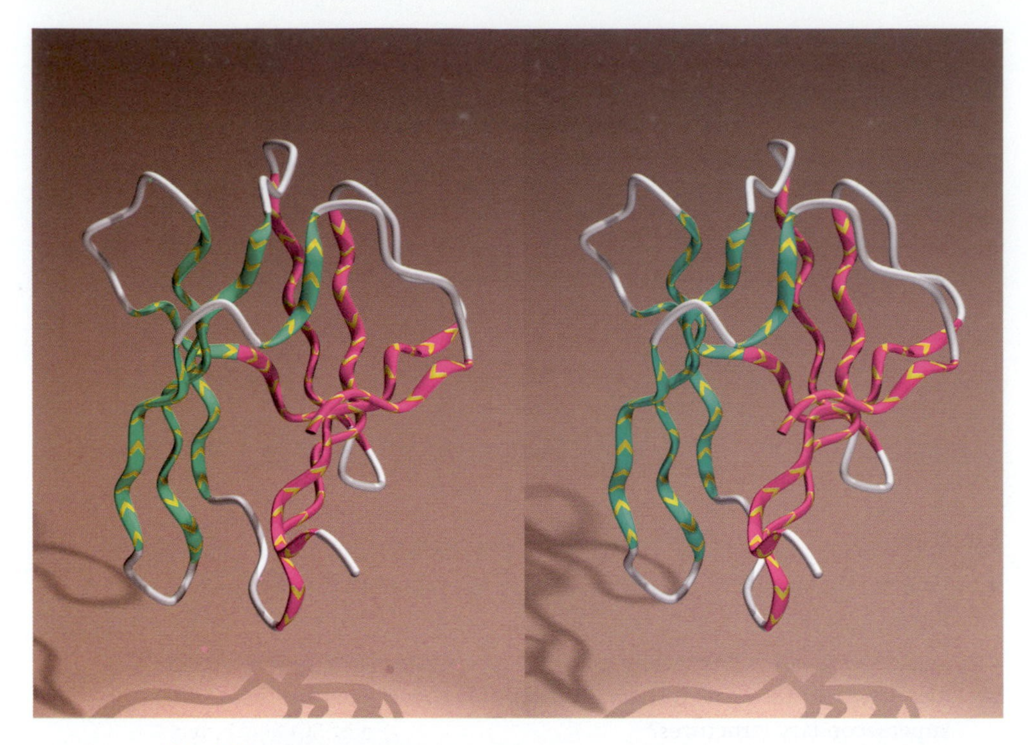

Fig. 4.23 β sandwich protein [1WHO].

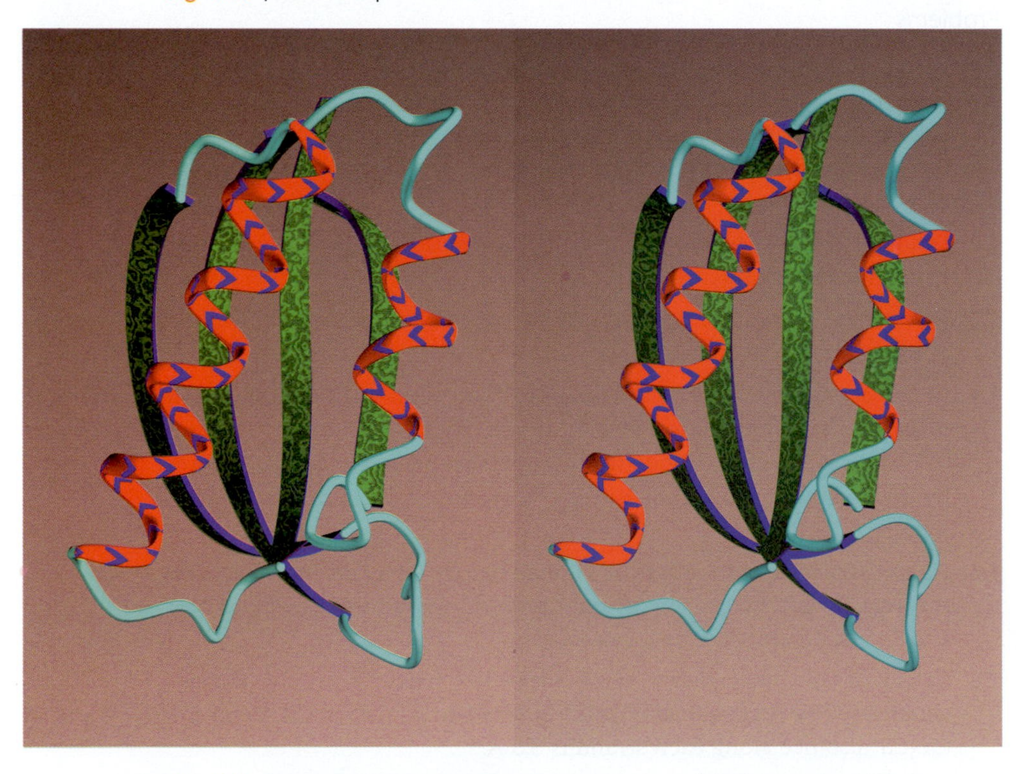

Fig. 4.24 Ribosomal protein S6 [1RIS]. N-terminus: top of third strand from left. C-terminus: top of fourth strand from left.

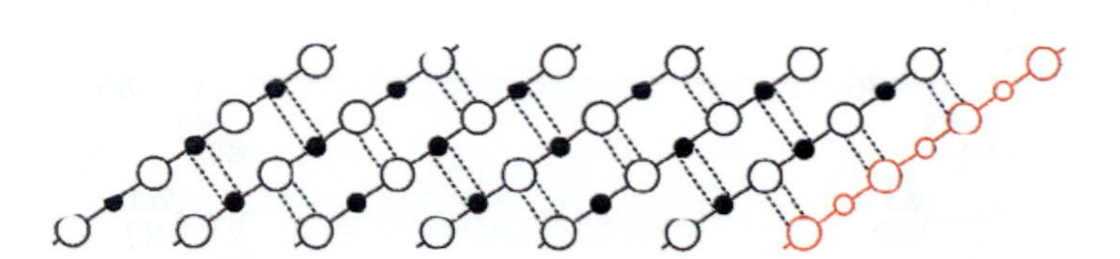

Fig. 4.25 Hydrogen-bonding pattern of β-barrel.

4.7. Show that the sidechains packed in the interior of a β-barrel will appear in layers perpendicular to the barrel axis if the ratio of the shear number to the numbers of strands (S/n) is integral.

Weblems

4.1. (a) In SCOP, find a protein in the same superfamily as enolase, but in a different family. (b) In CATH, find a protein with the same topology as dihydrolipoamide acetyltransferase, but in a different homologous superfamily.

4.2. Send the structure of acylphosphatase to the DALI server. What other proteins have the same topology?

4.3. (a) To what other enzymes is the family (in the sense of SCOP) of glyceraldehyde-3-phosphate dehydrogenases most closely related? (b) Are the substrates of these enzymes chemically similar? (c) Do the reactions catalysed by these enzymes have analogous mechanisms?

4.4. How is chloramphenicol acetyltransferase classified in SCOP and CATH?

4.5. Find another four-helix bundle and draw a diagram of its topology similar to that appearing on page 140.

4.6. Figure 4.26 shows a helix–helix packing. What is the nature of the ridges and grooves?

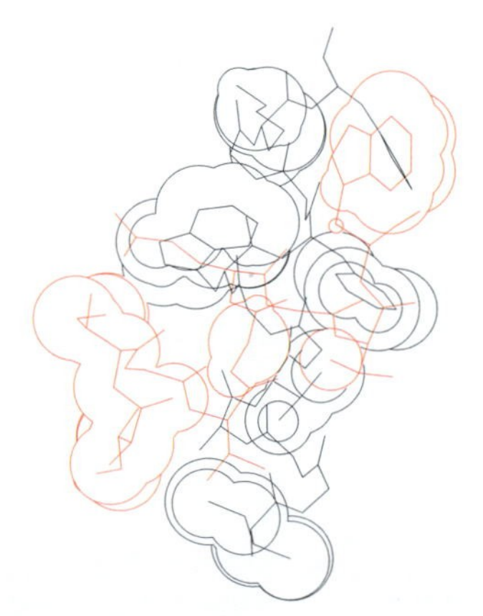

Fig. 4.26a

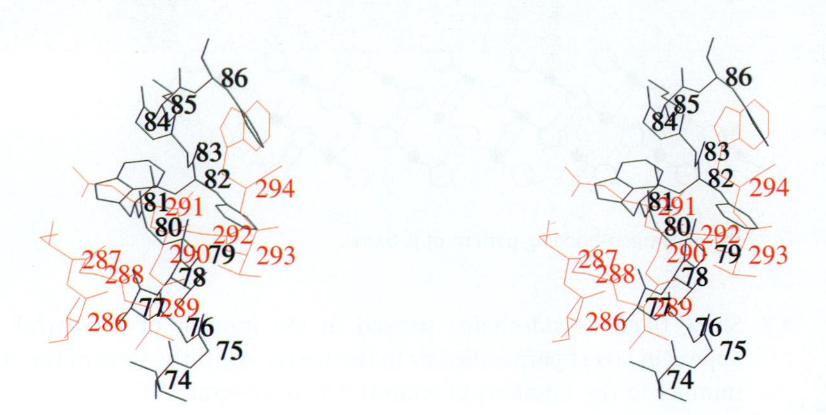

Fig. 4.26b

Fig. 4.26 A helix–helix contact from carboxypeptidase [5CPA]. (a) van der Waals slices through the interface. (b) A wire model with labels. Use the slices to see the pattern of the interaction, and the wire model only to check the residue numbers.

CHAPTER 5
Molecular evolution

Within the broad unity of life on Earth at the molecular level, there is also a great diversity, the result of evolution. This has been brought into its sharpest focus by studies of the relationships among nucleotide sequences of genes, amino acid sequences of proteins, and the three-dimensional structures and functions of proteins. DNA sequencing techniques now provide a direct window to the genome; and X-ray crystallography and NMR a direct view of the atomic structures and mechanisms of function of proteins. The availability of very large amounts of quantitative information in each of these categories has permitted very detailed studies of the processes of evolutionary change at the molecular level.

Some studies of molecular evolution relate molecular data to the results of traditional macroscopic evolutionary studies. We may ask: in what ways do molecular data confirm, supplement or modify our understanding of traditional phylogenetic relationships derived from comparative anatomy, embryology and the fossil record? Other questions are meaningful only in terms of molecular data: How do nucleotide sequences of DNA and amino acid sequences of proteins vary within and between individuals, populations and species? What kinds of selective forces and constraints are observed to apply to DNA and protein sequences and structures?

Evolution of DNA and proteins

We now recognize that the primary events in the generation of biological diversity are the mutation, insertion and deletion of nucleotides in DNA sequences, or larger-scale transposition of pieces of genetic material; and that selection reacts to protein function as determined by protein structure. If a 'wild-type' gene produces a functional protein product, a mutant gene may produce either an alternative protein of equivalent function—a *neutral mutation*—or a protein that carries out the same function but with an altered rate or specificity profile, or a protein with an altered function, or a protein that does not function—or even fold—at all. Evolution in a population—that is, the change in distribution of DNA sequences among the individual organisms—may occur through positive or negative selection, or by random fixation of variants without selective advantage.

Examination of homologous genes and proteins in different species have shown that evolutionary variation and divergence occur very generally at the molecular level. To a great extent, this has taken place in parallel with the divergence of species observed at the macroscopic level. Proteins from related species have similar but not identical amino acid sequences. These sequences determine similar but not identical protein structures. Even 40 years ago, the comparison of the haemoglobin and myoglobin structures showed that although the amino acid sequences had diverged, the basic qualitative 'fold' of the polypeptide chain was retained, with some modifications in structural details. Subsequent studies have extended these observations and established quantitative relationships between divergence of sequence and divergence of structure.

Direct access to the genome—nucleotide sequences

The determination of DNA sequences has revealed many new and unsuspected features of the contents and organization of the genome.

Most eukaryotic genes are interrupted by stretches of non-translated DNA, called intervening sequences or introns; the regions expressed are called exons. Introns appear to be fairly stable components of genes. For example, all functional vertebrate globin genes—including those of man, rabbit, mouse, chicken and *Xenopus laevis*, and also that of the clam *Scapharca inaequivalvis*—contain two introns at homologous positions. The intron–exon pattern of the globin gene has been constant since before the divergence of vertebrates and invertebrates, about 600 million years ago. However, the gene for leghaemoglobin in soybean has three intervening sequences, two of which correspond in position to those of the vertebrate globin genes; and the genes for insect and bacterial globins have none.

There has been considerable discussion of a suggestion of Gilbert, that exons represent structural components of proteins that can be recombined in different contexts, as a mechanism of development of new protein folds. Although this is observed in many modular proteins, the correlation between the intron–exon structures of genes and the structures of proteins has failed to support this hypothesis below the domain level.

Evolutionary changes in protein sequences

Tables of aligned amino acid sequences are a basic tool of molecular evolutionary studies. The patterns of conservation and variation at the individual positions provide clues to the nature of the selective constraints on the molecule, even more explicitly than a structure does. Table 5.1 illustrates aligned sequences from the globin family. Table 5.1a is limited to five mammalian globins: human and horse α and β chains, and sperm whale myoglobin. Table 5.1b contains globins from much more diverse species: it

One sequence keeps silent about its three-dimensional structure
Two aligned sequences whisper
But tables of many aligned sequences shout out loud

includes the mammalian globins plus globins from an insect, a plant, and a bacterium.

Each of the five mammalian globin sequences shown in Table 5.1a contains approximately 150 residues. At 25 positions the same amino acid is conserved in all five sequences, including the two histidine residues that interact with the haem iron. (In the line below the tabulation, upper-case letters indicate residues that are conserved in all five sequences, and lower-case letters indicate residues that are conserved in all but sperm whale myoglobin.) Other positions contain only amino acids with very similar properties: for example, position 3 contains only Ser or Thr; position 119 contains only Val, Ile or Leu. Still other positions show very wide variations in sidechain size and polarity; for example, position 32 contains Glu, Gly and Ile.

The many similarities strongly suggest that the sequences are related. Moreover, on the basis of the patterns of residue conservation and change at different positions, there is an obvious hierarchical classification. Position 32, for example, contains Glu in the human and horse α chains, Gly in the human and horse β chains, and Ile in myoglobin. The reader can easily identify other such positions. These show that the α chains are more similar to each other than to the β chains or to myoglobin, the β chains are more similar to each other than to the α chains or to myoglobin, but the α and β chains are more similar to each other than either is to myoglobin. Given another mammalian globin sequence, the reader would easily be able to identify it as a haemoglobin α chain, a haemoglobin β chain, or a myoglobin. This hierarchy is consistent with the evolutionary divergence of these three classes of globins. Another interesting feature of the patterns of conservation is the numerous pairs of conserved residues separated by 3, 4 or 7 in the sequences. This in an indication that they are in α-helices. Note that this inference is available only through the juxtaposition of aligned related sequences.

The sequences in Table 5.1b include the same five mammalian globins, and additional homologues from an insect, a plant and a bacterium. These are much more diverse—the bacterial globin is the most distant from the others—and indeed only four positions are conserved in all eight sequences. (In the line below the tabulation, upper-case letters indicate residues that are conserved in all eight sequences, and lower-case letters indicate residues that are conserved in all but the bacterial globin.) A residue interacting with the haem iron—known as the distal histidine—is one of the residues conserved in all the sequences shown except the bacterial globin; however, there are numerous eukaryotic globins known in which this residue has changed.

Table 5.1 Some aligned sequences from the globin family

Table 5.1a Mammalian globin sequences

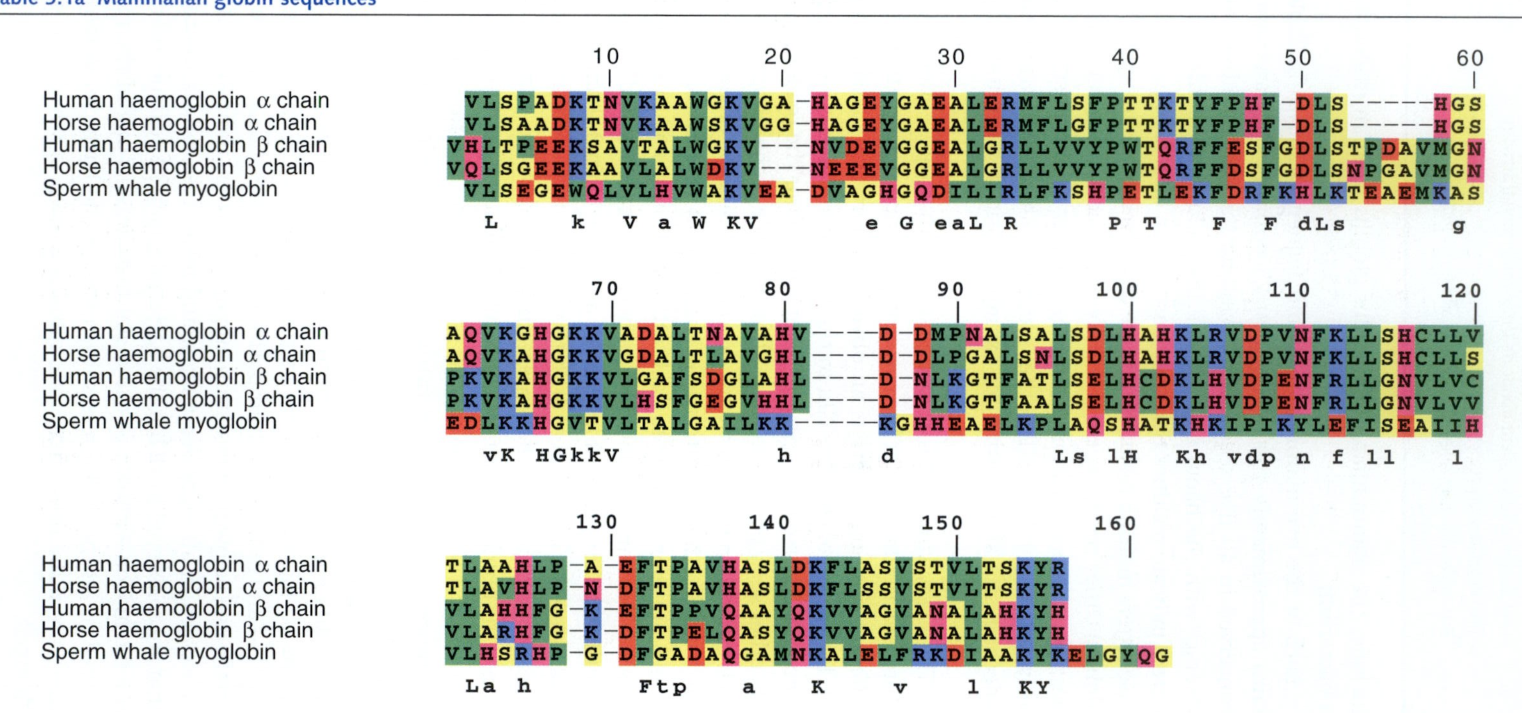

```
                                     10        20        30        40        50        60
                                      |         |         |         |         |         |
Human haemoglobin α chain         VLSPADKTNVKAAWGKVGA-HAGEYGAEALERMFLSFPTTKTYFPHF-DLS-----HGS
Horse haemoglobin α chain         VLSAADKTNVKAAWSKVGG-HAGEYGAEALERMFLGFPTTKTYFPHF-DLS-----HGS
Human haemoglobin β chain        VHLTPEEKSAVTALWGKV---NVDEVGGEALGRLLVVYPWTQRFFESFGDLSTPDAVMGN
Horse haemoglobin β chain        VQLSGEEKAAVLALWDKV---NEEEVGGEALGRLLVVYPWTQRFFDSFGDLSNPGAVMGN
Sperm whale myoglobin             VLSEGEWQLVLHVWAKVEA-DVAGHGQDILIRLFKSHPETLEKFDRFKHLKTEAEMKAS
                                   L    k  V a W KV      e G eaL R      P T   F F dLs      g

                                     70        80        90       100       110       120
                                      |         |         |         |         |         |
Human haemoglobin α chain        AQVKGHGKKVADALTNAVAHV----D-DMPNALSALSDLHAHKLRVDPVNFKLLSHCLLV
Horse haemoglobin α chain        AQVKAHGKKVGDALTLAVGHL----D-DLPGALSNLSDLHAHKLRVDPVNFKLLSHCLLS
Human haemoglobin β chain        PKVKAHGKKVLGAFSDGLAHL----D-NLKGTFATLSELHCDKLHVDPENFRLLGNVLVC
Horse haemoglobin β chain        PKVKAHGKKVLHSFGEGVHHL----D-NLKGTFAALSELHCDKLHVDPENFRLLGNVLVV
Sperm whale myoglobin            EDLKKHGVTVLTALGAILKK-----KGHHEAELKPLAQSHATKHKIPIKYLEFISEAIIH
                                  vK HGkkV        h    d        Ls lH  Kh vdp n f ll  l

                                    130       140       150       160
                                      |         |         |         |
Human haemoglobin α chain        TLAAHLP-A-EFTPAVHASLDKFLASVSTVLTSKYR
Horse haemoglobin α chain        TLAVHLP-N-DFTPAVHASLDKFLSSVSTVLTSKYR
Human haemoglobin β chain        VLAHHFG-K-EFTPPVQAAYQKVVAGVANALAHKYH
Horse haemoglobin β chain        VLARHFG-K-DFTPELQASYQKVVAGVANALAHKYH
Sperm whale myoglobin            VLHSRHP-G-DFGADAQGAMNKALELFRKDIAAKYKELGYQG
                                  La h     Ftp  a  K   v  l  KY
```

Table 5.1b Eukaryotic and prokaryotic globin sequences

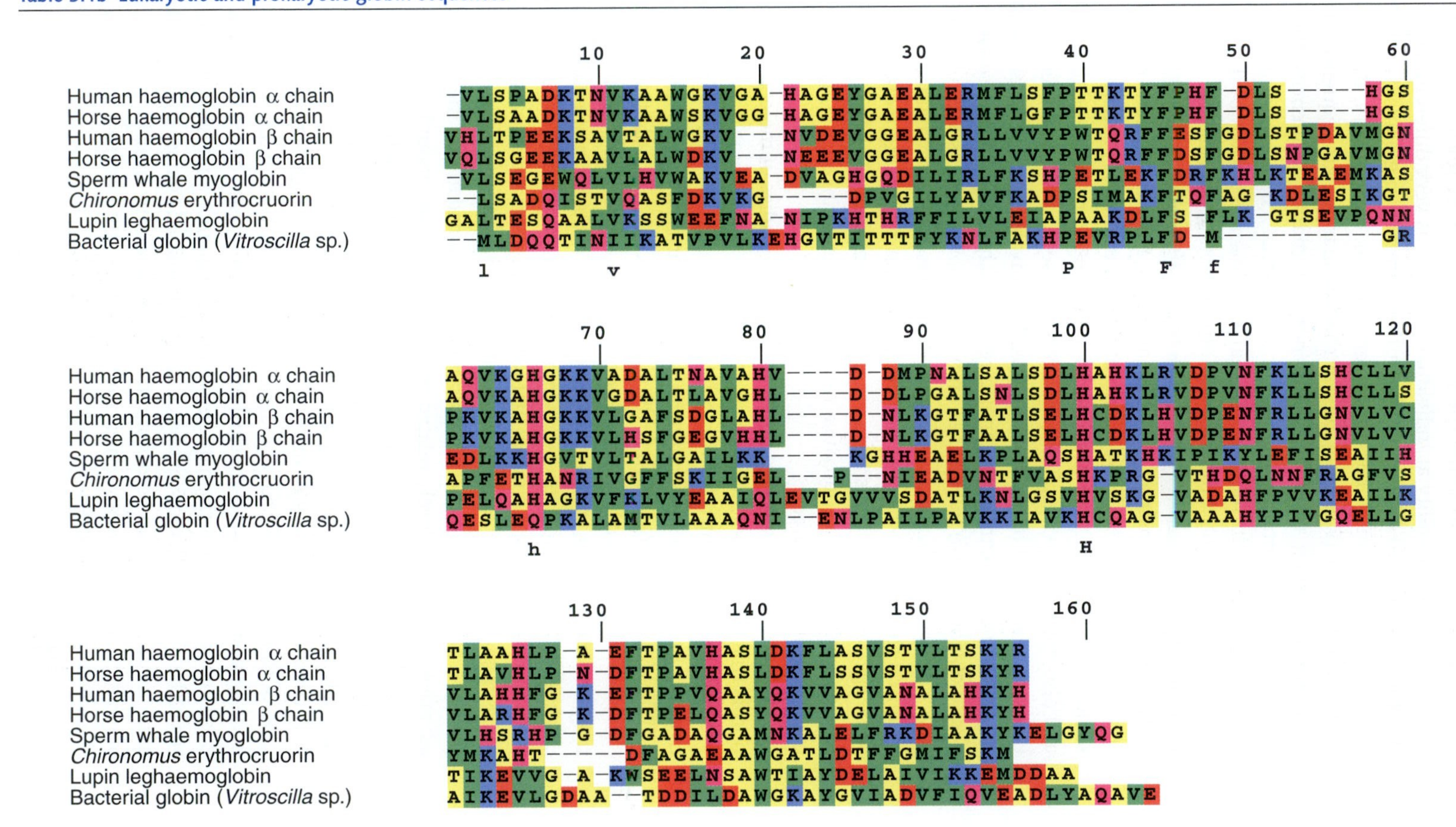

```
                                          10        20        30        40        50        60
                                           |         |         |         |         |         |
Human haemoglobin α chain           -VLSPADKTNVKAAWGKVGA-HAGEYGAEALERMFLSFPTTKTYFPHF-DLS-----HGS
Horse haemoglobin α chain           -VLSAADKTNVKAAWSKVGG-HAGEYGAEALERMFLGFPTTKTYFPHF-DLS-----HGS
Human haemoglobin β chain           VHLTPEEKSAVTALWGKV---NVDEVGGEALGRLLVVYPWTQRFFESFGDLSTPDAVMGN
Horse haemoglobin β chain           VQLSGEEKAAVLALWDKV---NEEEVGGEALGRLLVVYPWTQRFFDSFGDLSNPGAVMGN
Sperm whale myoglobin               -VLSEGEWQLVLHVWAKVEA-DVAGHGQDILIRLFKSHPETLEKFDRFKHLKTEAEMKAS
Chironomus erythrocruorin           --LSADQISTVQASFDKVKG-----DPVGILYAVFKADPSIMAKFTQFAG-KDLESIKGT
Lupin leghaemoglobin                GALTESQAALVKSSWEEFNA-NIPKHTHRFFILVLEIAPAAKDLFS-FLK-GTSEVPQNN
Bacterial globin (Vitroscilla sp.)  --MLDQQTINIIKATVPVLKEHGVTITTTFYKNLFAKHPEVRPLFD-M----------GR
                                       l       v                           P     F  f

                                          70        80        90       100       110       120
                                           |         |         |         |         |         |
Human haemoglobin α chain           AQVKGHGKKVADALTNAVAHV----D-DMPNALSALSDLHAHKLRVDPVNFKLLSHCLLV
Horse haemoglobin α chain           AQVKAHGKKVGDALTLAVGHL----D-DLPGALSNLSDLHAHKLRVDPVNFKLLSHCLLS
Human haemoglobin β chain           PKVKAHGKKVLGAFSDGLAHL----D-NLKGTFATLSELHCDKLHVDPENFRLLGNVLVC
Horse haemoglobin β chain           PKVKAHGKKVLHSFGEGVHHL----D-NLKGTFAALSELHCDKLHVDPENFRLLGNVLVV
Sperm whale myoglobin               EDLKKHGVTVLTALGAILKK-----KGHHEAELKPLAQSHATKHKIPIKYLEFISEAIIH
Chironomus erythrocruorin           APFETHANRIVGFFSKIIGEL---P--NIEADVNTFVASHKPRG-VTHDQLNNFRAGFVS
Lupin leghaemoglobin                PELQAHAGKVFKLVYEAAIQLEVTGVVVSDATLKNLGSVHVSKG-VADAHFPVVKEAILK
Bacterial globin (Vitroscilla sp.)  QESLEQPKALAMTVLAAAQNI--ENLPAILPAVKKIAVKHCQAG-VAAAHYPIVGQELLG
                                         h                                        H

                                          130       140       150       160
                                           |         |         |         |
Human haemoglobin α chain           TLAAHLP-A-EFTPAVHASLDKFLASVSTVLTSKYR
Horse haemoglobin α chain           TLAVHLP-N-DFTPAVHASLDKFLSSVSTVLTSKYR
Human haemoglobin β chain           VLAHHFG-K-EFTPPVQAAYQKVVAGVANALAHKYH
Horse haemoglobin β chain           VLARHFG-K-DFTPELQASYQKVVAGVANALAHKYH
Sperm whale myoglobin               VLHSRHP-G-DFGADAQGAMNKALELFRKDIAAKYKELGYQG
Chironomus erythrocruorin           YMKAHT-----DFAGAEAAWGATLDTFFGMIFSKM
Lupin leghaemoglobin                TIKEVVG-A-KWSEELNSAWTIAYDELAIVIKKEMDDAA
Bacterial globin (Vitroscilla sp.)  AIKEVLGDAA--TDDILDAWGKAYGVIADVFIQVEADLYAQAVE
```

It is interesting to examine the residue variation at several positions in Table 5.1b. The variability at individual positions is very different. Position 3, which was limited to Ser and Thr in the mammalian globins, is still limited to Ser and Thr in all but the bacterial globin. Position 119, that varied only conservatively (Val, Ile or Leu) in the five mammalian sequences, contains no additional residues in the nonmammalian sequences shown. Position 32, that contained Glu, Gly and Ile (already showing minimal constraint), is occupied in the nonmammalian sequences shown by Tyr, Ile or Lys. The variability at different positions reflects functional or structural importance.

It is possible to construct 'evolutionary trees' from tabulations of related sequences. Phylogenies derived from different families of proteins from the same range of species are usually mutually consistent in branching order, and also consistent with phylogenetic trees based on classical methods. Of course it is essential to choose functionally equivalent proteins. An attempt to derive mammalian phylogenetic relationships from globin sequences would obviously have to be based on haemoglobin α-chains taken by themselves or haemoglobin β-chains taken by themselves or myoglobin chains taken by themselves. It could not be carried out by mixing α-chains from some species with β-chains from others. 'Horizontal' gene transfer between species would, of course, produce results completely inconsistent with true overall evolutionary trees.

A subject that has received much attention in studies of molecular evolution is that of the 'molecular clock': the idea that amino acid substitution proceeds at a constant rate within individual protein families. If this were true, not only could we infer from substitution patterns such as those shown in Table 5.1 the topology of the evolutionary tree, but we could translate the extent of the divergence of the sequences into actual times of divergence from the most recent common ancestor. Such a calibration of evolutionary clocks could be used to date events in biological history prior to the available fossil record. It would also imply that rates of amino acid substitution appear to be independent of rates of morphological change and of variations in selective pressure, and therefore appear to be nonadaptive. Indeed, some adaptive changes in closely related species are brought about by only a few amino acid substitutions, which amount to so small a fraction of the total number of substitutions between species that they do not significantly affect the statistics of amino acid sequence divergence.

A conspicuous success of molecular clocks was the correct dating of the time of the human–African ape divergence by A.C. Wilson and co-workers, in disagreement with the conclusions of paleontologists based on fossils. It was later accepted that the date derived from sequence analysis was correct. Nevertheless, considerable evidence has accumulated against the constancy of molecular clock rates across an entire protein family over longer periods of geological time. Of course, it is for the very distant past that a molecular clock would be most useful.

Variability in selective constraints in protein molecules

What is undoubtedly true is that evolution runs at different rates in different protein families, and even in different regions in a single protein structure. Residues in proteins that are subject to strict functional constraints or that play crucial structural roles can accommodate mutations less easily than other residues. Different overall rates of amino acid substitution depend on the fraction of residues to which the structure and function are sensitive. For example, fibrinopeptides, regions exised and discarded during the maturation of fibrinogen and therefore subject to minimal functional constraint, have the fastest rate of evolution of any known family of proteins. The sequences of mammalian proinsulins show a similar phenomenon (See Box). The C-peptide sequences, which are excised and do not appear in the mature functional hormone, show much higher variability than the A and B chains, which make up the active molecule.

Sequences of mammalian insulins

```
B chain sequences
        Human         FVNQHLCGSHLVEALYLVCGERGFFYTPKT
        Pig           FVNQHLCGSHLVEALYLVCGERGFFYTPKA
        Cow           FVNQHLCGSHLVEALYLVCGERGFFYTPKA
        Guinea pig    FVSRHLCGSNLVETLYSVCQDDGFFYIPKD
        Rat           FVKQHLCGPHLVEALYLVCGERGFFYTPKS
C peptide sequences
        Human         RREAEDLQVGQVELGGGPGAGSLQPLALEGSLQKR
        Pig           RREAENPQAGAVELGGG--LGGLQALALEGPPQKR
        Cow           RREVEGPQVGALELAGGPGAGGL-----EGPPQKR
        Guinea pig    RRELEDPQVEQTELGMGLGAGGLQPLALEMALQKR
        Rat           RREVEDPQVPQLELGGGPEAGDLQTLALEVARQKR
A chain sequences
        Human         GIVEQCCTSICSLYQLENYCN
        Pig           GIVEQCCTSICSLYQLENYCN
        Cow           GIVEQCCASVCSLYQLENYCN
        Guinea pig    GIVDQCCTGTCTRHQLQSYCN
        Rat           GIVDQCCTSICSLYQLENYCN
```

Alignment of the amino acid sequences of the proinsulins of human, pig, cow, guinea pig and rat. The proinsulin chain, a single polypeptide comprising the B, C and A sequences (in that order in the gene), folds within the lumen of the endoplasmic reticulum, and disulphide bonds are formed between its A and B chains. The double basic residues at the ends (RR and KR) are the signals for cleavage of the (connecting) C peptide, which has no hormonal activity. The alignment shows the close similarity of the A and B chains which make up the mature and functional hormone, and the somewhat lower similarity in the C peptides.

These cases are fairly straightforward. Of course, it is not always possible to say, *a priori*, what the effect of a mutation will be. The mutation in human haemoglobin responsible for sickle cell anemia is a change from Lys to Val on the molecular surface, the clinical consequences of which could not be inferred from knowledge of the sequence and the structure.

The existence in genomes of multiple copies of genes, capable of independent evolutionary variation, also affects rates of change. Multiple copies may include different alleles in diploid organisms, and gene families. Gene families are closely linked multiplets of similar DNA sequences which may be differently expressed in different tissues or at different stages of development.

Evolution of protein structures

Included in the 10 000 protein structures now known are several families within which the molecules maintain the same basic folding pattern over ranges of sequence homology from near-identity down to below 20%. How is structure maintained in the face of such drastic sequence change? The answer is that structural details are not maintained; rather it is function on which selection is acting. It is function that is maintained, or modified. In general only the overall folding pattern is conserved within families of proteins, and as the sequences diverge the structures progressively deform.

In both closely and distantly related proteins the general response to mutation is conformational change. Variations in conformation in families of homologous proteins that retain a common function reveal how the structures accommodate changes in amino acid sequence. Residues active in function, such as the proximal histidine of the globins or the catalytic serine, histidine and aspartate of the serine proteinases, are resistant to mutation because changing them would interfere, explicitly and directly, with function. It is the ability of protein structures to accommodate mutations in nonfunctional residues that permits a large amount of apparently nonadaptive change to occur. Surface residues not involved in function are usually free to mutate. Loops on the surface can often accommodate changes by local refolding, provided that they are not involved directly in function.

Mutations that change the volumes of buried residues generally do not change the conformations of individual helices or sheets, but produce distortions of their spatial assembly. These tend to take the form of rigid-body shifts and rotations, typically 3–5 Å but occasionally larger. The nature of the forces that stabilize protein structures sets general limitations on these conformational changes; other constraints derived from function vary from case to case. In sets of related proteins, the general folding pattern is preserved, with distortions that increase in severity as the amino acid sequences progressively diverge. How, then, is function maintained in widely divergent sequences? It requires the *integration* of the response to mutations over all or at least a large portion of the molecule. The divergence of the structures is not an introduction of random independent noise. Instead, the distortions throughout the molecule are coupled to preserve, or modify, function.

The evolution of proteins with altered function

There is some information available about the way in which protein structures alter existing functions or develop new ones, although it is more anecdotal than systematic.

In families of closely related proteins, mutations generally conserve function but modulate specificity. For instance, the chymotrypsin family of serine proteinases contains a *specificity pocket*: a surface cleft complementary in shape and charge distribution to the sidechain adjacent to the scissile bond. Mutations tend to leave the backbone conformation of the pocket unchanged but to affect the shape and charge of the lining, altering the specificity (see Chapter 6).

Haemoglobin provides another example of modulation of function by a small number of point mutations. Most amino acid substitutions in vertebrate haemoglobin evolution appear to be functionally neutral. The allosteric properties of haemoglobin, especially the regulatory responses to ligands other than oxygen, are brought about by substitutions of a few amino acid residues in key positions. For example, human adult and foetal haemoglobins differ by the replacement of 39 out of 287 residues, including a substitution of Ser (143β) for His. Primarily as a result of this mutation, foetal haemoglobin has a lower affinity than adult haemoglobin for the regulatory ligand diphosphoglycerate (DPG). This promotes the transfer of oxygen across the placenta to the foetus. It is interesting that the intrinsic oxygen affinity of foetal hemoglobin (in the absence of DPG) is *lower* than that of normal adult haemoglobin.

Recruitment of proteins as lens crystallins illustrates another mode of evolution: a novel function *preceding* divergence. In the duck, an active lactate dehydrogenase and an enolase serve as crystallins, although they do not encounter these substrates *in situ*. In other cases crystallins are closely related to enzymes, but some divergence has already occurred, with loss of catalytic activity.

There are numerous examples of recruitment of structures for altered functions. The TIM-barrel structure, or very similar variants, has now appeared in over 40 enzymes. In many cases the sequence similarity is so low that it is impossible to say whether the proteins are genuinely related, or whether evolution has discovered this very stable and useful fold more than once. However, certain enzymes sharing the TIM-barrel fold, and which are similar enough for us to be confident of their homology, clearly show the divergent evolution of new functions.

For instance, the enolase superfamily of TIM-barrel enzymes contains several enzymes that catalyse different reactions with shared features of their mechanism. The group includes enolase itself, mandelate racemase, muconate lactonizing enzyme I, and D-glucarate dehydratase. Each acts by abstracting an α proton from a carboxylic acid to form an enolate intermediate. The subsequent reaction pathway, and the nature of the product, vary from enzyme to enzyme. These enzymes have very similar overall structures, a variant of the TIM-barrel fold. Different residues in the active site produce enzymes that catalyse different reactions.

In the enolase superfamily, function has changed by modification of a binding site. Although these enzymes catalyse different reactions, some aspects of the mechanism have been preserved. In contrast, the chymotrypsin family of serine proteinases offers examples of emergence of completely novel functions.

1. Haptoglobin is a chymotrypsin homologue that has lost its proteolytic activity. It acts as a chaperone, preventing unwanted aggregation of proteins. Haptoglobin forms a tight complex with haemoglobin fragments released from erythrocytes, with several useful effects including preventing the loss of iron. Haptoglobin has a number of other functions, including mediating immune responses.
2. The serine proteinase of rhinovirus has developed a separate, independent function, of forming the initiation complex in RNA synthesis, using residues on the opposite side of the molecule from the active site for proteolysis. This is not a modification of an active site; it is the creation of a new one.
3. Subunits homologous to serine proteinases appear in plasminogen-related growth factors. The role of these subunits in growth factor activity is not yet known, but it cannot be a proteolytic function because essential catalytic residues have been lost.
4. The insect 'immune' protein scolexin is a distant homologue of serine proteinases that induces coagulation of haemolymph in response to infection.

Neutral mutations

Over the short term, proteins can evolve by selection of one or a few sequence changes that confer altered or novel functions. After establishment of novel function the sequences will continue to diverge. Many of the changes will be 'neutral'; that is, new sequences created by random mutational processes may replace their predecessors by chance, *without necessarily* having conferred selective advantage on the organisms that bear them.

Protein engineering can give some idea of the extent of sequence change that is apparently 'neutral' in terms of its selective advantage. Lactate dehydrogenase (LDH) and malate dehydrogenase (MDH) provide an extreme example. LDH interconverts lactate $\longleftrightarrow$ pyruvate, and MDH interconverts malate $\longleftrightarrow$ oxalacetate. The two enzymes are very distant homologues, sharing a common tertiary structure and coenzyme, but with below 20% identical residues in the aligned sequences. J.J. Holbrook and his collaborators found that alteration of a *single* residue—Gly102 to Arg—changed wild-type *B. stearothermophilus* lactate dehydrogenase to a malate dehydrogenase. The ratio of the values of k_{cat}/K_M for pyruvate and oxalacetate is a measure of the relative specificity of the proteins as lactate and malate dehydrogenases. For the wild-type enzyme, the ratio is ~10^3; for the single-site mutant the ratio is ~10^{-4}. The mutant enzyme has a value of k_{cat}/K_M for its substrate, oxalacetate, as high as that of the wild type for its substrate, pyruvate. In fact, the mutant enzyme has a maximal velocity twice as fast as that of the natural *B. stearothermophilus* malate dehydrogenase!

The conclusion is that many of the sequence differences between homologous proteins are the result of random drift rather than selection. Of course not all changes are neutral. Some are advantageous, and their proliferation follows Darwinian principles. Some are deleterious and selection rejects them. However, an altered gene *can* take over in a population even if neither the sequence changes nor their effects on the structure have affected function to provide selective advantage. Proteins are rough-hewn as well as shaped by destiny.

Domain combination and recombination

Proteins can also achieve functional variety by combining domains in different ways. Many dehydrogenases share homologous NAD- or NADP-binding domains, coupled with unrelated domains that vary with the reaction catalysed.

A common mechanism for generating different partners is gene duplication followed by divergence. In some cases, such as the chymotrypsin-like serine proteinases, the duplicated material forms two domains tightly integrated into a functional unit, with the catalytic site split across the two domains. No known protein contains a single isolated serine proteinase domain.

In other cases, gene duplication provides a mechanism for generating regulatory control over function, by development of an oligomeric protein. In haemoglobin the tetrameric allosteric structure makes the transfer of oxygen from lungs to muscles more efficient, and creates binding sites between the subunits that can respond to regulatory ligands such as hydrogen or chloride ions and diphosphoglycerate. Mutations can achieve fine tuning of the control mechanism, as in the differential response of maternal and foetal haemoglobin to DPG.

Bovine glutamate dehydrogenase, an enzyme containing distinct catalytic and regulatory subunits, has an internal homology that suggests its descent from an earlier molecule containing only a catalytic site.

Proteins can *combine* gene duplication or fusion with generation of partners by *domain swapping*. Suppose that protein A undergoes gene duplication and divergence, or gene fusion, to form a two-domain structure A–B, and that an extensive A–B interface develops. Then there is the potential of forming a dimeric (four-domain) structure that contains two A–B interfaces from different monomers (Figure 5.1). Interleukin-5 is an example of what such a protein might look like (Figure 5.2).

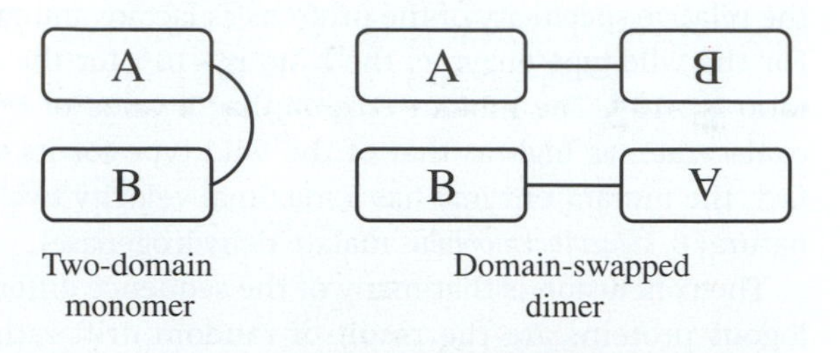

Fig. 5.1 Schematic diagram of domain swapping. A monomer contains two domains, A and B, connected by a flexible linker. Assume that there is a well-developed interface between A and B, in the relative orientation that exists in the monomer. The dimer contains four domains. The reader should verify that the two A–B interfaces in the dimer, as shown here, are formed from domains in the same relative position and orientation as the single interface in the monomer.

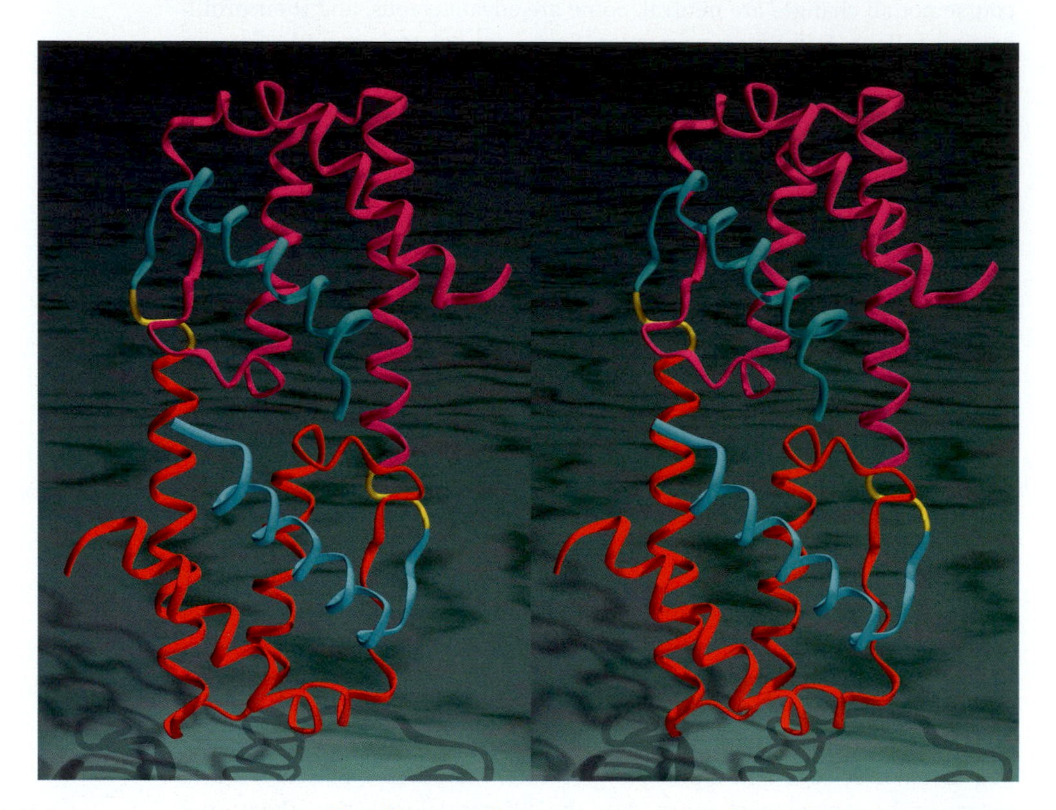

Fig. 5.2 Interleukin-5, a dimeric protein containing two equivalent domain–domain interfaces [1HUL].

Structural relationships among related molecules

Families of related proteins tend to retain similar folding patterns. Figure 5.3 shows sperm whale myoglobin and its distant relative lupin leghaemoglobin. Figure 5.4 shows the two sulphydryl proteases actinidin and papain. Figure 5.5 shows two related β-sheet proteins, plastocyanin and azurin. The reader is urged to spend some time comparing these sets of structures, cataloguing features that remain the same and features that change. Which pair of structures appears to be most closely related?

A general relationship between divergence of amino acid sequence and protein conformation in families of related proteins

If one examines sets of related proteins, it is clear that the general folding pattern is preserved, but that there are distortions which increase in

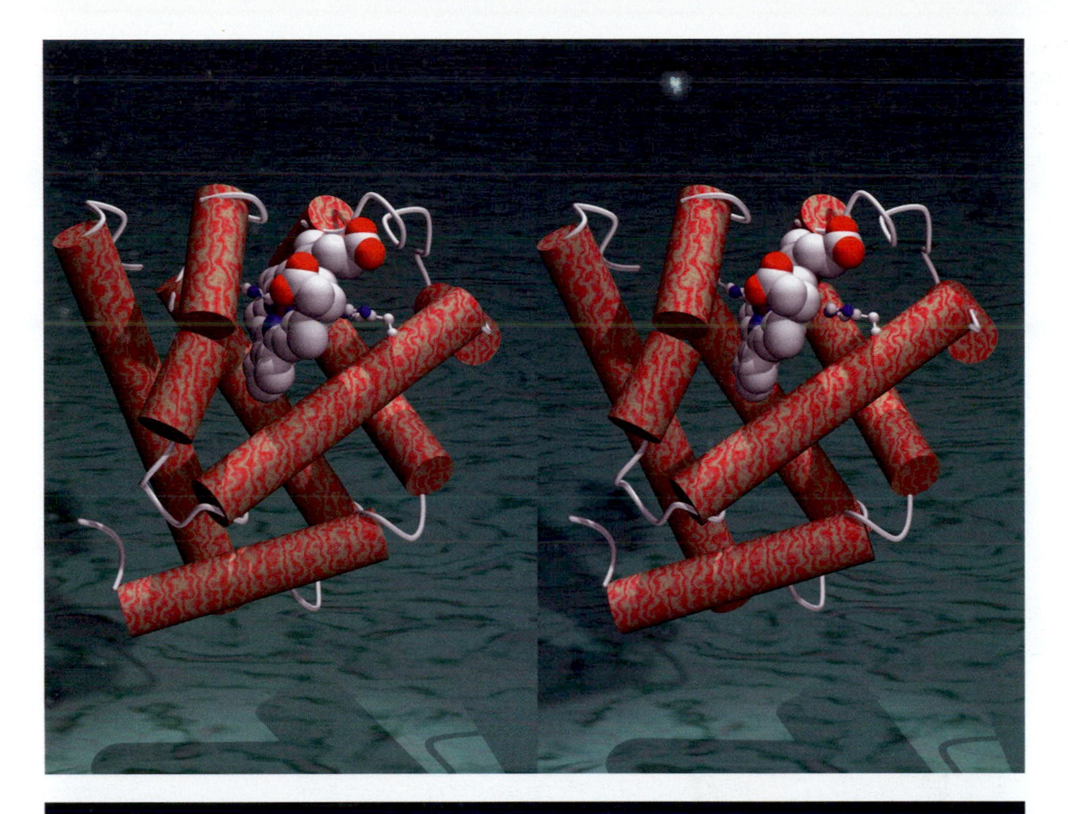

Fig. 5.3a

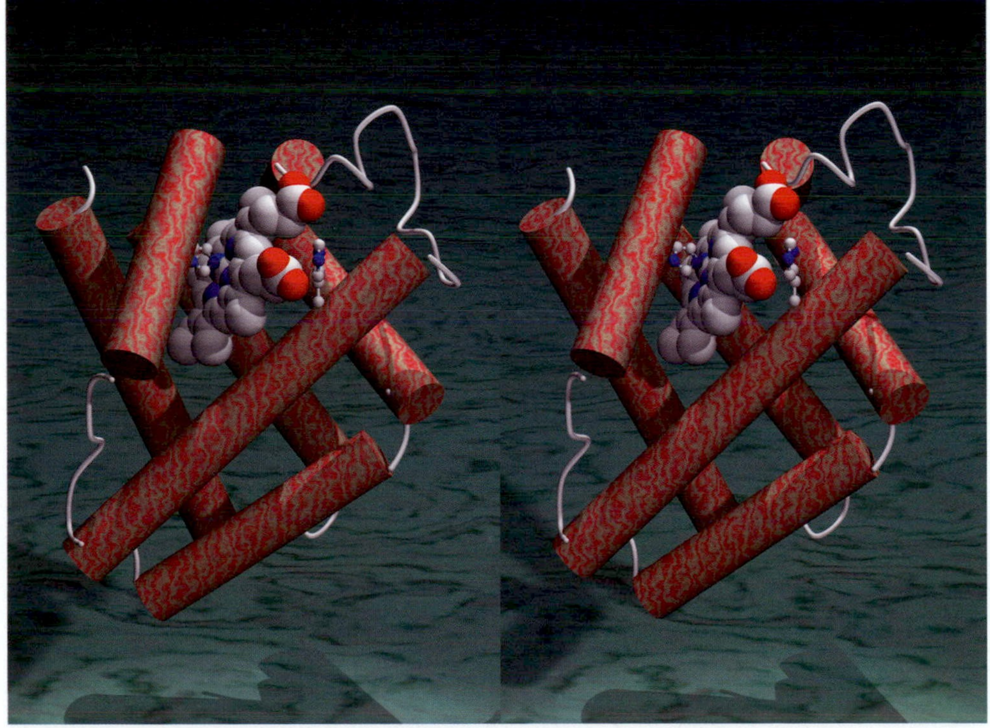

Fig. 5.3b

Fig. 5.3 (a) Sperm whale myoglobin [1MBD]. (b) Lupin leghaemoglobin [2LH7].

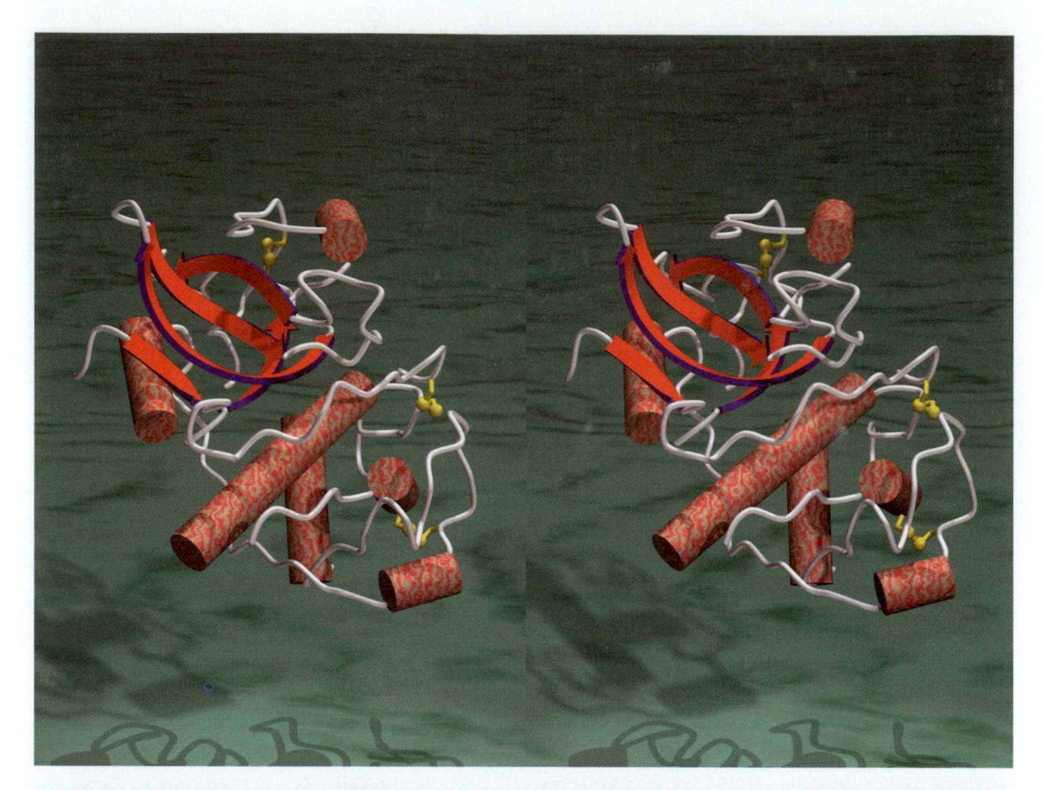

Fig. 5.4a

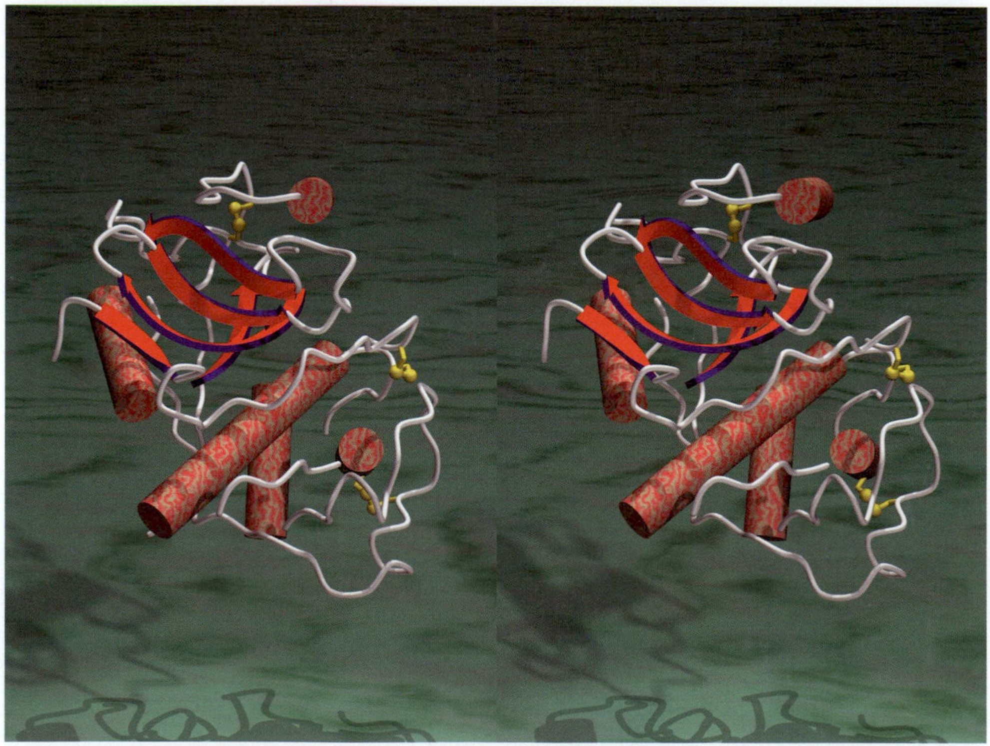

Fig. 5.4b

Fig. 5.4 (a) Actinidin [2ACT]. (b) Papain [9PAP].

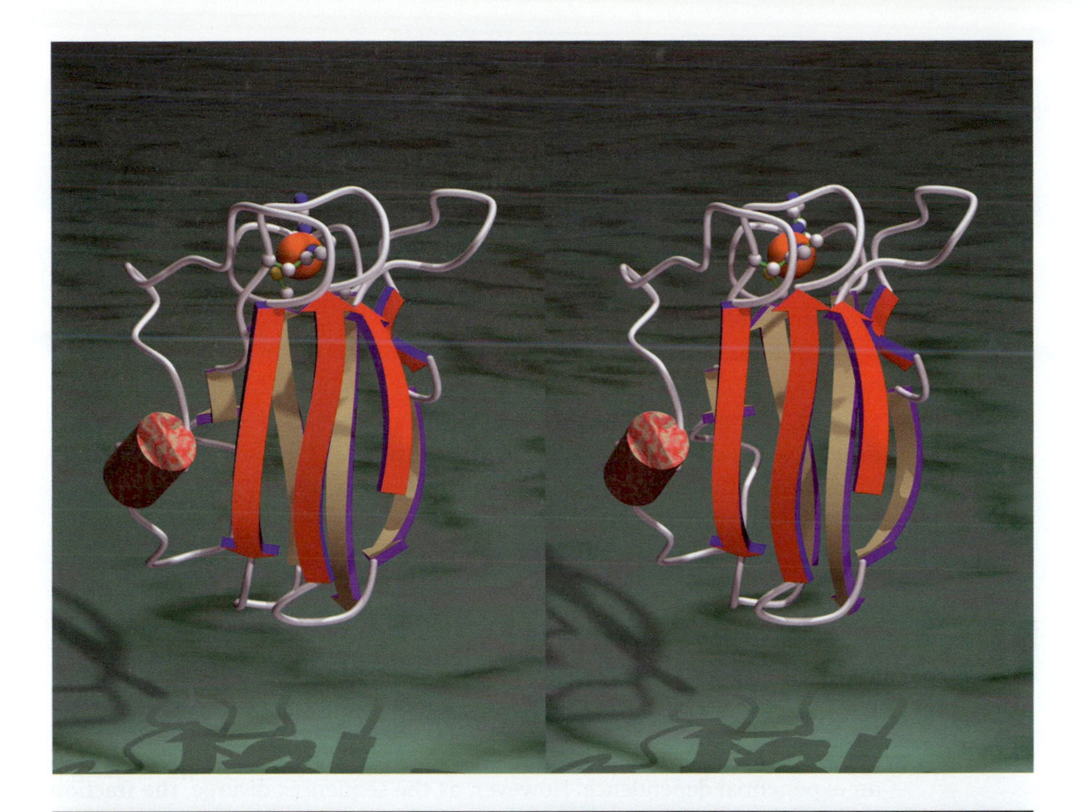

Fig. 5.5a

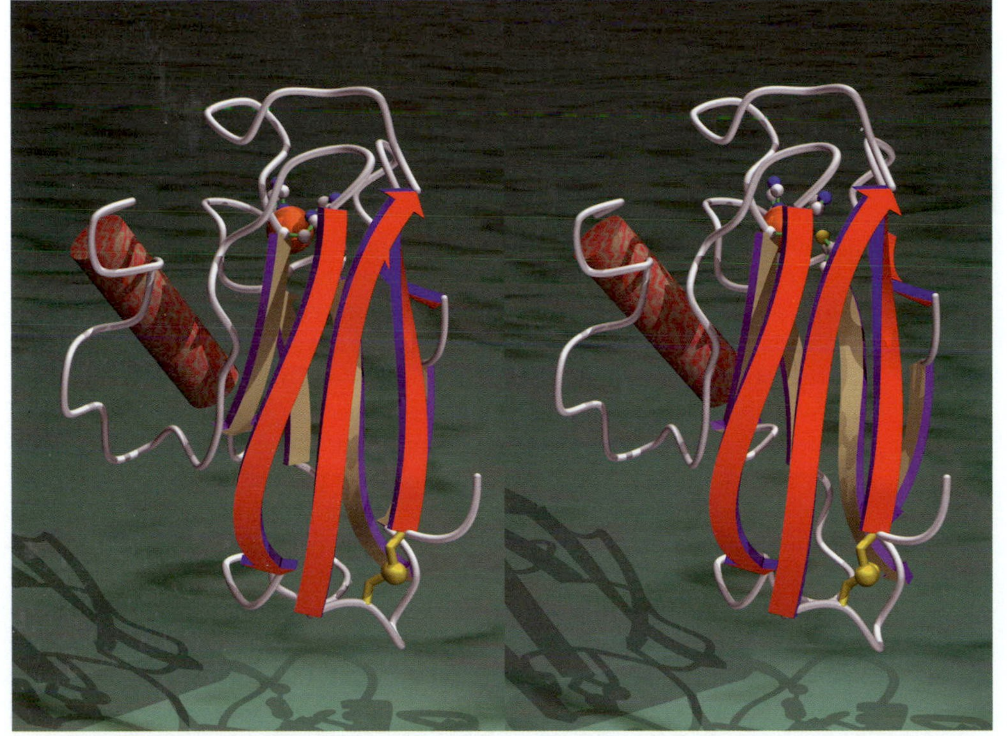

Fig. 5.5b

Fig. 5.5 (a) Plastocyanin [5PCY]. (b) Azurin [2AZA].

magnitude as the amino acid sequences diverge. A closer look reveals that the distortion is not uniformly distributed. Instead, in each family a core of the structure retains the same qualitative fold, with other parts of the structure changing conformation radically. An analogy: if one were to compare the capital letters R and B, the common core would correspond to those regions of the letters represented by the letter P. Outside this common core, R has a diagonal stroke and B has a loop. Comparison of a roman R and an italic *B* would illustrate a common core that is distorted between the two structures.

In proteins, the common core generally contains the major elements of secondary structure and peptides flanking them, including active site peptides. In globins and in actinidin/papain, the core includes almost all of the structure. In plastocyanin/azurin, the core includes the double β-sheet structure but not the long loop at the left; here the core is no more than about half the structure.

The r.m.s. deviation of the core measures the divergence of the structure. The fraction of identical residues in the core measures the divergence of the sequences. Figure 5.6 shows the relationship between divergence of sequence and structure for 32 pairs of proteins from eight different families. It is of the greatest interest and significance that there is a common relationship, that holds for all the different families studied.

The results show that as the sequence diverges, the structure diverges, with an exponential dependence. However, as the sequences diverge, the fraction of residues in the core—that is, the fraction of the structure that retains the same qualitative fold—may also decrease substantially (see Figure 5.7). For sequences more closely related than 60% residue identity, the core is observed to contain at least 90% of the residues, and the refolding of the remaining 10% will involve only minor surface loops. But for very distantly related structures, with residue identities below 20%, often evolution has had (and has taken) the opportunity to alter far more of the structure. In such cases the core can amount to as little as 40% of the structure or as much as 80–90%.

Let us consider representative points on the curve in Figure 5.6. First, the points at 100% residue identity represent independent structure determinations of the same protein. The observed differences in structure arise primarily from crystal-packing forces—the molecules have different environments in the crystals and are subjected to different patterns of intermolecular contacts. The average value of the root-mean-square difference in mainchain atomic positions in these pairs of structures is 0.33 Å. The implication is this: protein conformation is determined primarily by amino acid sequence, and modified by subsidiary factors, mainly crystal-packing forces, but also the conditions of solvent and temperature. The value 0.33 Å estimates the magnitude of the effects of these secondary factors. Because this value is much smaller than the structural differences observed in related but non-identical proteins, those structural differences must genuinely arise primarily from the changes in amino acid sequences.

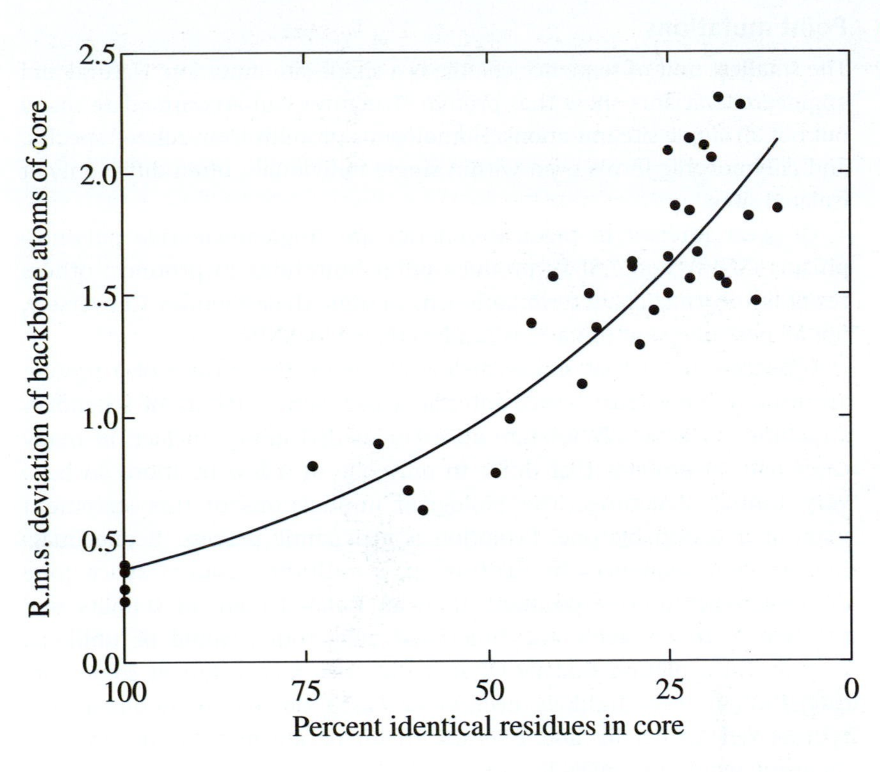

Fig. 5.6 Relationship between divergence of sequence and structure.

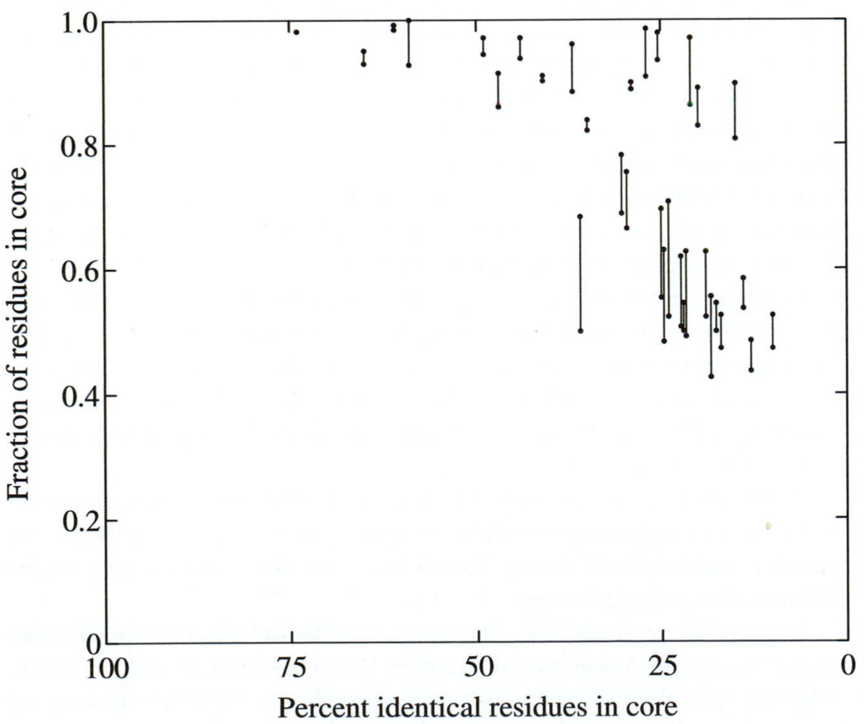

Fig. 5.7 Relationship between divergence of sequence and size of core.

Point mutations

The smallest unit of sequence change is a single-site mutation. Natural and engineered variants show that protein structures can accommodate many but not all single-site mutations. Homologous proteins from related species, and polymorphic forms even within single individuals, often differ only at isolated sites.

Of great interest in medical genetics are single-nucleotide polymorphisms (SNPs). Some SNPs produce point mutations in proteins; others result in incorrect chain termination, as in some thalassaemias. Conversely, not all possible point mutations can be created by SNPs.

Folklore tells us that single-site mutations on the surface of a protein are usually innocuous; buried interfaces between elements of secondary structure can generally tolerate an extra methyl group. In fact, in many cases natural proteins that differ in only one or a few positions do have very similar structures. The biological implications of this statement make it a reasonable one. Evolution is a dynamic process. If one imagines protein sequences as drifting in a multidimensional space (one dimension per residue position), then an isolated point of stability and function, with no stable and functional neighbours, would be unlikely: how would evolution find it? Or if it did arise as a result of some concatenation of very unlikely events—or based on a gene designed and synthesized artificially and released in an organism into the environment—it would be unstable.

Folklore notwithstanding, many single-site variants are not healthy proteins. Of course, if a mutation arising in nature caused total absence of function in an essential protein, it would be lethal and we should never see it. Many mutants that function poorly show up clinically, of which those of haemoglobin have been the most comprehensively and carefully studied. Most involve interior residues, especially those interacting with the haem group. However, the mutation that produces HbS, associated with sickle-cell anaemia, produces a correctly-folded protein which causes clinical problems because its deoxy form aggregates easily.

Of course, natural variants are a subset of all possible variants that have been subjected to natural selection. Artificial variants produced by site-directed mutagenesis extend our knowledge of what is tolerable. One technique is what we have called the *allumwandlung*: the replacement of a single residue by all 19 others, testing of functional properties, and the solution of the crystal structures, of all 20 proteins.

For example, M. Matsumura, W.J. Becktel and B.W. Matthews have carried out a study of replacements of Ile3 in bacteriophage T4 lysozyme. Figure 5.8 shows the effect of the change Ile3 $\rightarrow$ Tyr. There is in this case only a local deformation of the structures.

Is it possible to predict the structures of single-site mutants, knowing the parent structure? A preliminary question is whether it is possible to decide whether the mutation will cause only a small structural deformation, or more drastic changes. This cannot be done entirely reliably *a priori*, but often

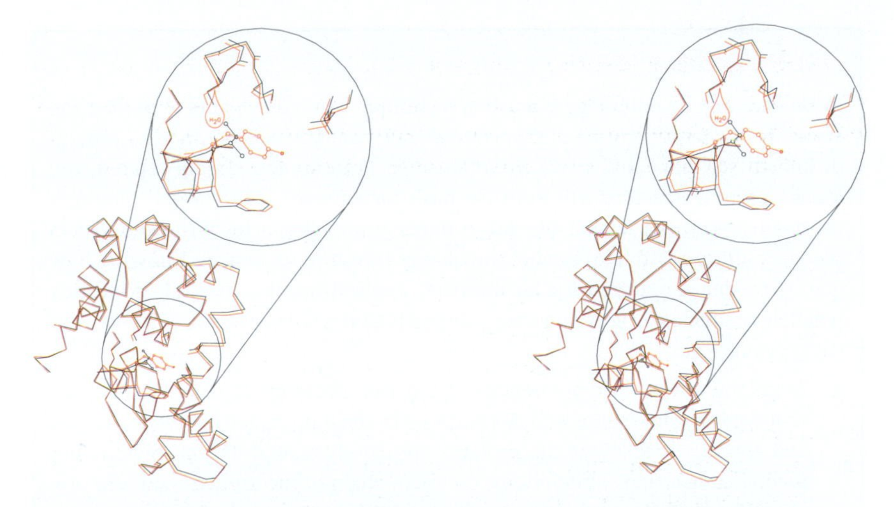

Fig. 5.8 Structural effect of single-site mutation Ile3 → Tyr in T4 lysozyme [3LZM] and [1L18]. The wild type (black) contains Ile, packing into the structure. The Tyr in the mutant (red) cannot fit into the Ile pocket, but it can reach the surface and points out into solution. This threatens to create a cavity in the space occupied by the Ile, but this is filled by a water molecule, which forms hydrogen bonds to the carbonyl oxygens of Tyr3 and Cys97 in the mutant, and also interacts with 6Met.

it is a justifiable assumption, especially if it is known experimentally that the mutant is functional.

The idea would be to use the parent structure, with one residue replaced, as an initial condition for energy minimization or molecular dynamics. But the energy models that underlie those systems are not sufficiently refined: if they are run on the native structures, with no mutations, they cause distortions which can be of the same order of magnitude or larger than the differences between native and mutant structures. If the calculations are artificially restrained—by tethering the atoms to their initial positions—one risks not allowing genuine changes to happen. It is therefore difficult to predict the small structural changes: one is trying to draw a fine line with a thick brush.

Closely related and distantly related proteins

Consider next the pair of sulphydryl proteases papain and actinidin (Figure 5.4). In this case the core contains 206 residues (out of a total of 212 in papain and 218 in actinidin). Thus the core includes almost all of both molecules. The residue identity is 49%, and the r.m.s. deviation of the core is 0.77 Å. Papain and actinidin are typical closely related proteins.

Finally, for pairs of proteins with below 20% residue identity, the disturbance to the structure is severe. Plastocyanin and azurin are typical examples (Figure 5.5). The core may be limited to no more than 40% of the sequence, with 60% of the chain adopting a quantitatively different fold. The reason for the onset of larger changes below 50% sequence identity of the core is that in closely related proteins the sequence changes primarily involve surface

Homology modelling

Model-building by homology is a useful technique when one wants to predict the structure of a target protein of known sequence, that is related to another protein of known sequence *and* structure. If the two proteins are closely related, the known protein structure will form the basis for a model of the target. Although the quality of the model will depend on the degree of similarity of the amino acid sequences, it is possible to predict the quality from the sequences. Knowing how good a model is required for an intended application, it is possible to decide whether a sufficiently accurate structure prediction will be available. The steps in the procedure are:

1. Align the amino acid sequences of the two proteins. It will generally be observed that insertions and deletions lie in the loop regions between helices and sheets, especially if the proteins are closely-related. For all regions not involving insertions or deletions, the main chain of the known structure provides the model for the corresponding residues of the target structure
2. Determine main chain segments to represent the regions subject to insertions and deletions, and insert them into the conserved main chain framework. Rules for modelling loops (Chapter 3) are useful at this stage.
3. Replace the sidechains of residues that have mutated. For residues that have not mutated, retain the sidechain conformation. For residues that have mutated, keep the same sidechain conformational angles as far as possible.
4. Examine the model—both by eye and by programs—to detect any serious collisions between atoms. Relieve these collisions, as far as possible, by manual manipulations.
5. Refine the model by limited energy-minimization. The role of this step is to fix up the exact geometrical relationships at places where regions of main chain have been joined together, and to allow the sidechains to wriggle around a bit to place themselves in comfortable positions. The effect is really only cosmetic—energy refinement will not fix serious errors in such a model.

In a sense, this procedure produces 'what you get for free' in that it defines the model of the protein of unknown structure by making minimal changes to its known relative. Unfortunately it is not easy to make substantial improvements. A rule of thumb (referring again to Figures 5.6 and 5.7) is that if the two sequences have at least 40–50% identical amino acids in an optimal alignment of their sequences, the procedure described will produce a model of sufficient accuracy to be useful for many applications. If the sequences are more distantly related, neither the procedure described nor any other currently available will produce a generally useful model of the target protein from the structure of its relative.

residues, which exert relatively little leverage on the structural framework. Distantly related proteins show mutations in buried residues, changes which have more serious effects on the structure of the core.

Application to homology modelling

Suppose that the sequence of an unknown protein is found to be homologous to that of a protein of known structure. Then, the results presented in the preceding section show that the core of the known protein can provide a model for the core of the unknown one, and that the minimum expected quality of this model can be predicted from the divergence of the sequences.

The level of 50% residue identity—or any closer relationship—provides a useful 'rule of thumb' for the utility of such a model. Thus, suppose one had determined the amino acid sequence of a new sulphydryl protease with 50% residue identity with actinidin, and built a model of this protein by taking the backbone of actinidin and replacing the mutated sidechains, retaining the sidechain conformation of the parent structure whenever possible. If the structure of the new protein was solved by X-ray crystallography, we should expect, on the basis of this relatively close relationship between the amino acid sequences, that:

1. A core of over 90% of the residues would retain a common fold. Fewer than 10% of the residues would be found in loops with radically different conformations.
2. The backbone atoms could be superposed with a r.m.s. deviation of about 1.0 Å or less. The binding site might show even less deviation, as evolution tends to alter binding sites relatively conservatively, provided that one is dealing with a family of proteins in which function is maintained.
3. The sidechains of 90% of the nonmutated residues, and of 50% of the mutated residues, would have similar conformations.

Such a model would give a reasonable picture of the unknown structure, at a level useful for analysis of its structural and functional properties. If the sequences were more closely related, the quality of the model would be correspondingly improved.

For proteins more distantly related than this threshold of 50% residue identity, the model-building procedure described could be less successful, and one should be discouraged from trying to build a full-blown three-dimensional model. The core might include no more than 40% of the sequence, with 60% of the chain adopting a qualitatively different fold. The geometry of the core residues would be altered more radically, and r.m.s. deviation would be much higher. The one consolation is that the active site might be relatively well conserved between the known and unknown structures, permitting some analysis of the effects on function of mutations within the active site itself.

Unfortunately, there is no better way of generating a model of an unknown protein from a very distant relative, nor is there any procedure—

such as those based on conformational energy calculations or molecular dynamics—that will reliably improve such a model. This is an active subject of current research.

Useful web sites

Homology modelling server:
http://www.expasy.ch/swissmod/SWISS-MODEL.html
Collections of aligned sequences:
http://croma.ebi.ac.uk/dali/fssp/fssp.html
http://www.sanger.ac.uk/Pfam *or* http://pfam.wustl.edu/
http://coot.embl-heidelberg.de/SMART/
Database of metabolic pathways:
http://www-c.mcs.anl.gov/home/compbio/PUMA/Production/puma_graphics.html

Recommended reading and references

Ayala, F.J. (1999). Molecular clock mirages. *BioEssays* **21**, 71–6.

Bowmaker, J.K. (1998). Evolution of colour vision in vertebrates. *Eye* **12**, 541–7.

Dean, A.M. (1998). The molecular anatomy of an ancient adaptive event. *Amer. Sci.* **86**, 26–37.

Hunt, D.M., Dulai, K.S., Cowing, J.A., Julliot, C., Mollon, J.D., Bowmaker, J.K., Li, W.H. and Hewett-Emmett, D. (1998). Molecular evolution of trichromacy in primates. *Vision Research* **38**, 3299–306.

Jeffery, C. J. (1999). Moonlighting proteins. *Trends Biochem. Sci.* **24**, 8–11.

Murzin, A.G. (1998). How far divergent evolution goes in proteins. *Curr. Opin. Struc. Biol.* **8**, 380–7.

Patthy, L. (1999). *Protein evolution.* Blackwell Science, Oxford.

Šali, A. (1995). Modelling mutations and homologous proteins. *Current Opinion Biotechnol.* **6**, 437/51.

Šali, A. (1995). Protein modeling by satisfaction of spatial restraints. *Molecular Medicine Today* **1**, 270–7.

Tramontano, A. (1998). Homology modeling with low sequence identity. *Methods* **14**, 293–300.

Wilks, H.M., Hart, K.W., Feeney, R., Dunn, C.R., Muirhead, H., Chia, W.N., Barstow, D.A., Atkinson, T., Clarke, A.R. and Holbrook, J.J. (1988). A specific, highly active malate dehydrogenase by redesign of a lactate dehydrogenase framework. *Science* **242**, 1541–4.

Wray, G.A. and Abouheif, E. (1998). When is homology not homology? *Curr. Opin. Gen. Devel.* **8**, 675–80.

Exercises, problems and weblems

Exercises

5.1. What uppercase letter corresponds to the 'core' of the following pairs of letters? (a) T and I. (b) F and L. (c) A and V. Note: (1) Here these letters do not stand for amino acids, but are to be considered as patterns themselves. (2) Don't forget that as patterns they may be rotated in *three* dimensions.

5.2. Why, before the availability of recombinant insulin, was pig insulin (rather than insulin from some other species) used for the clinical treatment of diabetes?

5.3. From the alignment of insulins from five species, what fraction of residues is conserved in all five sequences in the (a) A chains, (b) B chains, (c) C-peptides.

5.4. From the aligned table of globin sequences, choose five positions that are likely to contain residues that are on the surface of monomeric globins, and five positions that are likely to be buried in the interior of monomeric globins.

5.5. On photocopies of Figures 5.4 and 5.5, showing the comparisons of actinidin and papain and plastocyanin and azurin, indicate the regions of conformational differences.

Problems

5.1. In the following table of aligned sequences, (a) What are the most similar and most distant members of the family? (b) What secondary structural elements would you expect? Explain your reasoning. (c) Suppose that an experimental structure is known only for the first sequence. For which others would you expect to be able to build a model with an overall r.m.s. deviation of ≤ 1.0 Å for 90% or more of the residues? (d) On a photocopy of the alignment table highlight all positively-charged residues. Then consider the conservation pattern of charged residues near the C-terminus of these sequences. Propose a reasonable guess about what kind of molecule these domains interact with.

```
TYLWEFLLKLLQDR.EYCPRFIKWTNREKGVFKLV..DSKAVSRLWGMHKN.KPD
VQLWQFLLEILTD..CEHTDVIEWVG.TEGEFKLT..DPDRVARLWGEKKN.KPA
IQLWQFLLELLTD..KDARDCISWVG.DEGEFKLN..QPELVAQKWGQRKN.KPT
IQLWQFLLELLSD..SSNSSCITWEG.TNGEFKMT..DPDEVARRWGERKS.KPN
IQLWQFLLELLTD..KSCQSFISWTG.DGWEFKLS..DPDEVARRWGKRKN.KPK
IQLWQFLLELLQD..GARSSCIRWTG.NSREFQLC..DPKEVARLWGERKR.KPG
IQLWHFILELLQK..EEFRHVIAWQQGEYGEFVIK..DPDEVARLWGRRKC.KPQ
VTLWQFLLQLLRE..QGNGHIISWTSRDGGEFKLV..DAEEVARLWGLRKN.KTN
ITLWQFLLHLLLD..QKHEHLICWTS.NDGEFKLL..KAEEVAKLWGLRKN.KTN
LQLWQFLVALLDD..PTNAHFIAWTG.RGMEFKLI..EPEEVARLWGIQKN.RPA
IHLWQFLKELLASP.QVNGTAIRWIDRSKGIFKIE..DSVRVAKLWGRRKN.RPA
RLLWDFLQQLLNDRNQKYSDLIAWKCRDTGVFKIV..DPAGLAKLWGIQKN.HLS
RLLWDYVYQLLSD..SRYENFIRWEDKESKIFRIV..DPNGLARLWGNHKN.RTN
IRLYQFLLDLLRS..GDMKDSIWWVDKDKGTFQFSSKHKEALAHRWGIQKGNRKK
LRLYQFLLGLLTR..GDMRECVWWVEPGAGVFQFSSKHKELLARRWGQQKGNRKR

  L fl  1L         i W      F        a  WG K
```

In the line below the table: uppercase letters signify residues conserved in all sequences; lowercase letters signify residues conserved in all but one of the sequences.

5.2. Humans achieve colour vision through the action of three visual pigments, with absorption maxima in the violet, green, and yellow-green regions of the spectrum. These pigments are homologous proteins, called opsins, conjugated to a common ligand, 11-*cis*-retinal. They are distinguished according to their spectral properties as S, M and L, for Short, Medium and Long wavelength absorbance maxima.

In humans and other old-world primates, S opsin is encoded by an autosomal gene, and M and L by genes on the X-chromosome. This is why certain kinds of colour-blindness in humans are sex-linked traits. In contrast, most species of new-world monkeys have only two loci, an autosomal gene for S opsin, and a polymorphic X-linked gene for M and L. In these species, females heterozygous at the opsin locus on the X chromosome can express three opsins (including the autosomal S), but males only two. As a result, monkeys of different sexes have uniformly different colour vision: female heterozygotes have full colour vision, but all males are partially colour blind.

Because the chromophore is common to all the proteins, the spectral shifts must arise from amino acid sequence differences. The difference between M and L is the result of substitutions at three sites, residues 180, 277 and 285 (see following Table).

Species	L opsin residue number			M opsin residue number		
	180	277	285	180	277	285
Old-world species:						
Human	S	Y	T	A	F	A
Chimpanzee	S	Y	T	A	F	A
Gorilla	S	Y	T	A	F	A
Diana monkey	S	Y	T	A	F	A
Macaque	S	Y	T	A	F	A
Talapoin monkey	S	Y	T	A	F	A
New-world species:						
Capuchin monkey	S	Y	T	A	F	A
Marmoset	S	Y	T	A	Y	A
Howler monkey	S	Y	T	A	F	A

These results show that spectral tuning is the result, in most cases, of the same set of mutations. They suggest the hypothesis (1) that the divergence between L and M opsins preceded the divergence of old-world and new-world primate species. This would require there to have been gene duplication and translocation of the gene on the X chromosome of old-world primates. The alternative hypothesis (2) is that the proteins are showing convergent evolution, under the same selective pressure.

(a) For each hypothesis, sketch the appearance of the implied evolutionary tree relating the following four sets of proteins: old-world primate L opsins, old-world primate M opsins, new-world primate L opsins, new-world primate M opsins.

(b) According to each hypothesis, which would be expected to resemble old-world L opsin more closely: the old-world M opsin or the new-world L opsin?

(c) To choose between these alternatives, consider the following alignment table of partial sequences of opsins. (The asterisks mark the three sites at which the L/M divergence is the result of selective pressure on spectral properties. A '.' indicates that the residue at that position is the same as that of the human L sequence. Humans are old-world and Howler monkeys new-world.)

```
Exon 3:
                                                  *
Human L       AIISWERWLVVCKPFGNVRFDAKLAIVGIAFSWIWSAVWTAPPI
Human M       ......M.....................I......A........
Howler L      ......R.....................V......S........
Howler M      ......R.....................V......A........

Exon 4:

Human L       GPDVFSGSSYPGVQSYMIVLMVTCCIIPLAIIMLCYLQVWLAIRA
Human M       .........................T...S..V............
Howler L      .....................I...FL..G..E............
Howler M      ....................VI...IL..S..V............

Exon 5:

                           *       *
Human L       KEVTRMVVVMIFAYCVCWGPYTFFACFAAANPGYAFHPLMAALPAYFAKS
Human M       ..........VL.F.F.....A............P...............
Howler L      ...........M.Y.......T............................
Howler M      ...........I.F.......A............................
```

Excluding the three sites at which selection was operating, determine the number of differences between each pair of partial sequences. Which hypothesis is supported by the results?

(d) What additional data and analysis would you want to adduce to confirm or refute your conclusion?

5.3. Figures 5.9b and 5.9e show the superposition of bovine liver phosphotyrosine protein phosphatase [1PHR] and one of two identical chains of *E. coli* IIb cellobiose [1IIB]. (Front and back views.) Figures 5.9a and 5.9d show the phosphatase and Figures 5.9c and 5.9f show the cellobiose structure individually. (a) On a photocopy of these figures, indicate with highlighter on the copies of parts a, c, d and f the core of

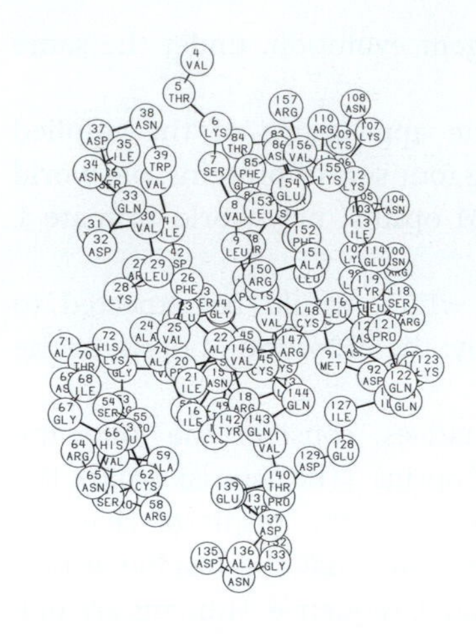

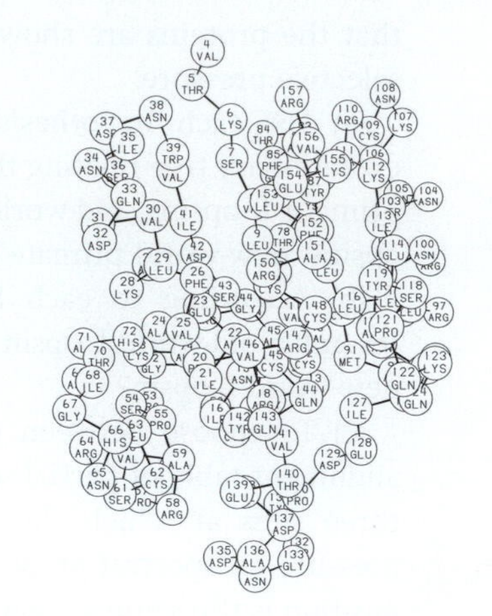

Fig. 5.9a

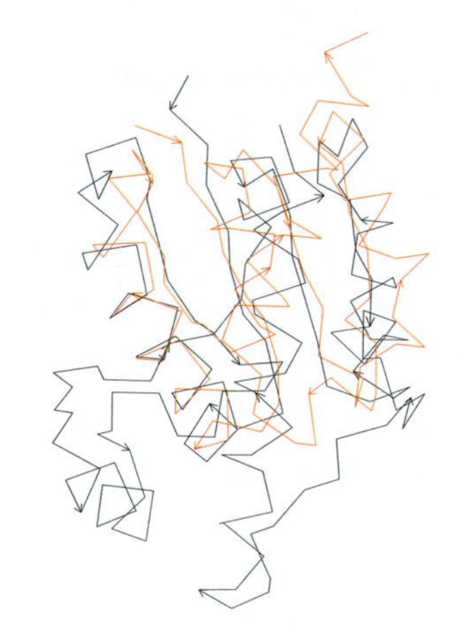

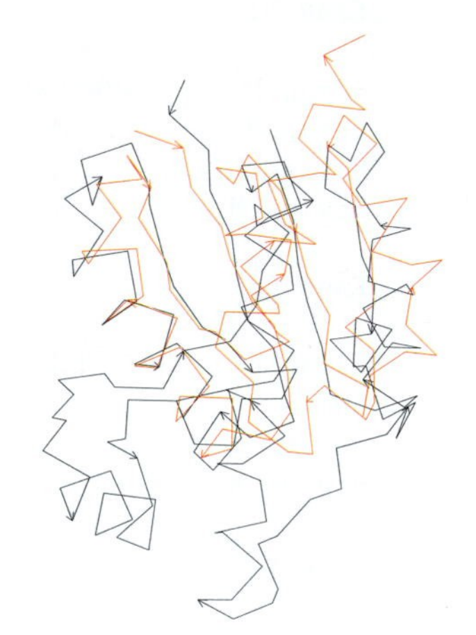

Fig. 5.9b

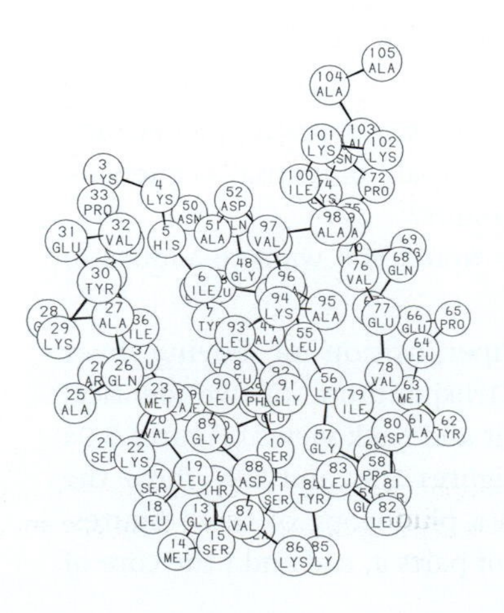

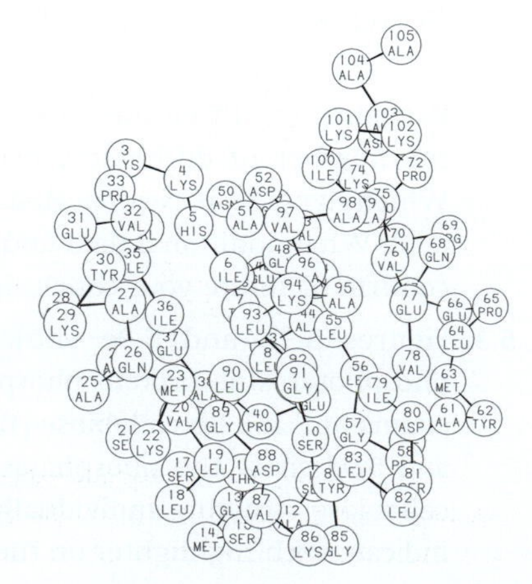

Fig. 5.9c

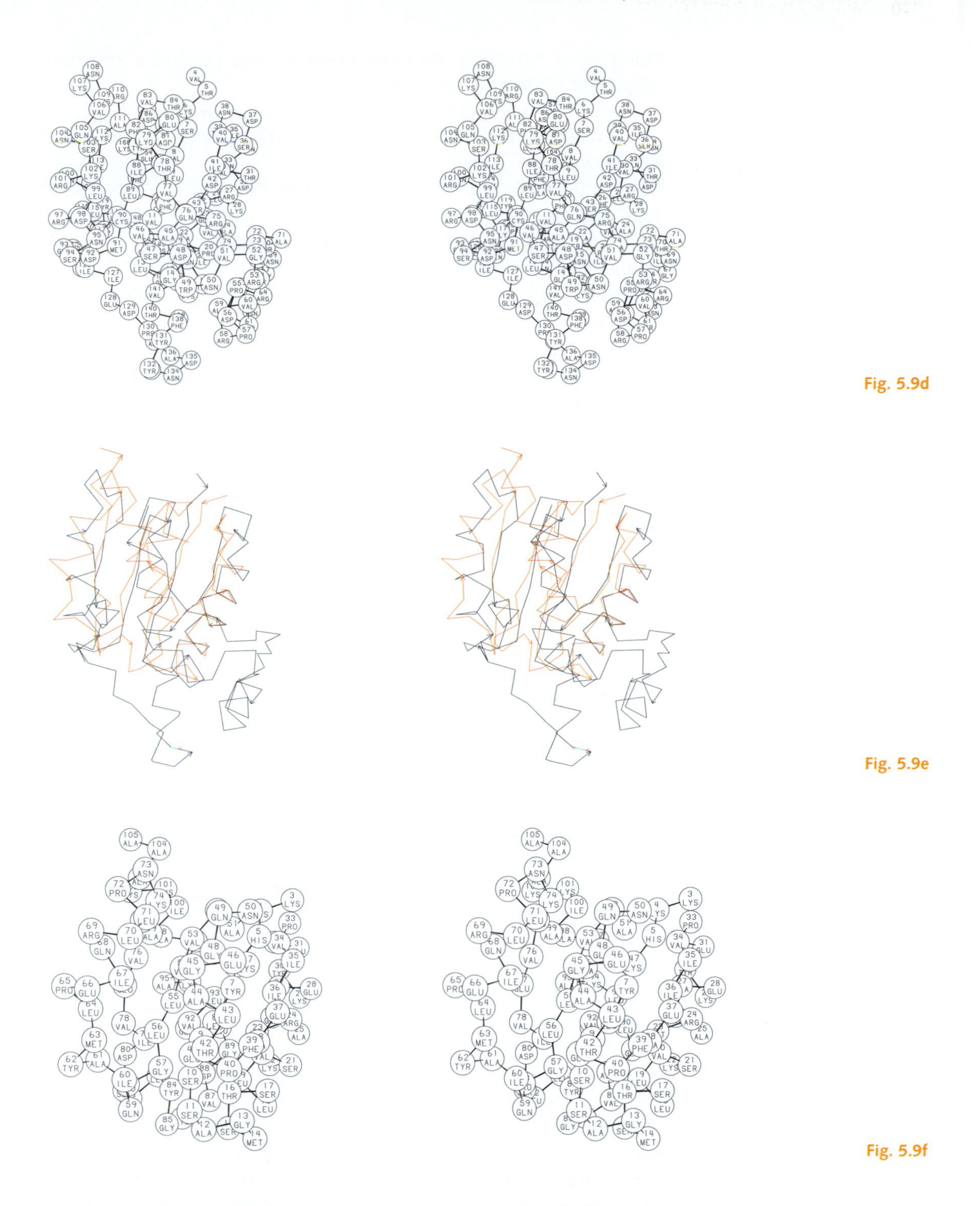

Fig. 5.9 (a,d) Bovine liver phosphotyrosine protein phosphatase [1PHR]. (b,e) Superposition of phosphatase and IIb cellobiose. (c,f) The A chain of *E. coli* IIb cellobiose [1IIB]. Parts a, b, and c differ from d, e and f by a 180° rotation, providing front and back views.

this pair of structures. (b) From these figures, prepare a structural alignment of the two sequences:

Bovine liver phosphotyrosine protein phosphatase [1PHR]

4 VTKSVLFVCL GNICRSPIAE AVFRKLVTDQ NISDNWVIDS GAVSDWNVGR SPDPRAVSCL 63

64 RNHGINTAHK ARQVTKEDFV TFDYILCMDE SNLRDLNRKS NQVKNCRAKI ELLGSYDPQK 123

124 QLIIEDPYYG NDADFETVYQ QCVRCCRAFL EKVR 157

The A chain of *E. coli* IIb cellobiose [1IIB].

3 KKHIYLFSSA GMSTSLLVSK MRAQAEKYEV PVIIEAFPETL AGEKGQNADV VLLGPQIAY 62

63 MLPEIQRLLP NKPVEVIDSL LYGKVDGLGV LKAAVAAIKKA AA 104

(These sequences are for convenience only; in fact it would be possible to transcribe the sequences from the figures.)

(c) Estimate the r.m.s. deviation of the mainchain atoms of the core.

5.4. Isozymes are enzymes that occur in related but not identical forms. In higher organisms they may appear in different tissues and/or different stages of development. Hexokinases and glucokinase are enzymes that convert glucose to glucose-6-phosphate. In humans, three forms of hexokinase occur in various organs, including liver, brain and muscle; glucokinase appears only in liver. Their enzymatic properties differ (see following table).

	M_r	K_M for glucose	Inhibited by glucose-6-P?
Hexokinases	100 000	20–170 μM	Yes
Glucokinase	50 000	6 mM	No

The region of normal plasma glucose concentrations, 5 mM, is below K_M for glucokinase but not for hexokinase. Therefore in this concentration range glucokinase activity increases with increasing glucose concentration. This, together with the absence of product inhibition, makes it suitable to 'soak up' excess plasma glucose for conversion to glycogen. In contrast, the entry of glucose into muscle cells is insulin-regulated, and intracellular glucose concentrations are about 0.1 mM. Hexokinases in muscle catalyse the first step of glycolysis for energy release.

(a) Do the relative sizes of glucokinase and the other hexokinases suggest any hypothesis about the relationship of their structures?

(b) What evidence for this hypothesis would you seek if you were given the amino acid sequences? What might you expect to find?

(c) Assuming that the amino acid sequences supported your

hypothesis, what additional support might be available from the DNA sequences of the regions containing the genes for these proteins?

(d) What would you expect the molecular weight of prokaryotic homologues to be?

(e) What would you expect to be the relationship between the sequences of mammalian hexokinases and glucokinases, and the sequences of yeast and prokaryotic homologues?

Weblems

5.1. Find the sequences of globins from an insect and from a bacterium. Submit these two sequences to a web server that aligns sequences pairwise. Compare the results with the alignment appearing in the table in this chapter.

5.2. Find the alignment of the globins in the Pfam data base. What amino acids can appear at the sites that correspond to positions 3, 32 and 119 in Table 5.1?

5.3. It is possible to form the intact insulin molecule from isolated A and B chains. Before the availability of recombinant insulin, would it have been possible to create human insulin by isolating and combining the A chain from one animal species with the B chain from another?

5.4. Carry out the research indicated in problem 5.2(d). What did you find? Was it what you expected? Report on any surprises.

5.5. A student too impatient to work out the answer to Problem 5.3 by studying the figures decided to find the answer by using tools available on the web. How would he or she go about this? Would it be possible to prevent students from taking this route to solving the problem by replacing the correct amino acid sequences in Figures 5.9 by random amino acid sequences? Could a truly determined student still find the answer via the web? Justify your answer.

CHAPTER 6

Evolution in selected protein families

Evolution has produced families of proteins in which the amino acid sequences have diverged widely, but the biological functions and three-dimensional structures have remained much the same. How can amino acid sequences that are very different form proteins that are very similar in three-dimensional structure? What is the mechanism by which proteins adapt to mutations during the course of their evolution? What is the role of the possible pathways of evolutionary change—as opposed to the requirements of the structure itself—in limiting the changes in structural features of homologous proteins?

In this chapter we shall examine several families of proteins in more detail. These include a family of all α-helical proteins, the globins; a family of all β-sheet proteins, the chymotrypsin-like serine proteinases; and a family of α/β proteins, the NAD-binding domains of dehydrogenases.

Evolution of the globins

Globins are a family of proteins found in prokaryotes, animals and plants, with the primary function of binding oxygen. The globin structure contains an assembly of helices surrounding a haem group. Crystal structures of globins from widely divergent species show how these molecules have changed conformation during evolution, and how their freedom to do so has been constrained by requirements of stability and function.

What remains conserved and what varies during the course of globin evolution? We had a glimpse of what goes on at the sequence level in the previous chapter. Now we look at the structures.

The secondary structure of sperm whale myoglobin, the first protein structure to be solved by X-ray crystallography, consists of 9 helices, labelled A, B, C, D, E, F, F′, G and H (Figure 6.1a and Exercise 6.2). The C helix is a 3_{10} helix. The helices, and their basic topological arrangement to create the haem pocket, are conserved in other globin structures, except for the D helix, which has unwound in some members of the family. For example,

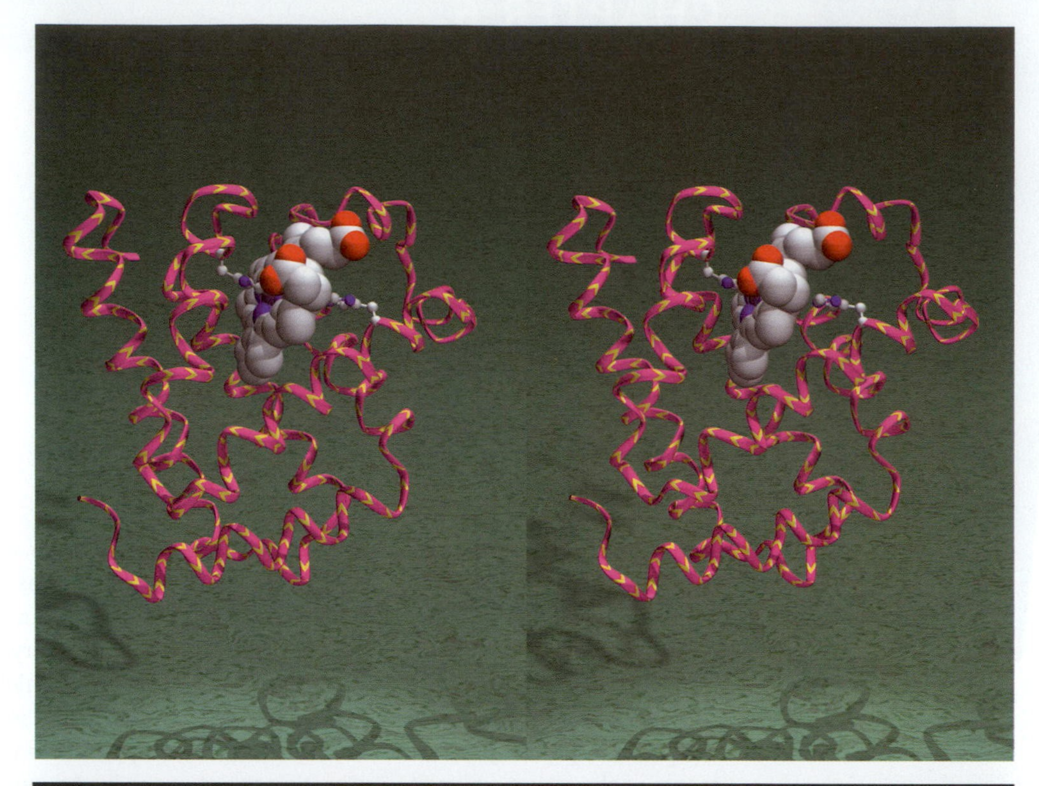

Fig. 6.1a

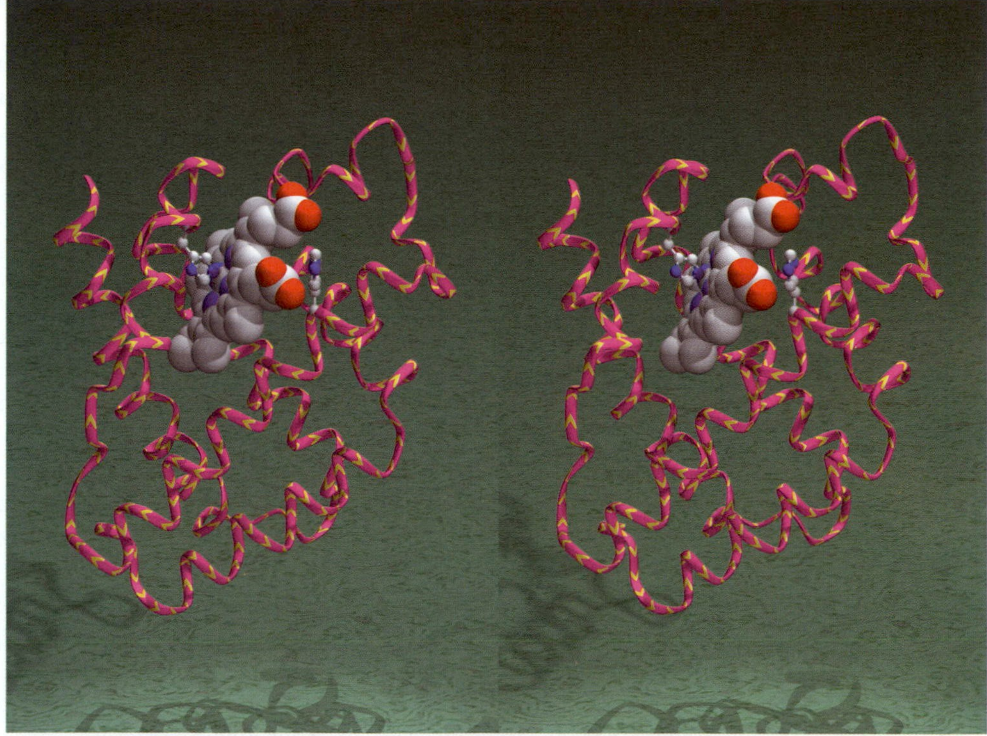

Fig. 6.1b

Fig. 6.1 (a) Sperm whale myoglobin [1MBO]. (b) Lupin leghaemoglobin [2LH7].

the β chain of human haemoglobin contains a D helix but the α chain does not. In mammalian globins the F helix has broken into two segments; in non-mammalian globins it is straight. (The F-helix is a particularly sensitive region of the structure because it contains the proximal histidine that binds the haem iron.)

Although the basic complement of 8 or 9 helices appears consistently, the lengths of the helices vary. Comparing sperm whale myoglobin with a plant globin, lupin leghaemoglobin (Figure 6.1b), the E helix (diagonal, in the front) is much longer in the plant globin than in myoglobin. The loops between helices are quite variable in sequence and structure.

The native states of globins are stabilized by the packing of about 60 residues at interfaces between helices. Approximately half of them are buried—in the sense of inaccessible to water molecules—and the other half are at the peripheries of interfaces.

Most of the helix–helix contacts follow the 'ridges into grooves' model of the structure of helix interfaces (see Chap 4). For instance, the F–H interface in lupin leghaemoglobin is a standard '$i\pm4/i\pm4$' packing (Figure 6.2). A ridge created by residues four apart in the sequence of the F helix pack against a groove created by sets of residues four apart in the H helix. However, the globins are rich in exceptions to the standard types of helix interface structures. The B–E and G–H contacts contain interfaces with unusual structures in which two ridges cross at a notch produced by a small sidechain flanked by two larger ones.

The tertiary structure, the set of interhelix contacts, is largely conserved both at the level of interactions between helices, and even at the level of interactions between residues. Five helix–helix contacts, between the A–H, B–E, B–G, F–H and G–H helices, are common to different globins. Other contacts, between the A–E, C–G, B–D and E–H helices, occur in some but not all globin structures. Homologous interfaces between pairs of helices in different globins are conserved in structure, in that they retain the basic type of ridge–groove packing. For example, the B–G helix contact is a '$i\pm4/i\pm4$' packing in different globins. Figure 6.3 compares the B–G interfaces in the α subunit of horse methaemoglobin and the larval insect *Chironomus* globin erythrocruorin. The B–E contact is a crossed-ridge structure in different globins.

Despite the retention of the structure of the interface, there are significant shifts and rotations between pairs of packed helices, These changes in structure are a consequence of the dense packing of the helix interfaces. When a mutation of a residue in an interface changes the volume of the sidechain, the helices must move to adjust the packing. The changes in relative geometry of packed helices can in some cases be substantial. The spatial disposition of the B and E helices in globins vary by shifts of up to 7 Å and rotations of up to 30°. Figure 6.4 shows the distribution of the interaxial distances and angles for the different helix interfaces in nine globins.

Despite the large shifts and rotations, a subtle feature of the interface structure is preserved. *The pattern of residue–residue contacts at interfaces tends to*

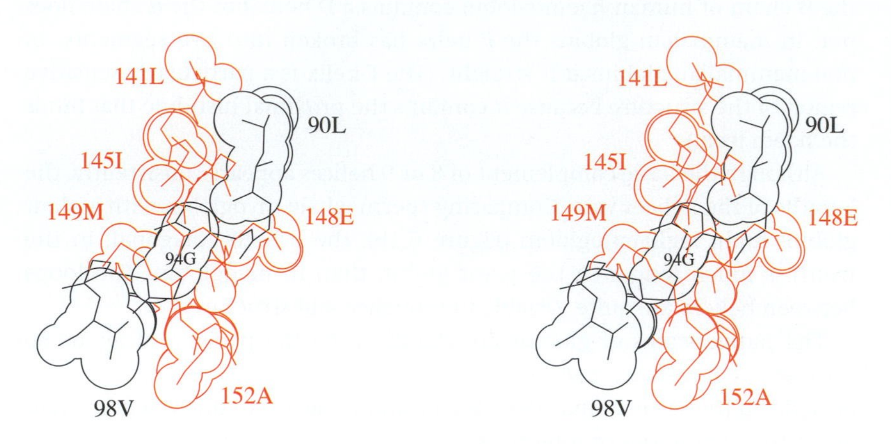

Fig. 6.2 The F–H helix contact in lupin leghaemoglobin [2LH7].

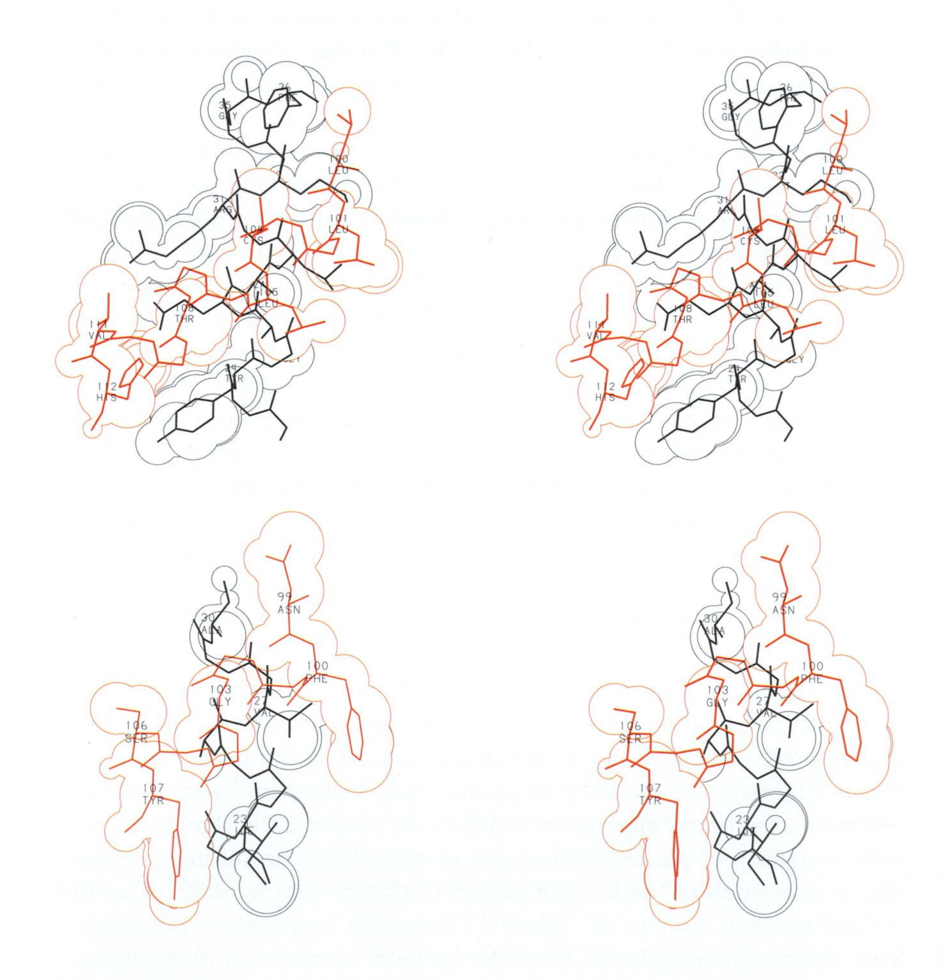

Fig. 6.3a

Fig. 6.3b

Fig. 6.3 The B–G helix contact in (a) horse methaemoglobin, α-chain [1EHB], (b) *Chironomus* erythrocruorin [1ECD].

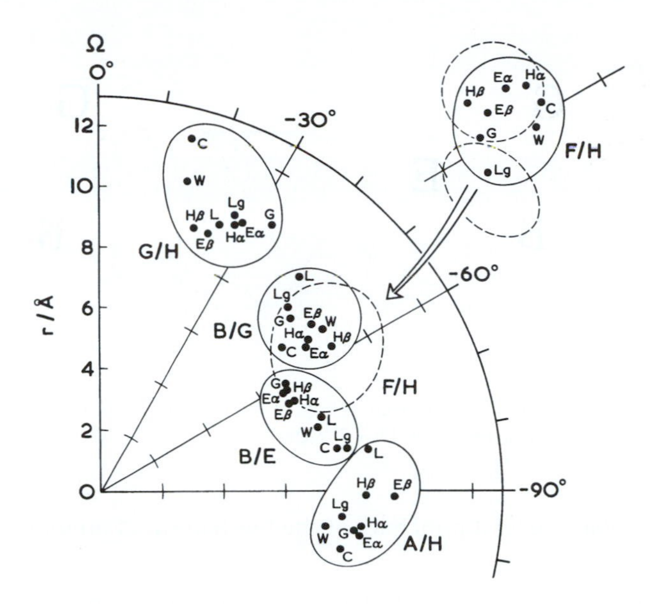

Fig. 6.4 Distribution of interaxial distances and angles of packed helices in nine globin structures.

remain, even if the residues that make the contact mutate. That is, if a pair of residues is in contact in one structure, the homologous residues in a related protein are likely to make a contact also. This conservation of the reticulation of the residues arises from the requirement for maintaining well-packed interfaces during evolution, as a condition for retaining stability. Point mutations in a helix–helix interface can be accommodated in the packing by a wriggling around of sidechains, and by the shifts and rotations of the helices with respect to each other. But if the reticulation were not conserved—if the interface were to jump to a completely new set of residue-residue contacts—the complementarity of the surfaces would be entirely destroyed, and the stability of the contact lost. This explains why insertions and deletions of amino acids do not appear in packed helices. Insertion of an amino acid would turn at least part of the interface by 100°, destroying structural complementarity.

How do the globins reconcile the large changes in individual helix-helix contacts to preserve function? The constraints on the structure appear to apply at the global level rather than to the individual helix–helix contacts. Figure 6.5 shows a superposition of the B, E and G helices from human haemoglobin, α-chain and lupin leghaemoglobin. The B and E helices and B and G helices are in contact, but the E and G helices are not. The figure shows that there have been large rotations of the helices at the B–E and B–G contacts. But these rotations have been coupled to leave the E and G helices in the same relative spatial position, in order to form the haem pocket.

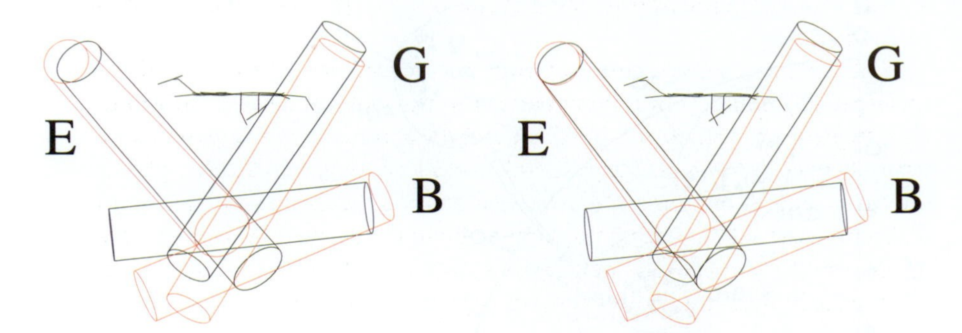

Fig. 6.5 The B, E and G helices surrounding the haem group, in human haemoglobin, α chain [1HHO] (black), and lupin leghaemoglobin [2LH7] (red).

The globins reveal some general principles of the mechanism of protein evolution:

1. Most of the major secondary structural elements—in the case of the globins, the helices—remain intact, although they may change their lengths.
2. Most of the major tertiary structural interactions—the helix–helix contacts—are preserved, although the relative positions and orientations of the helices in contact may change by large amounts.
3. These large changes must be coupled to preserve function. In the case of the globins, the haem pocket, and the ability to bind oxygen, must be retained.
4. The requirement for stability during the course of evolution constrains the changes in the amino acid sequence. The tertiary structural interactions—the packing at the helix–helix interfaces—can accommodate extensive sequence changes provided that they are made one by one, but a completely novel well-packed interface cannot form in a single step. In consequence:
   (a) insertions and deletions in helices are confined to their ends;
   (b) the pattern of residue contacts at interfaces between helices and sheets tends to be conserved.
5. These considerations apply to natural proteins, and not necessarily to engineered proteins. To any two natural homologous proteins, there must be continuous evolutionary paths from their common ancestor during which all intermediates were stable and functional. This constraint does not apply to engineered proteins.

Phycocyanin and the globins

Phycocyanins are the major constituents of the phycobilisome, the supramolecular light-harvesting complexes of cyanobacteria and red algae. They bind a chromophore which is an open-chain tetrapyrrole structure similar to a bile pigment (the haem group is a tetrapyrrole structure *closed* into a macro-ring.)

When the first phycocyanin structure was solved, it showed an entirely unsuspected similarity in folding pattern to the globins, despite the difference in prosthetic group and function, and the absence of any apparent similarity between globin and phycocyanin sequences (Figure 6.6). (The phycocyanin structure contains an N-terminal extension containing two helices in addition to the set displaying the same structural relationships that the globin helices do. These 'extra' helices are involved in oligomerization to form the phycobilisome, a process that phycocyanins do not share with the globins. Ignore them.)

What is the origin of the similarity of the folds? Do globins and phycocyanins share a common ancestor, or did their very different functions somehow require the independent evolution of the same folding pattern? It is difficult to answer these questions because:

1. the sequences are so different that it is not easy to align them with any confidence;
2. some of the structural differences between globins and phycocyanins are much larger than the differences between distantly-related globin structures.

To try to distinguish between true evolutionary relationship or convergence, we can only observe the similarities of the structures and try to interpret them. The basic dilemma is the distinction between points of similarity —which may be consistent with either homology or convergence—and criteria for true relationship. W.E. Le Gros Clark, the palaeontologist, wrote:

> While it may be broadly accepted that, as a general proposition, degrees of genetic relationship can be assessed by noting degrees of resemblance in anatomical details, it needs to be emphasized that morphological characters vary considerably in their significance for this assessment.

For us, the implication is that we must give different weight to different sorts of structural similarities, disregarding features common to many classes of proteins, and emphasizing the unusual ones. We must look for structural similarities that are not specifically required by structure or function. That is the principle. The problem with putting it into practice is that it is impossible to be *sure* that any feature of a protein structure is not required for structure or function.

The C-helices in globins and phycocyanins are both of the rare 3_{10} structure; the rest are all α-helices. The C-helix does not appear to play a role in the function of monomeric globins. (It is important in the allosteric change in haemoglobin, but that was surely a much later development.) If, as seems likely, a sequence compatible with an α-helix in this region could produce a viable globin or phycocyanin, then the fact that the C-helices in both globins and phycocyanins are of the unusual 3_{10} type is a structural similarity not specifically required by structure or function.

What about the pattern of interactions of residues at the helix interfaces?

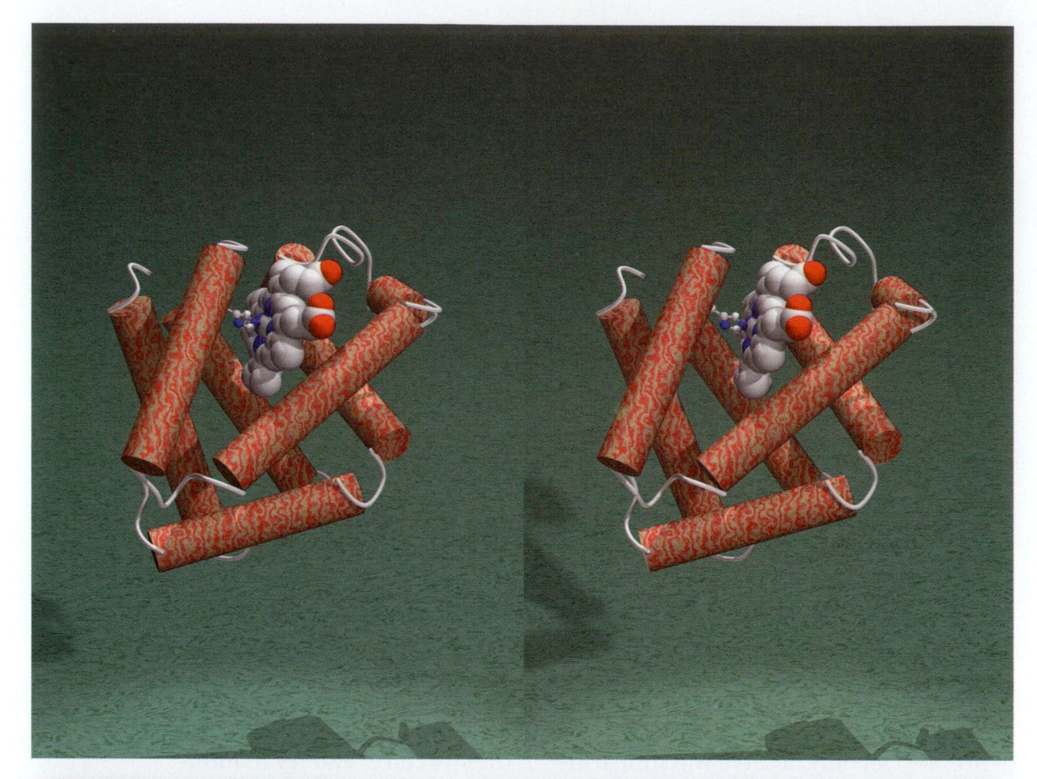

Fig. 6.6a

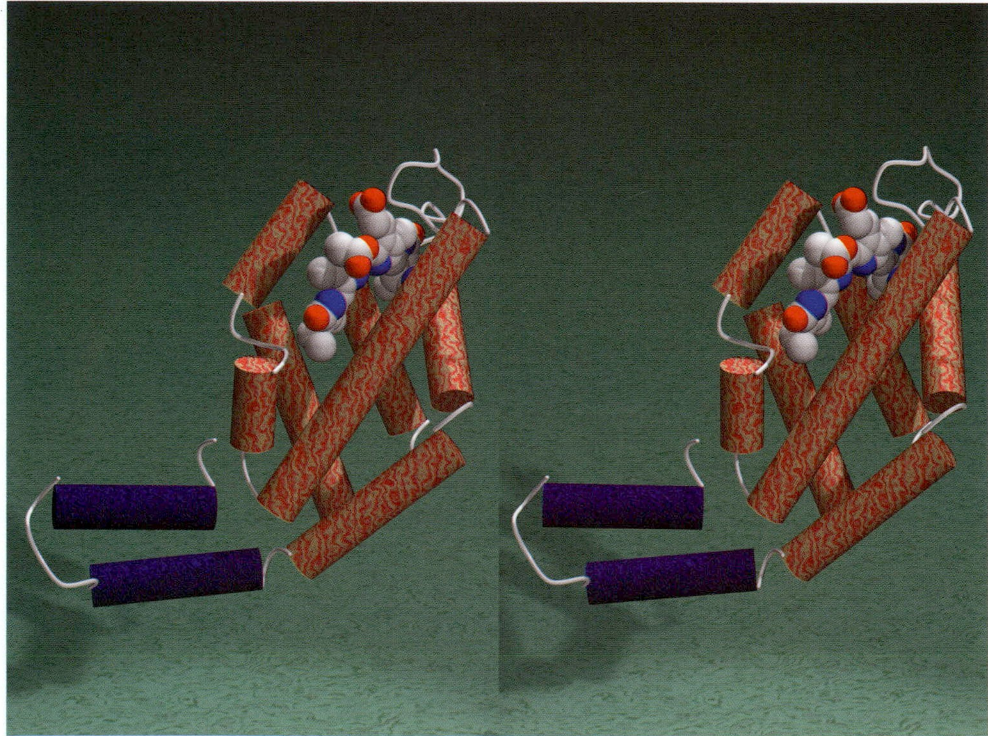

Fig. 6.6b

Fig. 6.6 (a) Sea hare (*Aplysia limacina*) myoglobin [1MBA]. (b) *Mastigocladus laminosus* phycocyanin, α-chain.

The observation of common types of packing, for example, the '$i\pm4/i\pm4$' structure, in interfaces between the same pairs of helices in globins and phycocyanins, is not evidence for evolutionary relationship, as this structure appears in many unrelated proteins. More significant are the interfaces with unusual structures: the B–E and G–H packings. In the globins these interfaces have the unusual 'crossed-ridge' structure. It is very interesting to see a similar crossed-ridge packing in the B–E interface in phycocyanin (Figure 6.7). A similar crossed-ridge structure is also observed in the G–H contact in phycocyanins. But can we argue that such an interface is not required by the structure? In phycocyanins the interaxial angles in the B/E interfaces are all in the region of −50° to −60° that would be expected from a normal '$i\pm4/i\pm4$' packing. Therefore the phycocyanin structure cannot require the special crossed-ridge structure to achieve an unusual inter-helix-axis angle. We can never really settle the question rigorously, because evolutionary events, which occurred in the distant past, are not directly observable. However, if we had to bet, the observation that the structural similarities between globins and phycocyanins extend down to the minute structural details, apparently not required by fold or function, suggests that the structures are homologous rather than convergent.

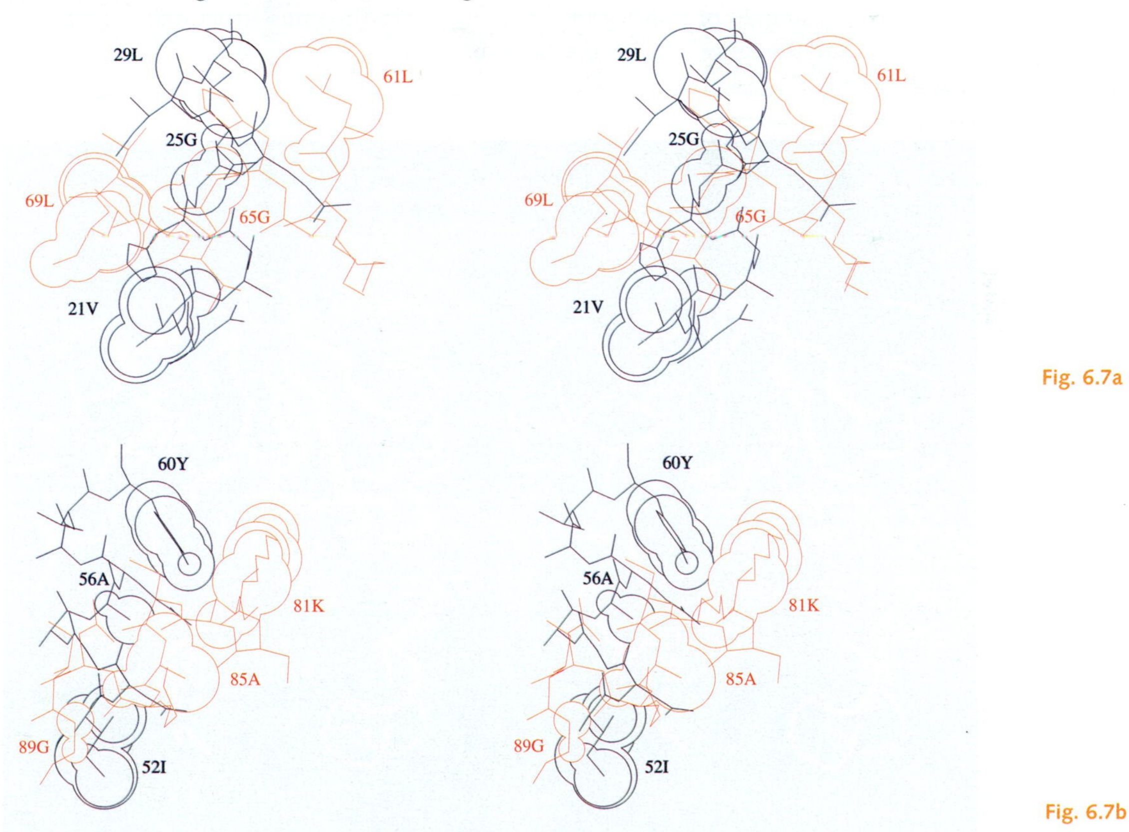

Fig. 6.7a

Fig. 6.7b

Fig. 6.7 B–E contact in (a) sperm whale myoglobin. (b) *Mastigocladus laminosus* phycocyanin, α-chain.

Evolution of serine proteinases of the chymotrypsin family

The chymotrypsin-like serine proteinases are a family of enzymes appearing in animals, plants, bacteria and viruses (Figure 6.8). Like other protein families, they diverged from a common ancestral protein in parallel with the organisms in which they are expressed, but not without exceptions: *Streptomyces griseus* trypsin is similar to mammalian enzymes and probably represents a gene transfer from a higher organism, prompting B.S. Hartley to speculate that 'the bacterium may have been infected by a cow'.

Mammalian serine proteinases participate in numerous physiological processes, including digestion, blood clotting, fertilization, and complement activation in the immune response. In several disease states, including emphysema, tumour metastasis and arthritis, the levels of the proteinases or inhibitors are elevated or out of balance. In *Drosophila*, serine proteinases are implicated in developmental control.

The mechanism of action involves the interaction of the substrate with a catalytic triad: Asp102–His57–Ser195 (Figure 6.9). This combination appears in many proteins, with very different folds (see Box on p. 207). These are examples of convergent evolution, not to the same structural topology, but to the same catalytic mechanism.

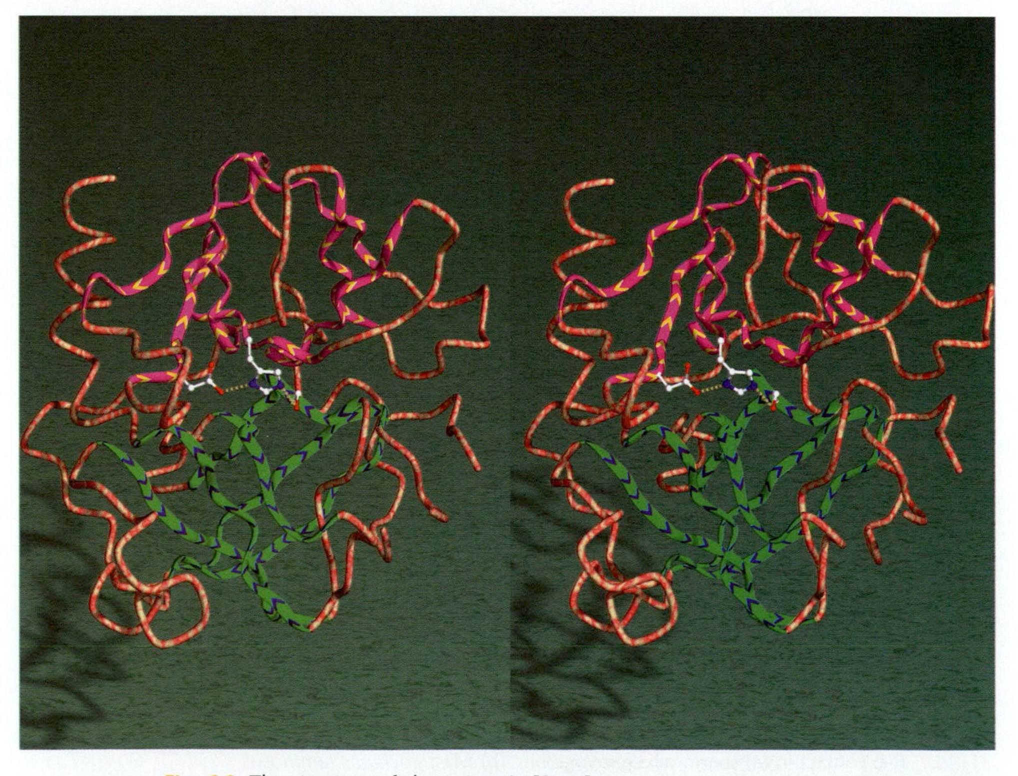

Fig. 6.8 The structure of chymotrypsin [8GCH].

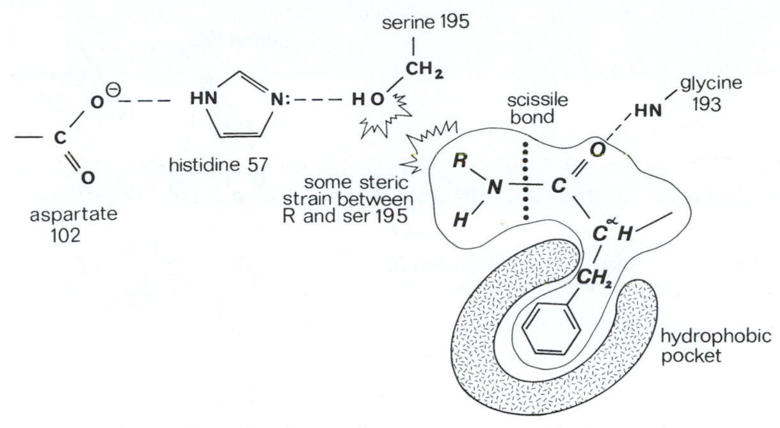

a) Formation of Michaelis complex

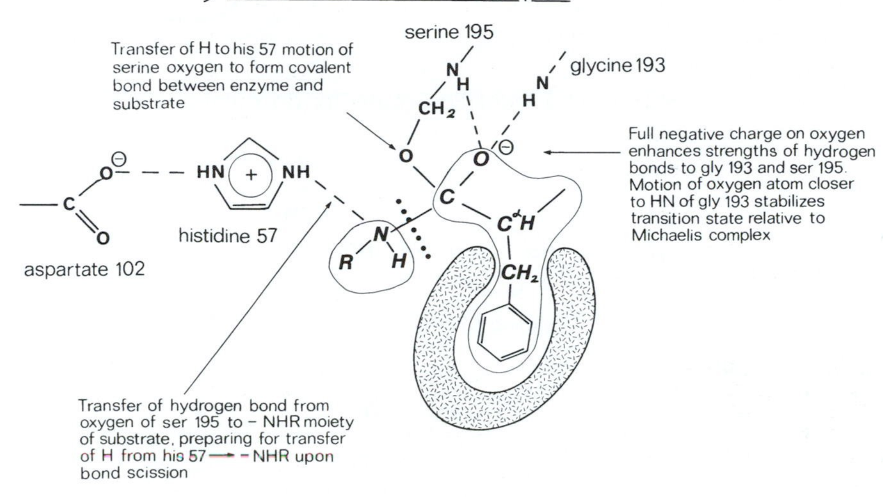

b) Transition state

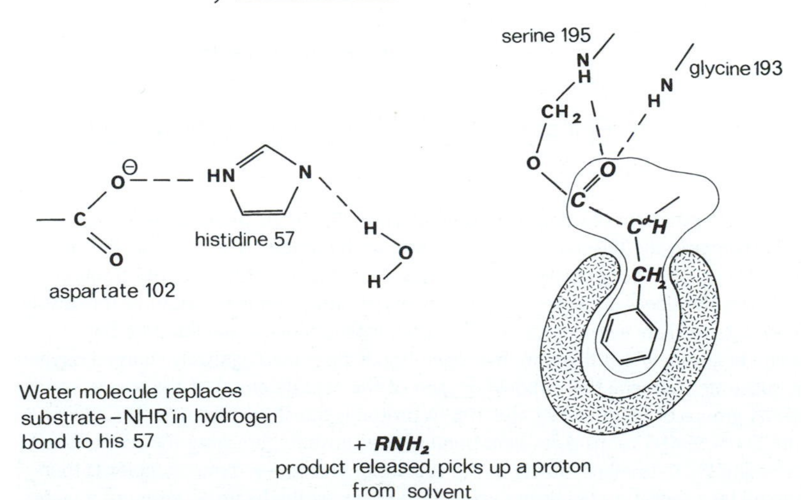

c) Acyl-enzyme intermediate

Fig. 6.9a–c

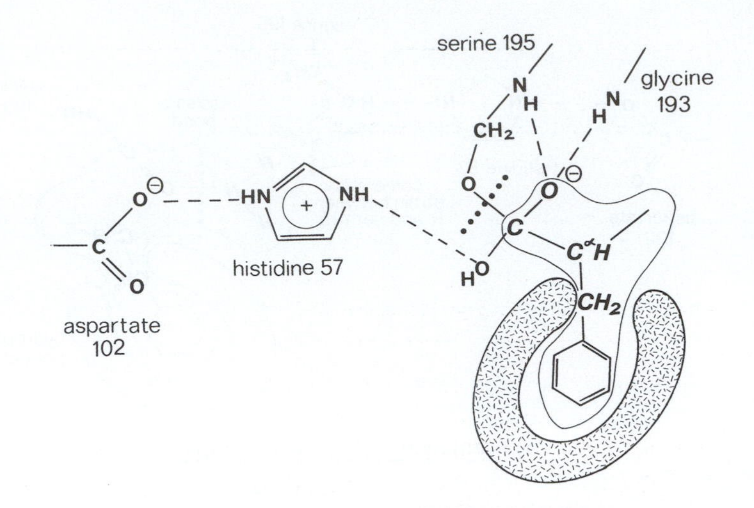

d) Transition state for deacylation

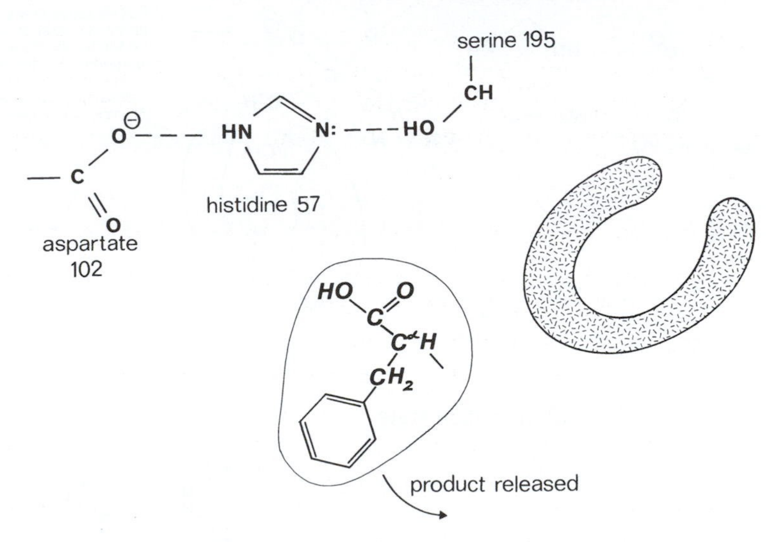

Fig. 6.9d–e

e) Release of product and return to initial state.

Fig. 6.9 The mechanism of catalysis of chymotrypsin-like serine proteinases. In the Michaelis complex, the sidechain of the substrate at the residue N-terminal to the scissile bond binds in a 'specificity pocket'. The catalytic triad Asp102–His57–Ser195 positions and polarizes the sidechain of Ser195 for nucleophilic attack on the C atom of the scissile bond. An intermediate in which this carbon atom is tetrahedral is stabilized by the 'oxyanion hole'—in which there are hydrogen bonds from the negatively charged oxygen of the substrate (formerly the carbonyl oxygen of the peptide group of the scissile bond) to the NH groups of residues 193 and 195. A proton is transferred from Ser195 to His57, and the Oγ of Ser195 forms a covalent bond to the substrate, breaking the scissile bond and releasing the C-terminal moiety of the substrate. The acyl–enzyme complex is then hydrolysed by a similar mechanism: nucleophilic attack by the hydroxyl group of a water molecule hydrogen bonded to His57 releases the carboxylic acid product, restoring the enzyme to its initial state.

Proteins containing the Ser–His–Asp catalytic triad

Chymotrypsin-like serine proteinases
Other proteinases: subtilisin, carboxypeptidase II
Lipases
Acetylcholinesterase
Natural catalytic antibody
Enoyl-CoA hydratase
4-chlorobenzoyl-CoA dehalogenase

Recent discoveries of distant relatives extend the scope of the family. Viral 3C proteinases do not retain the classic Ser–His–Asp catalytic triad; they have Cys instead of Ser, and, in some cases, Glu instead of Asp. Some homologues have developed different functions: haptoglobin acts as a chaperone, preventing unwanted aggregation of proteins, and shows several other functions including mediation of immune responses. The rhinovirus 3C proteinase has a second function as an RNA-binding protein. Vertebrate plasminogen-related growth factors contain domains homologous to chymotrypsin-like serine proteinases, that have lost their catalytic triad, and function in receptor activation. The insect 'immune' protein scolexin is a distant homologue that induces coagulation of haemolymph in response to infection.

The structure of serine proteinases of the chymotrypsin family

The structures of serine proteinases have been of interest since determination of the crystal structure of chymotrypsin in 1969. From now, for brevity we shall use the term serine proteinase to refer only to this family, even though certain other families of proteinases also contain catalytic serines, and the viral 3C proteinases, homologues of chymotrypsin, do not.

What are the structural constraints to which these proteins have been subject during their evolution, and how have they explored these limits?

Each serine proteinase contains two domains of similar structure, likely to have arisen by gene duplication and divergence. The domains pack together, with the active site between them. In most members of the family, intradomain disulphide bridges help to keep the molecule intact and to maintain the structure of the active site. The region of the substrate that includes the scissile bond binds in the cleft between the domains (Figure 6.10). The N-terminal domain contains the His and Asp of the catalytic triad. The C-terminal domain contains the Ser of the catalytic triad, the 'oxyanion hole' that stabilizes the transition state, and the specificity pocket. Domain 1 is composed primarily of residues from the N-terminal portion of the molecule, and domain 2 primarily of residues from the C-terminal portion.

Structures of individual domains

Each domain contains a six-stranded antiparallel β-sheet, folded into a β-barrel. Figures 6.11 and 6.12 illustrate the domains of *S. griseus* proteinase b,

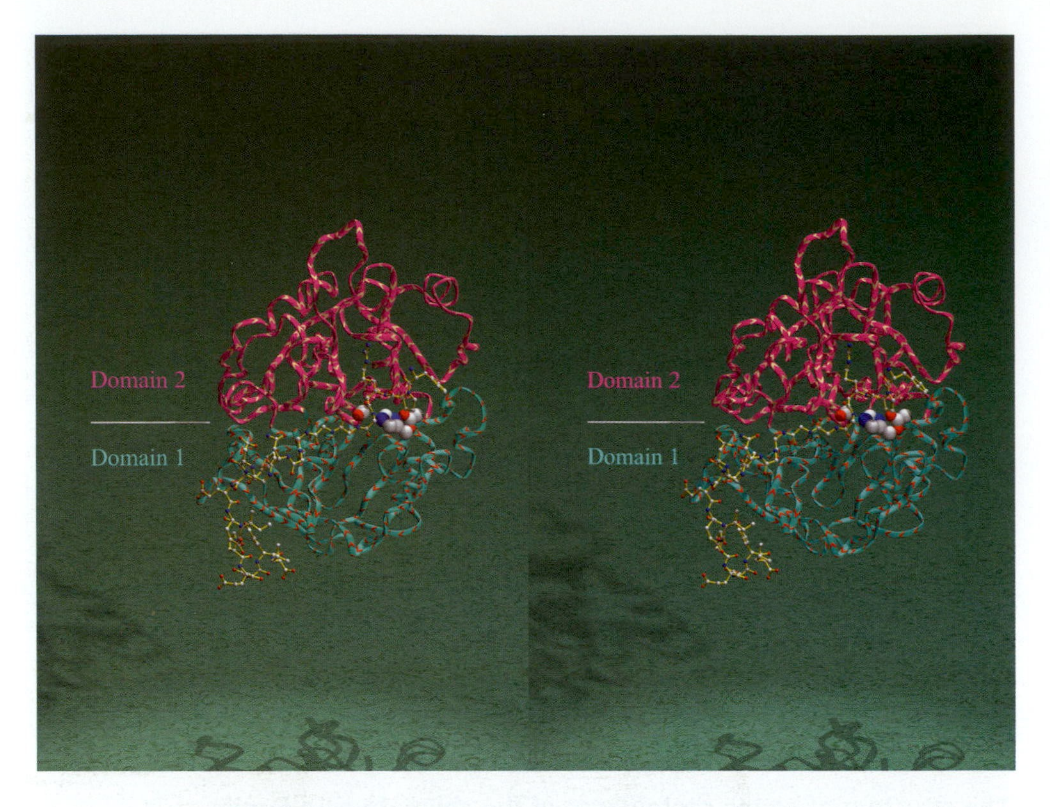

Fig. 6.10 Distribution between the two domains of elements of the catalytic mechanism of serine proteinases. This figure show human thrombin binding the artificial inhibitor hirulog 3, a 20-residue synthetic peptide related to the natural inhibitor hirudin from the leech [1ABI]. Domain 1 is below domain 2. An arginine is inserted into the specificity pocket, and the C-terminus of the molecule interacts with the anion-binding exosite.

one of the smaller structures. In the sheet diagrams, the edge strand appears twice, at the left and—in red—at the right. It is useful, although somewhat fanciful, to think of the stands of sheet as the stripes on the shirt of an American football referee, and the long hairpins as his arms folded across his chest as in the signal for 'illegal shift' (Figure 6.13).

Domains from different serine proteinases differ in the lengths of the strands of β-sheet making up the common parts of the fold, and in the lengths and structures of the regions connecting the strands. The proteins vie with one another for the gaudiest decorations on their common core.

Each domain contains two repetitions of a motif consisting of three strands of antiparallel β-sheet connected by two hairpins (Figure 6.14). (The entire protein contains four copies, two from each domain.) The first hairpin is bent, through an angle of approximately 90°. The two motifs are consecutive along the chain with a linking segment between them. This linking segment connects the only two consecutive strands that are not hydrogen-bonded to each other; that is, it is the only connection between consecutive strands that is not a hairpin. (This segment corresponds to the arching line linking strands three and four in Figure 6.12.) Each motif contains two

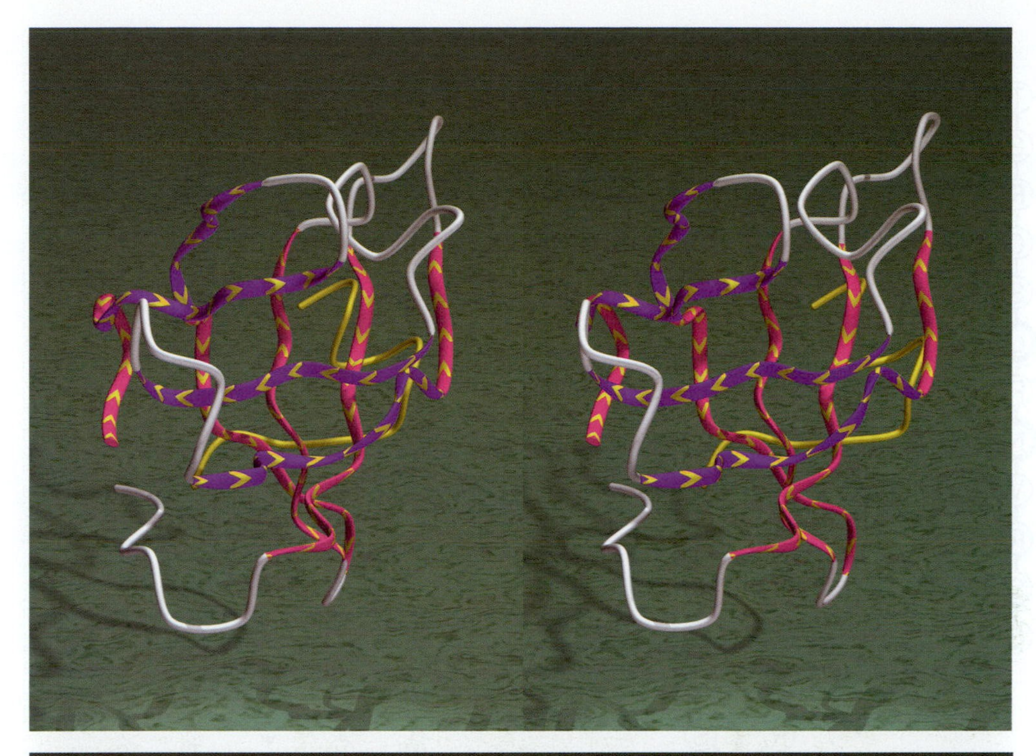

Fig. 6.11a

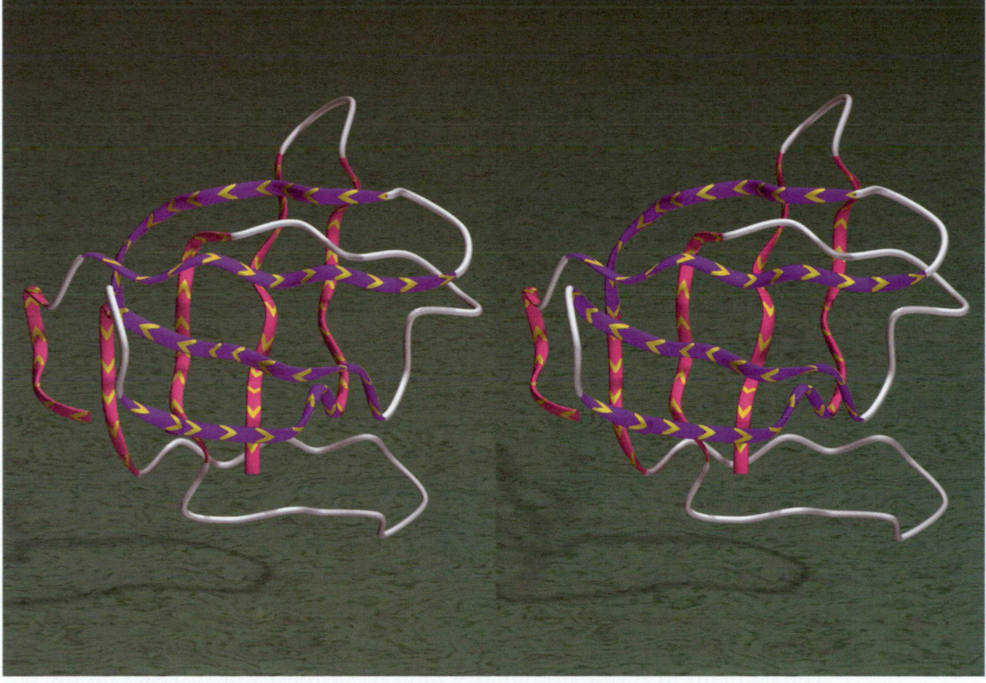

Fig. 6.11b

Fig. 6.11 Chain tracings of the domains of *S. griseus* proteinase b [3SGB]. (a) Domain 1, (b) domain 2. Domain 1 contains a short helix in the connection between the central vertical strand and the lower folded-over loop; a helix is present in this region in several but not all of the structures. For a view parallel to barrel axis, see upper domain of Figure 6.8.

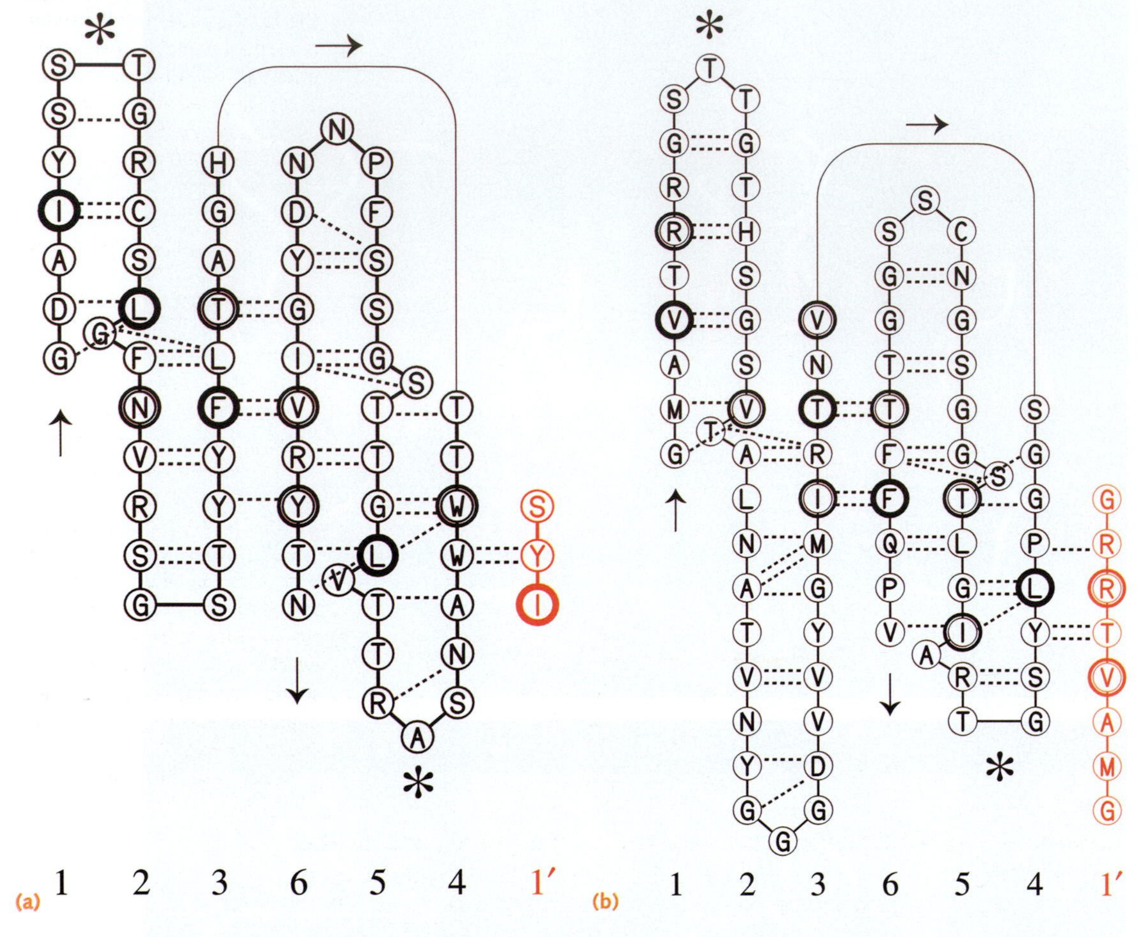

Fig. 6.12 Hydrogen-bonding patterns of the β-sheets in *S. griseus* proteinase b [3SGB]. (a) Domain 1. (b) Domain 2. The structure has been unfolded to lie in a plane. Each domain contains a six-stranded β-sheet. Strands are numbered in order of their appearance in the chain. The N-terminal strand is repeated, appearing once (in black) at the left and, again (in red) at the right. To form the three-dimensional structure, two of the hairpins – those marked by * – must be folded out of the page, and twisted to lie horizontally across the sheet.

Most domains from other serine proteinases have the same general pattern but differ in the lengths of the hairpins and of the connections between them. Some of the longer connections contain α-helices. In domain 1 of Sindbis virus capsid proteinase the general pattern is incomplete. The residues of the sheet that are packed inside the β-barrel are ringed.

hairpins. The first hairpins of the two motifs form β-sheet hydrogen bonds to each other, and the second hairpins of each form β-sheet hydrogen bonds to each other; this closes the barrel.

The domain–domain interface

The domains in serine proteinases interact primarily at residues on the outside of several strands of the β-barrel of each domain, but also in the loops connecting these strands. The interactions between the domains are mostly van der Waals contacts, with a few H-bonds. The structure of the interface is characterized by a central and conserved region near the catalytic triad, sur-

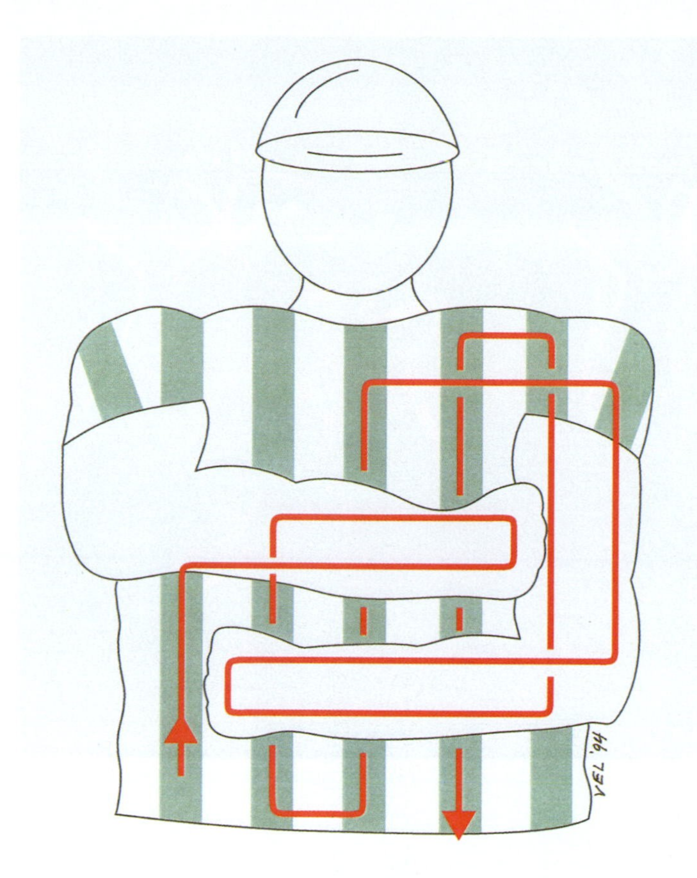

Fig. 6.13 An idealized tracing of the folding pattern of the serine proteinase domain superposed onto a picture of an American-football referee. This figure exaggerates the approximate symmetries within the domain: there are two axes of approximate two-fold symmetry. One is perpendicular to the page and, if the direction of the chain is ignored, another at an angle of 45° to the horizontal (compass direction Northeast–Southwest).

In most cases, the two hairpins folded over the sheet (the referee's arms) form hydrogen bonds to each other, closing the sheet into a barrel. These hydrogen bonds appear in Figure 6.12 between the second strand from the right and the strand at the extreme right which is a copy of the leftmost strand. The axis of the barrel is approximately along the NE–SW line of Figure 6.13.

Drawing by V.E. Lesk.

rounded by additional regions that vary among the proteinases. In the mammalian proteinases, the residues at the core of the domain–domain interface are all either absolutely conserved or vary within fairly narrow limits, with a few exceptions. In most proteinases, a Gly–Gly contact across the domain–domain interface facilitates the close approach of the domains.

The specificity pocket

The primary specificity is determined by the sidechain of the residue adjacent to the scissile bond, which fits into a pocket next to the catalytic site. Figure 6.15 shows the structure of this pocket in chymotrypsin, including the sidechain of a substrate tryptophan residue. The sides of the pocket are formed from two loops joining successive strands of the β-sheet of domain 2; they connect strands 4–5 and 5–6. The base of the pocket is occupied by Ser189 in chymotrypsin; the residue at this position contributes to the specificity. Glycines at positions 216 and 226 produce a narrow, slotlike pocket in chymotrypsin, suitable for binding a flat hydrophobic sidechain.

Comparing the specificity pockets in different proteinases, the mainchain is rather rigid, and the change in specificity is achieved primarily by mutation. To the extent that this is true, it should be possible to model the structure of the specificity pocket in other proteinases, and from this to predict the specificity.

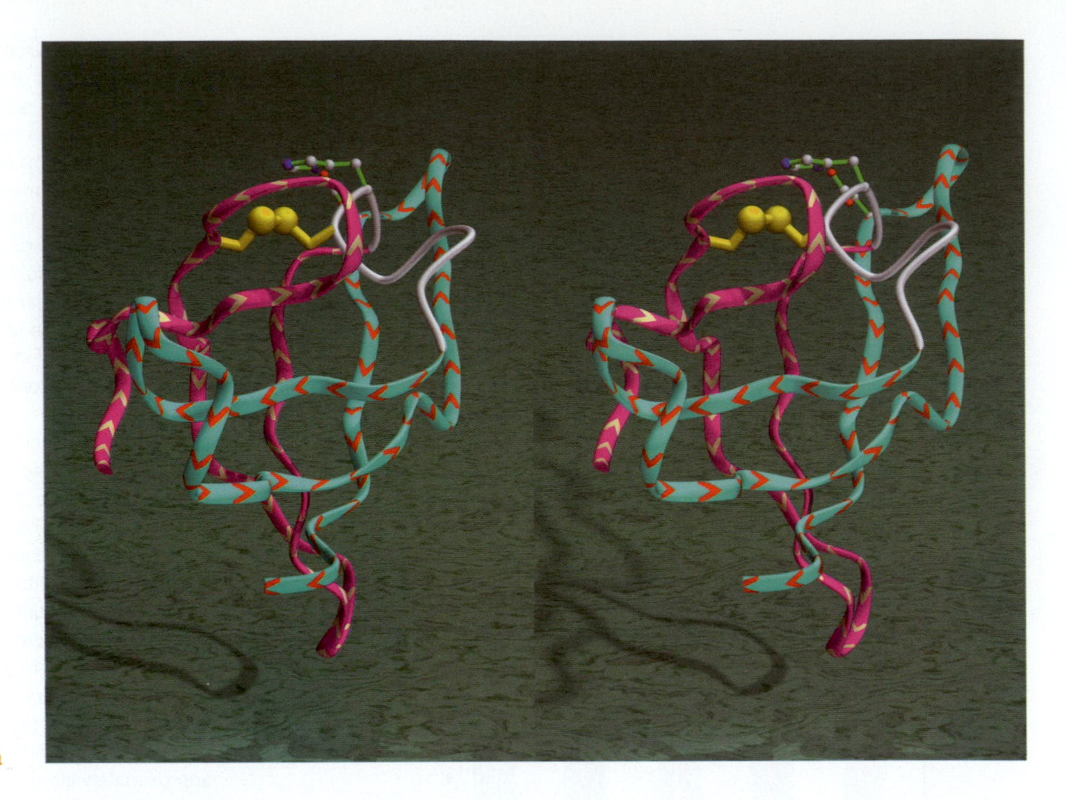

Fig. 6.14a

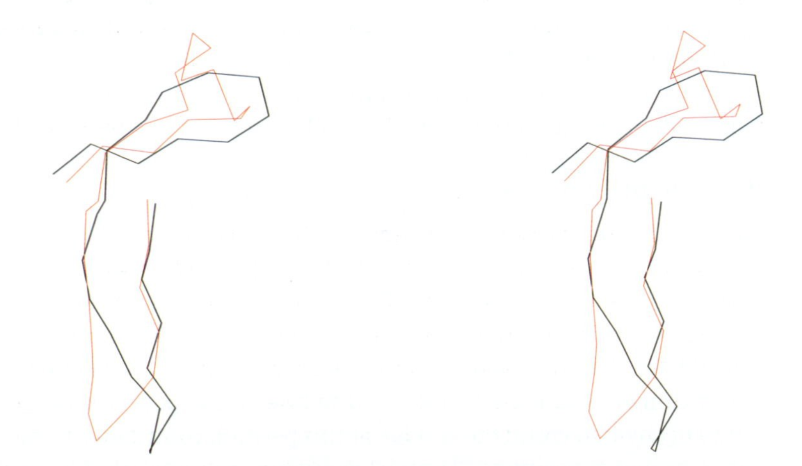

Fig. 6.14b

Fig. 6.14 (a) Domain 1 of *S. griseus* proteinase b, showing the relationship between the two similar motifs [3SGB]. The N-terminal motif (residues 1–30) is shown in pink/yellow ribbon with dense cross-hatching, the linking region (residues 31–39) in white, and the C-terminal moiety (residues 39–71) in green/orange. (b) Superposition of residues from the two motifs of domain 1 of *S. griseus* proteinase b.

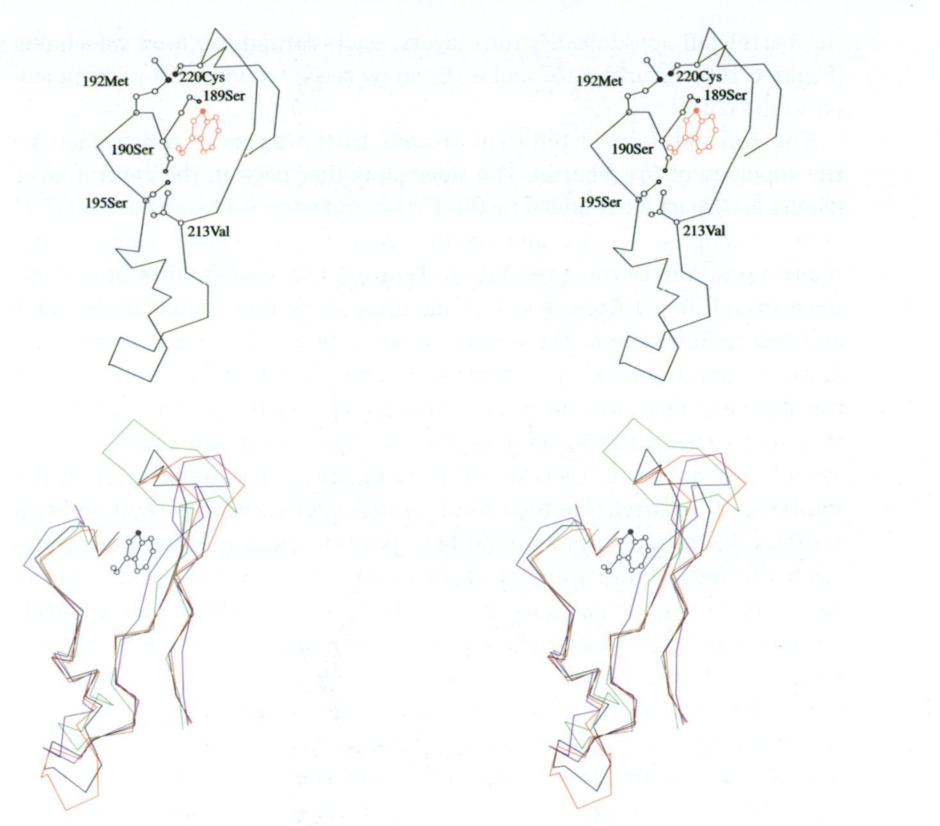

Fig. 6.15 (a) Specificity pocket in chymotrypsin, occupied by a tryptophan sidechain from an autolysis product [8GCH]. (b) Superposition of the mainchains of the residues creating the specificity pocket in chymotrypsin [5CHA] (black, including Trp sidechain from product), thrombin [1PPB] (red), *Achromobacter* proteinase [1ARB] (green) and Sindbis virus capsid proteinase [2SNV] (blue).

The β-barrels in serine proteinase domains and the packing of residues in their interiors

Recall that any β-barrel topology can be characterized by two *discrete* parameters, the strand number, n, and the 'shear number', S, which measures the stagger of the strands (Chapter 4). The β-barrels in serine proteinases contain $n = 6$ strands and have a shear number $S = 8$. For barrels in which S/n is integral, as in the triose phosphate isomerase fold ($S = n = 8$; $S/n = 1$) or in interleukin-1β ($S = 12$, $n = 6$; $S/n = 2$), there is a regular packing of sidechains in the interior of the barrels to form natural layers, as a consequence of the geometry of the assembly of the strands. In chymotrypsin-like serine proteinases, the incommensurability between S and n produces a less regular structure in the sidechains packed inside the barrel.

Figure 6.16a shows the hydrogen-bonding pattern in the β-barrel of domain 1 of elastase. The edge strand is repeated and appears at both the right and the left; the actual structure contains only six strands. Note the catalytic His and Asp at the top of the third and fourth strands from the left. Although the ratio S/n is non-integral, the residues packed in the interior of the barrel fall *approximately* into layers, each containing four sidechains (Figure 6.16b). Figures 6.16c and 6.16e show serial sections, cut perpendicular to the barrel axis.

The contributions of different strands to the layers is constrained by the topology of the β-barrel. The sidechains that pack in the central layer (Figure 6.16d) are surrounded by thick rings in Figure 6.16a, and are marked as I♠ (two copies), V♡, L◇ and V♣, in strands 1, 3, 6, 4 and 1′. Consider the relative positions of these residues in Figure 6.16a: start from I♠ in the leftmost strand. To get from I♠ to V♡ one must move two strands to the right and two residues down. The residue between I♠ and V♡ (the Thr in strand 2, *one* column right and *one* residue down) has its sidechain pointing out of the sheet and does not contribute to the packing of the interior. If this pattern were repeated three additional times, four sidechains from alternate strands and alternate levels would pack together on a single layer in the interior of the barrel. The final result of moving two strands right and two residues down, four times, would be a position eight residues down, precisely the vertical displacement of the two copies of I♠, *but with a horizontal displacement of eight strands rather than six.* This pattern is observed in β-barrels with the regular triose phosphate isomerase topology, for which $S = 8$, $n = 8$, and $S/n = 1$ (Figure 4.20).

Consider in contrast the relationship between the next pair of residues in the central layer: V♡ and Leu◇. The relationship between these two can be described as 'one over, two down'. If the 'one over, two down' pattern were repeated five more times, six inward-pointing sidechains, from adjacent strands but alternate levels, would pack in a layer in the interior of the barrel. The final result of moving one strand to the right and two residues down, six times, would be a position six strands over, precisely the horizontal displacement of the two copies of the edge strand, *but with a vertical displacement of 12 residues rather than eight.* This pattern is observed in β-barrels with the β-trefoil topology of interleukin-1β, for which $S = 12$, $n = 6$, and $S/n = 2$ (Figure 4.25).

The problem faced by serine proteinases in forming their barrels is that moving two strands right and two residues down, four times, and ending up at the repeated copy of the residue from which one started, requires eight strands, but only six are present. Alternatively, moving one strand right and two residues down, six times, gives a shear number of 12, but the actual shear number is eight. The incommensurability of the strand and shear numbers means that an irregular pattern is required, and what the molecule does is to mix—indeed, in the case of the central layer of domain 1 of elastase, to alternate—the patterns of the two types of regular β-barrels: $S/n = 1$ and $S/n = 2$.

A useful shorthand suggests itself if one thinks of the residues as laid out on a chessboard. The 'two over, two down' pattern connecting the positions of Ile♠ and Val♡ in strands one and three is like a bishop moving two squares along a diagonal. The 'one over, two down' pattern connecting the positions Val♡ and Leu◇ in strands three and four is a move like that of a

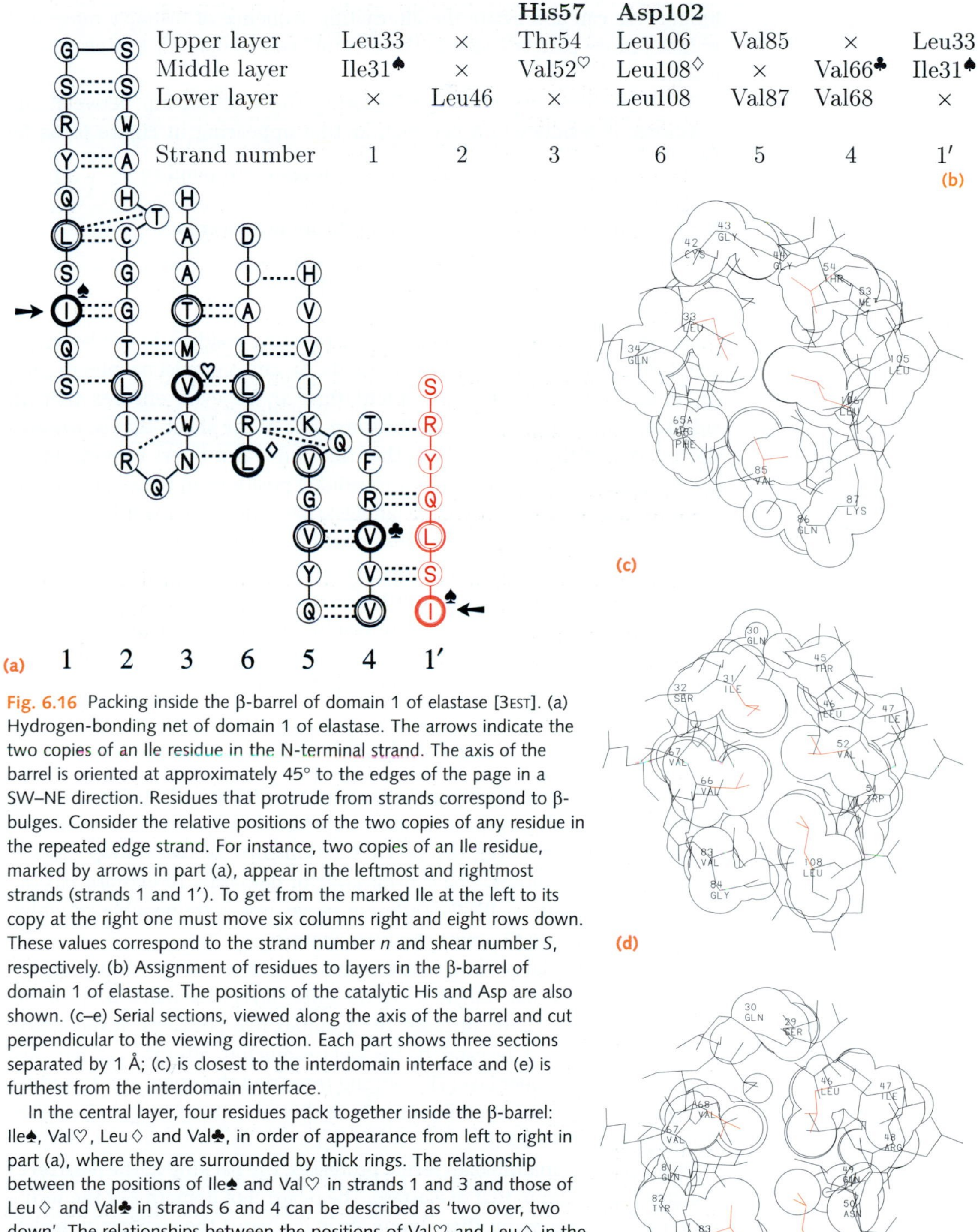

His57 Asp102
Upper layer Leu33 × Thr54 Leu106 Val85 × Leu33
Middle layer Ile31♠ × Val52♡ Leu108◇ × Val66♣ Ile31♠
Lower layer × Leu46 × Leu108 Val87 Val68 ×
Strand number 1 2 3 6 5 4 1′
(b)
(a) 1 2 3 6 5 4 1′
(c)
(d)
(e)

Fig. 6.16 Packing inside the β-barrel of domain 1 of elastase [3EST]. (a) Hydrogen-bonding net of domain 1 of elastase. The arrows indicate the two copies of an Ile residue in the N-terminal strand. The axis of the barrel is oriented at approximately 45° to the edges of the page in a SW–NE direction. Residues that protrude from strands correspond to β-bulges. Consider the relative positions of the two copies of any residue in the repeated edge strand. For instance, two copies of an Ile residue, marked by arrows in part (a), appear in the leftmost and rightmost strands (strands 1 and 1′). To get from the marked Ile at the left to its copy at the right one must move six columns right and eight rows down. These values correspond to the strand number n and shear number S, respectively. (b) Assignment of residues to layers in the β-barrel of domain 1 of elastase. The positions of the catalytic His and Asp are also shown. (c–e) Serial sections, viewed along the axis of the barrel and cut perpendicular to the viewing direction. Each part shows three sections separated by 1 Å; (c) is closest to the interdomain interface and (e) is furthest from the interdomain interface.

In the central layer, four residues pack together inside the β-barrel: Ile♠, Val♡, Leu◇ and Val♣, in order of appearance from left to right in part (a), where they are surrounded by thick rings. The relationship between the positions of Ile♠ and Val♡ in strands 1 and 3 and those of Leu◇ and Val♣ in strands 6 and 4 can be described as 'two over, two down'. The relationships between the positions of Val♡ and Leu◇ in the third and fourth strands from the left, and those of Val♣ and Ile♠, in strands 4 and 1′, can be described as 'one over, two down'. The combination of these patterns is required because exclusive use of either pattern is inconsistent with the non-integrality of S/n.

knight. We can abbreviate the alternating sequence of bishop's move and knight's move patterns seen in the central layer of domain 1 of elastase as BNBN.

What patterns are possible? Denoting the relationship between the positions of residues in a net such as that appearing in Figure 6.16a by $\begin{pmatrix}\Delta x_i\\ \Delta y_i\end{pmatrix}$ (where the bishop's and knight's moves correspond to $B = \begin{pmatrix}2\\2\end{pmatrix}$ and $N = \begin{pmatrix}1\\2\end{pmatrix}$), the constraint on the residues forming a layer is:

$$\Sigma \begin{pmatrix}\Delta x_i\\ \Delta y_i\end{pmatrix} = \begin{pmatrix}n\\ S\end{pmatrix}$$

where the sum is taken over the inward-pointing residues in the layer, S is the shear number, n the number of strands, Δx_i and Δy_i must be integral, and Δy_i must be even. (The requirement that Δy_i be even ensures that all sidechains are pointing towards the same side of the sheet; that is, towards the interior of the β-barrel. Note that this constraint refers to positions in the hydrogen-bonding net and not to residue positions in the sequence; the distinction arises where there are β-bulges. Exceptions to this rule can occur at the ends of strands where residues might have conformations outside the β-strand region of the Sasisekharan–Ramakrishnan–Ramachandran diagram.) For regular barrels—S/n integral—these equations are satisfied by $\Delta x_i = n/L = S/2$, where L is the number of residues in the layer (if $S/n = 1$, $L = n/2$; and if $S/n = 2$, $L = n$) and $\Delta y_i = (S/n)\Delta x_i$. For the serine proteinase domains with four residues per layer, the equations take the form:

$$\Sigma_{i=1}^{4} \begin{pmatrix}\Delta x_i\\ \Delta y_i\end{pmatrix} = \begin{pmatrix}6\\ 8\end{pmatrix}$$

which has as its smallest integral solutions $\Delta y_i = 2$, for $i = 1, 2, 3, 4$ and $\Delta x_i =$ 1 (twice) and 2 (twice), to give some combination of two 'two over, two down', and two 'one over, two down' relationships; or two bishop's moves and two knight's moves. The central layer of domain 1 of elastase has the most common pattern, BNBN. Different move orders are possible, of which BNBN and BNNB are the most common choices.

The other two layers in elastase domain 1 can be analysed in a similar way. The residues that contribute to the outer layers are marked by double rings in Figure 6.16a. In all cases, the residues five positions in the amino acid sequence before the catalytic His and residue four after the catalytic Asp contribute to the upper layer (T in strand three and the upper L in strand six in Figure 6.16a.) Leu ◇ contributes to both the central and bottom layer (see Figures 6.16d and 6.16e.)

In most of the mammalian serine proteases, the packing in the β-barrel of domain 1 is similar to that of elastase. The β-barrel of domain 2 of the serine proteinases also contains three layers. As in domain 1 the packing in the layers shows preferred but not unique patterns among the mammalian proteinases, with greater variability and irregularity in the bacterial and viral molecules.

Conclusions

Evolution of the serine proteinases, like that of other protein families, shows a qualitative conservation of conformation at the secondary and tertiary structural level.

1. The secondary and tertiary structure—β-barrels of similar topology—is conserved in mammalian and bacterial proteinases. (Greater variations appear in viral enzymes.) The loops joining the strands of β-sheet can and do change in length and conformation.
2. Mutations in and around the binding site must preserve the active site. Indeed, residues in and around the active site, including catalytic residues, the specificity pocket, and residues in the interface, are well conserved.
3. Preservation of the topology of the β-barrel requires that mutations be consistent with good packing of its interior. The structure of the β-barrel cannot vary continuously, because its geometry is determined by two discrete parameters—number of strands $n = 6$ and shear number $S = 8$. The backbone of the β-barrel is therefore relatively rigid, and the molecule can only escape this structure by giving up the closure of the barrel; this occurs in domain 1 of Sindbis virus capsid proteinase. On the other hand, the pattern of packing inside these β-barrels is more variable than those in TIM-like or β-trefoil proteins for which S/n is integral, because there is no clearly preferred arrangement of sidechains. This packing shows somewhat more consistent and regular patterns in the mammalian proteinases than in the others.

NAD-binding domains of dehydrogenases

In 1973, knowing the structures of lactate, malate and alcohol dehydrogenases, C.-I. Brändén, H. Eklund, B. Nordström, T. Boiwe, G. Söderlund, E. Zeppezauer, I. Ohlsson and Å. Åkeson wrote:

> The coenzyme binding region [of horse liver alcohol dehydrogenase] has a main-chain conformation very similar to a corresponding region in lactate and malate dehydrogenase. It is suggested that this substructure is a general one for binding of nucleotides and, in particular, the coenzyme NAD.

Since then, many additional crystal structures have confirmed and extended this insight. There are now over 30 different enzymes that contain a common NAD-binding domain. The paradigm nucleotide-binding domain in horse liver alcohol dehydrogenase contains two sets of β–α–β–α–β units, together forming a single parallel β-sheet flanked by α-helices (Figure 6.17). From the point of view of Figure 6.17, the strands appearing from left to right are in the order 6–5–4–1–2–3. There is a long loop, or crossover, between strands three and four. As Brändén described in 1980, this feature of the fold creates a natural cavity contributing to the binding of the adenine ring of the NAD and, in other molecules with similar supersecondary structures, of other nucleotide-containing fragments.

Fig. 6.17a

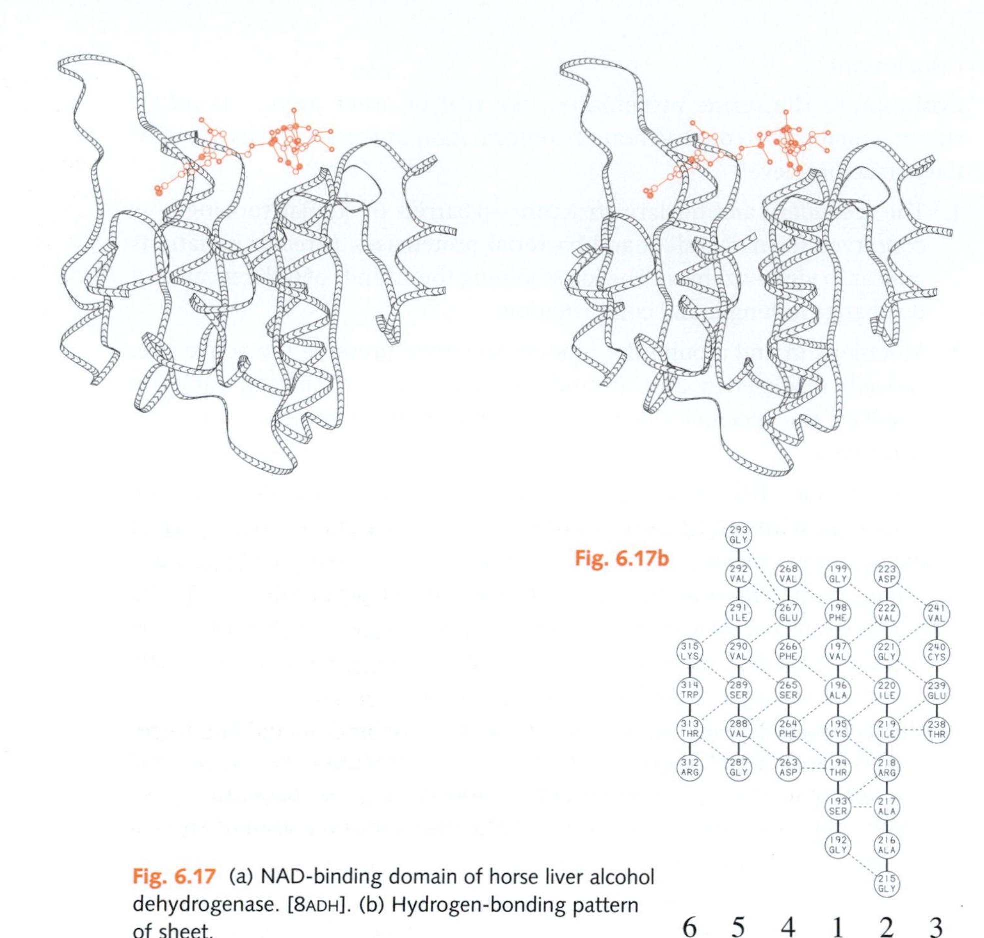

Fig. 6.17 (a) NAD-binding domain of horse liver alcohol dehydrogenase. [8ADH]. (b) Hydrogen-bonding pattern of sheet.

NAD-binding domains combine with very different other domains to form enzymes with very different substrate specificities. They occur as oligomers of different sizes; many are tetramers. Figure 6.18 shows formate dehydrogenase, which is a dimer. The NAD-binding domains in each monomer are highlighted.

Other proteins than dehydrogenases bind nucleotide-containing cofactors in a manner similar to the binding of NAD to dehydrogenases, and create the binding site from domains with generally similar secondary and tertiary structure. Examples include the FMN in flavodoxin and the FAD in pyruvate oxidase. Conversely, other structures have extended the repertoire of modes of binding NAD and related ligands. Some have folding patterns very different from the dehydrogenase NAD-binding domains. These include proteins from other general topological classes: all β and $\alpha + \beta$ as well as proteins in the α/β class but unrelated to the dehydrogenase NAD-binding fold.

The sequence motif G*G**G

The binding of NAD to dehydrogenases involves numerous hydrogen bonds and van der Waals contacts between the cofactor and the enzyme. In partic-

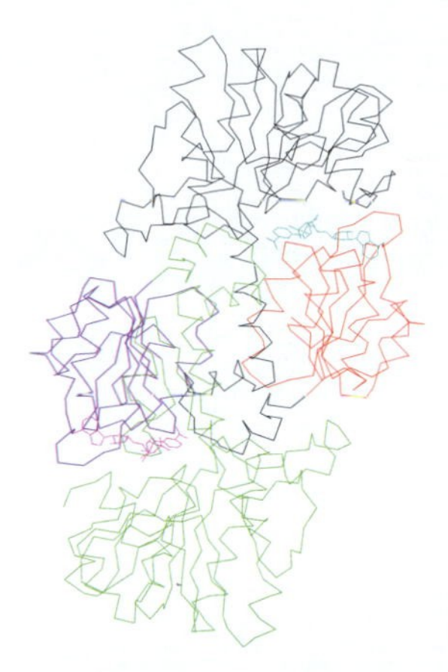
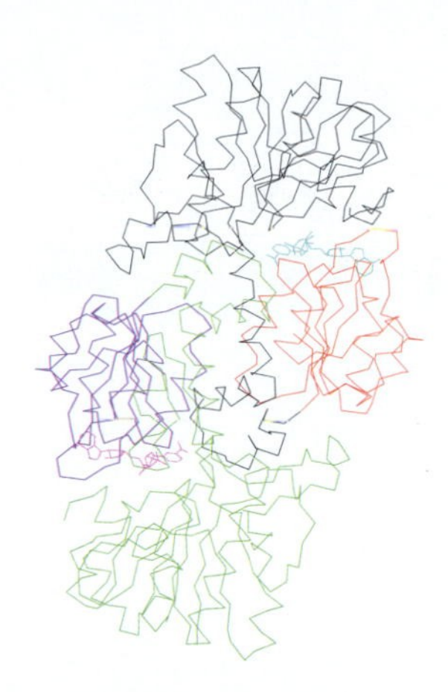

Fig. 6.18 The dimeric structure of formate dehydrogenase, viewed down the axis of symmetry [2NAD]. One subunit is black, with its NAD-binding domain highlighted in red. The other is green, with its NAD-binding domain highlighted in blue.

ular, there are usually hydrogen bonds from the residues in the turn between the first strand and the helix that follows it, to one of the phosphate groups of the cofactor. These interactions give rise to a consensus sequence in this region containing the three-glycine pattern G*G**G characteristic of the first β–α–β unit in the dehydrogenase domain. (This motif signifies a hexapeptide: Gly–xxx–Gly–xxx–xxx–Gly, where xxx represents any residue.)

The first two glycines are involved in nucleotide binding and the third, which is in the helix following the first strand, is involved in the packing of the helix against the sheet. The first glycine is in an α_L conformation integral to the structure of the turn. A Cβ at the position of the second glycine would collide with the cofactor. The third glycine is in the helix following the first strand, and allows for packing of the helix against the sheet. In horse liver alcohol dehydrogenase, a Cβ in the residue at the position of the third glycine would clash with the carbonyl of the first glycine.

In most dehydrogenases there is a well (but not absolutely) conserved aspartate approximately 20 residues C-terminal to the G*G**G motif. This aspartate appears near the C-terminus of the second strand, and forms hydrogen bonds to the ribose of the adenosine moiety of the NAD.

Comparison of NAD-binding domains of dehydrogenases

Figures 6.19 through 6.24 illustrate six NAD-binding domains, showing the folding of the chain and the hydrogen-bonding net of the sheet. These give some idea of the observed structural variety.

In five of these domains, the sheets contain the canonical six strands, but are extended by additional strands. Dihydropteridine reductase has a seventh strand adjacent and parallel to the sixth strand and an eighth strand forming a hairpin with the seventh. 6-phosphogluconate dehydrogenase and formate dehydrogenase each have a seventh strand adjacent but parallel to the sixth, and no eighth strand. Glyceraldehyde-3-phosphate dehydrogenase has a short stretch of antiparallel sheet between the third and fourth strand, before the 'crossover'. 3α,20β-hydroxysteroid dehydrogenase has a seventh strand adjacent and parallel to the sixth.

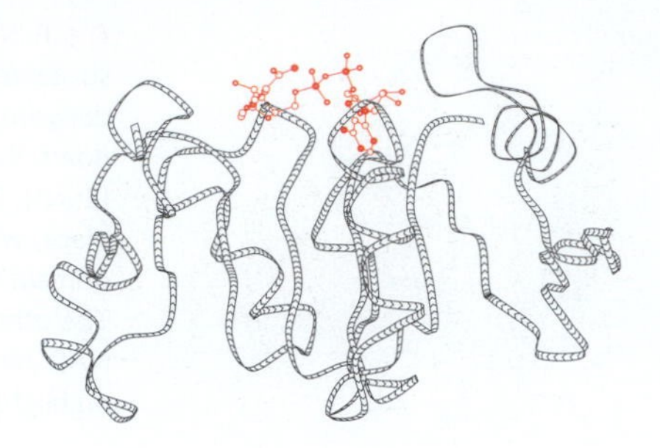

Fig. 6.19a

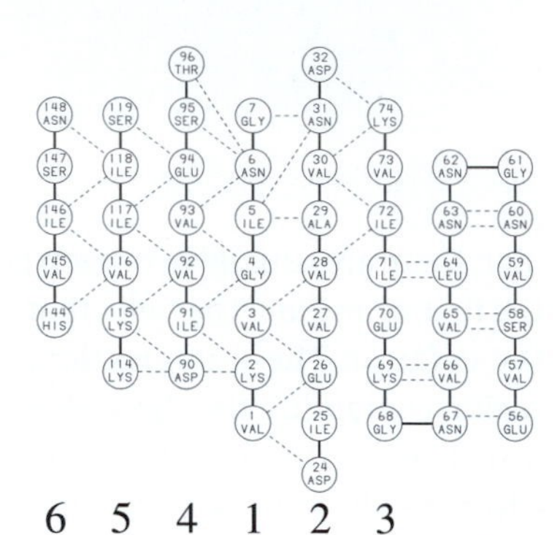

Fig. 6.19 (a) NAD-binding domain of glyceraldehyde-3-phosphate dehydrogenase [1GD1]. (b) Hydrogen-bonding pattern of sheet.

Fig. 6.19b

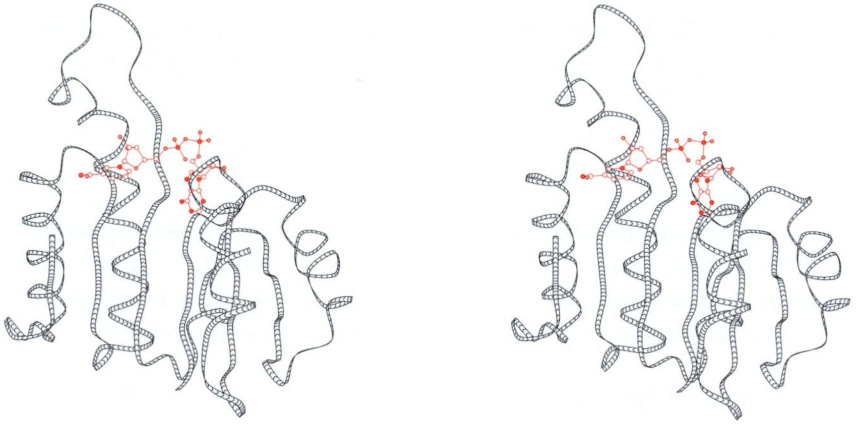

Fig. 6.20a

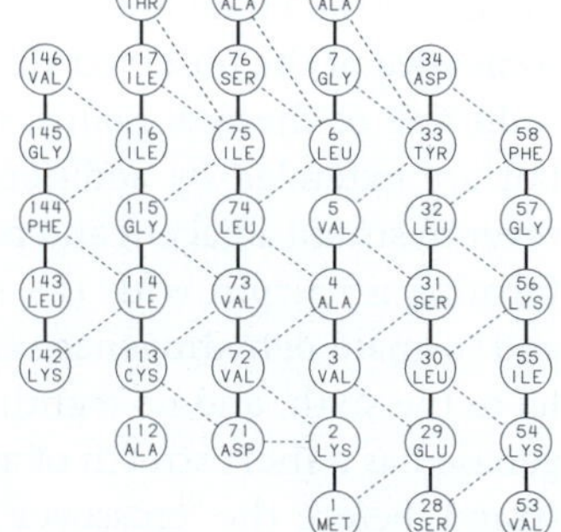

Fig. 6.20 (a) NAD-binding domain of malate dehydrogenase [1EMD]. (b) Hydrogen-bonding pattern of sheet.

Fig. 6.20b

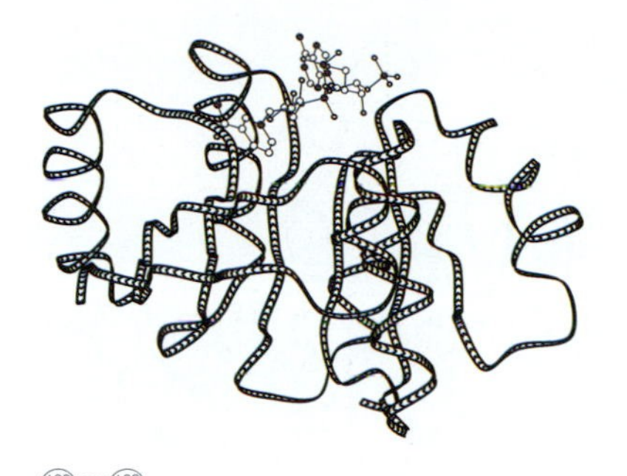

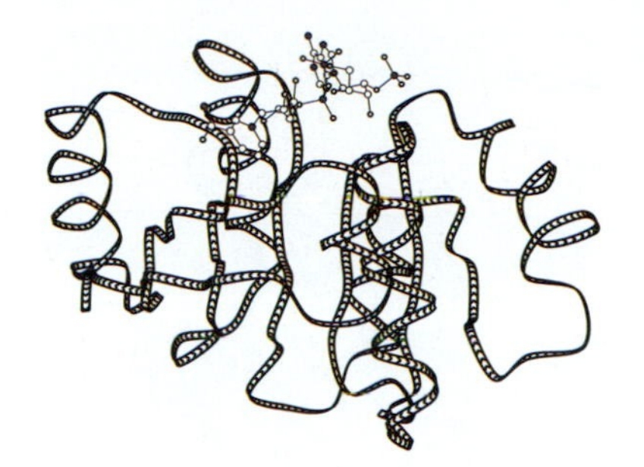

Fig. 6.21a

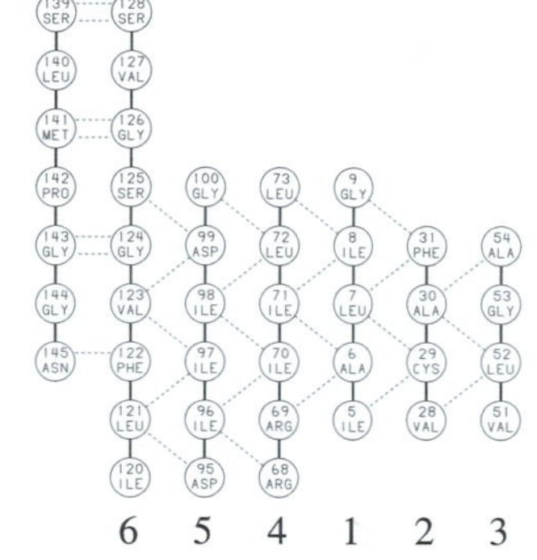

Fig. 6.21b

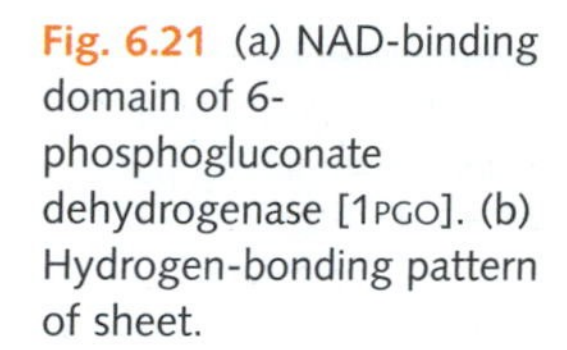

Fig. 6.21 (a) NAD-binding domain of 6-phosphogluconate dehydrogenase [1PGO]. (b) Hydrogen-bonding pattern of sheet.

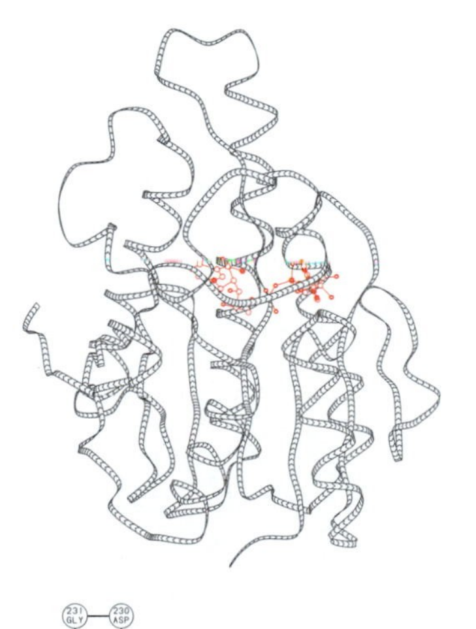

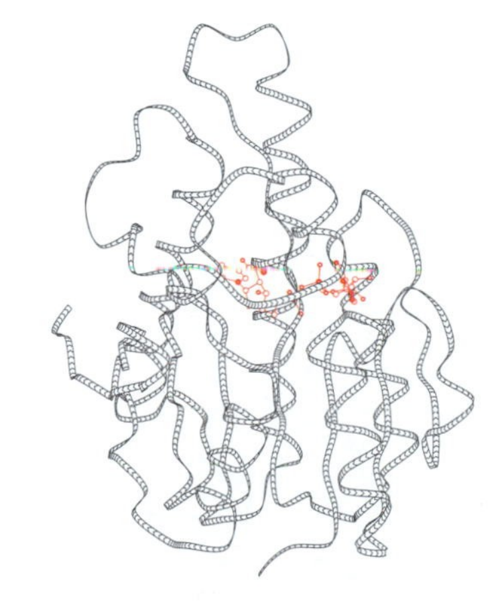

Fig. 6.22a

6 5 4 1 2 3

Fig. 6.22b

Fig. 6.22 (a) NAD-binding domain of dihydropteridine reductase [1DHR]. (b) Hydrogen-bonding pattern of sheet.

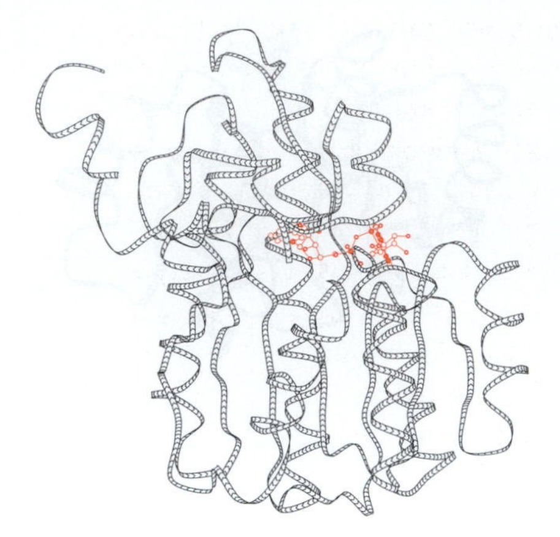

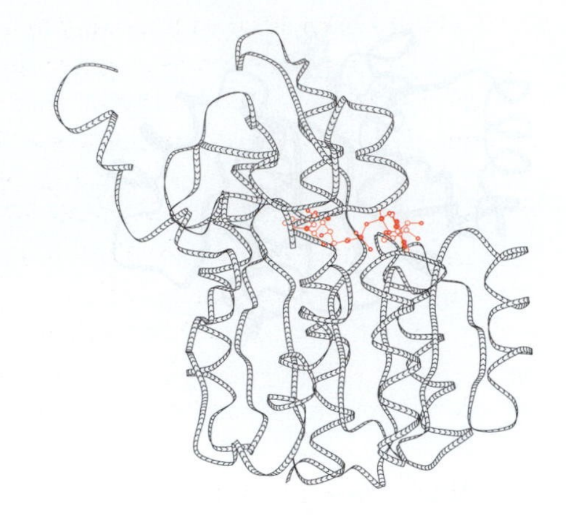

Fig. 6.23a

Fig. 6.23 (a) NAD-binding domain of 3α,20β-hydroxysteroid dehydrogenase [2HSD]. (b) Hydrogen-bonding pattern of sheet.

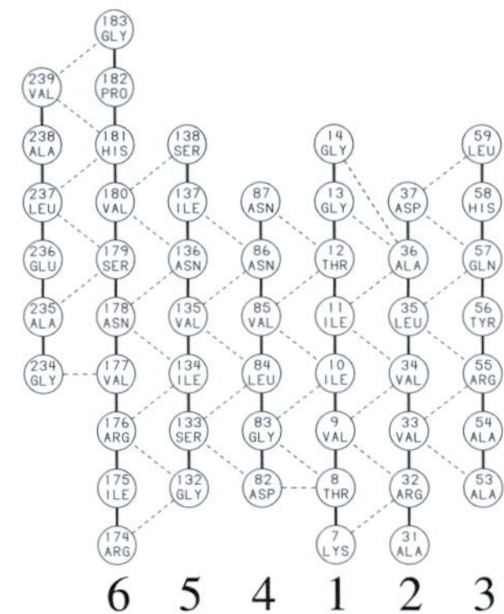

Fig. 6.23b

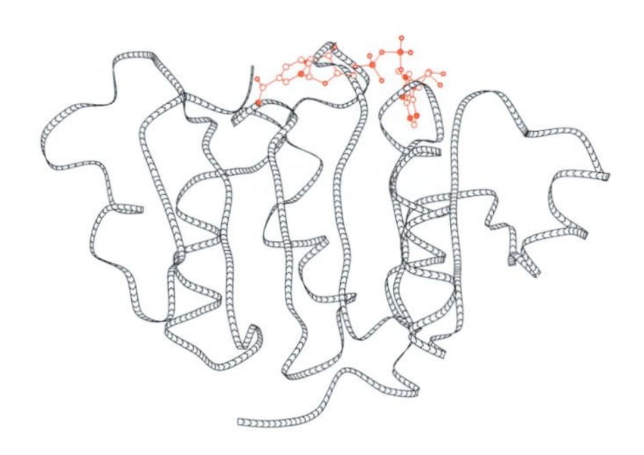

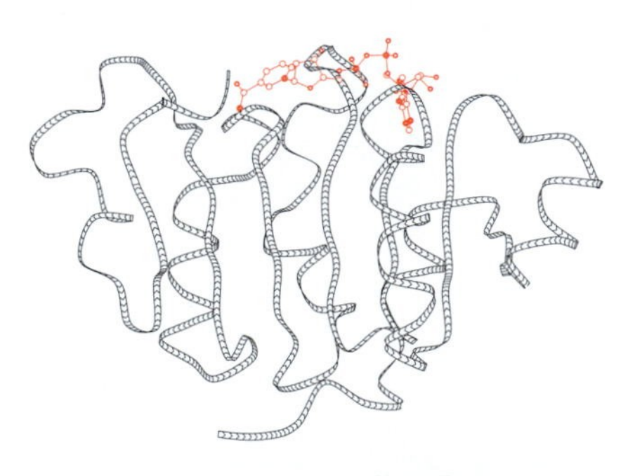

Fig. 6.24a

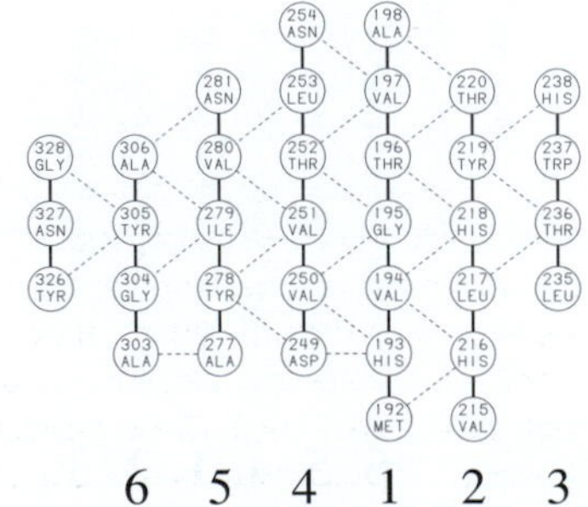

Fig. 6.24 (a) NAD-binding domain of formate dehydrogenase [2NAD]. (b) Hydrogen-bonding pattern of sheet.

Fig. 6.24b

The helices also differ considerably in length, and can shift relative to the sheet. In 3α,20β-hydroxysteroid dehydrogenase and dihydropteridine reductase the two helices in the C-terminal portion of the domain (between strands four and five, and between strands five and six) are elongated. There is also a helix in the crossover region, which appears in many of these structures. Dihydropteridine reductase has lost the helix between strands two and three.

What is the conserved core of the family? Figure 6.25 shows the superposition of NAD-binding domains of a closely related pair of enzymes (lactate and malate dehydrogenases) and a more distantly related pair (alcohol dehydrogenase and dihydropteridine reductase). The crossover region—the loop between strands three and four—is especially variable in structure among the different enzymes. The maximal common substructure of the distantly-related pair corresponds to the core of this family of domains.

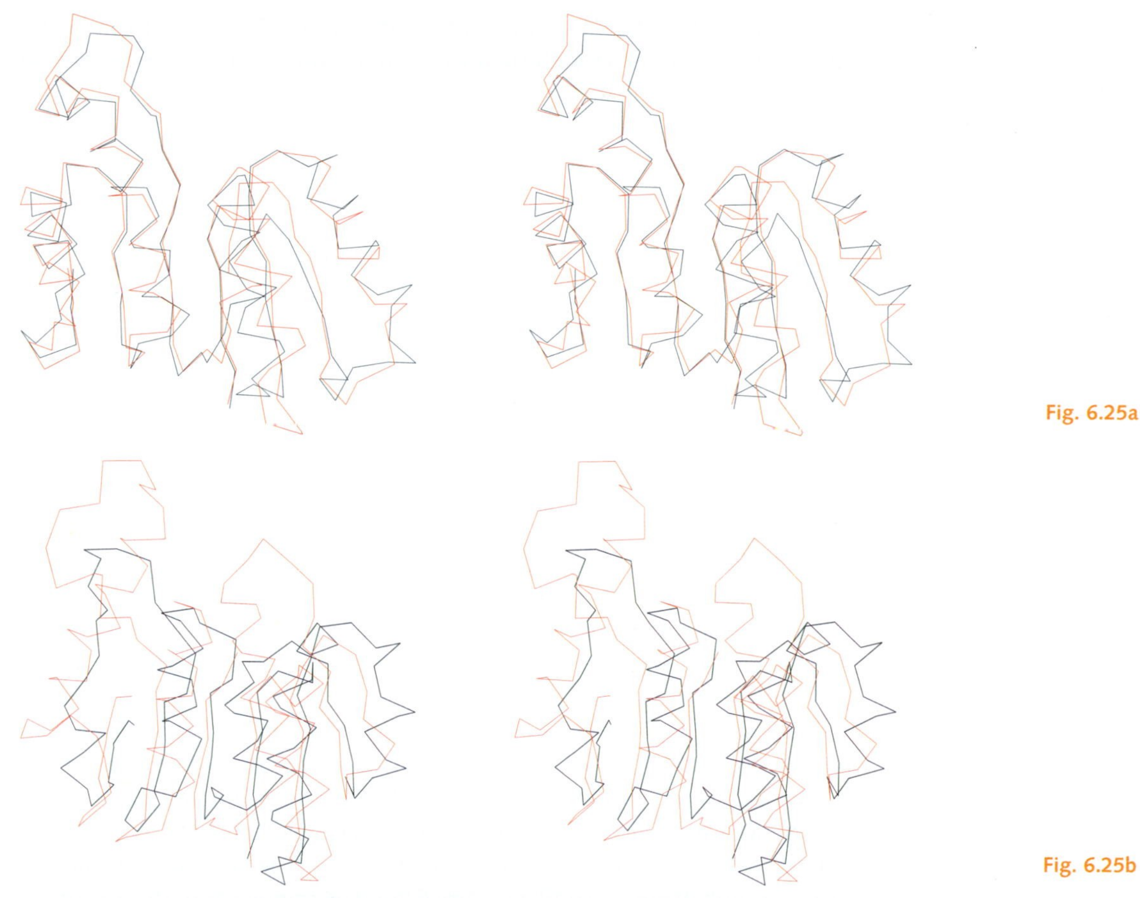

Fig. 6.25a

Fig. 6.25b

Fig. 6.25 (a) Superposition of NAD-binding domains of lactate dehydrogenase [9LDT] (black) and malate dehydrogenase [1EMD] (red). The sequences of these regions have 23% identical residues upon optimal alignment. Although these molecules have developed different functions, they are still fairly similar. (b) Superposition of NAD-binding domains of horse liver alcohol dehydrogenase [2OHX] (black) and dihydropteridine reductase [1DHR] (red). These molecules have diverged more radically. There are only 14% identical residues in an optimal alignment.

Four of the structures deviate from the G*G**G motif. Malate and formate dehydrogenases contain an alanine instead of the first glycine: In formate dehydrogenase, the alanine is in a ++ conformation. In malate dehydrogenase, the alanine is preceded by a glycine in a ++ conformation, and a structural superposition shows that the alanine is an insertion in the sequence of the loop causing only local deformation of the chain. 6-phosphogluconate dehydrogenase contains an alanine at the position of the second glycine. In this structure, which binds NADP, the usual interaction between the first β–α–β unit and the pyrophosphate is absent. In 3α,20β-hydroxysteroid dehydrogenase, there is also an insertion in the loop; a structural superposition onto horse liver alcohol dehydrogenase gives the alignment:

```
horse liver alcohol dehydrogenase       TCAVFGL-GGVGL
3α,20β-hydroxysteroid dehydrogenase     TVIITGGARGLGA
```

Residues from the C-terminus of the fifth strand (the second from the left in Figure 6.17) interact with the nicotinamide ring of the cofactor and usually form hydrogen bonds to the amide group.

Cofactor binding

The conformation and interactions of the cofactor differ among the structures. Most but not all show the common interactions between the pyrophosphate and the region of conserved sequence between the first strand and helix (see Figure 2.7). However, in 6-phosphogluconate dehydrogenase, these hydrogen bonds are absent. Nevertheless the structure and sequence pattern is largely conserved. In addition, the asparate that forms hydrogen bonds to the ribose is asparagine in 6-phosphogluconate dehydrogenase. This is significant for the mechanism of discrimination between NAD and NADP binding sites.

Binding of NAD vs NADP

Several domains in this family are now known that, although containing most elements of the standard NAD binding site, bind NADP instead. In dehydrogenase domains that bind NADP, the extra phosphate of NADP occupies the space which, in NAD-binding domains, contains the conserved hydrogen bond between the aspartate (223 in horse liver alcohol dehydrogenase) and the adenosine ribose. Dehydrogenases appear to use several different structural mechanisms for effecting the switch in cofactor specificity.

A comparison of the structures of *E. coli* quinone oxidoreductase (cofactor NADP) and horse liver alcohol dehydrogenase (cofactor NAD) shows the adjustment in this region in this particular case. The structures of the first β–α–β–α unit are quite similar (Figure 6.26). The sequences in this region are also similar:

```
                                                      *
alcohol dehydrogenase   TCAVFGL-GGVGLSVIMGCKAAGAARIIGVDINK-DKFAKAKEVGA
quinone oxidoreductase  QFLFHAAAGGVGLIACQWAKALGA-KLIGTV-GTAQKAQSALKAGA
common                          GGVGL      KA GA   IG       K   A   GA
```

The * marks the position of the Asp conserved in NAD-binding enzymes.

Figure 6.26 shows that there has been a reconformation of the loop that connects the second strand and the second helix. In place of the Asp223–ribose hydrogen bond in alcohol dehydrogenase, there is a hydrogen bond between the N of Gly173 and a phosphate oxygen. Note that the phosphate oxygen occupies the same region of space as the asparate in the NAD-binding enzyme.

An alternative mechanism of accommodation of NADP is seen in the structure of 6-phosphogluconate dehydrogenase (Figure 6.27). An Asn appears at the position occupied by the conserved Asp in NAD-binding dehydrogenases, and forms hydrogen bonds to the phosphate as well as to

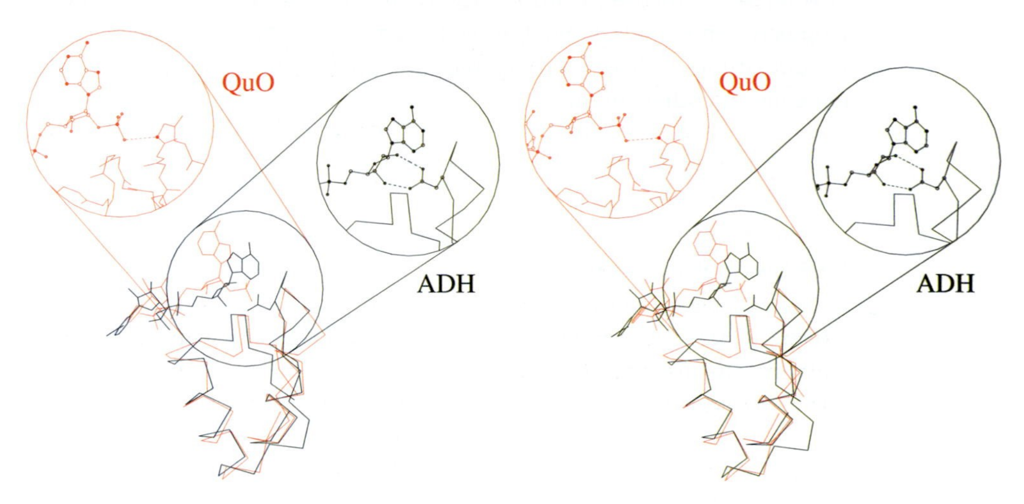

Fig. 6.26 Superposition of regions interacting with adenosine unit in horse liver alcohol dehydrogenase [2OHX] (black), binding NAD and quinone oxidoreductase [1QOR] (red), binding NADP.

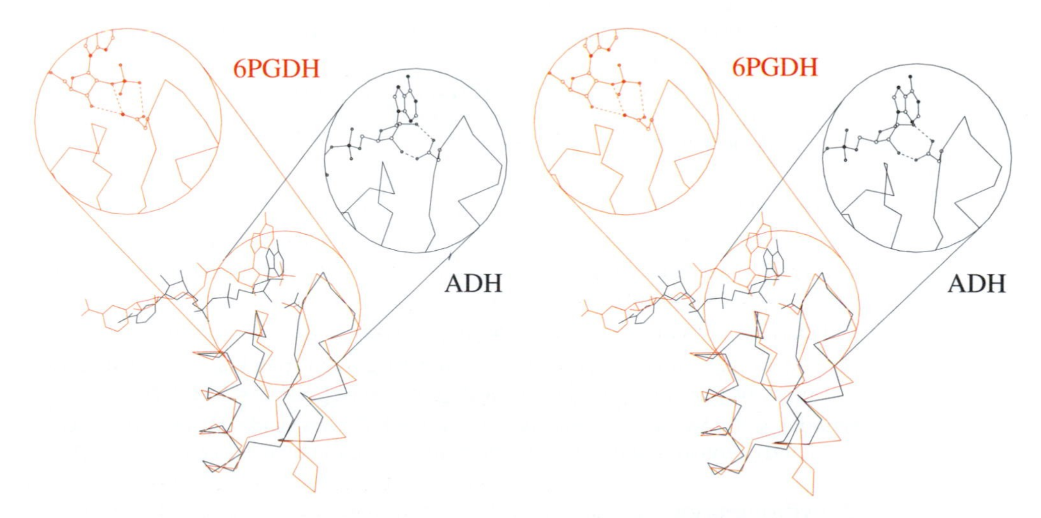

Fig. 6.27 Superposition of regions interacting with adenosine unit in horse liver alcohol dehydrogenase [2OHX] (black), binding NAD and 6-phosphogluconate dehydrogenase [1PGO] (red), binding NADP.

the sugar. The cofactor has moved away from the domain, breaking the interaction between the pyrophosphate and the loop at the end of the first β-strand.

Conclusions

Great versatility is required of NAD-binding domains. They combine with unrelated second domains. They form enzymes in different oligomeric states, which catalyse different reactions. The cofactor can vary—some bind NADP rather than NAD. Nevertheless the family retains a core of the domain, the six-stranded parallel β-sheet flanked by helices on both sides. This tertiary structure, featuring the cavity created by the long 'crossover' loop between strands 3 and 4, provides a generic cofactor binding site, and is compatible with the variety required in the interactions with other domains and other subunits.

Useful web sites

Globins: http://bmbsgi11.leeds.ac.uk/promise/GLOBINS.html
MEROPS database of peptidases:
http://www.bi.bbsrc.ac.uk/Merops/Merops.htm
PROLYSIS: proteinases and inhibitors: http://delphi.phys.univ-tours.fr/Prolysis
IUBMB EC list for peptidases:
http://www.chem.qmw.ac.uk/iubmb/enzyme/EC34

Recommended reading and references

Globins

Lesk, A.M. and Chothia, C. (1980). How different amino acid sequences determine similar protein structures: the structure and evolutionary dynamics of the globins. *J. Mol. Biol.* **136**, 225–70.

Pastore, A. and Lesk, A.M. (1990). Comparison of the structures of globins and phycocyanins: evidence for evolutionary relationship. *Proteins: Structure, Function and Genetics* **8**, 133–55.

Proteinases

Barrett, A.J., Rawlings, N.D. and Woessner, J.F. (ed.) (1998). *Handbook of proteolytic enzymes*. Academic Press: London. (A CD-ROM version is available which permits searching for substrate specificity.)

Keil, B. (1992). *Specificity of proteolysis*. Springer-Verlag, Berlin and New York.

Lesk, A.M. and Fordham, W.D. (1996). Conservation and variability in the structures of serine proteinases of the chymotrypsin family. *J. Mol. Biol.* **258**, 501–37.

Scharpe, S., De Meester, I., Hendriks, D., Vanhoof, G., van Sande, M. and Vriend, G. (1991). Proteases and their inhibitors: today and tomorrow. *Biochemie* **73**, 121–6.

Dehydrogenases

Brändén, C.-I. (1980). Relation between structure and function of α/β-proteins. *Quart. Revs. Biophys.* **13**, 317–38.

Brenner, S.E., Chothia, C., Hubbard, T.J.P. and Murzin, A.G. (1996). Understanding protein structure: using scop for fold interpretation. *Methods in Enzymology* **266**, 635–43.

Carugo, O. and Argos, P. (1997). NADP-dependent enzymes. I. Conserved stereochemistry of cofactor binding. II. Evolution of the mono- and dinucleotide binding domains. *Proteins: Structure, Function and Genetics* **28**, 10–28 and 29–40.

Fothergill-Gilmore, L.A. and Michels, P.A. (1993). Evolution of glycolysis. *Prog. Biophys. Mol. Biol.* **59**, 105–235.

Persson, B. (1997). Alcohol dehydrogenases. *Adv. Exp. Med. Biol.* **1997**, 591–4.

Exercises, problems and weblems

Exercises

6.1. Describe the basic symmetry of the NAD-binding domain. Which parts of the structure do not follow this symmetry?

6.2. On photocopies of Figure 6.1a and 6.1b, label the helices according to the standard alphabetic notation (A . . . H). If you felt you needed a hint, which other figure appearing in this book would you look at?

6.3. Suggest two pairs of helices that are *unlikely* to be in contact in the structure of any globin monomer.

6.4. Make two photocopies of Figure 6.14a. (a) Assume that one copy represents the N-terminal domain of a serine proteinase, and label the approximate positions of the catalytic histidine and aspartate. (b) Assume that the other copy represents the C-terminal domain of a serine proteinase, and label the approximate position of the catalytic serine.

6.5. On a photocopy of Figure 6.16d, write the symbols ♠, ♡, ◇ and ♣ on the residues labelled by these symbols in Figure 6.16a.

Problems

6.1. Photocopy Figure 6.12a, cut it out, and paste strands 1 and 1′ on top of each other to form a barrel.

6.2. Would it be possible for a globin structure to exist with 'normal' $i\pm4/i\pm4$ ridges-into-grooves packing at its B–E helix contact, instead of the observed crossed-ridge structure? Indicate briefly how would you go about designing an amino acid sequence, differing minimally from that of sperm whale myoglobin, to test this.

6.3. The residues packed in the central layer of elastase [3EST] are distributed among the strands of β-sheet according to the scheme BNBN. What are the equivalent schemes for the top and bottom layers? (See Figure 6.16.)

Weblems

6.1. Compare the lengths of the α-helices in sperm whale myoglobin with those in globins from lamprey and *Glycera*.

6.2. Are there any proteins other than chymotrypsin-like serine proteinases that contain β-barrels with six strands and shear number 8?

6.3. How much surface area is buried in the complex between trypsin and bovine pancreatic trypsin inhibitor?

6.4. In sperm whale myoglobin, how much surface area is buried in the G–H contact?

6.5. In sperm whale myoglobin, which helices make contact with the haem group?

6.6. Find a known inhibitor of human neutrophil elastase.

6.7. Identify a serine proteinase of the chymotrypsin family that cleaves the peptide bond after Phe but not after Trp.

6.8. The channel-activating proteinase (CAP1) regulates the activity of the epithelial sodium channel. (a) To what protease of known structure is CAP1 from *Xenopus laevis* most closely related? (b) Using a server available on the web build a model of *X. laevis* CAP1. Make a reasonable prediction of its specificity.

6.9. Mineralocorticoid receptors show relatively weak specificity, and are regulated in different tissues by local metabolic control of concentrations of different possible ligands. In the human kidney, 11β-hydroxysteroid dehydrogenase-type 2 catalyses the rapid conversion of cortisol to an inactive derivative, as a result of which the receptor is sensitive to aldosterone, which the enzyme does not alter. (Mutations impairing the activity of the enzyme—or ingestion of large amounts of licorice, which contains an inhibitor—cause cortisol to swamp the receptors.)

(a) Find proteins in *C. elegans* that most closely resemble human 11β-hydroxysteroid dehydrogenase-type 2, These proteins are putatively similar to ancestors of the human enzyme.

(b) Find other mammalian homologues of the *C. elegans* enzymes. What is the substrate specificity of the closest mammalian relative of the the *C. elegans* enzymes? What does this suggest about the evolutionary origin of human 11β-hydroxysteroid dehydrogenase-type 2?

(c) Find an enzyme in *E. coli* that is homologous to the *C. elegans* enzymes identified in (a).

(d) Compare the amino acid sequences of human 11β-hydroxysteroid dehydrogenase-type 2 and the homologues you have found. Do the results suggest that the *E. coli* enzyme is a prokaryotic precursor of the eukaryotic ones? What alternative hypothesis might you consider to explain the results?

CHAPTER 7

Some proteins of the immune system

Antibody structure

The vertebrate immune system has the job of identifying foreign substances, and defending the body against them. It has evolved to be able to recognize the entire organic world.

Upon challenge by a pathogen, the initial *primary immune response* mobilizes antibodies with dissociation constants in the micromolar range. This is relatively weak binding, compared with the nanomolar dissociation constants of the antibodies of the secondary response, produced within a few days. These results are impressive, and—as the AIDS epidemic has forcibly demonstrated—the immune system is essential for survival.

The generation of diversity in the immune system differs in several ways from the Darwinian model of random generation of variation followed by selection. For most proteins, evolution has produced a unique molecule (or a relatively small number) within each species, adapted for each particular role. For antibodies, in contrast, nature has produced a system with a premium on generation and *maintenance* of diversity. During our lifetimes, most of us will synthesize three major haemoglobin molecules—embryonic and foetal versions followed by an adult form—but 10^{10} antibodies.

The mechanism by which antibody diversity is created is now understood in considerable detail, both at the level of gene sequences and at the level of protein structure. The genes for antibodies produced in the primary immune response arise by assembly from a combination of DNA segments. Each light chain gene contains a V segment, a J segment and a C segment; each heavy chain gene contains a V segment, a D segment, a J segment, and several C segments. The total number of possible genes is at least as high as the product of the number of possible choices for each segment. In fact it is much higher because of extra variability introduced in joining the V, D and J segments. These genes then undergo somatic mutation to tune the affinity and specificity of the antibodies for the secondary response.

Like enzymes and many other proteins, immunoglobulins have specific binding sites that interact with ligands, usually with lock-and-key complementarity. Molecules related to antibodies appear on cell surfaces, to mediate cell–cell recognition and signalling processes, or triggering proliferation of particular cells in response to antigenic challenge. Many other related proteins are known; antibodies are members of a large superfamily of proteins, not all of which participate in the immune response (see following Box).

Some proteins of the immunoglobulin superfamily

Protein	Function
Immunoglobulins	Antibodies B lymphocyte antigen receptors
T subset antigens	
CD4, CD8α,CD8β	Interact with peptide-presenting MHC proteins
CTLA4, CD28	Coreceptors involved in T-cell activation
MHC complex	
MHC Class I α chain	Present peptides to T-cell receptors
MHC Class I β_2 microglobulin	
MHC Class II α chain	
MHC Class II β chain	
T-cell receptor complex	
TCR α, β, γ, δ chains	Recognition of MHC–peptide complexes
CD3 γ, δ, ϵ chains	
T-cell adhesion molecules CD2, LFA-3	CD2 of T-cells interacts with LFA-3, mediating cell adhesion
Immunoglobulin receptors	
FcϵRI	IgE receptor
FcγRI	binds aggregated IgG
Molecules of the nervous system	
Neural adhesion molecule (NCAM)	Mediates neural cell adhesion
Myelin associated gp (MAG)	Myelination (?)
Growth factor receptors	
Platelet-derived growth factor receptor Colony stimulating factor-1 receptor	Interact with growth factors to trigger cell division, etc.
Muscle proteins	
Telokin	Domain of myosin light chain kinase
Titin	Muscle ultrastructure and elasticity

For a longer list, see: Halaby D.M. and Mornon, J. (1998). *J. Mol. Evol.* **45**, 389–400.

The structures of immunoglobulins

Common features of immunoglobulins were first recognized physico-chemically and serologically, then described in terms of the amino acid sequences, and only later revealed in atomic detail by X-ray crystallography. Limited proteolytic digestion and cleavage of disulphide bonds showed that antibodies contain multiple polypeptide chains (Figure 7.1). The chains were distinguished by size into light (L) chains (M_r about 23 000) and heavy (H) chains (M_r about 50 000–70 000). Systematic internal homologies in the amino acid sequences suggested that immunoglobulins are composed of multiple copies of related units, each approximately 100 residues long. These similarities indicated that each of the basic units forms an individual, quasi-independent, three-dimensional structure, or domain, with a common folding pattern. Light chains contain two domains, and heavy chains

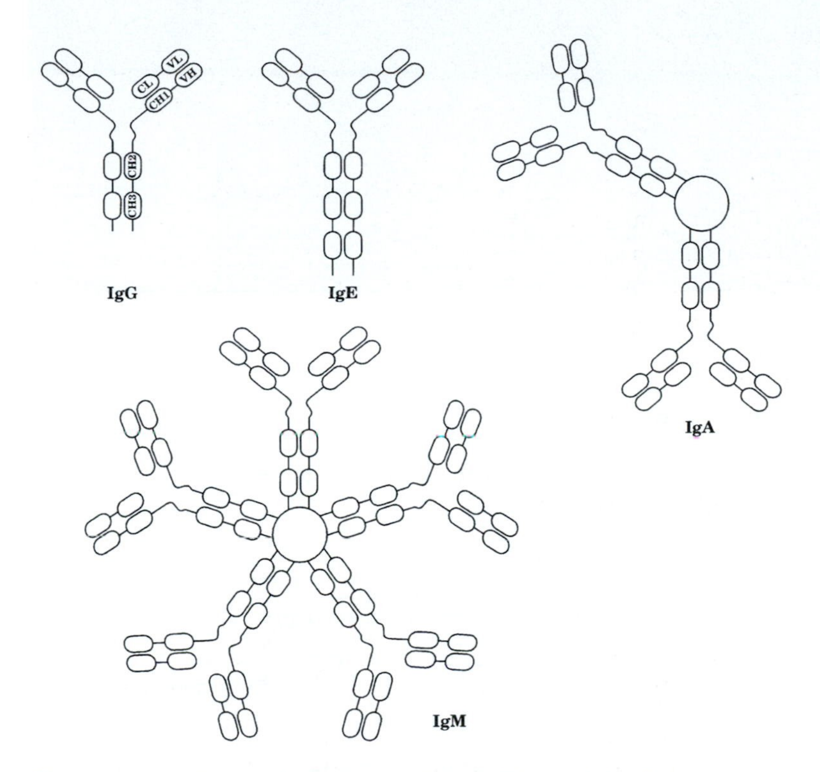

Fig. 7.1 Most antibody molecules contain light and heavy chains. Each comprises one variable domain and different numbers of constant domains. The combination of 2 light + 2 heavy chains is a higher-order building block. Immunoglobulins of different classes (IgG, IgA, IgM, IgD and IgE) show various states of oligomerization, with additional chains where necessary serving as linkers. Soluble IgDs are like IgGs.

The IgG molecule contains four polypeptide chains: two identical light chains each containing one variable and one constant domain, denoted VL and CL; and two identical heavy chains each containing one variable and three constant domains, denoted VH, CH1, CH2, and CH3. The CH1 and CH2 domains are linked by a peptide called the hinge region. The angle between the VL–VH domain pair and the CL–CH1 domain pair is called the elbow angle. In solution, these joints are probably flexible. Immunoglobulins also contain carbohydrate moieties not shown in this figure.

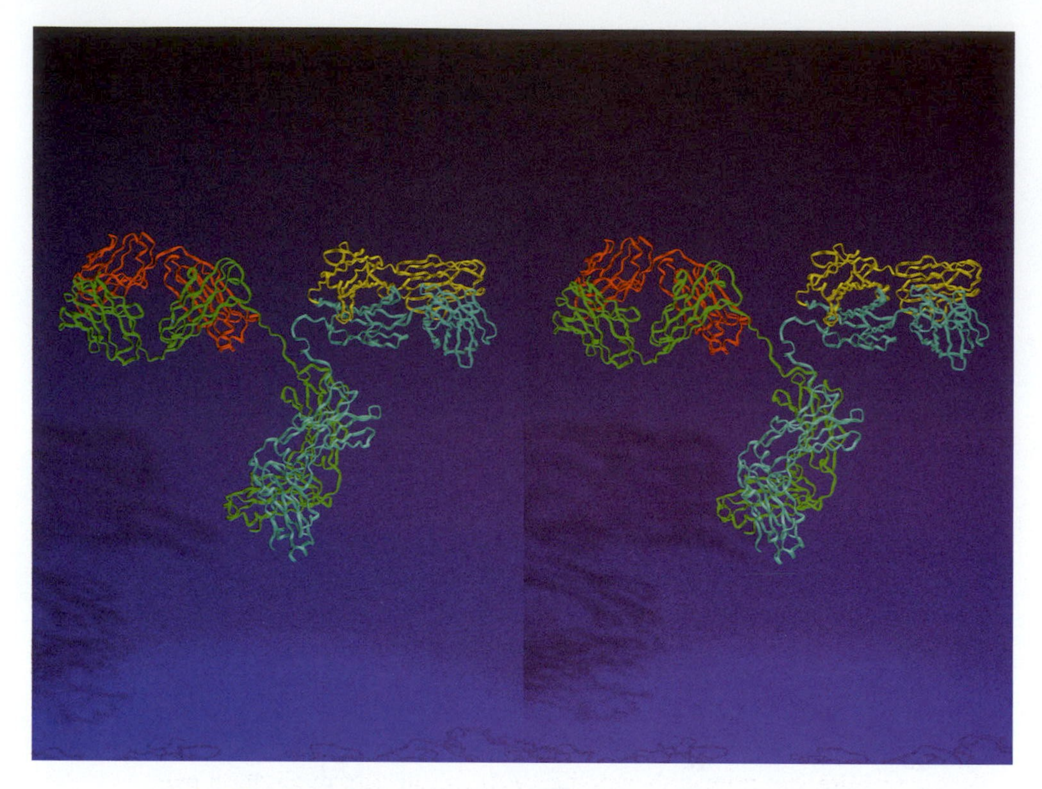

Fig. 7.2 Structure of complete immunoglobulin G [1IGT].

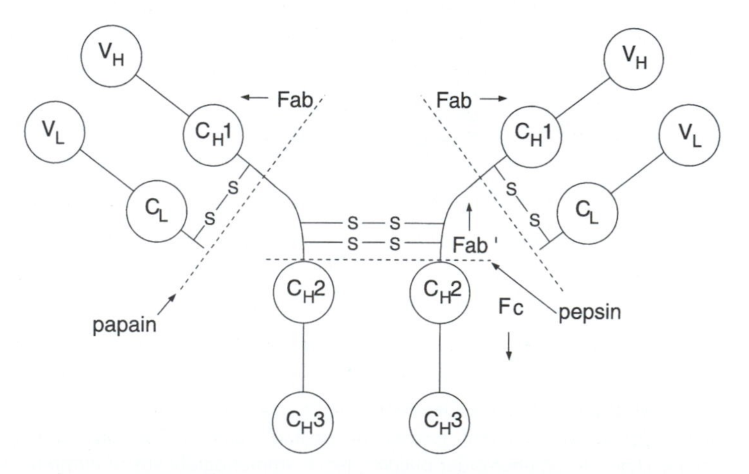

Fig. 7.3 A schematic diagram of the structure of an IgG, showing the distribution of domains in the heavy and light chains, the interchain disulphide bridges, and the definitions of the fragments—Fab, Fab′ and Fc—produced by limited proteolytic cleavage.

contain four or five domains. Light chains were distinguished, originally on the basis of their amino acid sequences, into κ and λ classes or isotypes. In the human, κ and λ light chains are present in comparable proportions; in the mouse κ light chains predominate.

Before the first crystal structure determinations of immunoglobulins, Kabat and co-workers were able to identify the regions involved in antigen-binding by analysis of the distribution of variability in aligned immunoglobulin sequences. They recognized two types of domains: variable (V) and constant (C); the variability in sequence being much lower in constant than in variable domains. Within the variable domains they noticed regions of still higher variability—called the hypervariable regions or complementarity determining regions (CDRs)—that they correctly and presciently identified as responsible for antibody specificity. The structurally-conserved regions of the variable domains outside the CDRs are called the framework.

When the first structures of immunoglobulin domains were determined by X-ray crystallography, it appeared that constant and variable domains had a similar double β-sheet structure in the framework, that the hypervariable regions corresponded to surface loops in the variable domains, and that these loops did indeed interact with antigens (see Figure 7.2.)

Different classes of immunoglobulins—IgG, IgA, IgM, IgD and IgE—differ in the assembly of their chains and domains. Figure 7.1 shows the domain structure of the different classes of immunoglobulins. Molecules in the class best known structurally, the IgGs, usually have two heavy chains each containing four domains and two light chains each containing two domains. The antigen-combining site in most IgGs is formed from three loops from the light chain and three from the heavy chain. Each IgG has two copies of the binding site, one from each light–heavy chain pair. This duplication permits multiple interactions with antigens to form aggregates.

Figure 7.3 defines the fragments produced by limited proteolytic cleavage. Most of the structural analysis has been carried out on Fab fragments. Useful molecules that have been produced by genetic engineering are single-chain Fv fragments, containing only VL and VH domains, with the C-terminus of the VH domain linked by a flexible peptide to the N-terminus of the VL domain, or vice versa.

The interactions between the different chains in an IgG include the disulphide bridges (see Figure 7.3); and interfaces between corresponding domains—VL-VH, CL-CH1, and CH3-CH3 but not CH2-CH2—that pack together to form extensive van der Waals contacts. Like other oligomeric proteins, the interior interfaces are formed by the packing of complementary surfaces. This complementarity of fit fixes the relative spatial disposition of the pieces that interact. The tendency to conserve the residues involved in these interfaces explains why different light and heavy chains can pair fairly freely to form complete immunoglobulins. In the case of the VL–VH interaction, the conservation of the relative geometry has the additional important consequence that, to a reasonable first approximation, *the double β-sheet frameworks of VL and VH domains form a scaffolding of nearly constant structure on which the antigen-binding site is erected.*

Figures 7.4 and 7.5 show the folding patterns and sheet diagrams of VL, VH and C domains. Each has the form of a double β-sheet 'sandwich.' In

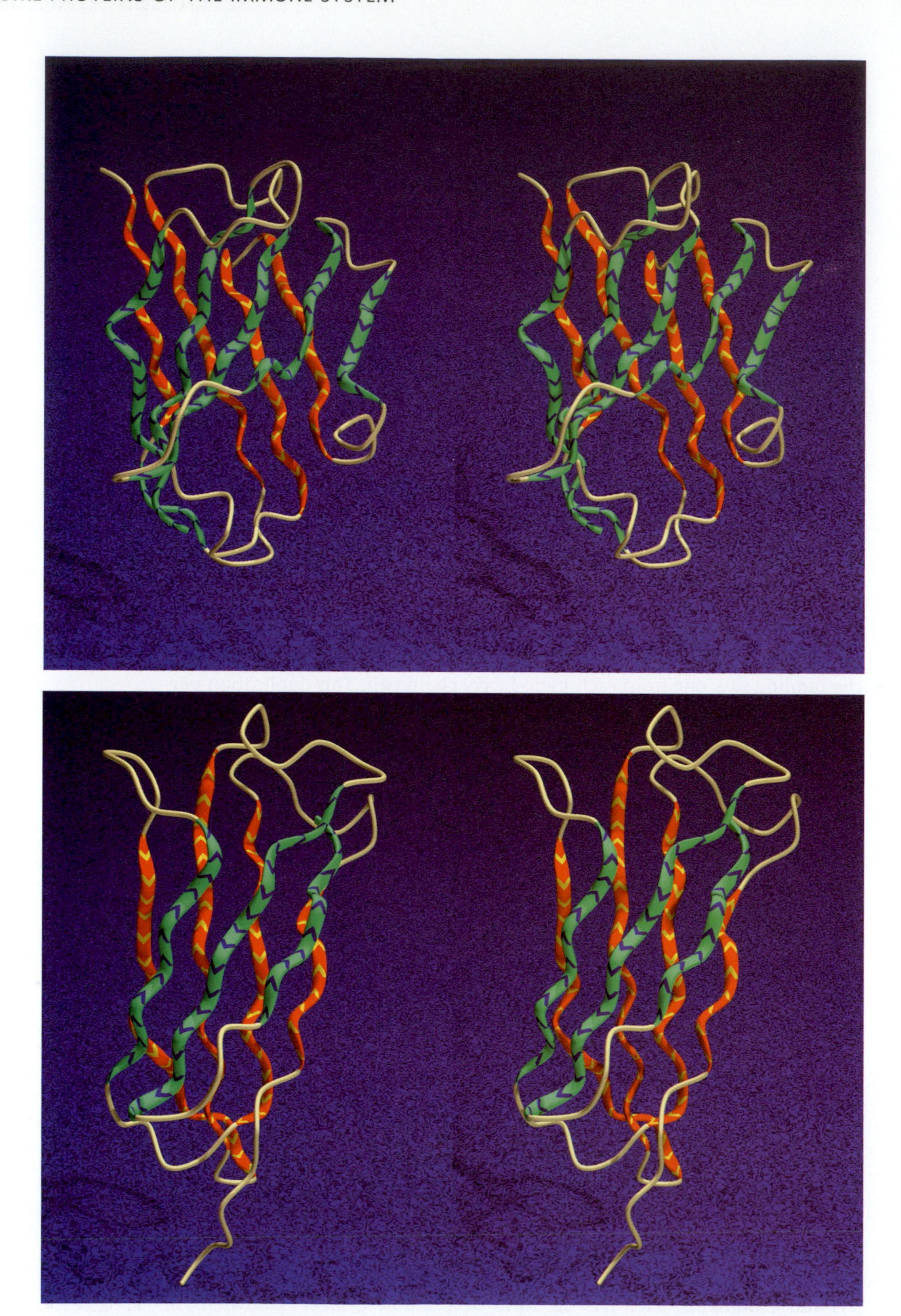

Fig. 7.4 The folding patterns of (a) variable domains and (b) constant domains. The domains shown here are the VL and CH1 domains of Fab KOL [2FB4]. Each domain contains two β-sheets; one is shown in green and the other in red.

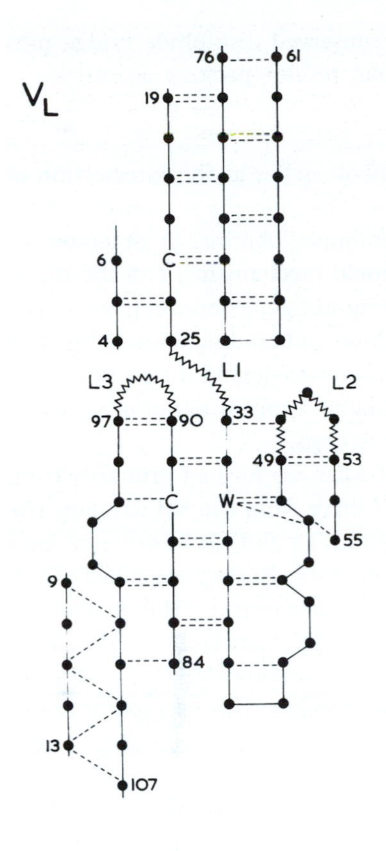

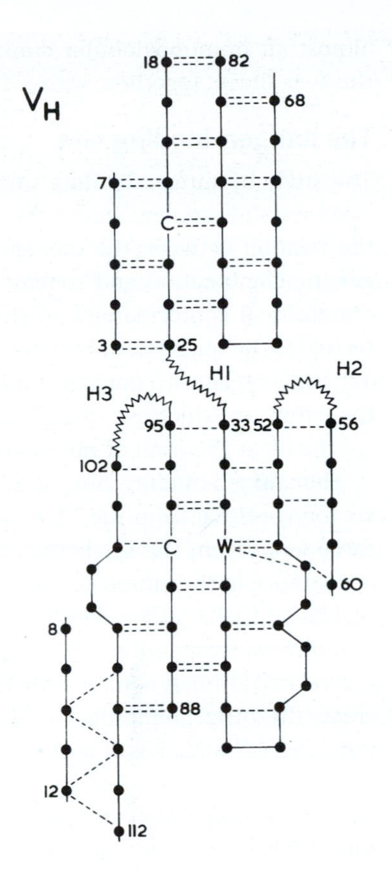

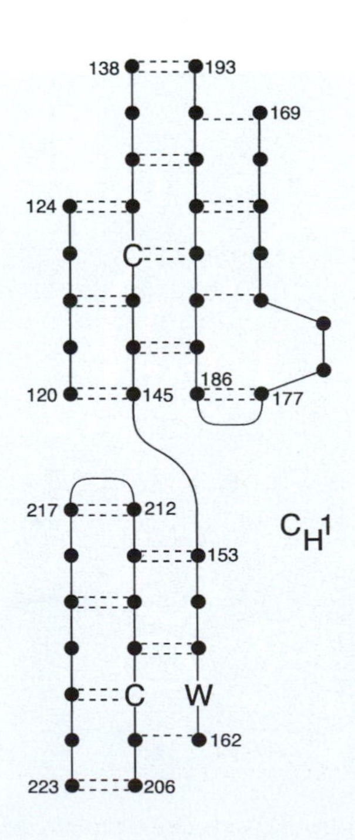

Fig. 7.5 Schematic representations of typical hydrogen-bonding patterns of VL, VH and C domains. The two sheets have been opened like a book with the spine horizontal. Strands in the upper sheets in this figure are drawn in red in Figure 7.4; strands in the lower sheets in this figure are drawn in green in Figure 7.4.

almost all immunoglobulin domains, a conserved disulphide bridge pins the two sheets together, with a tryptophan residue packing against it.

The antigen-binding site

The study of antigen-binding sites of antibodies lies at the intersection of two topics of interest. One is the biology of the immune response: what is the relation between the sequences of immunoglobulins, as generated by genetic combination and somatic mutational mechanisms, and the three-dimensional conformations of the resulting antigen-combining sites? The second is the question of how the recognition and binding of an antigen by antibodies relates to our understanding of protein–ligand interactions, and the extent to which we can give a quantitative explanation of affinity and specificity at the level of interatomic interactions.

The antigen-binding sites of most antibodies are formed primarily from six loops—three from the VL domain and three from the VH domain. The three loops from the VL domain are called L1, L2 and L3, in order of their appearance in the amino acid sequence; alternatively, they are called CDR1, CDR2 and CDR3 (*CDR = complementarity-determining region*). VH domains contain three corresponding CDRs: H1, H2 and H3. (The antibodies of the camel and related animals contain only heavy chains, and three VH CDRs suffice to create the antigen-binding site. This observation has obvious applications to the design or selection of artificial antibodies for therapy.) Four of the loops, L2, H2, L3 and H3, are β-hairpins. (Recall that a β-hairpin is a loop that links successive antiparallel strands of a single β-sheet.) In contrast, L1 and H1 form bridges from a strand in one of the two β-sheets to a strand in the other (see Figure 7.6).

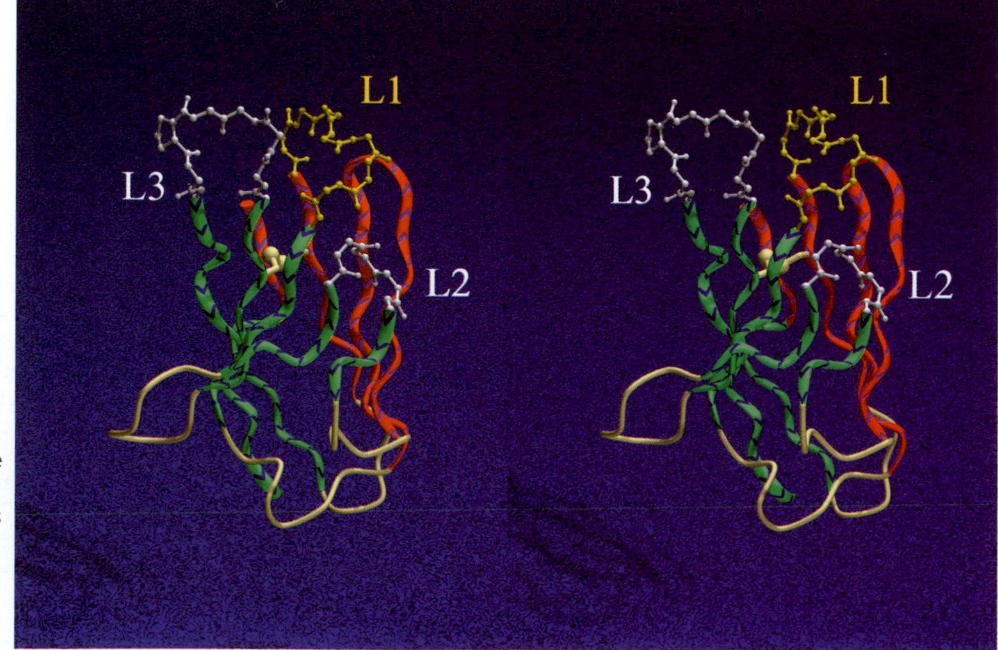

Fig. 7.6 The Vκ domain REI, indicating the antigen-binding loops L1, L2, and L3 [1REI].

Figures 7.7 and 7.8 illustrate the construction of the antigen-binding site. Figure 7.7 shows the binding of a small organic ligand, or hapten—phosphorylcholine—by the immunoglobulin McPC603. An orientation was chosen looking down onto the antigen-binding site—an 'antigen's-eye view.' Figure 7.8 shows the complex of Fab HyHEL-5 and hen egg white lysozyme.

In another complex, between Fab D1.3 and hen egg white lysozyme, the interface is a relatively flat patch of dimensions 20×30 Å, although one glutamine sidechain from the lysozyme inserts into a pocket in the antibody. A total of 17 residues from the antibody make contact with a total of 17 residues from the lysozyme. There are 32 pairs of residues in contact. Of the 17 residues from the antibody, 14 are from the CDRs. The other three are framework residues, but they are adjacent to CDRs.

In the lysozyme–antibody complexes of known structure, there appears to be rather little change in structure of either protein relative to the unliganded states. These complexes therefore fit the picture of recognition by fairly rigid bodies with preformed complementarity. This is not observed however in some complexes of antibodies with oligopeptides clipped out of protein structures. For example, in the complex between Fab 17/9 and a eight-residue region from influenza virus haemagglutinin, the peptide in the complex has a conformation different from its conformation in the native haemagglutinin structure. Comparison of the structure of the Fab fragment of the antibody in free and liganded forms shows that there has also been some structural change in the antibody.

Somatic mutation and the maturation of the antibody response

Somatic mutation is the process whereby antibodies active in the primary response are tuned in affinity and specificity by tinkering with their sequences to produce the 'mature' antibodies of the secondary response. Whereas the diversity in the primary antibodies is highest in H3 and L3, around the centre of the antigen-binding site, somatic mutations spread the diversity to its periphery. Usually, somatic mutations are isolated point substitutions, and not insertions or deletions.

P.G. Schultz and co-workers have studied maturation of an anti-nitrophenyl phosphonate catalytic antibody by X-ray crystallography. They solved the structures of the primary (germ-line) and somatically mutated Fab fragment, each with and without ligand. Their results reveal both the sites and the structural effects of mutations.

There are nine amino-acid sequence changes between germ-line and mature antibody; three in the light chain and six in the heavy chain. One of the light chain mutants appears within the L1 loop; the other two are in positions in the sequence near L1 and L2. One of the heavy-chain mutants appears in H2, three are in regions adjacent to or in contact with the antigen-binding loops, and two are surface residues at the opposite ends of the domains, and would appear to have little effect on the antigen-binding site. No mutant is at a position directly in contact with the antigen. No mutant

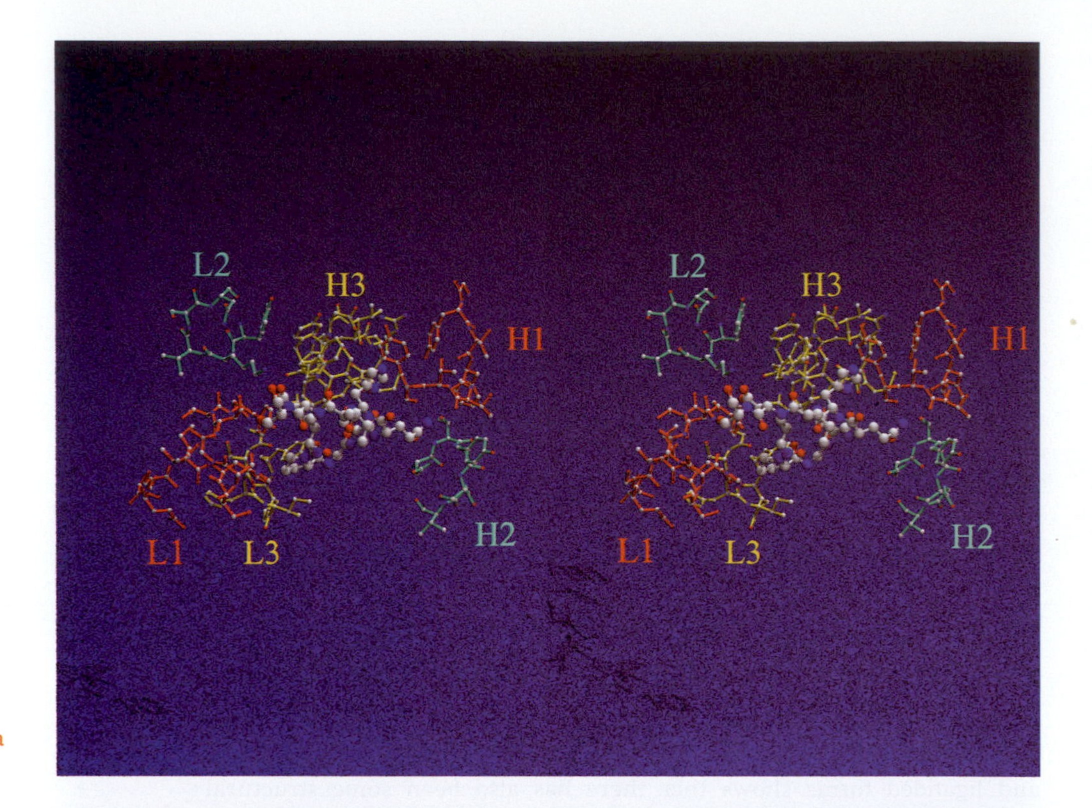

Fig. 7.7a

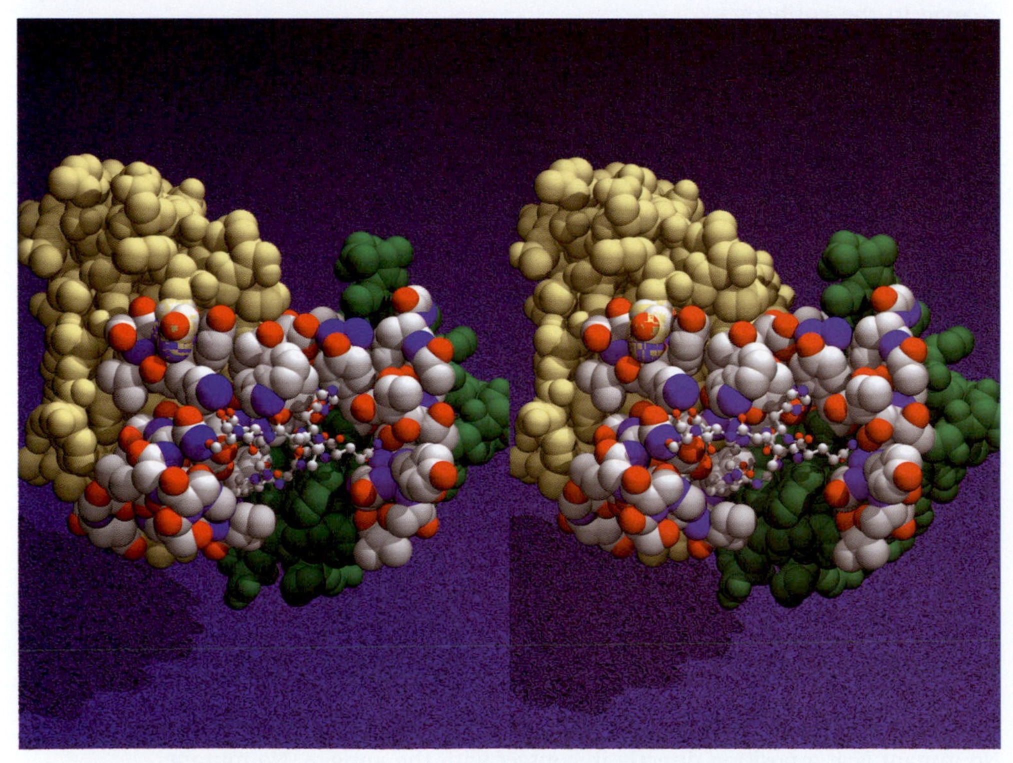

Fig. 7.7b

Fig. 7.7 (Facing page) The spatial distribution of antigen-binding loops, or CDRs, in the antigen-binding site of McPC603 [2MCP]. The ligand is phosphorylcholine. The orientation is chosen to give a view down onto the antigen-binding site, an 'antigen's-eye view'.

The light chain loops are at the left, the heavy chain loops are at the right. Reading clockwise starting at '7 o'clock', the loops appear in the order L1, L2, H3, H1, H2, L3. There is a rough symmetry in the arrangment of the loops: L1 is opposite H1, L2 is opposite H2, and L3 opposite H3. The central position of H3 is noteworthy.

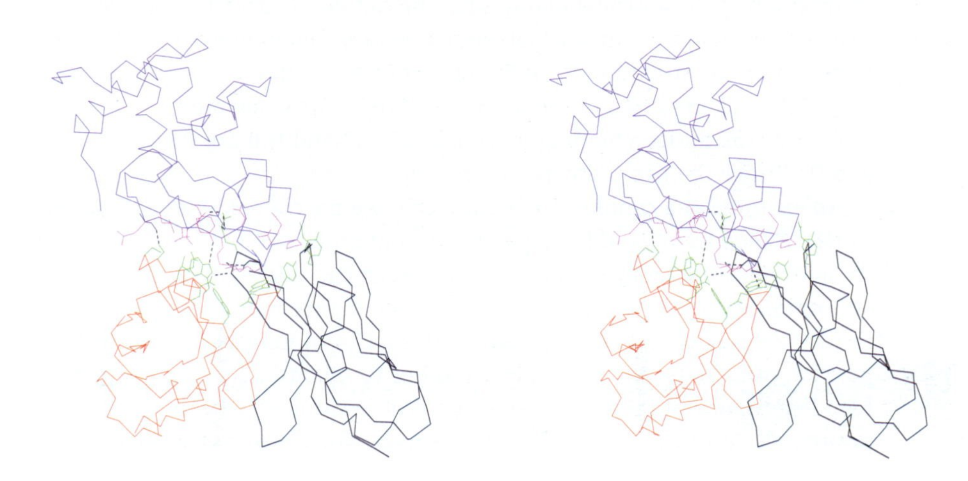

Fig. 7.8 Interactions between HyHEL-5 (VL, black; VH, red) and hen egg white lysozyme (blue) [1BQL]. Residues in the lysozyme that make contact with the antibody are shown in magenta. Residues in the antibody that make contact with the lysozyme are shown in green. Broken lines indicate hydrogen bonds between antigen and antibody.

appears in H3, even though there are extensive contacts between H3 and the antigen. The conformations of two of the antigen-binding loops (H1 and H2) differ between germ-line and mature antibody.

The antigen-binding site of the mature antibody has a similar conformation in ligated and unligated states; a fixed conformation complementary to the antigen. This structure is shared by the ligated state of the germ-line antibody, but the unligated state of the germ-line antibody shows differences in conformation. That is, the germ-line antibody *can* adopt the antigen-binding conformation, and is *induced* to do so by the ligand; the mature antibody adopts this conformation even in the absence of ligand. Who among readers has not sat at a table that irritatingly rocked between two states, and corrected it by inserting an empty book of matches under the short leg? Who has realized that he or she was mimicking part of the mechanism of antibody maturation?

Canonical structures of antigen-binding loops of antibodies

Five of the six antigen-binding loops of immunoglobulin structures have only a small discrete repertoire of mainchain conformations, called *canonical structures*. These conformations are determined by a few particular residues within the loop, or outside the loop but interacting with it. Among corresponding loops of the same length, only these residues need to be conserved to maintain the conformation of the loop. Other residues in the sequences of the loops are thus left free to vary, to modulate the surface topography and charge distribution of the antigen-binding site. The canonical structure specifies the main chain conformation of a loop quite precisely. Different exemplars of the same canonical structure in different antibodies differ by no more than 0.4–0.5 Å r.m.s. deviation of the positions of the main chain atoms.

Four of the antigen-binding loops (L2, L3, H2 and H3) are β-hairpins. L1 and H1 bridge the two β-sheets (Figure 7.9).

An example of a canonical structure that we have already seen (Chapter 3), the L3 loop from Vκ REI, is a β-hairpin, containing a *cis* proline at position 95, and stabilized by hydrogen bonds between the sidechain of the residue at position 90, just N-terminal to the loop, and main chain atoms of residues in the loop. The sidechain at position 90 is a Gln in REI (the residue at this position can also be an Asn or a His in other Vκ chains). The combination of loop length, one of these polar sidechains at position 90 and the proline at position 95 constitute the 'signature' of this conformation of this loop, from which it can be recognized in a sequence of an immunoglobulin for which the structure has not been experimentally determined.

B. Al-Lazikani, C. Chothia and the author have published a roster of canonical structures (see Figure 7.10). The results have been tested and refined by predictions of the atomic structure of binding sites in new immunoglobulin structures *before* they have been determined by X-ray crystallography.

What is the biological significance of this robustness of the main chain conformation of the antigen-binding loops? Perhaps it plays a role in the tuning of antibodies by somatic mutation. That is, in the system as it has evolved, a change in a residue in the antigen-combining site will in most cases produce only a conservative structural change in the main chain—this is what one would want in a system that already shows affinity and requires only 'fine-tuning'. Consider the alternative: if most mutations completely altered the main chain conformation of the loops, the effect would be to produce a succession of independent primary responses, rather than a secondary response with structures perturbed in only minor (but crucially important) ways from a set of already selected primary antibodies. The reader should not underestimate the power of very tiny structural changes to produce very large effects on binding affinity.

Greater variability in the H3 loop

H3, the third hypervariable region of the heavy chain, is far more variable in length, sequence and structure than the other antigen-binding loops. It

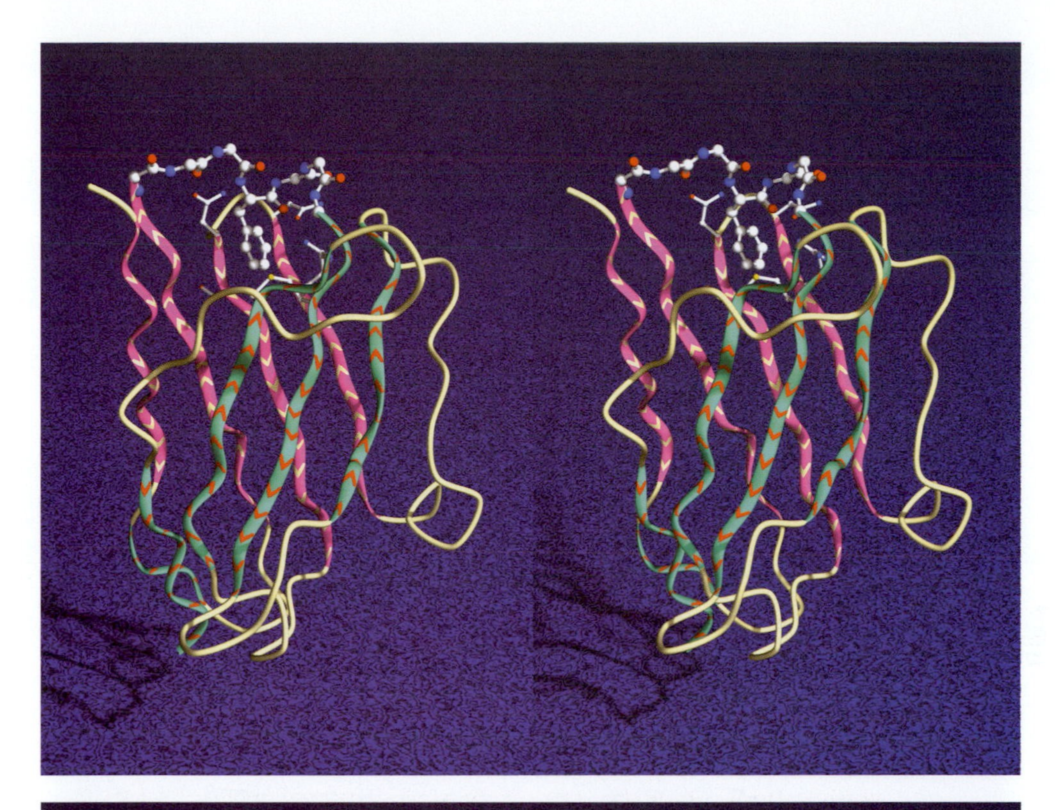

Fig. 7.9a

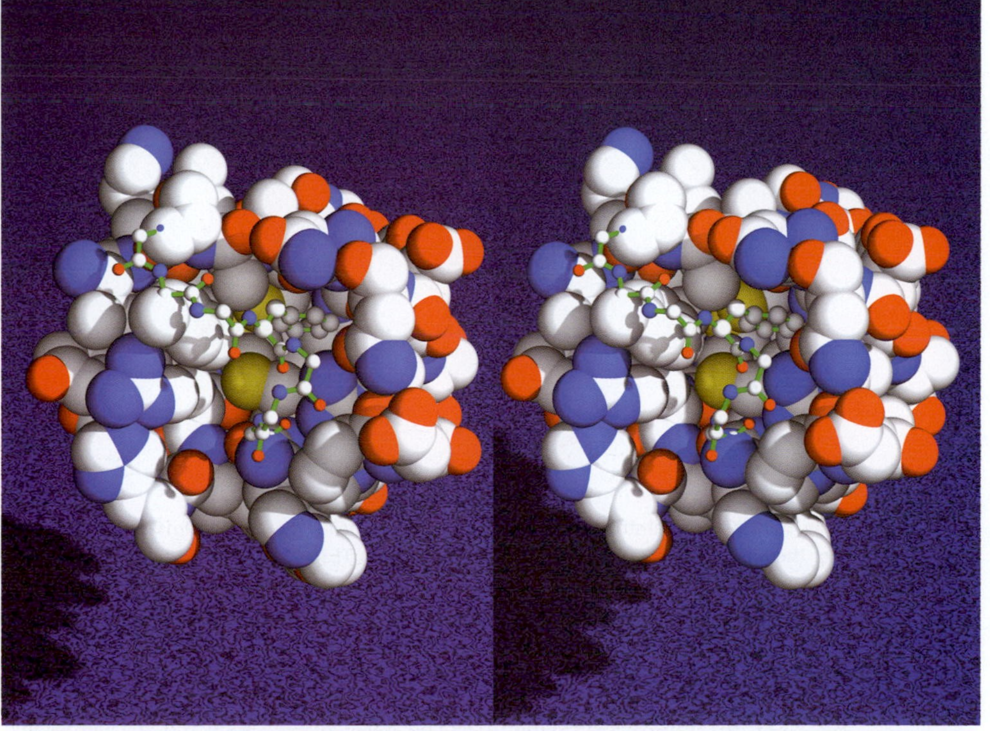

Fig. 7.9b

Fig. 7.9 The H1 loop of Fab J539 [2FBJ]. An important contribution to the stabilization of the conformation of this loop is the packing of residue Phe29 into the region between the two sheets. (a) View perpendicular to sheets of VH domain; H1 loop at top. (b) Looking down at H1 loop.

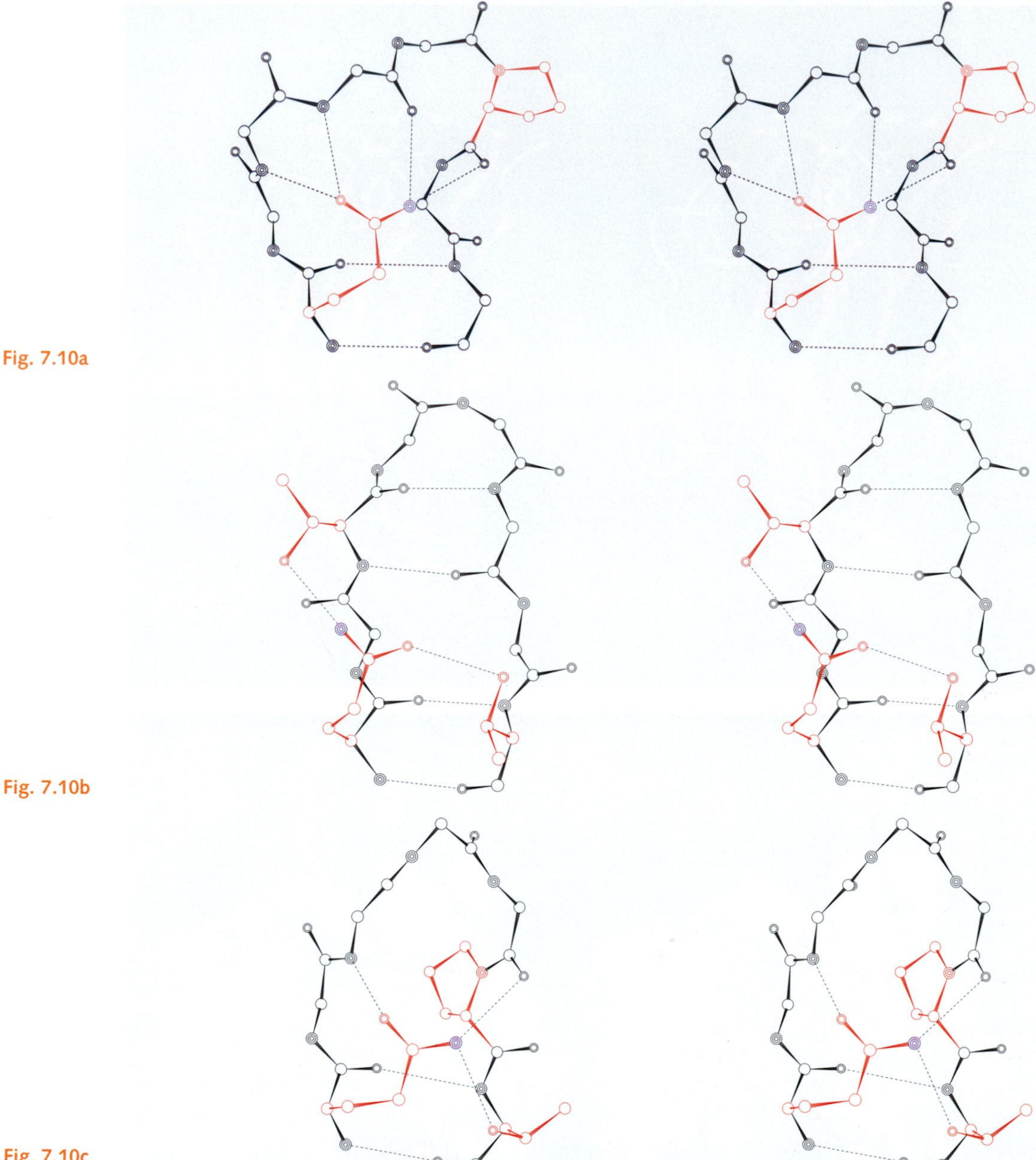

Fig. 7.10 Canonical structures for Vκ L3 loops and their signature patterns in the sequences of the region. (a) Vκ L3 canonical structure 1, from TE33 [1TET]. This is the most common canonical structure of L3 loops from Vκ domains. It has a characteristic *cis*-proline near the C-terminus. A polar sidechain at the residue just before the loop forms hydrogen bonds to stabilize the conformation of the region. (b) Vκ L3 canonical structure 2, from J539 [2FBJ]. (c) Vκ L3 canonical structure 3, from HyHEL-5 [3HFL]. (d) Vκ L3 canonical structure 4, from AN02 [1BAF]. (e) Vκ L3 canonical structure 5, from anti-gp41 monoclonal antibody [1DFB]. (f) Vκ L3 canonical structure 6, from 17E8 [1EAP]. (g) Signature sequences of Vκ L3 canonical structures. Residues in parentheses are in the framework.

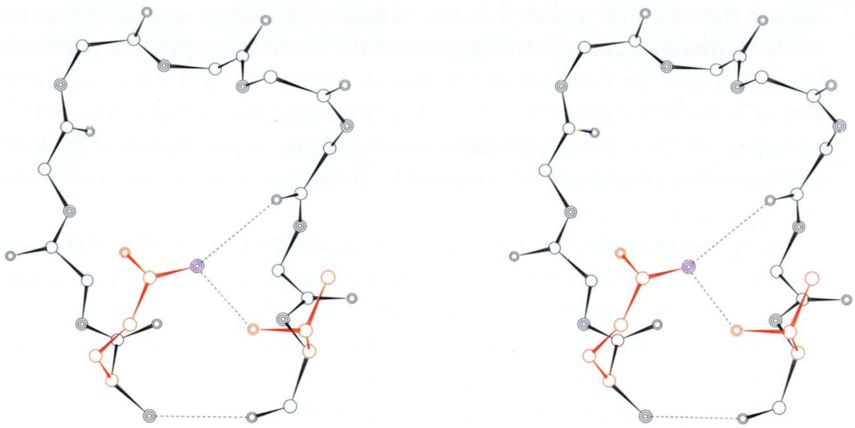

Fig. 7.10d

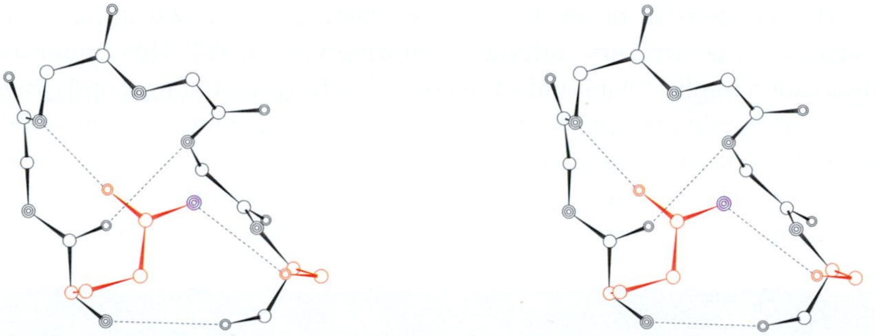

Fig. 7.10e

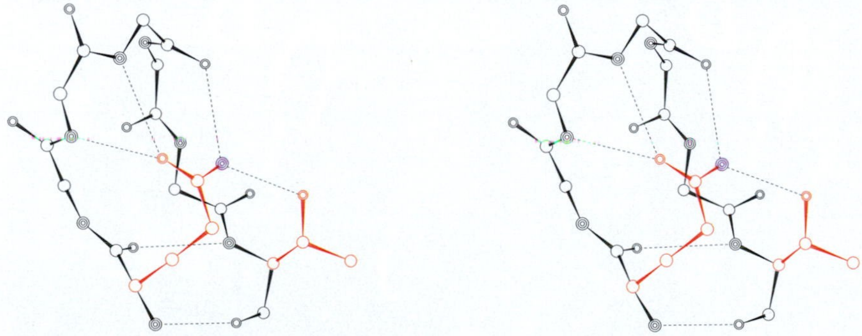

Fig. 7.10f

Canonical structure	Length	Residues Sequence pattern								
1	6	(90)	91	92	93	94	95	96		(97)
		QNH					P			ST
2	6	(90)	91	92	93	94	95	96		(97)
		QNH				P				ST
3	5	(90)	91	92	93	94	95			(97)
		QNH					P			ST
4	4	(90)	91	92	93	94				(97)
		QNH								ST
5	7	(90)	91	92	93	94	95	96	96a	(97)
		QNH						P		ST
6	5	(90)	91	92	93	94	95			(97)
		QNH				L	Y			ST

Fig. 7.10g

cannot therefore be included in the canonical-structure description of the conformational repertoire that applies to the three hypervariable regions of VL chains and the first two of VH chains. Because the H3 loop falls in the region of the V–D–J join in the assembly of the immunoglobulin heavy chain gene (Figure 7.11), several mechanisms contribute to generation of its diversity, including combinatorial choice of V, D and J gene segments, and imprecise splicing at the junctions.

In expressed antibodies, H3 appears prominently at the centre of the antigen-binding site (Figure 7.12). Given this central position, H3 makes significant interactions—with other loops, with the framework, with the light-chain partner, and with ligands—that influence its conformation. Thus H3, in contrast to the other five antigen-binding loops, has a conformation that depends strongly on its molecular environment. Indeed, structures containing the VH domain of antibody B1-8 combined with two different VL domains show two very different conformations of H3. This important observation implies that (unlike the other five antigen-binding loops) general rules governing the conformation of H3 *must* involve interactions outside its local region in the sequence.

It has proved convenient to divide the H3 region into a torso section comprising residues proximal to the framework—four residues starting from the

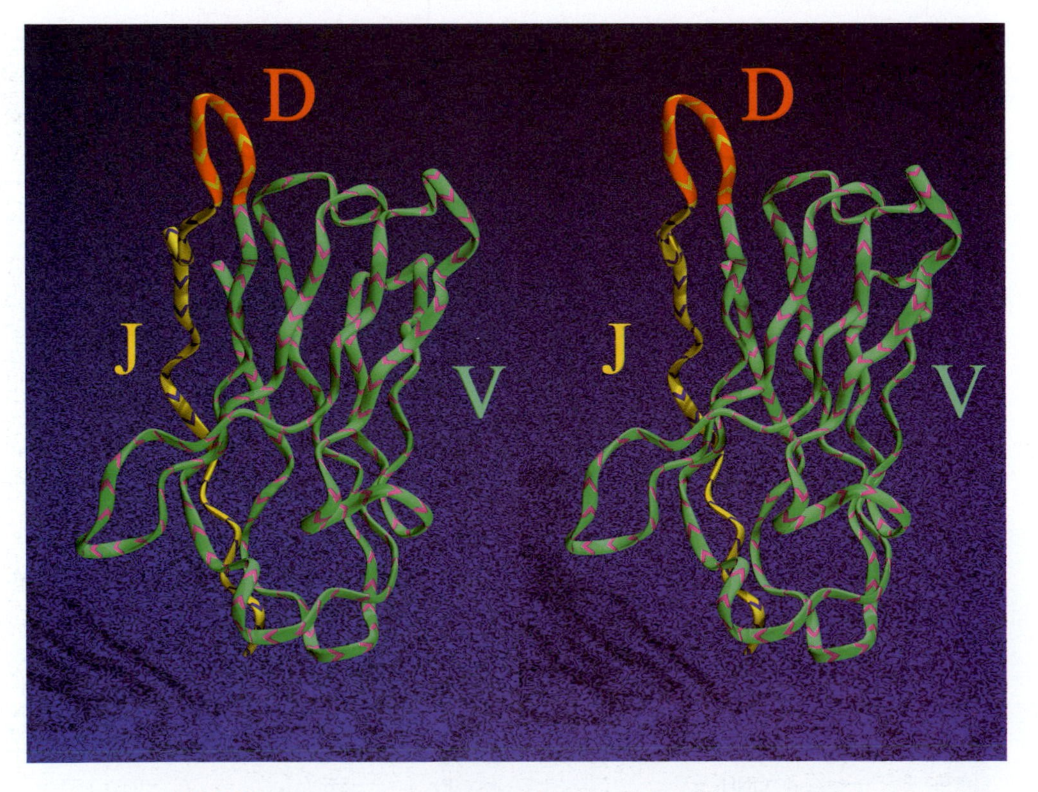

Fig. 7.11 H3 appears in the region of the V–D–J join in the assembly of the heavy chain gene. In the VH domain of McPC603, shown in this figure, the region encoded by the D segment coincides almost exactly with the H3 region.

conserved Cys at position 92 from the N-terminus, and six residues from the C-terminus—and a head, the apex of the loop. For H3 structures that contain more than 10 residues, there are two main classes of torso conformations (Figure 7.13). In the major class, the conformation of the torso has a β-bulge at residue 101. For a few H3 regions, the torso does not contain a bulge, but continues the regular hydrogen-bonding pattern of the β-sheet.

The choice of bulged or non-bulged torso conformation is dictated primarily by the sequence: a bulged torso will be formed when residues are present that permit formation of a salt bridge between the sidechains of an Arg or Lys at position 94 and an Asp at position 101, unless a residue at position 93 can form the salt bridge. Indeed, the limits of the torso region were chosen as the largest set of residues proximal to the framework that appeared to have a limited repertoire of conformations, as in the canonical structure model of other antigen-binding loops.

In contrast, the heads or apices of the H3 loops have a very wide variety of conformations. In shorter H3 regions, and in those containing the non-bulged torso conformation, the heads follow the rules relating sequence to structure in short hairpins. For longer H3 regions containing the bulged torso conformation, there are many very different conformations of the head, which can be catalogued but which are difficult to classify.

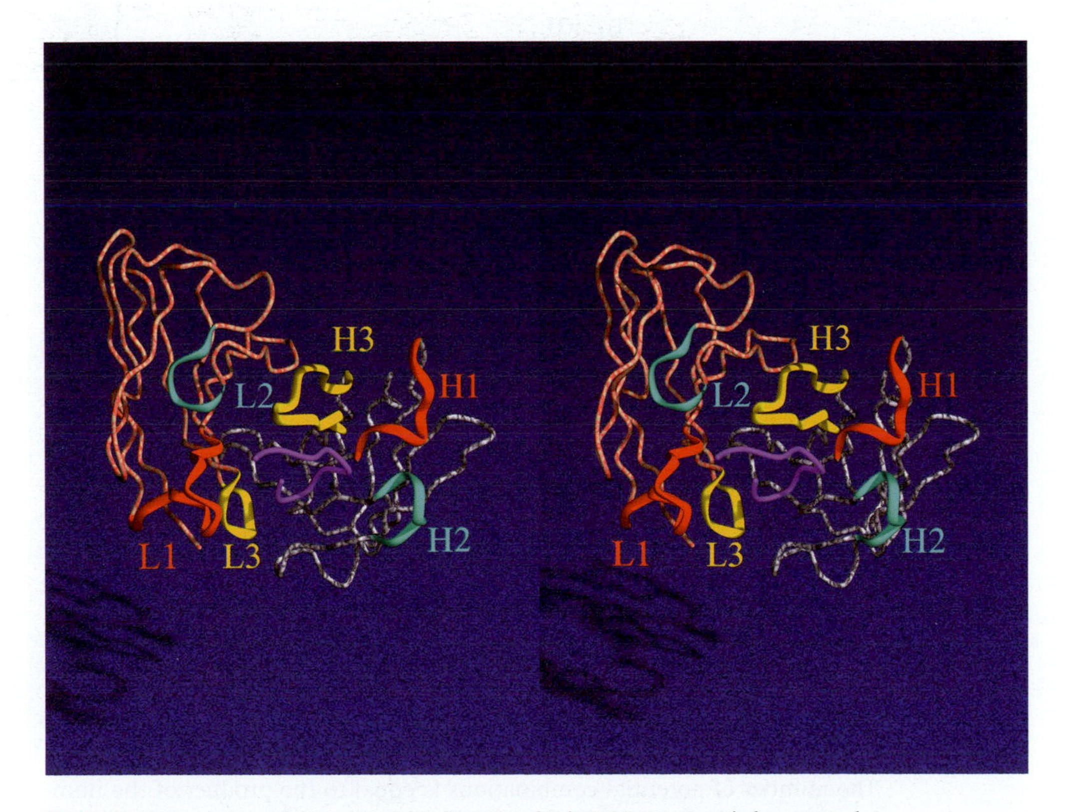

Fig. 7.12 General view of the antigen-binding site of Fab TE33 [1TET], with fragment of cholera toxin (purple). H3 appears in a central position within the antigen-binding site.

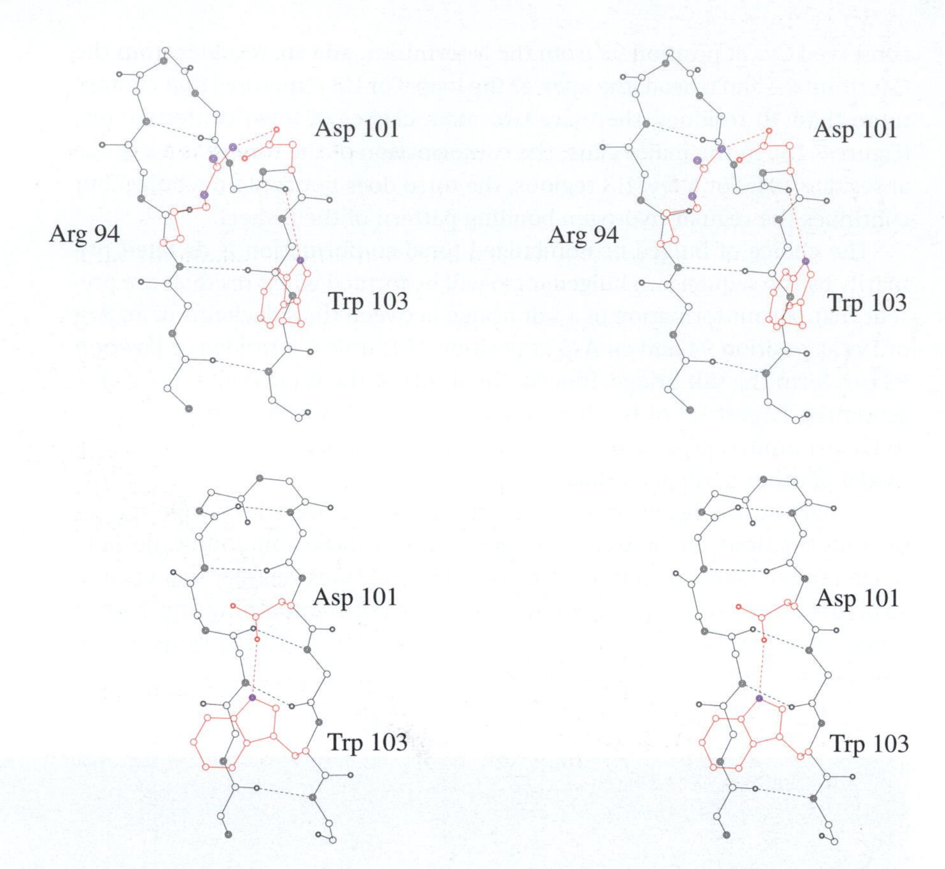

Fig. 7.13 Two H3 loops with alternative torso conformations: (a) McPC603, the bulged torso conformation [1MCP], (b) 4–4–20, the non-bulged torso conformation [1FLR].

Conclusions: how does the immune system generate and then refine molecules of such a wide range of specificity?

Sequence

At the genetic level, the many genes coding for the variable domains of light and heavy chains are put together by combining selected gene segments or exons. The variable domain of the light chain is formed by linking a V segment and a J segment; that of the heavy chain by linking a V segment, a D segment and a J segment. The human genome contains, for the light chain: 40 Vκ segments, five Jκ segments, 31 Vλ segments, four Jλ segments; and for the heavy chain: about 51 VH segments, 25 D segments, and six JH segments. Imprecise joining at the borders between the V, D, and J segments is a source of further variability at the genetic level, contributing to a large diversity in the vicinity of hypervariable region H3.

Light and heavy chains combine to form different pairwise combinations. The number of potential combinations is equal to the product of the number of light chain genes and the number of heavy chain genes. It is estimated that the genetic repertoire contains over 10^3–10^4 different Vκ genes and over 10^6–10^9 different VH genes. The product exceeds, by far, the number of B cells in the body. (Indeed, as the diameter of a B cell is about 20 μm, a back of the envelope calculation shows that the total weight of $10^4 \times 10^9 = 10^{13}$ B cells would be 40 kg; if the laboratory mouse had a body weight sufficient to accommodate so many B cells, the history of immunology would undoubtedly have been very different.)

Somatic mutation during antibody maturation provides additional variation, perturbing the germ-line structures rather than effecting major reconfiguration. Only minor structural changes are necessary, to have very large effects on affinities.

Structure

The antigen-combining site has a mainchain conformation determined by the canonical structures of loops L1, L2, L3, H1, and H2, and the more variable H3. The number of potential mainchain structures in the antigen-binding site is equal to the product of the number of possible canonical structures for loops L1, L2, L3, H1, and H2, and of the number of possible structures of H3. Putting numbers into these statements gives very large numbers of potential antibody sequences. However, there appear to be biases in the distribution of combinations of canonical structures in populations of expressed antibodies.

Because only a relatively small number of residues is required to fix the canonical structures of the main chain, most of the antigen-binding loops can be freely decorated with sidechains to vary the surface topography and charge distribution of the binding site.

The immune system is one of many biological examples of molecular recognition. Individual antibody–antigen complexes fit the classical ideas of lock-and-key complementarity and induced fit; but the system as a whole is more complex than, for example, enzymes, in which the lock and key are in most cases fixed and unchanging within the lifetime of the organism. In the primary response, the immune system achieves affinity for many keys by providing many locks; but achievement of the spectacular affinity of the secondary response requires the mechanism for perturbing the locks for better fit. It is the integration of the immune response over the system as a whole, involving both genetic and structural diversity, that adds the dimensions of complexity.

Proteins of the Major Histocompatibility Complex

Surgical patients, if not immunosuppressed by drugs, will reject transplanted organs—unless the donor is an identical twin—because the transplant is recognized as foreign. The immunological distinction between 'self' and 'non-self' resides in the proteins of the Major Histocompatibility Complex (MHC) and their interaction with T-cell receptors. MHC proteins bind intracellularly-produced peptides and present them on cell surfaces. The triggering event in alerting the immune system to the presence of a foreign protein is the recognition, by a T-cell receptor, of a complex between an MHC protein and a peptide derived from the foreign protein.

MHC proteins fall into two classes, with related but different structure and function. The two classes function as parallel systems, to produce different immune responses appropriate to intracellular and extracellular pathogens, respectively (Figure 7.14). Class I MHC molecules appear on the surfaces of most cells of the body, and present peptides derived from pro-

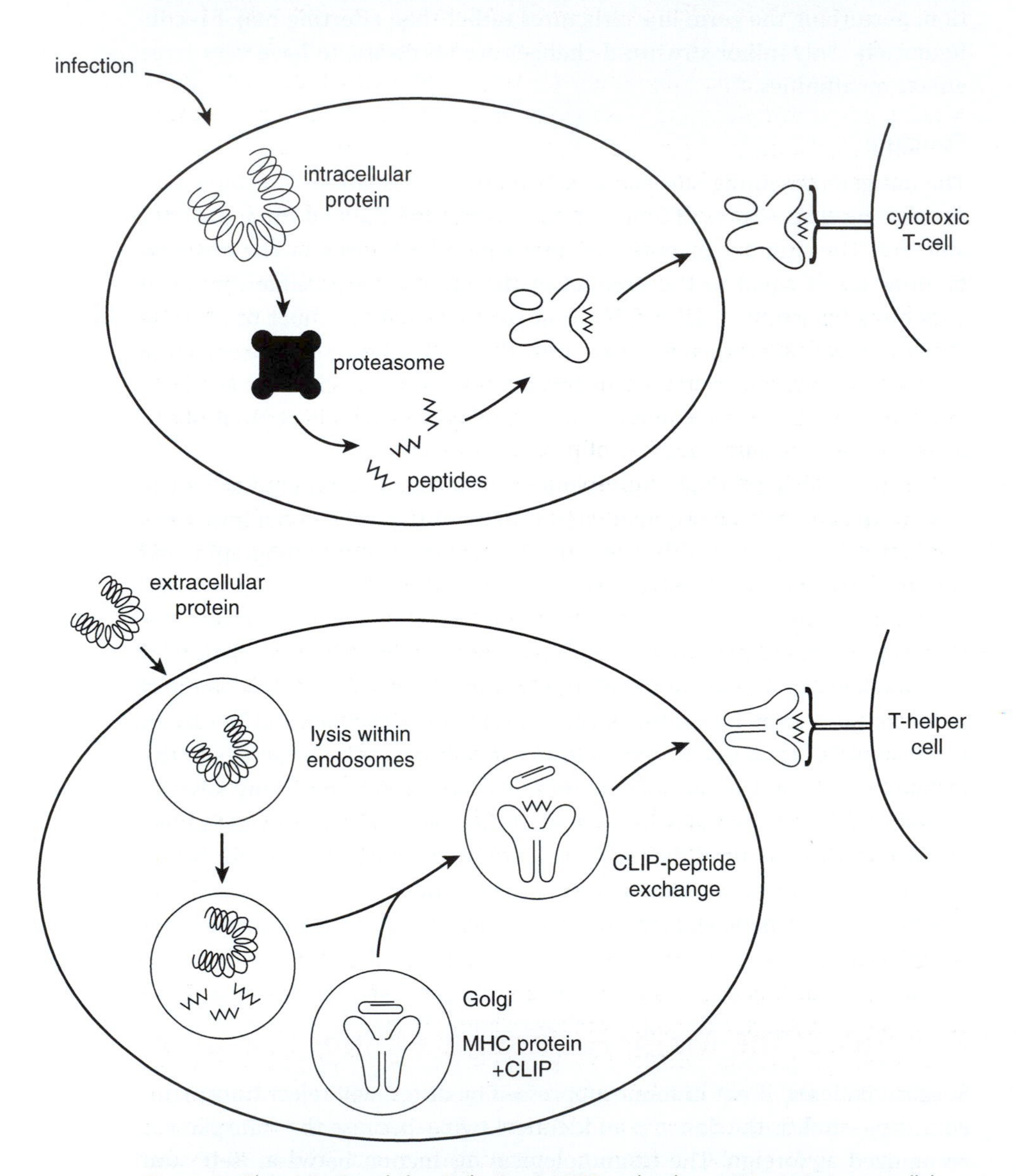

Fig. 7.14 Class I (top) and class II (bottom) MHC molecules participate in two parallel systems to trigger the immune response to foreign proteins originating inside and outside cells. Peptides derived from foreign proteins are loaded intracellularly and transported to the surface where they are presented to T cells. The representation of the class II invariant peptide (CLIP) is an icon not a drawing of its structure.

teins degraded in the cytosol. These peptide–MHC complexes alert the body to intracellular pathogens. They interact with cytotoxic T-cells, and direct the immune response to the presenting cell and those in its vicinity. Class II MHC molecules appear on the surfaces of specialized cells of the immune system: B-lymphocytes and antigen-presenting macrophages. They present oligopeptides derived from exogenous antigens (which have been endocytosed and chopped into peptides). These peptide–MHC complexes interact with helper T-cells, mediating the proliferation of cells synthesizing antibodies that circulate in the blood, and the activation of macrophages.

In addition to triggering immune responses in mature individuals, MHC–peptide complexes are also involved in the removal of self-complementary T cells in the thymus during development, at the stage when the distinction between self and non-self is 'learnt'.

Each individual in vertebrate species expresses a set of MHC proteins selected from a diverse genetic repertoire in the species. In humans the MHC complex is a set of linked genes on chromosome six. The system is highly polymorphic, with 50–150 alleles per locus, showing greater sequence variation than most polymorphic proteins. Each of us produces six class I molecules and a somewhat higher complement of class II. Each MHC protein must therefore be able to bind many peptides, if ~30 MHC proteins are to present the large number of possible antigens.

The set of MHC proteins expressed defines the haplotype of an individual. The number of possible haplotypes has been estimated to be of the order of 10^{12} although, because of linkage, the combinations are non-random, and because of selective pressure fewer combinations appear than expected. Like fingerprints and restriction fragment length polymorphisms, our haplotypes are a personal identification code. Haplotypes have been used extensively in anthropology in measuring quantitatively the relationships between human populations, and in tracing paths of migration.

In addition to their roles in peptide presentation, which are fairly well understood at the molecular level, it has been suggested that the contribution of MHC proteins to our individual identities also influences mate selection (apparently correlated with subtle body odours).

Structures of MHC proteins

MHC proteins have modular structures containing several domains. These include peptide-binding domains with folds characteristic of the MHC system, and immunoglobulin-like domains (Figure 7.15).

Class I MHC proteins contain two polypeptide chains (Figure 7.16.) The longer chain (M_r 44 000 in humans; 47 000 in mice) has a modular structure, inherited from the exons of the gene, of the form α_1–α_2–α_3, followed by a short hydrophobic membrane-spanning segment, and a 30-residue cytoplasmic tail. The α domains are approximately 90 residues in length. The second chain is β_2 microglobulin, a non-polymorphic structure (that is, constant within the species), the gene for which is unlinked from the MHC complex.

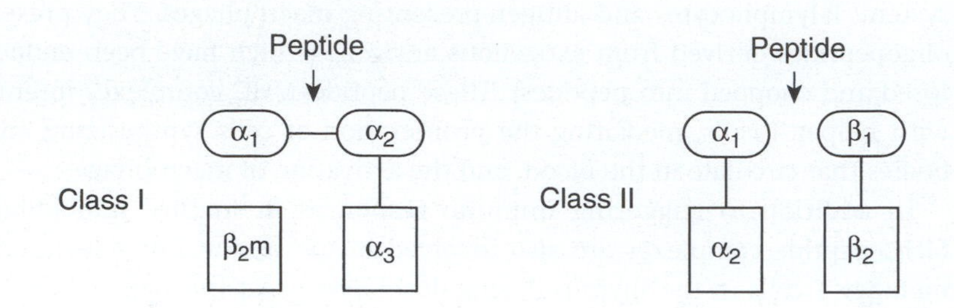

Fig. 7.15 MHC proteins are modular proteins containing two peptide-binding domains (cartouches) and two immunoglobulin-like domains (rectangles). Peptides bind in a groove created by the α_1/α_2 domains in class I MHC proteins and by the α_1/β_1 domains in class II MHC proteins.

The α_1 and α_2 domains of class I MHC proteins have a common fold, and interact to form a symmetric combined structure. They bind peptides in a groove between them, created by two long curved α-helices (Figure 7.17). The variability in the amino acid sequence of MHC proteins is high in the regions that surround the groove, to create variety in specificity. The α_3 domain and β_2 microglobulin do not interact with bound peptides. They are double β-sheet proteins with topologies of the immunoglobulin superfamily. Each has the disulphide bridge, and the tryptophan packed against it, in common with immunoglobulin domains.

The α_1 and α_2 domains of class I MHC proteins, and the α_1 and β_1 domains of class II, have a common folding pattern. Each has a four-stranded β-sheet at the N-terminus, followed by a short bridging helix and then a long C-terminal helix that lies across the β-sheet. The strands from each domain interact laterally to form an 8-stranded β-sheet, positioning the C-terminal helices from the two domains to form the sides of the peptide-binding cleft. Each helix has a pronounced curvature. Because of the approximate twofold symmetry of the α_1–α_2 unit, the long C-terminal helices run antiparallel to each other. Peptides bind in an orientation parallel to that of the α_1 helix.

Class II MHC proteins contain an α chain (M_r 34 000) containing two domains, $\alpha_1 + \alpha_2$, a homologous β chain (M_r 29 000) containing two domains, $\beta_1 + \beta_2$, and a invariant chain I1 (M_r 31 000), which is not present in the mature cell-surface form (Figure 7.18; the invariant chain does not appear). The α_1 and β_1 domains pack together to make a structure similar to that formed by the α_1 and α_2 domains of Class I MHC proteins, and they bind peptides in a similar mode (Figure 7.19).

Figure 7.20 shows a superposition of the peptide-binding domains from class I and class II MHC proteins, each containing a peptide ligand. The peptide binding domains have a similar structure, and the peptides occupy a similar region of space relative to the MHC protein, running parallel to the α_1 helix in both classes. Note that in the class I molecules, the cleft is pinched off at both ends: at the right by sidechains, and at the left by the unwinding of two turns at the N-terminus of the helix in the α_1 domain, resulting from a deletion in this region in class I molecules relative to

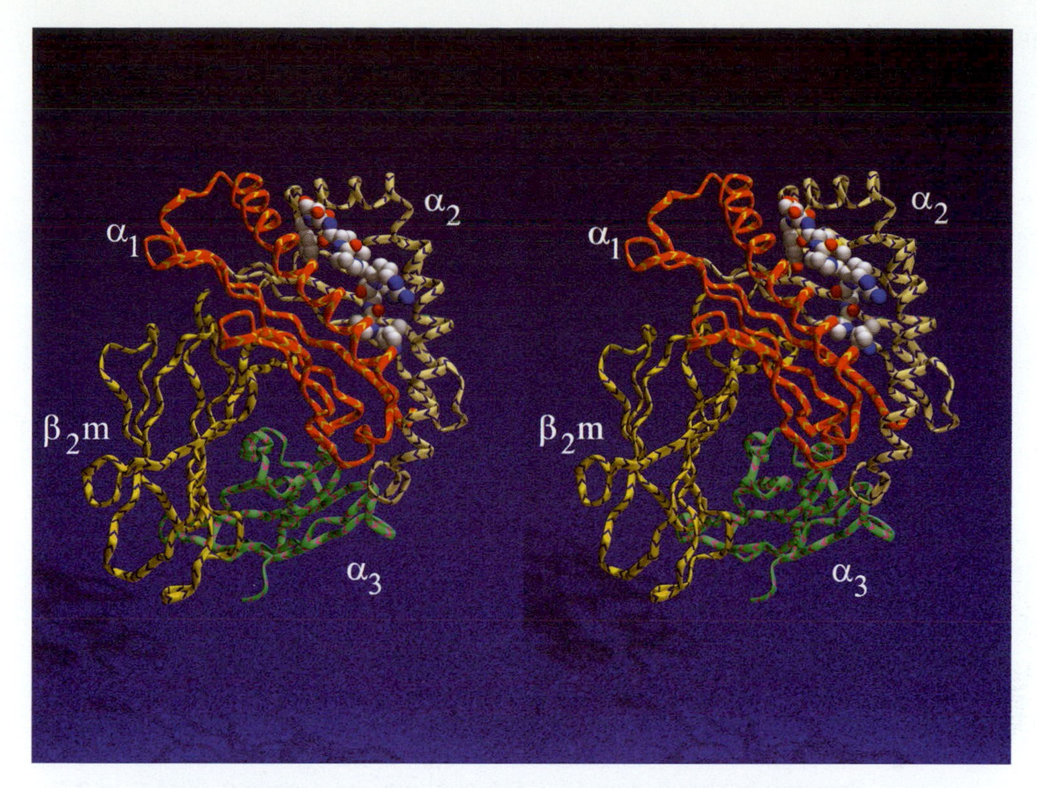

Fig. 7.16 The structure of a class I MHC protein, B35, binding the peptide VPLRPMTY from the nef protein of HIV-1 [1A1N]. The peptide ligand is represented by spheres.

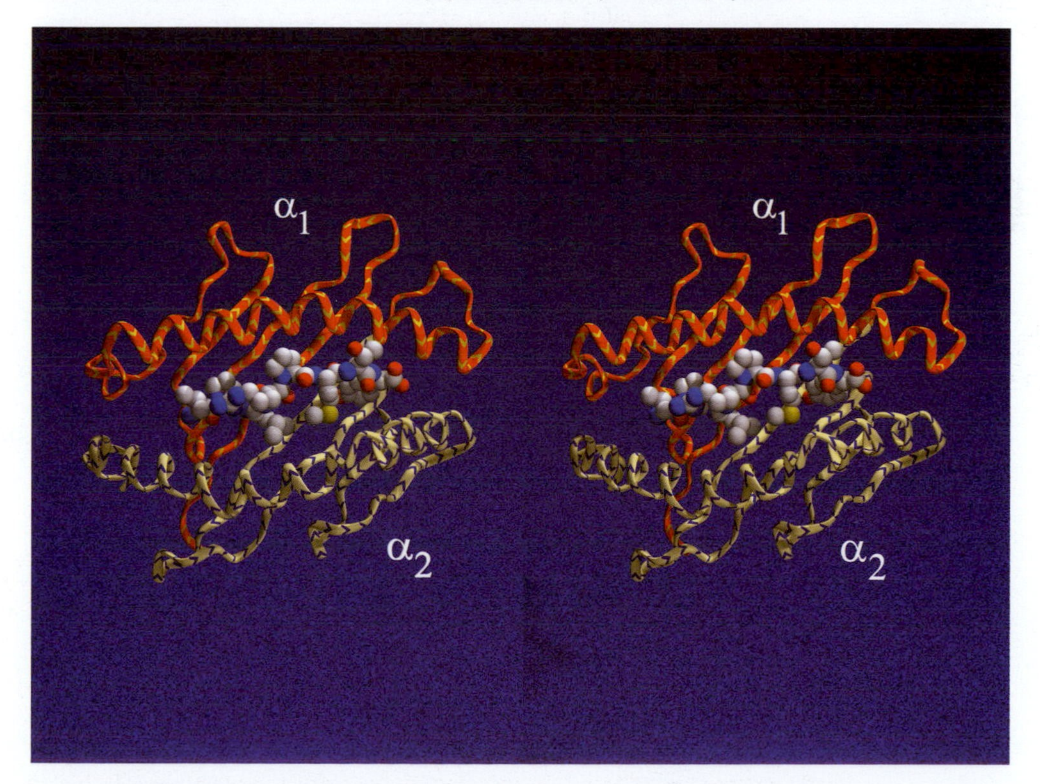

Fig. 7.17 The peptide-binding domains and ligand from class I MHC protein B35 [1A1N].

class II. The MHC protein sidechains that bind the chain termini of the peptide tend to be conserved; in class I these include tyrosines at positions 7, 59, 159, and 171 that interact with the N-terminus of the peptide, and residues 84Tyr, 146Lys and 147Trp that interact with the C-terminus of the peptide. In class II also the cleft is closed at the right, but open at the left.

Class I molecules can bind peptides of limited length—from about 8–11 residues—but most commonly 8–9. Two factors impose the limits: the closure of the cleft at both ends, and a salt bridge to the C-terminal carboxyl group of the peptide. The cleft will accommodate nine-residue peptides in a nearly-extended conformation; longer peptides can bulge out or zig-zag or their C-termini can extend out beyond the end of the pocket.

The α_2 and β_2 domains of class II MHC proteins, like the α_3 and β_2 microglobulin domains of Class I, resemble immunoglobulin constant domains. A fit of the Cα atoms of the α_2 domain of I-AK to the CH2 domain of human immunoglobulins superposed 92 Cαs with r.m.s. deviation 1.27 Å.

The nomenclature is somewhat unfortunate, because the Greek letter in the domain designation indicates the chain in which it appears and not the folding pattern, which is more common usage. The two peptide-binding domains are called α_1 and α_2 in the class I molecules, and α_1 and β_1 in class II. In class I molecules, domain α_3 is *not* homologous to the other two α domains. In class II, domain α_1 is homologous to domain β_1, and α_2 to β_2!

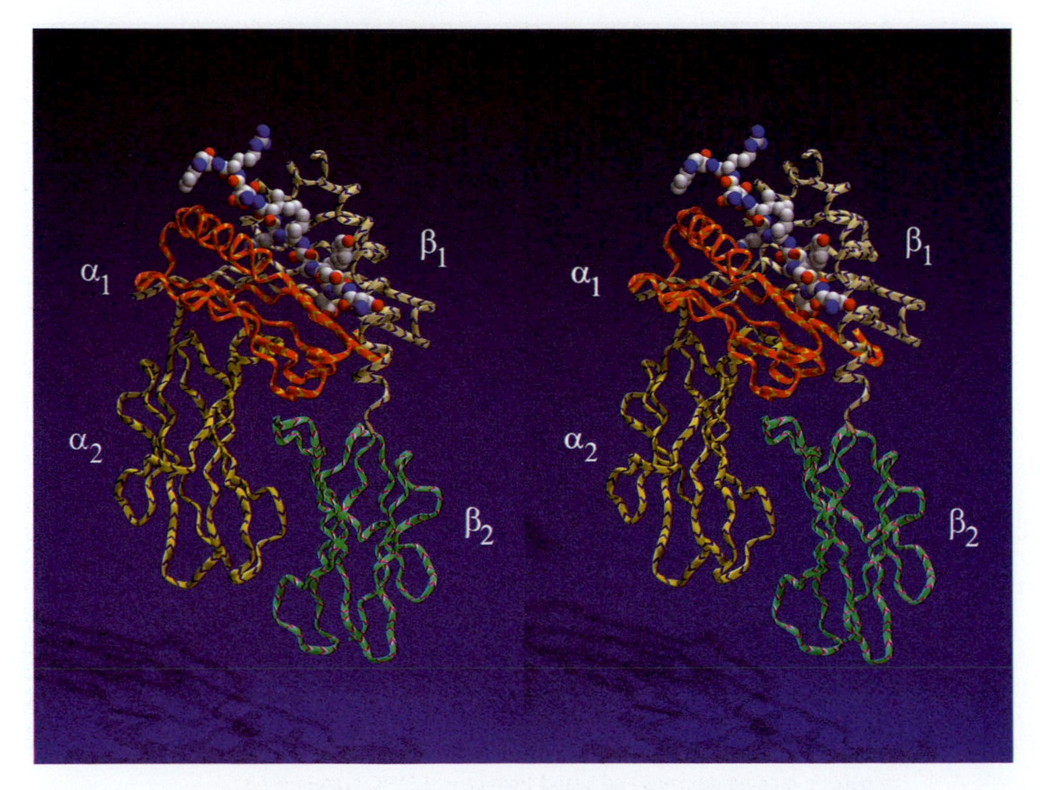

Fig. 7.18 The structure of a class II MHC protein, I-Ak, binding the peptide STDYGILQINSRW from hen egg white lysozyme [1IAK].

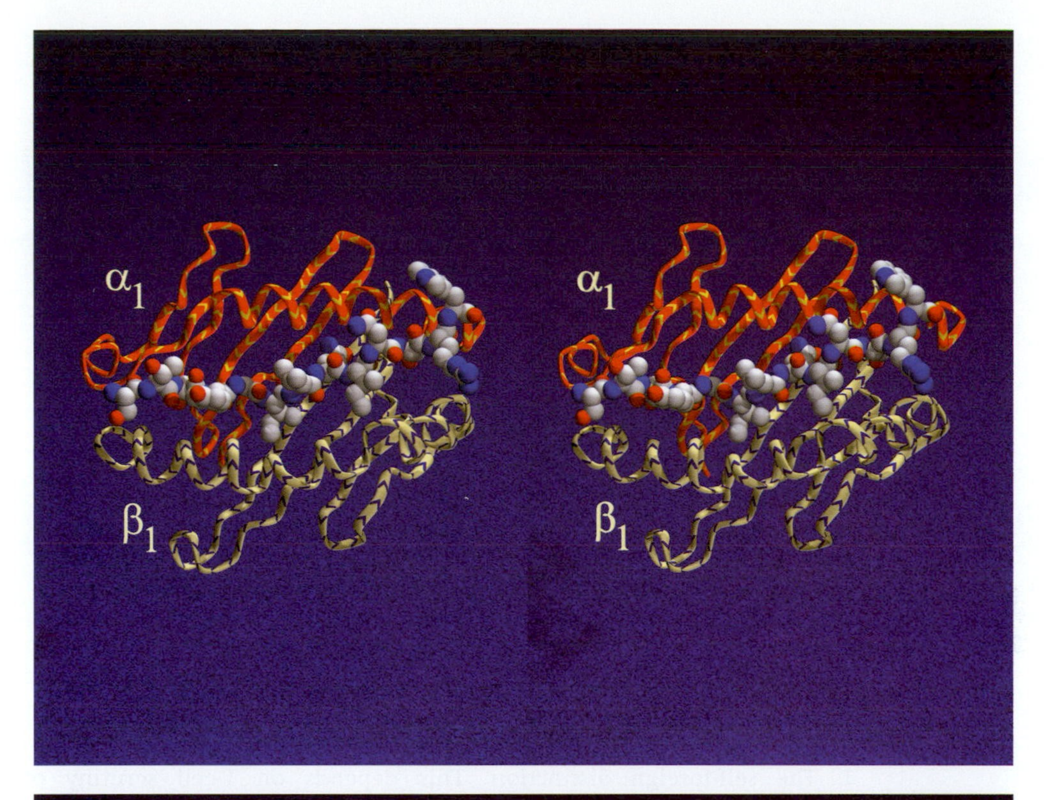

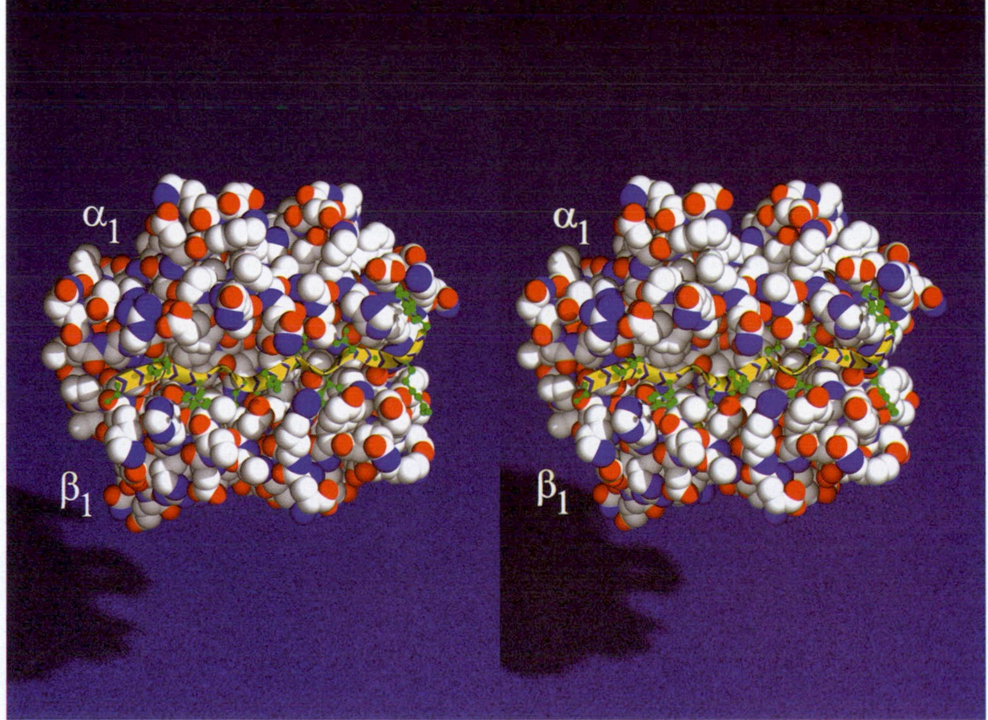

Fig. 7.19 The peptide-binding domains and ligand from class II MHC protein (I-AK) [1IAK]. Two representations.

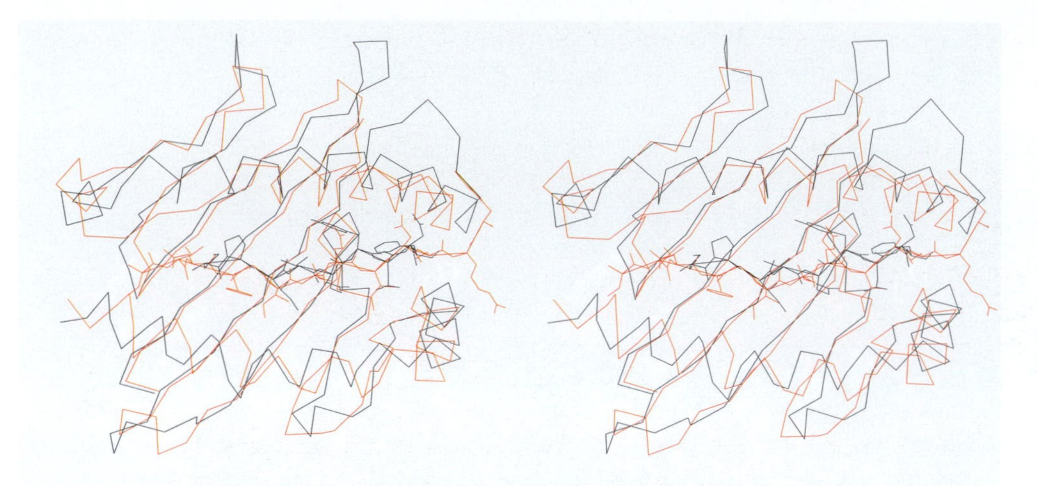

Fig. 7.20 Superposition of peptide-binding domains and ligand from class I (B35) [1A1N] (black) and class II (I-Ak) [1IAK] (red) MHC proteins, showing the overall similarity in structure. The ligands occupy similar regions of molecular space, in the groove between the two helices.

Specificities of the MHC system

Two aspects of MHC specificity are:

1. The self-foreign distinction. This depends on T-cell scrutiny of MHC–peptide complexes.
2. The selection of different types of immune reponse to different categories of threats—extracellular vs. intracellular. This is accomplished by directing peptides derived from extracellular and intracellular sources through the class I and class II systems, respectively, to form complexes recognized by different types of T-cells.

The MHC proteins themselves have broad specificity, each binding many peptides including those of self and non-self origin. Cell surfaces display large numbers of MHC–peptide complexes, among which those binding foreign peptides are a small minority. T-cell receptors, in contrast, have narrow specificity and pick out the complexes containing foreign peptides, like an antiques dealer spotting a valuable item in a rummage sale. T-cell diversity is created by the same types of genetic recombination mechanisms that produce the diversity of antibodies in the primary immune response. (There is no analogue of somatic mutation in the T-cell system, however.)

T cells 'see' both the MHC and peptide components of the complex they recognize. The variability in MHC proteins extends to the residues that interact with T cells but do not interact with the bound peptides. Two MHC proteins binding the same peptide could be recognized by different T cells.

Class I and class II MHC proteins function in parallel, selecting different immune responses to extracellular and intracellular pathogens

Challenge by internally-synthesized foreign proteins, as in virus-infected cells, leads to cleavage by cytosolic proteasomes. The peptides encounter class I MHC proteins in the endoplasmic reticulum, and the peptide complexes are moved to the cell surface.

Challenge by external foreign proteins leads to cleavage in endosomes. Class II MHC proteins start out in the endoplasmic reticulum associated with the Invariant (Ii) chain. In this complex the peptide-binding site is blocked, to suppress flooding the system with self-derived peptides, and to prevent picking up peptide fragments from internally-synthesized proteins. (Internally-synthesized proteins are the responsibility of class I MHC molecules.) The complex moves to specialized organelles where it is exposed to peptide fragments of imported proteins. Cleavage of the I chain leaves the CLIP (CLass II Invariant chain Peptide) in the binding site. A molecule related to the MHC proteins—HLA-DM in humans and H-2M in mice—catalyses peptide exchange, removing the CLIP and enhancing the rate of binding. Peptide binding stabilizes the α chain–β chain dimers, which are brought to the cell surface.

Peptide binding

Crystal structures of both class I and class II MHC proteins, with and without bound peptides, reveal:

1. the nature of the peptide–MHC protein interactions;
2. the mechanism of the broad specificity;
3. the nature of the surface presented to T-cell receptors.

The complexes of the class I MHC protein HLA-B53 show the conformation and interactions of the ligand. Figure 7.21 shows the shape of the cleft from the complex of HLA-B53 with the peptide LS6 (KPIVQYDNF) from the malarial parasite *Plasmodium falciparum.* The cleft is a deep groove, with a hump in the middle, and pockets at either end which receive inward-pointing sidechains from near the chain termini of the peptide. It is interesting that the middle part of the floor of the groove is lined by polar sidechains: 9Tyr, 70Asn, 74Tyr, and 97Arg. The peptide is in an extended conformation; the conformational angles of all residues are in the β region of the Sasisekharan–Ramakrishnan–Ramachandran plot. The main chain residues (N, Cα, C, O) of all but two central peptide residues make contact with the MHC protein, and every sidechain of the peptide makes contact with the MHC protein.

Interactions between the peptide and the MHC protein can be divided into those that involve the main chain of the peptide—and are therefore independent of the sequence of the ligand—and those that involve the sidechains—and depend on the sequence of the ligand. The ligand-sequence-independent interactions explain the affinity and the broad specificity of the interaction; the others explain its selectivity.

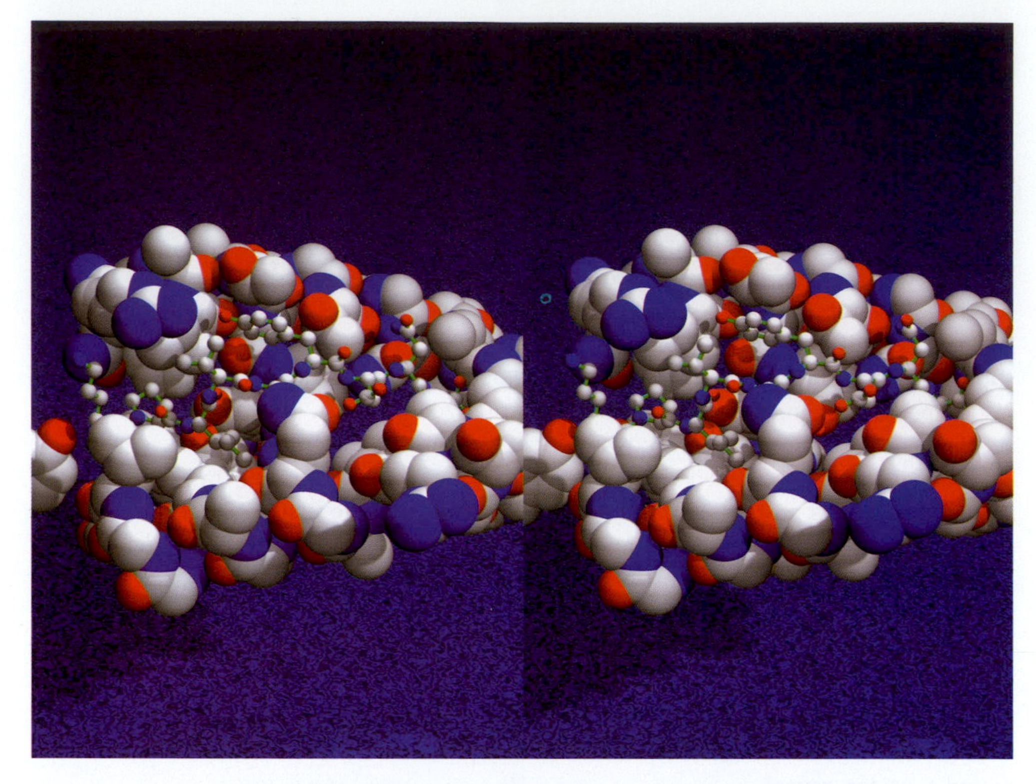

Fig. 7.21 Space-filling picture of the cleft in class I MHC protein B53 [1A1O].

The terminal residues of the peptide in the LS6–HLA-B53 complex are anchored by networks of hydrogen bonds (Lys is the N-terminal residue and Phe the C-terminal residue):

Peptide residue			MHC protein residue	
Lys	N	...	OH	7Tyr
Lys	N	...	OH	171Tyr
Lys	O	...	OH	159 Tyr
Phe	N	...	OD1	77 Asn
Phe	O	...	NZ	146 Lys
Phe	O	...	OG1	143 Thr

The residues from the MHC protein that make these hydrogen bonds are highly conserved. Residues 3 and 4 of the peptide also make backbone hydrogen bonds to the MHC protein. All these hydrogen bonds are peptide-sequence independent.

Two of the buried peptide residues, 2Pro and 9Phe, lie in crevices at the base of the interhelix cleft. The other two, 3Ile and 5Gln, are buried between the α_2 domain C-terminal helix and the peptide itself. Figure 7.22 shows two sets of slices through the complex, (a) at a relative high 'altitude' in the cleft, and (b) nearer the base of the cleft. In (a) it is seen that the sidechains of 3Ile and 5Gln are not pointing down into the MHC protein but are nevertheless buried between it and the peptide backbone. In (b) it is seen that the sidechains 2Pro and 9Phe project through the lowest slice, to be buried in a pocket in the MHC protein. Figure 7.23 shows the good packing of the two buried residues 2Pro and 9Phe in the pockets they

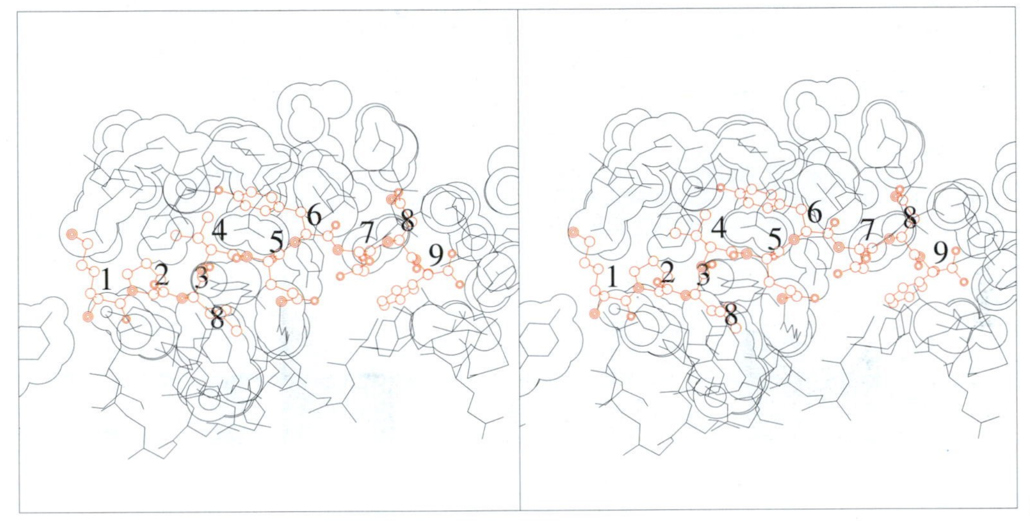

Fig. 7.22 a

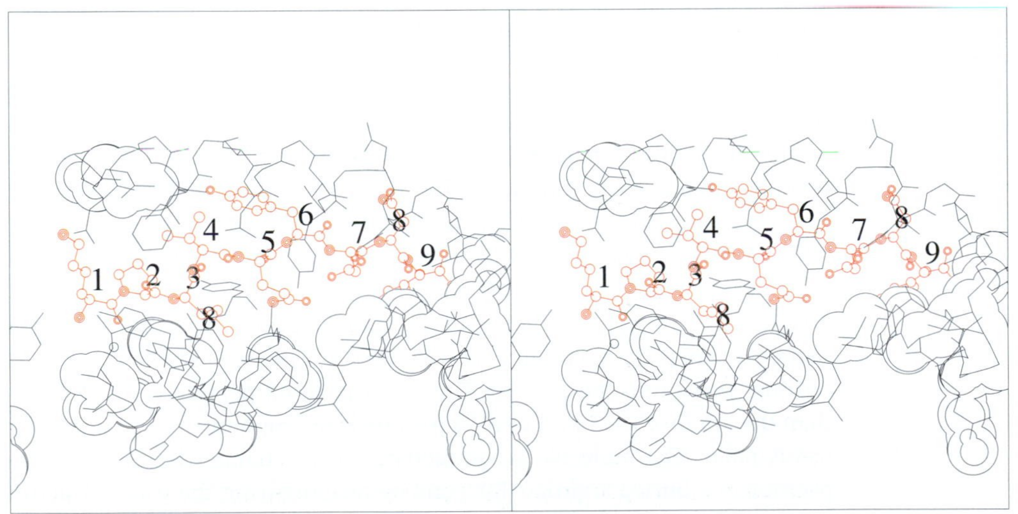

Fig. 7.22 b

Fig. 7.22 Slices through the cleft in Class I protein [1A1O]. (a) Fairly low down within the cleft. Note that the ligand sidechains Pro and Phe protrude through these slices into the bottom of the cleft. (b) Nearer the top of the cleft (oblique view). Note that the ligand sidechains are buried between the MHC protein and the peptide backbone.

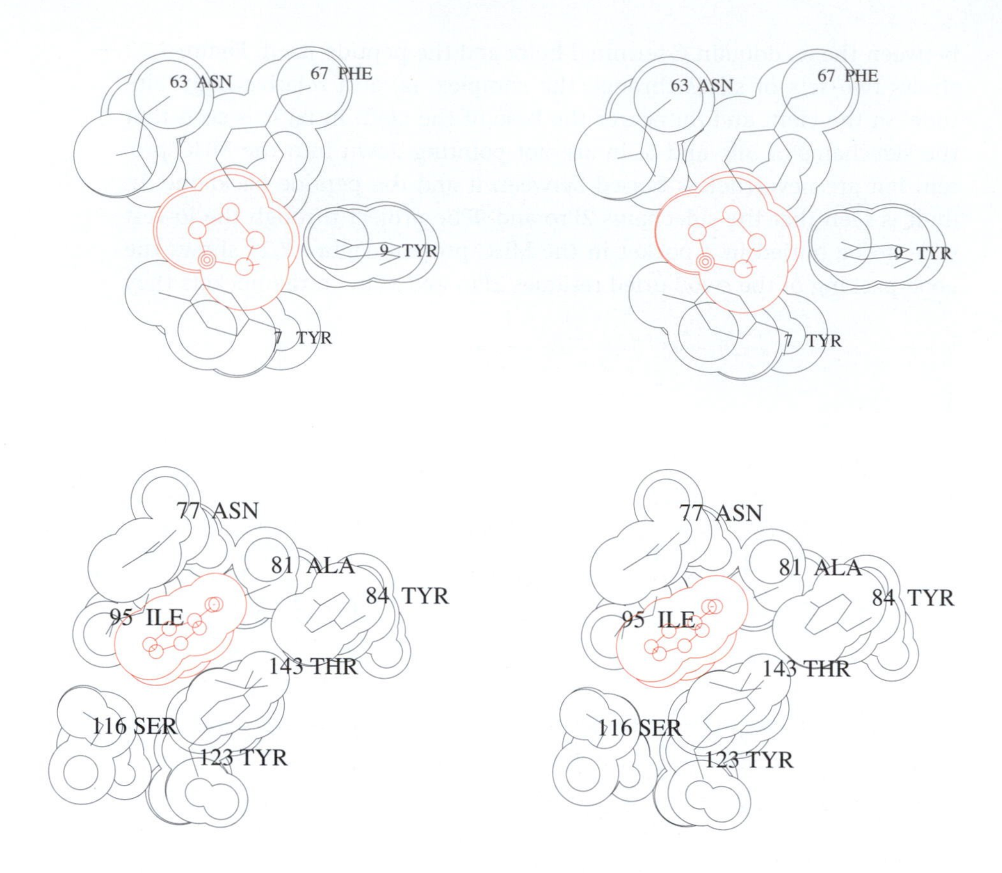

Fig. 7.23 Tight fitting of buried sidechains from the bound peptide KPIVQYDNF from the malarial parasite, into pockets in the MHC protein [1A1O]. (a) 2 Pro. (b) 9 Phe. Residue numbers in the figure refer to the MHC protein.

occupy. It is not surprising that Pro is at this position in the consensus sequence of peptides binding to HLA-B53.

The HLA-B53 system reveals how the MHC protein can bind to alternative peptides. It has been solved binding the LS6 peptide (KPIVQYDNF) from *P. falciparum*, and also with a peptide from the gag protein of HIV-2 (TPYDINQML). Some of the buried residues have mutated; how does the structure accommodate this? Figure 7.24 compares the binding of ligand residues 4–7 in the complexes of the two peptides with HLA-B53. The sequences are IVQY in LS6 and YDIN in the HIV-2 peptide. The first and third residues point into the protein. The Y in the HIV-2 peptide is much larger than the I in LS6. The I in the HIV-2 peptide is hydrophobic whereas the Q in LS6 is polar. The molecule accommodates these changes by changing the position of a buried arginine (R97) and by reconfiguring the water structure in the cleft.

Another degree of freedom, used by the class I protein HLA-B35, is 'induced fit': conformational changes in the MHC protein accommodate changes in sequence in the ligand. These changes appear in comparing the structure of HLA-B53, binding the 9-mer KPIVQYDNF, with that of the closely

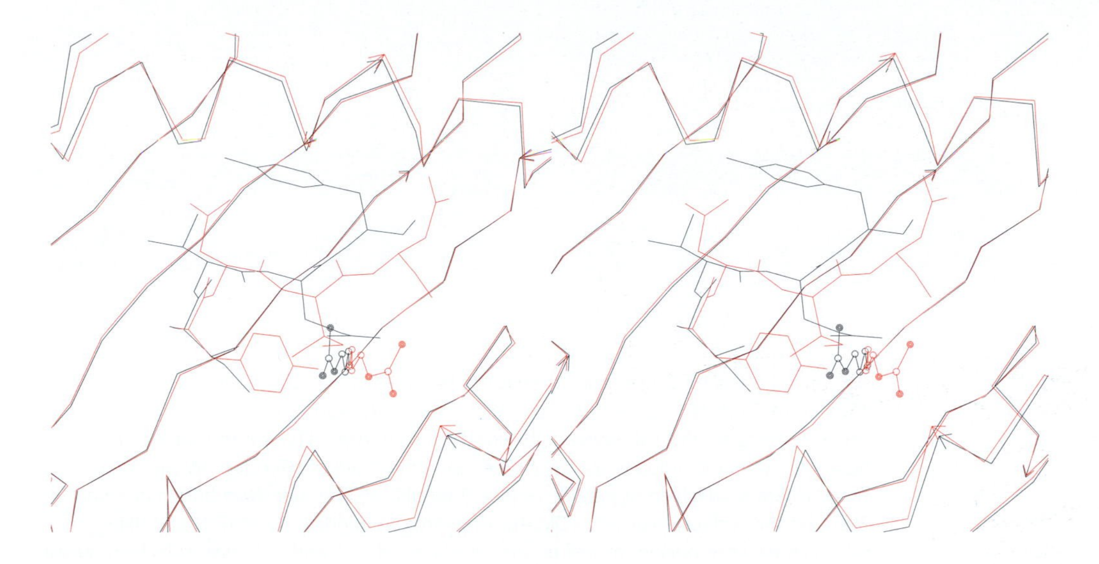

Fig. 7.24 Binding of two peptides with different sequences to the same MHC protein, HLA-B53 [1A1O] (black) and [1A1M] (red). Note the change in conformation of the Arg residue in the MHC proteins.

related HLA-B35, binding the 8-mer VPLRPMTY. There appear to be two main causes of conformational change. First, the few sequence changes between the two MHC proteins alter the pocket that binds the C-terminal aromatic residue of the peptide, giving HLA-B35 specificity for tyrosine and changing the position of the C-terminal region of the ligand backbone. The position of the C-terminal residue of the peptide bound to B35 would clash sterically with the position of the short helix in B53, as would the peptide residue 7Asp. Conformational changes in the MHC protein avoid these clashes. Both ligands in these complexes contain Pro in the second position. This sidechain is locked into a crevice at one end of the groove; the C-terminal residue (Y or F) is locked into a crevice at the other. To fit in an extra residue, something has to give—the longer peptides buckle in the middle.

The reader may wonder why the question of ligand-induced conformational changes is addressed by comparison of different peptides binding to the same, or very similar, MHC molecules, rather than by comparing one MHC molecule in free and ligated states. Peptide binding is important to stabilize the structure of MHC proteins, and it has not been possible to crystallize the unligated state.

Binding of peptides to class II MHC proteins

In the structure of the class II MHC protein I-Ak binding the peptide STDYGILQINSRW, the peptide lies in the cleft with its ends protruding from both sides of the MHC protein (Figure 7.25). All residues of the ligand are in the β conformation. As the binding groove is pinched off at the right (in the orientation of this picture), the ligand turns up, out of the cleft. There is a difference between class I and class II in the interactions between peptide

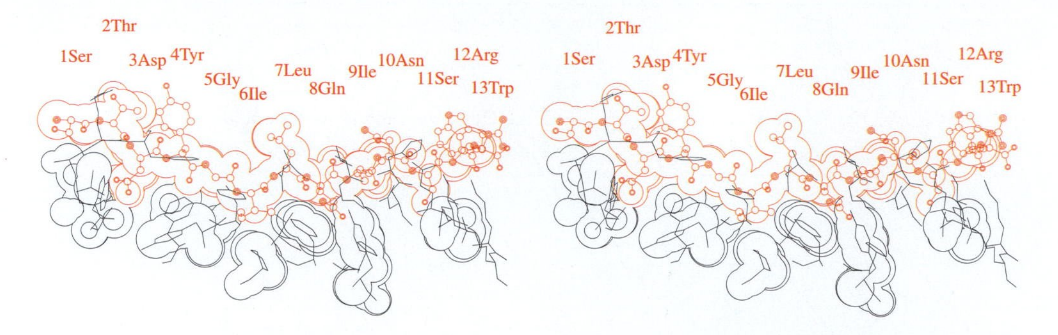

Fig. 7.25 Ligand binding to class II MHC protein I-Ak [1IAK] (black). Peptide residue (red): 1Ser, on the surface, exposed; 2Thr, on the surface, exposed; 3Asp, sidechain buried in a cavity, forming double salt bridge with sidechain of 52Arg.; 4Tyr, on the surface, exposed; 5Gly, two aromatic residues (22Tyr, 24Phe) and mainchain of 9Tyr and 9A Gly occupy space where a sidechain of peptide residue 5 would be; 6Ile, sidechain buried in a pocket; 7Leu, on the surface, exposed; 8Gln, forms hydrogen bonds to O of 62N and 9H, sidechain packs in pocket formed by the sidechains of 65T and 11F; 9Ile, sidechain buried in a pocket formed by sidechains of 61W and 47Y, covered by sidechain of 67Y; 10Asn, on the surface, exposed, forms hydrogen bond to 65T; 11Ser, sidechain sits in polar pocket formed by sidechains of 57D and 76R; 12Arg, on the surface, exposed; 13Trp, on the surface, exposed—sidechain slots into pocket formed by 71E and 75K.

sidechains and binding groove. In Class I, sidechains at or adjacent to the termini of the ligand sit in pockets at the bottom of the groove, and the central sidechains do not point down into the groove but at right angles to it. In I-Ak, class II, the central sidechains are buried in pockets in the base of the groove.

The peptide ligand makes numerous hydrogen bonds to the MHC protein. In the accompanying box the hydrogen bonds are classified according to the participation of main chain and sidechain atoms from peptide and MHC protein. Entries in columns one and three, involving peptide main chain atoms, are peptide-sequence independent and account for the broad specificity. Nine of the 13 peptide residues make main chain hydrogen bonds, mostly but not exclusively with MHC protein sidechains. The two hydrogen bonds forming the interaction between residues 1Ser–3Asp of the peptide with 54Arg of the MHC protein are like those of a parallel β-sheet. Because the peptide has a nearly extended conformation, to a first approximation alternate sidechains in the central region of the peptide point into the groove and out towards the surface.

Conclusions

1. Class I and class II molecules bind peptides in ways that are generally similar but different in detail. (The peptide-binding domains of MHC proteins have been compared to a mouth, the two long helices corresponding to the lips. In binding peptides, Class I and class II molecules adopt a different embouchure.) The most obvious difference is the interaction of the terminal residues of peptide ligands with class I molecules in a way that limits the length of peptides that can be bound.

Hydrogen bonds between class II MHC protein I-Ak and peptide ligand

MHC mainchain peptide mainchain	MHC mainchain peptide sidechain	MHC sidechain peptide mainchain	MHC sidechain peptide sidechain
1 O . . . 54 N			
	1 OG . . . 279 O		
		2 O . . . 275 NE2	
3 N . . . 54 O			
			3 OD1 . . . 53 NH2
			3 OD1 . . . 280 OG1
			3 OD2 . . . 53 NE
		4 N . . . 276 OD1	
		4 O . . . 276 ND2	
6N . . . 9 O			
		7N . . . 268 OE2	
		8N . . . 63 OD1	
	8 NE2 . . . 63 O		
	8 NE2 . . . 67 N		
			8 OE1 . . . 205 ND1
		9N . . . 226 OH	
		9O . . . 70 ND2	
		11N . . . 70 OD1	
		11O . . . 69 NE2	
			11 OG . . . 253 OD1

Peptide sequence: 1S 2T 3D 4Y 5G 6I 7L 8Q 9I 10N 11S 12R 13W

2. MHC proteins achieve the goal of broad specificity but fairly tight binding by numerous contacts with the main chain of the peptide, which are peptide-sequence independent, and by conformational changes that allow tuning of the crevice to bind several peptides.

T-cell receptors

T-cell receptors (TCRs) are proteins related to immunoglobulins, that interact with MHC-presented peptides to trigger immune responses. Like immunoglobulins, they show a diversity generated by combinatorial assembly of genes from segments, but there is no subsequent development by somatic mutation. The binding sites of TCRs are created by six loops, three from each V segment, homologous to the CDRs of antibodies. However, unlike antibodies which must bind molecules of very different sizes and shapes, the targets of TCRs are exclusively MHC–peptide complexes, the *overall* sizes and shapes of which are fairly well circumscribed.

Binding of an MHC–peptide complex by a TCR on the surface of either cytotoxic T cells or helper T cells initiates a signal transduction cascade.

Because of the high concentration of MHC proteins presenting self peptides, T cells that would bind them are destroyed in the thymus early in life, to prevent autoimmune responses.

TCRs are dimers of proteins built of domains in a manner similar to immunoglobulins. Most TCRs contain an α and a β chain, each consisting of a variable (V) domain (bearing the binding loops), a constant (C) domain, a transmembrane segment and a cytoplasmic tail. (The α, β notation is independent of the nomenclature of the chains in MHC proteins.) The α chains have M_r 40 000–50 000, and the β chains have M_r 40 000–45 000. Neither chain has additional constant domains, as do immunoglobulin heavy chains.

Like other domains of the immunoglobulin superfamily, TCR V and C domains contain two β-sheets packed face to face, linked by a disulphide bridge with a conserved tryptophan packed against it. Immunoglobulins and TCRs share most of their secondary and tertiary structure, including the placement in the domain of the loops that create the binding site. However, in most TCR Vα domains and one known Vβ domain, there is a difference in the topology of the sheets, relative to immunoglobulin V domains, one strand having 'crossed the aisle', to appear in the other sheet (Figure 7.26). Also, although the canonical structure model applies to the first two CDRs of each TCR V domain, TCRs and immunoglobulins have different repertoires of canonical structures.

Canonical structures for TCR binding loops

Compared to immunoglobulins, relatively few crystal structures of TCRs are available. Our glimpse of the structural repertoire of the CDRs is limited. However, for the first two loops of the Vα and Vβ domains, it has been pos-

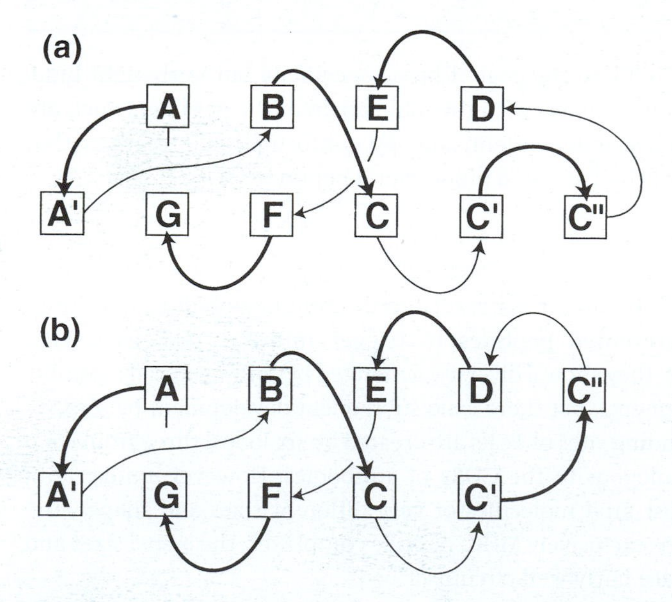

Fig. 7.26 Schematic diagrams of the β-sheet structures of the (a) standard immunoglobulin V domain and (b) TCR Vα domain. Note the change in the position of the C'' strand.

sible to define canonical structures from the available structures. The CDR3s of both chains are more variable (like CDR3 of VH of immunoglobulins) because of the complexity of gene assembly. Like antibody light chains, the Vα chains of TCRs are products of genes formed by linking V and J segments for the variable domain, followed by a gene segment for the constant domain. Like antibody heavy chains, Vβ chains of TCRs are products of genes formed by linking V, D, and J segments for the variable domain, followed by a single gene segment for the constant domain.

Examination of the germ-line genes for TCRs suggests that most of the Vα and Vβ domains will have CDR1 and CDR2 loops that follow one of the canonical structures already seen in a solved structure. Other canonical structures have been defined in terms of their sequence patterns, and await structure determination.

The TCR–MHC–peptide complex

The ternary complex of a T-cell receptor, with a class I MHC protein presenting a nine-residue viral peptide, show how the TCR recognizes the MHC–peptide complex (Figure 7.27). Figure 7.28 opens the complex out like a book, showing in part (a) the TCR plus the peptide, and the sidechains of the TCR that interact with the peptide and the MHC protein, and in part (b) the MHC protein plus the peptide, and the sidechains of the MHC that interact with the peptide and the TCR.

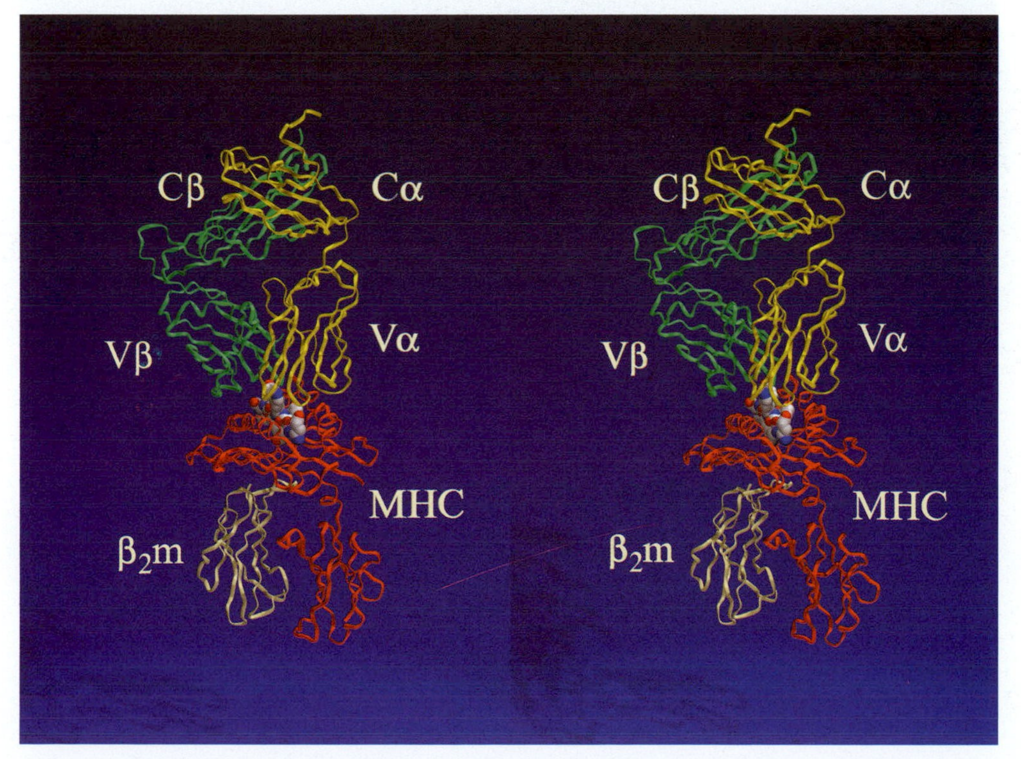

Fig. 7.27 a

Fig. 7.27 b

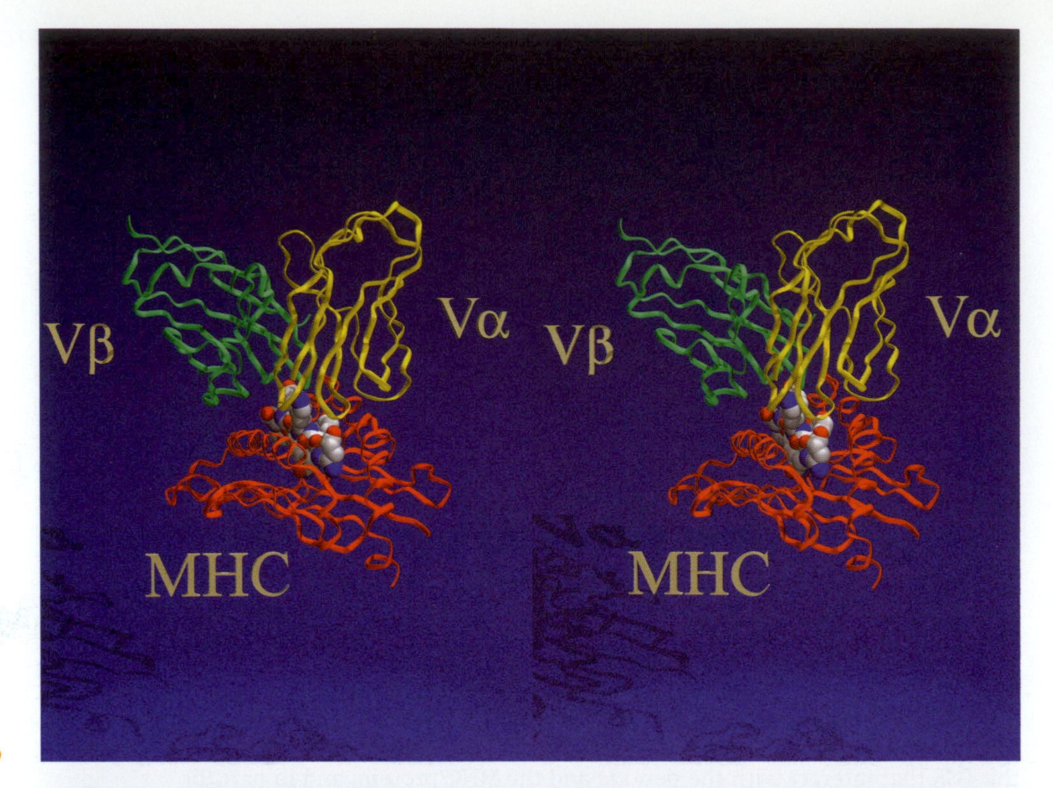

Fig. 7.27 c

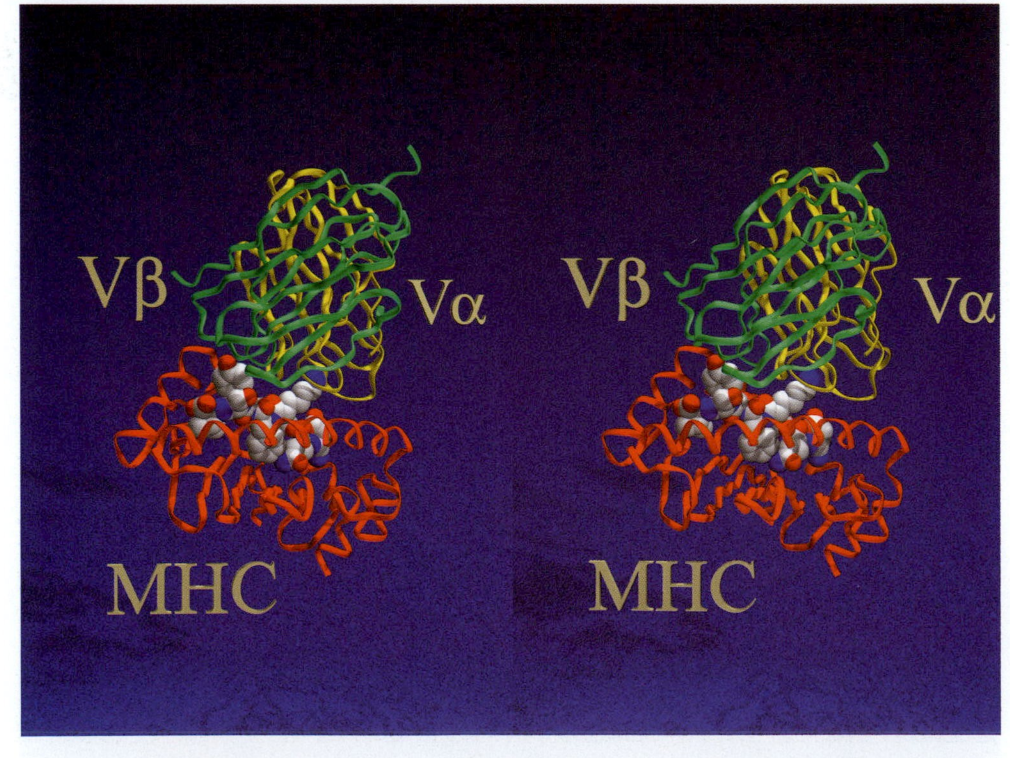

Fig. 7.27 (a) A human TCR in complex with a class I MHC molecule and a viral peptide [2ckb]. (b, c) The V domains of the TCR and the α_1 and α_2 domains of the MHC, plus the peptide. Two orientations.

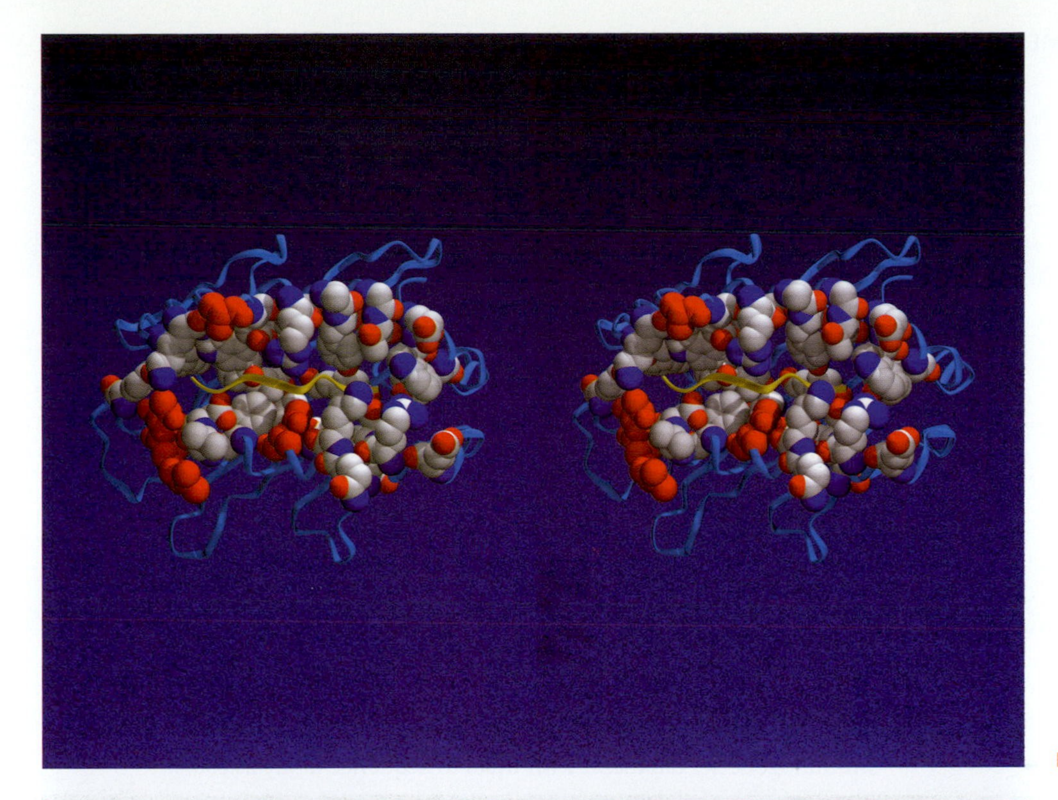

Fig. 7.28 a

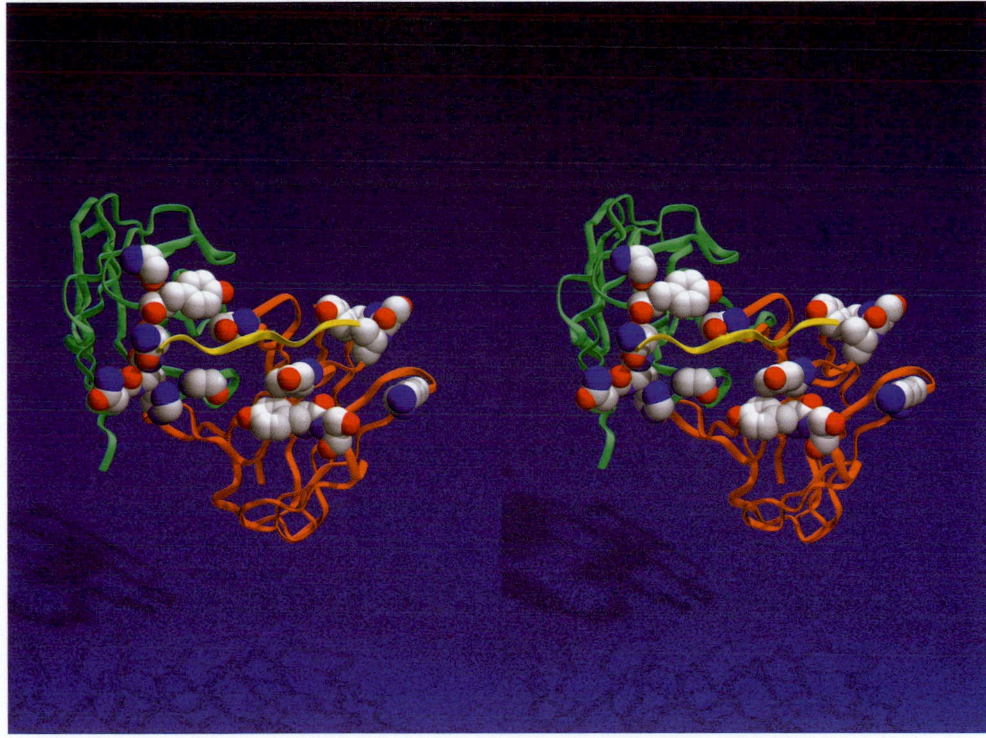

Fig. 7.28 b

Fig. 7.28 (a) The surface of the TCR that is in contact with the MHC–peptide complex. (b) The surface of the MHC molecule that is in contact with the TCR. The structure shown in the previous figure is prised apart and the contact residues coloured [2CKB].

A total of 11 residues from the TCR are in contact with the peptide. 17 residues from the TCR are in contact with the MHC, only six of which also make TCR–peptide contacts. The remaining 11 recognize the MHC, or, at least, the conformation of the MHC that appears in this peptide complex. Conversely, 26 residues from the MHC are in contact with the peptide. 16 residues from the MHC are in contact with the TCR, of which five are also in contact with the peptide.

Useful web sites

General collection of antibody-related material:
http://www.antibodyresource.com

Searching the Kabat databank of antibody sequences and specificities:
http://immuno.bme.nwu.edu/

Compilation of links to on-line databases and resources of immunological interest, as well as much other material related to computational molecular biology: http://www.infobiogen.fr/services/deambulum/english/db5.html

Immunogenetics database, with links to sequence analysis tools:
http://www.genetik.uni-koeln.de/dnaplot/

Recommended reading and references

Immunoglobulin structure and superfamily

Al-Lazikani, B., Lesk, A.M. and Chothia, C. (1997). Standard conformations for the canonical structures of immunoglobulins. J. *Mol. Biol.* **273**, 927–48.

Bork, P., Holm, L. and Sander, C. (1994). The immunoglobulin fold. Structural classification, sequence patterns and common core. J. *Mol. Biol.* **242**, 309–20.

Gallagher, R.B., Gilder, T. and Nossal, G.J.V. (1995). *Immunology / The Making of a Modern Science.* Academic Press, New York and London.

Halaby, D.M. and Mornon, J.P. (1998). The immunoglobulin superfamily: an insight on its tissular, species, and functional diversity. *J. Mol. Evol.* **46**, 389–400.

Harpaz, Y. and Chothia, C. (1994). Many of the immunoglobulin superfamily domains in cell adhesion molecules and surface receptors belong to a new structural set which is close to that containing variable domains. *J. Mol. Biol.* **238**, 528–39.

Padlan, E. (1996). X-ray crystallography of antibodies. *Adv. Prot. Chem.* **49**, 57–133.

Wedemayer, G.J., Patten, P.A., Wang, L.H, Schultz, P.G. and Stevens, R.C. (1997). Structural insights into the evolution of an antibody combining site. *Science* **276**, 1665–9.

Proteins of the MHC and TCRs

Batalia, M.A. and Collins, E.J. (1997). Peptide binding by class I and class II MHC molecules. *Biopolymers* **43**, 281–302.

Bongrand, P. and Mallissen, B. (1998). Quantitative aspects of T-cell recognition: from within the antigen-presenting cell to within the T cell. *BioEssays* **20**, 412–20.

Campbell, R.D. and Trowsdale, J. (1993). Map of the human MHC. *Immunol. Today* **14**, 349–52.

Iminishi, T., Wakisaka, A. and Gojobori, T. (1992). Genetic relationships among various human populations indicated by MHC polymorphisms. In: *HLA 1991*, Vol. 1, p. 627. (K. Tsugi, M.Aizawa, T. Sasazuki, ed.) Oxford University Press, Oxford.

Jones, E. Y., Tormo, J., Reid, S. W. and Stuart, D. I. (1998). Recognition surfaces of MHC class I. *Immunol. Res.* **163**, 121–8.

Klein, J., Takahata, N. and Ayala, F.J. (1993). MHC polymorphism and human origins. *Sci. Amer.* **269**(6), 46–51.

Robertson, M. (1998). Antigen presentation. *Curr. Biol.* **8**, R829–31.

Exercises, problems and weblems

Exercises

7.1. A complete IgG—as shown in Figure 7.2—has an M_r of about 170 000. Estimate the M_r of (a) a Fab fragment, (b) the Fc fragment, (c) an Fv fragment.

7.2. Estimate the minimum number of amino acids in a peptide antigen that could make contacts with all six CDRs of the antibody shown in Figure 7.7. (The total width of the figure is 40 Å.)

7.3. A complete IgG comprises two VL–VH dimers. What kind of approximate symmetry relates the two halves of the molecule shown in Figure 7.2?

7.4. The frameworks of the variable domains of an antibody create a scaffolding of fairly constant structure for the antigen-binding loops. Which of the following contributes to the constancy of this structure, as opposed to making a general contribution to the stability of the protein? (a) The disulphide bridge between the sheets of each domain. (b) The packing of the conserved Trp residue against the disulphide bridge. (c) Conservation of residues at the interface between the two domains. (c) Burial of large amounts of surface area between the sheets of the two domains. (d) Lateral hydrogen bonding between strands of β-sheet.

7.5. Here is the sequence of the Vκ light chain of an antibody. Which canonical structure of the L3 region would you expect?

```
        10         20         30         40         50         60
DIQMTQSPSS LSASVGDRVT ITCQASQDII KYLNWYQQTP GKAPKLLIYE ASNLQAGVPS
        70         80         90        100
RFSGSGSGTD YTFTISSLQP EDIATYYCQQ YQSLPYTFGQ GTKLQIT
```

7.6. The antibody D11.15 has the sequence CTRDDNYGAMDYWG in its H3 region. Would you expect the bulged or non-bulged structure of this region? The C is residue 92 and the final G is residue 104.

7.7. The observation of two carbonyl groups in consecutive residues in a protein pointing approximately parallel to each other suggests the presence of a residue outside the allowed regions of the

Sasisekharan–Ramakrishnan–Ramachandran plot. One of the L3 loops in Figure 7.9 has this feature. Which immunoglobulin, and at which residue in its L3 loop? Which conformational angle (ϕ, ψ or ω) is in a nearly *cis* conformation?

7.8. The end-to-end distance of a N-residue peptide in a nearly extended conformation is about 3.8 Å × N. The rise per residue of an α-helix is 1.5 Å. Estimate the number of residues in an α-helix needed to flank a groove designed to bind a nine-residue peptide in a nearly-extended conformation. Compare with the lengths of the long helices in MHC molecule peptide-binding domains.

7.9. How would acetylation of the N-terminus, and formation of an amide at the C-terminus, be expected to affect the binding of peptides to (a) class I and (b) class II MHC proteins?

7.10. The MHC protein I-Ak binds the peptide STDYGILQINSRW. Which of the following peptides would you expect it to bind also? (a) STDYGIIQINSRW. (b) STDYGRLQINSRW. (c) STKYGILQINSRW.

7.11. Antiserum to human haemoglobin is useful in diagnosis. For example, relatively early identification of bowel cancer is possible by detection of 'occult blood' in faeces. It was reported in 1981 that it is very difficult to raise antisera to human haemoglobin in rabbits, although antisera to human haemoglobin can be raised in other species, such as goats and sheep. (a) Suggest an explanation for the difficulty of raising antisera to human haemoglobin in rabbits. (b) What experiments might confirm this hypothesis? (c) Against what possible 'false-positive' results, that might arise in using goat and sheep anti-human haemoglobin antisera to detect blood in faeces, would one have to take precautions? What specific recommendation should be made to patients before testing for occult blood is carried out? (d) How could selected monoclonal antibodies reduce or eliminate false positives, relative to the use of polyclonal antibodies from goats or sheep?

Problems

7.1. The homologous antigen-binding loops of Vκ and Vλ domains of immunoglobulins have different repertoires of canonical structures. Suppose the antigen binding loops from a Vκ antibody are 'transplanted' into the homologous positions in a Vλ domains. Would you expect affinity to be retained? Explain your reasons.

7.2. Class I and class II MHC proteins contain similar domains arranged differently on polypeptide chains. How might these have arisen during evolution?

7.3. (a) Assume the following estimates: (1) Stimulation of a T cell requires approximately 100 MHC–peptide complexes per cell (as a consequence of the kinetics of dissociation of MHC–peptide complexes and of cell–cell encounters), and (2) an antigen-presenting cell expresses on the order of

10^5 MHC molecules on its surface. How many peptide species can an antigen-presenting cell effectively display at any time? (b) A typical eukaryotic cell synthesizes significant amounts of about 2000 proteins, average length 300 residues. How many 8-residue peptides can theoretically be generated from them? (c) What fraction of these peptides can a single antigen-presenting cell effectively display at any time?

Weblems

7.1. (a) Find the sequence of an antibody or antibody fragment against the HIV gp120 protein. Use: http://immuno.bme.nwu.edu/antibody_spec.html Report the name of the antibody. (b) What species is it from?

7.2. In what tissues are proteins of the immunoglobulin superfamily found in (a) vertebrates, (b) invertebrates?

7.3. Find PDB entries of 10 immunoglobulins, or fragments of immunoglobulins, determined by X-ray crystallography at better than 2.3 Å resolution.

7.4. Retrieve the DNA sequence of the heavy chain of the human anti-rabies antibody MAb105 (EMBL Data Library accession number L08089). Submit the sequence to a search of the databank V BASE of human antibody genes. Use: http://www.genetik.uni-koeln.de/dnaplot/ (a) From what germ-line V sequence is this domain derived? (b) What are the differences between the amino acid sequence coded by the germ-line segments and that of the corresponding region of the expressed antibody?

7.5. (a) Calculate the surface area buried in forming the hen egg white lysozyme–D1.3 complex. (Use PDB data set 1vfb.) (b) What fraction of the area of the lysozyme buried in the complex is non-polar? (c) What fraction of the area of the antibody buried in the complex is non-polar?

7.6. The Protein Data Bank contains structures of influenza haemagglutinin, and of a complex between Fab 17/9 and an heptapeptide cut out from influenza haemagglutinin. What is the r.m.s. difference in mainchain atom positions between the heptapeptide in the Fab complex and within the complete haemagglutinin structure?

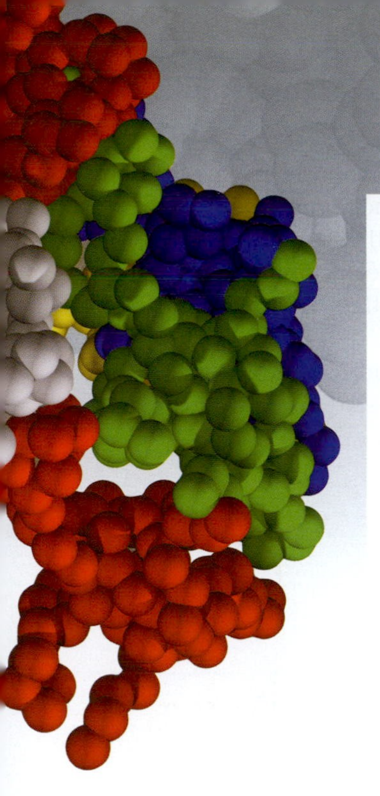

CHAPTER 8

Conformational changes in proteins

In previous chapters we have been talking of THE structures of protein, their unique native states. For many proteins, however, the mechanism of action requires changes between different conformational states.

Structural changes arising from change in state of ligation

There are now numerous cases of proteins for which structures have been determined in more than one state of ligation. In some cases, the structure undergoes little change, except perhaps for specific and localized changes associated with particular catalytic residues, as in triose phosphate isomerase (Figure 3.14). In sperm whale myoglobin the oxy and deoxy forms are very similar in conformation. In contrast, in haemoglobin, oxygen binding leads to changes in tertiary and quaternary structure. In some proteins, such as insulin, there are specific changes associated with the formation of a new binding site. Some enzymes, such as citrate synthase, show conformational changes in which the binding of ligands in a site between two domains leads to a closure of the interdomain cleft around them. Allosteric transitions involve long-range integrated conformational changes, as in haemoglobin, aspartate carbamoyltransferase, phosphorylase, phosphofructokinase, and others. Some very large proteins function as motors or pumps through changes in conformation; examples include myosin of muscle, the chaperonin GroEL, and ATPase.

Sperm whale myoglobin

The allosteric change of haemoglobin upon binding oxygen is so famous that many people are surprised to learn that the structures of oxymyoglobin and deoxymyoglobin are extremely similar. The most accurate coordinate sets are based on crystal-structure analyses of sperm whale myoglobin by J. Vojtěchovský, K. Chu, J. Brendzen, R. M. Sweet and I. Schlichting, who studied crystals of sperm whale myoglobin in different states of ligation under carefully controlled conditions of solvent and temperature. They collected data to near atomic resolution—1.15 Å for the deoxy state and 1.0 Å for the oxy state. The mainchains of the oxy and deoxy forms of myoglobin are extremely similar except for three residues at the C-terminus. The optimal r.m.s. deviation of all backbone atoms—N, Cα, C, O—of residues 1–150 is 0.15 Å! (The entire chain is 153 residues long.) Indeed, the optimal r.m.s. deviation of *all* atoms in these residues is 0.28 Å, the large differences arising

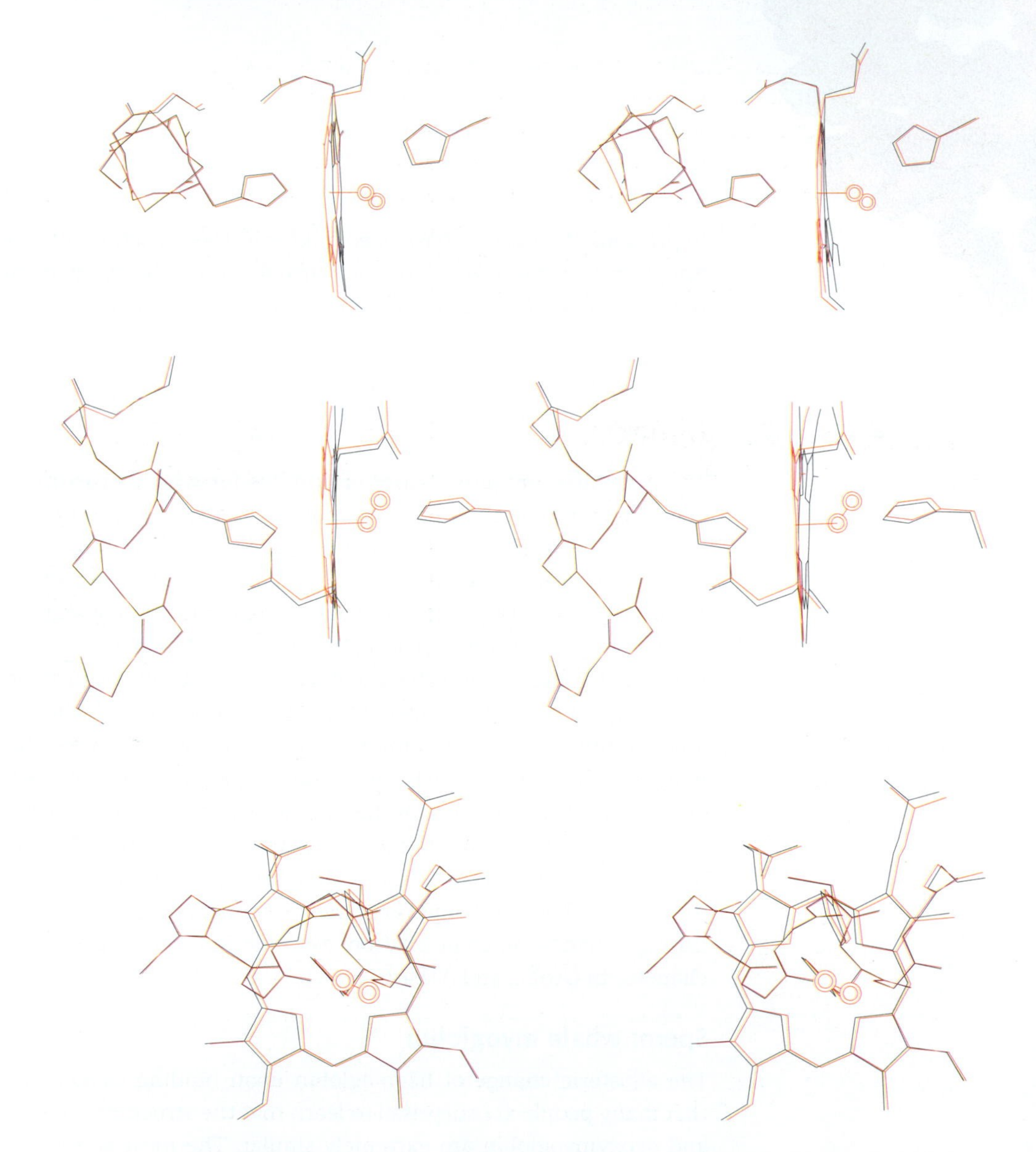

Fig. 8.1a

Fig. 8.1b

Fig. 8.1c

Fig. 8.1 (a–c) Oxy (red) and deoxy (black) forms of sperm whale myoglobin: haem group and F helix [1A6M, 1A6N]. (Multiple sidechain conformations not shown.) Three orientations.

from atoms at the ends of long sidechains. The optimal r.m.s. deviation of all N, Cα, C, O, Cβ, Cγ, and Oγ atoms in residues 1–150 is 0.16 Å. (These numbers arise from simplified calculations in which multiple conformers of certain sidechains are ignored.)

Figures 8.1a to 8.1c show a superposition of part of oxymyoglobin (red lines) and deoxymyoglobin (black lines). This picture includes the haem group, the F helix and the iron-bound proximal histidine, the distal histidine, and the O_2 of the ligated structure. The lengths of bonds provide a scale: the reader will see that the largest shifts are about a fifth of a bond length, or ~ 0.3 Å.

It is interesting that, although the ligated and unligated states have very similar conformations, the molecule must open up *during* the process of oxygen capture or release, because access to the oxygen-binding site is blocked in both the initial and final states (Figure 8.2). Readers may recall the scenes in *The Marriage of Figaro* in which the Count, having locked the closet before leaving the stage, returns to find all apparently as he left it, incognizant of the frantic intervening action that took place out of his view. Only rarely can crystallography reveal the dynamics of transient processes.

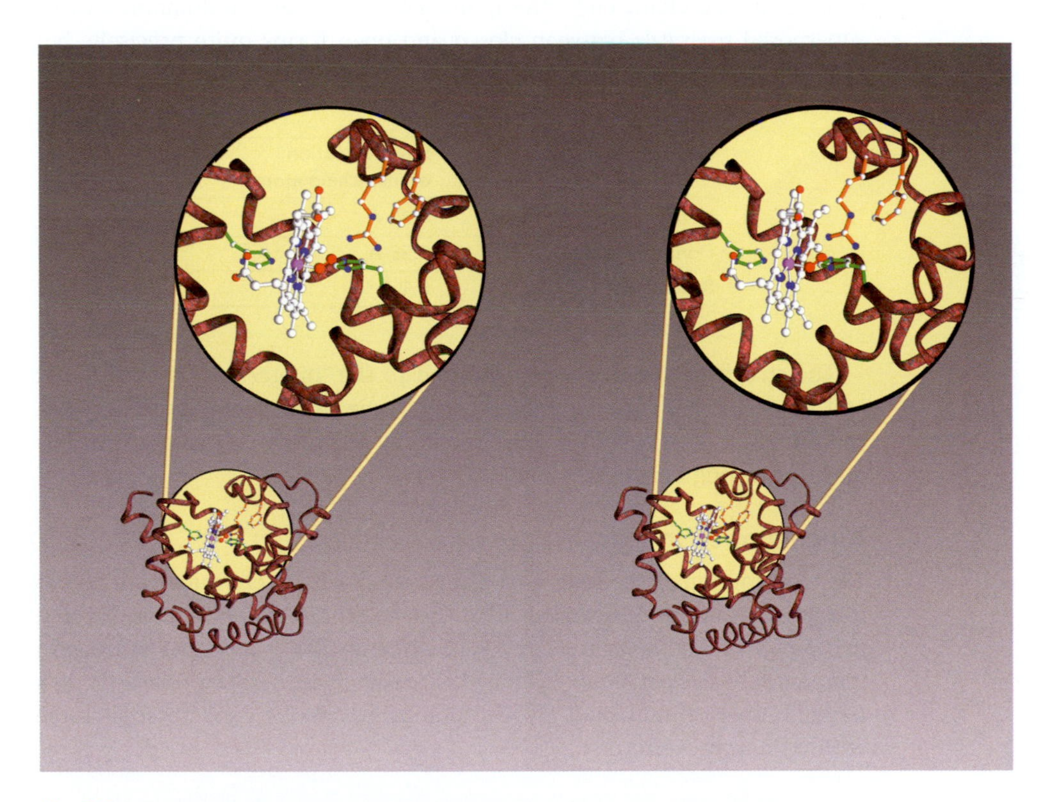

Fig. 8.2 Sperm whale myoglobin: the blocking of access to oxygen-binding site. Sidechains 45Arg and 46Phe are shown in orange, distal and proximal histidines in green, oxygen ligand in red [1A6M].

Hinge motions in proteins

The simplest possible mechanism of conformational change in proteins is a hinge motion, in which two parts of a structure move rigidly with respect to each other.

Hinge motion in lactoferrin

Lactoferrin is an iron-binding protein found in secretions such as milk and tears. It contains a single polypeptide chain approximately 700 residues long, folded into two large lobes of related sequence and similar structure. Each lobe contains two domains, with the chain passing between the domains twice. In the presence of iron, both lobes bind iron and have a 'closed' conformation. In the absence of bound iron the N-terminal lobe forms an open conformation, but the C-terminal lobe stays closed although it releases its iron (Figure 8.3). The conformational change in lactoferrin is a simple 'hinge' motion. A structure—a protein or a door—exhibits hinge motion if it consists of rigid objects that change their spatial relationship while maintaining their individual structures. The motion is permitted by a change in structure of a small amount of material linking the rigid bodies.

In the N-terminal lobe of lactoferrin, except for a small number of residues at the chain ends, the mainchain atoms of the domains can be superposed *separately* between closed and open forms quite precisely, but cannot of course be superposed together:

Residues superposed	r.m.s. deviation of mainchain atoms
Domain 1 alone:	0.82 Å
Domain 2 alone:	0.58 Å
Both domains together:	6.1 Å

The structural differences are limited to the 'hinges' at residues 89–92 and 249–252. There are two hinges because the chain passes twice between the domains. The change from open to closed form is a rotation of the domains through an angle of 54°.

Hinge motion in myosin

The contraction of muscle is a transformation of chemical energy to mechanical energy. It is carried out at the molecular level by a hinge motion in a protein called myosin (Figure 8.4), while myosin is attached to an actin filament. The cycle of *attach to actin–change conformation–release from actin*, in a large number of individual myosin molecules, creates a macroscopic force within the muscle fibre.

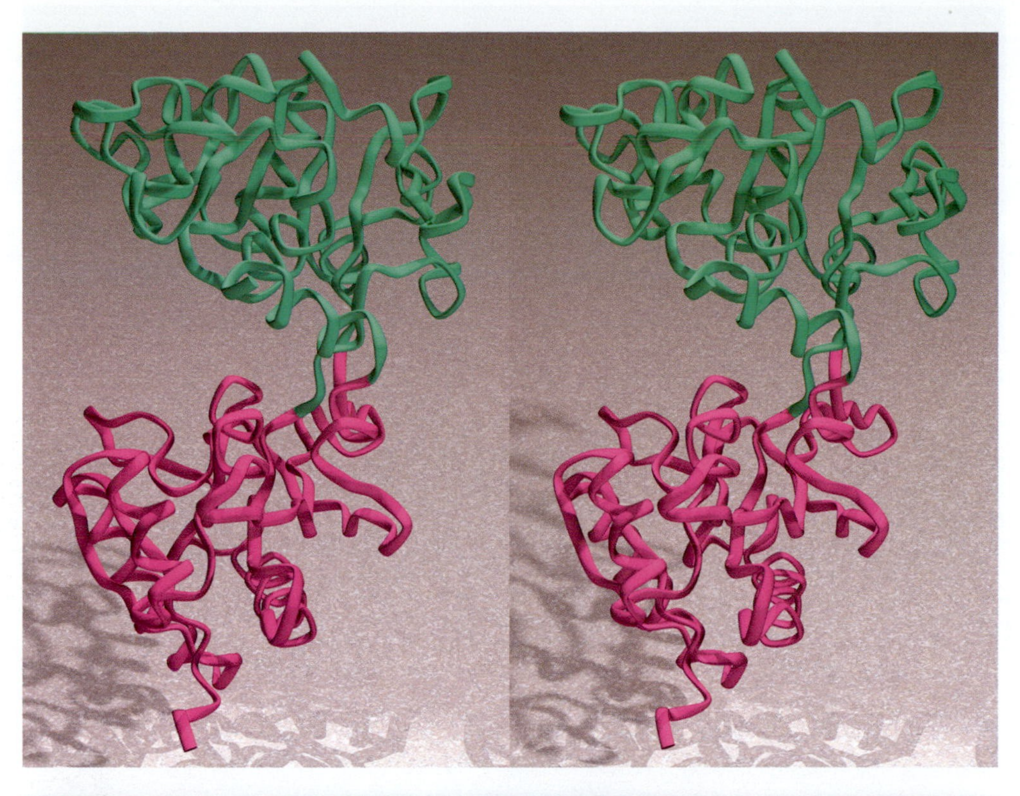

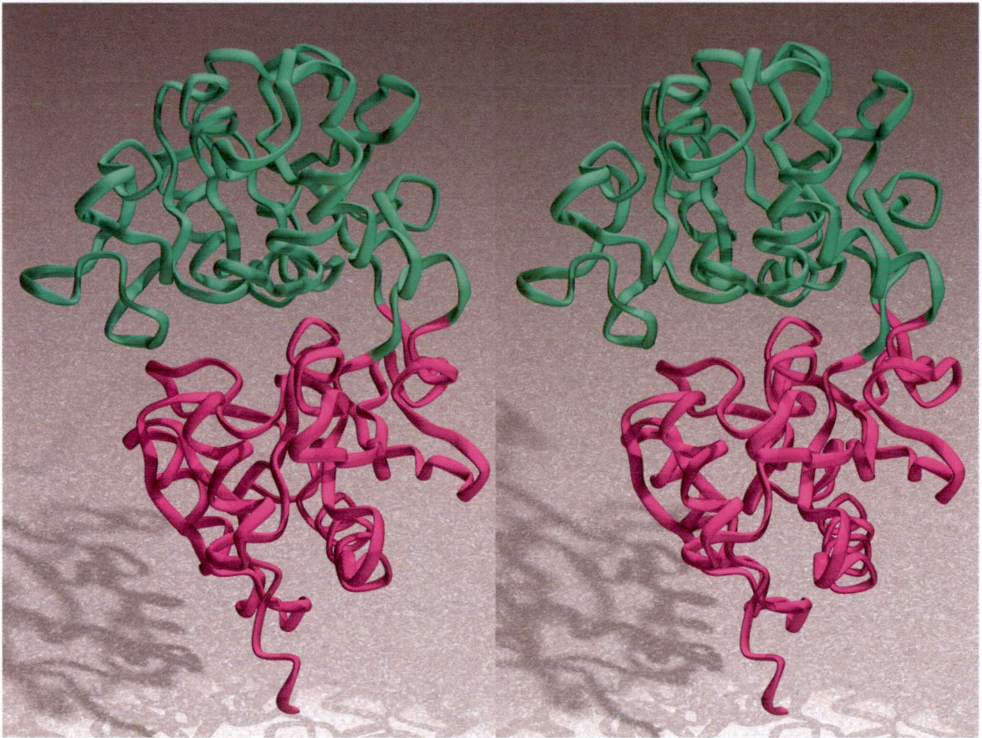

Fig. 8.3 Open and closed forms of lactoferrin [1LFG] and [1LFH].

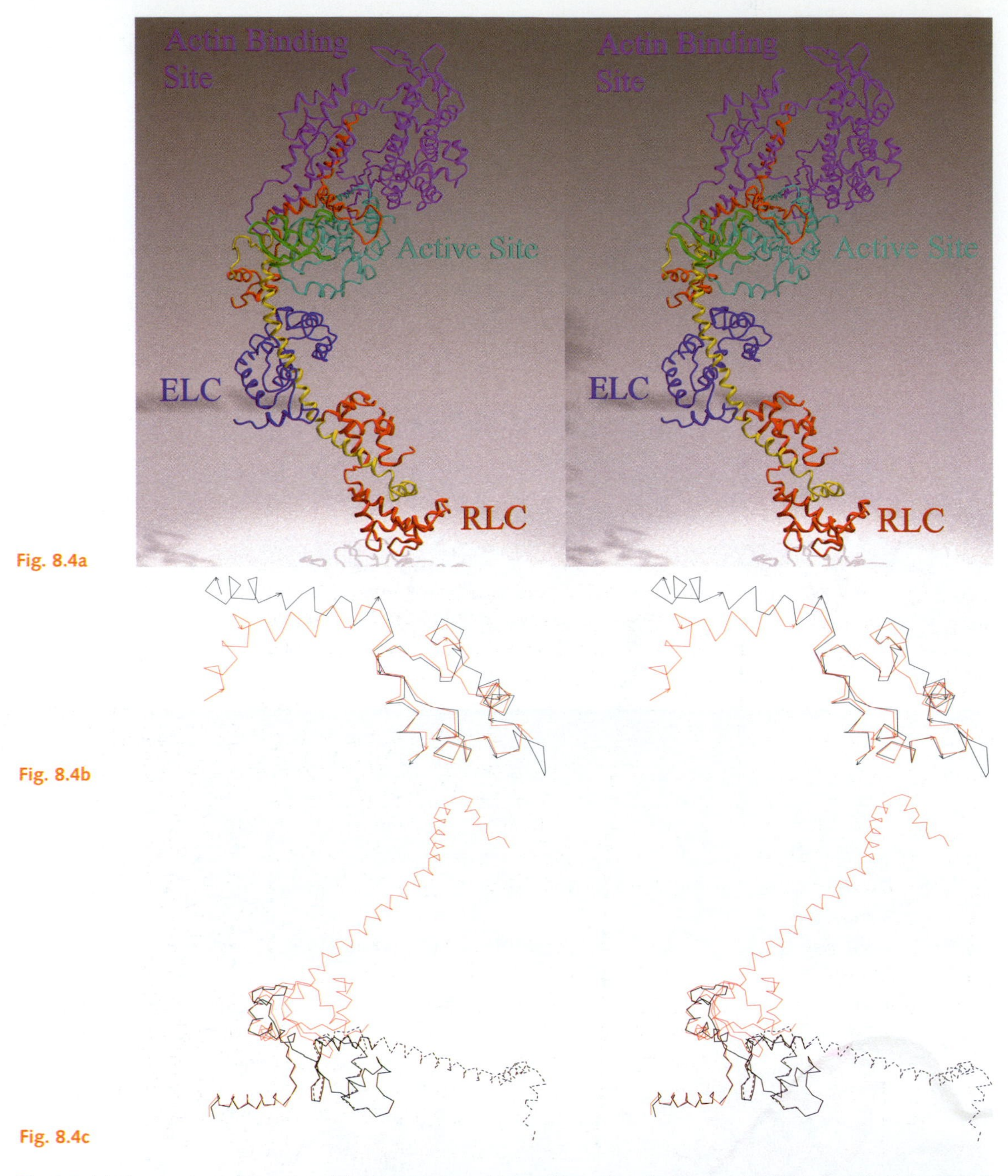

Fig. 8.4 (a) The structure of myosin subfragment 1 from chicken. The active site binds and hydrolyses ATP. ELC and RLC are the essential and regulatory light chains [2MYS]. (b) Hinge motion in myosin. Comparison of parts of chicken myosin open form [2MYS] (no nucleotide bound) and closed form binding ATP analogue ADP · AlF_4^- [1BR2]. This shows the segments of the structure that surround the hinge region. The segments shown in part b appear within the regions coloured red and yellow in part (a). (c) Model of the swinging of the long helical region in myosin as a result of the hinge motion. The broken line shows a model of the position that the complete long helix would occupy in the closed form, [2MYS] and [1BR1]. This conformational change is coupled to hydrolysis. It takes place while myosin is bound to actin, providing the power stroke for muscle contraction. In the context of the assembly and mechanism of function of a muscle filament, it is arguable that one should regard the helix as fixed and the head as swinging. But that would not show the magnitude of the conformational change as dramatically.

The 'helix interface shear' mechanism of conformational change

Hinge motion requires that the subunits be free to move with respect to each other. Sometimes close packing at interfaces between domains hinders their free movement. In these cases the mechanism of conformational change can be more complex.

Insulin

The pig insulin monomer consists of two chains, A (21 residues) and B (30 residues). The chains are linked by two disulphide bridges; there is a third disulphide bridge within the A chain. The A chain contains two helices; the B chain contains a helix and a strand of sheet (Figure 8.5). Two monomers form a dimer, held together by hydrogen bonding between strands of β-sheet and by van der Waals contacts (Figures 8.5b–8.5c). In the presence of Zn^{2+} the dimers assemble into hexamers. Crystals grown at low salt concentration contain the 2Zn form, with two zinc ions per hexamer. High salt concentration produces an alternative form containing four zinc ions per hexamer (Figure 8.6). Symmetry conditions dictate that the two zinc ions in 2Zn insulin fall on the three-fold axis, and that at least one of the four zinc ions in 4Zn insulin must be on the axis. One on-axial zinc binding site is very similar in 2Zn and 4Zn forms. The 4Zn form also contains three off-axial zinc binding sites related by the threefold symmetry.

Both the 2Zn and 4Zn forms of pig insulin have been solved to high resolution. The asymmetric unit of each crystal contains two monomers—each hexamer can be thought of as a trimer of dimers. The three dimers are related by a crystallographic, i.e. an exact, threefold symmetry and are identical in structure. The two monomers of each dimer are related by an approximate twofold axis. Therefore, in each crystal form, there are two different monomer environments; in each, two different monomeric conformational states are produced.

The 2Zn and 4Zn crystal structures present four independent monomer structures. Of these, one monomer of the 2Zn form is very similar in conformation to the corresponding monomer in the 4Zn form, and can serve as a 'reference structure' for the analysis of conformational changes and the search for their origin. In the 2Zn form, the second monomer is quite similar to the reference monomer (Figure 8.7a). In the second monomer of the 4Zn form the N-terminus of the B chain has changed its conformation entirely in the process of forming the off-axial zinc binding site (Figure 8.7b). Here we do not focus on the large change in the N-terminus of the B chain, but on the structural adjustments of the core of the protein.

A detailed comparison of the mainchain conformations of the insulin monomers suggests a physical picture of the deformations. The insulin structure contains a set of nearly-rigid segments of chain connected by more flexible regions, mobile with respect to one another. The chains contain many regions (3–10 residues in length) which have very similar mainchain

Fig. 8.5 (a) The secondary structure and disulphide bridges in the 2Zn insulin monomers. The strand in the B chain forms a β-sheet with the corresponding strand in another monomer. (b) The 2Zn insulin dimer [3INS]. The threefold axis relating this dimer to the other two dimers in the hexamer is vertical in the plane of the page, passing through the two zinc ions (spheres). (c) The 4Zn insulin dimer [1ZNI]. (Zn ions not shown.)

Fig. 8.5a

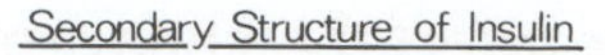

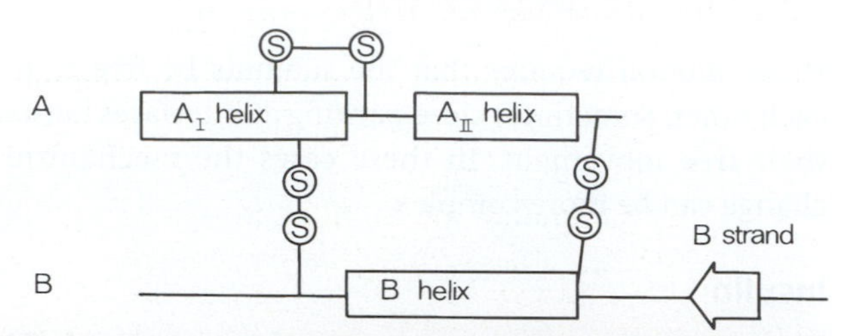

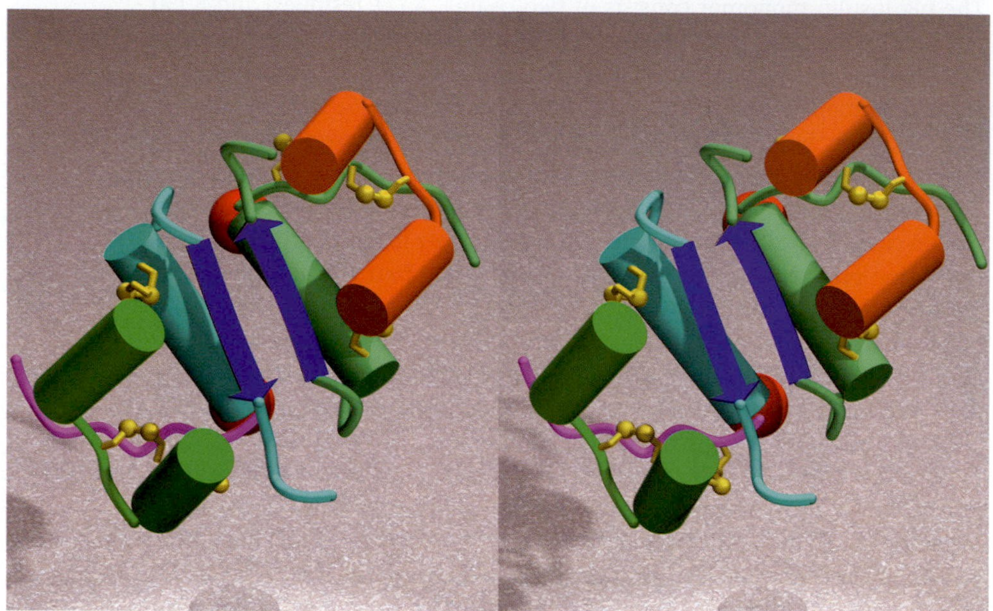

Fig. 8.5b

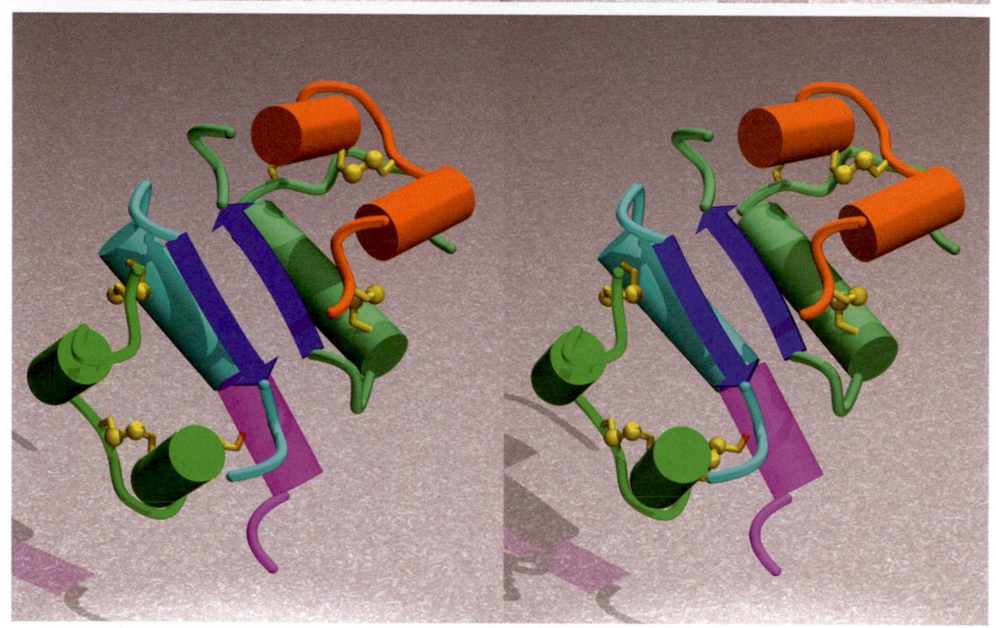

Fig. 8.5c

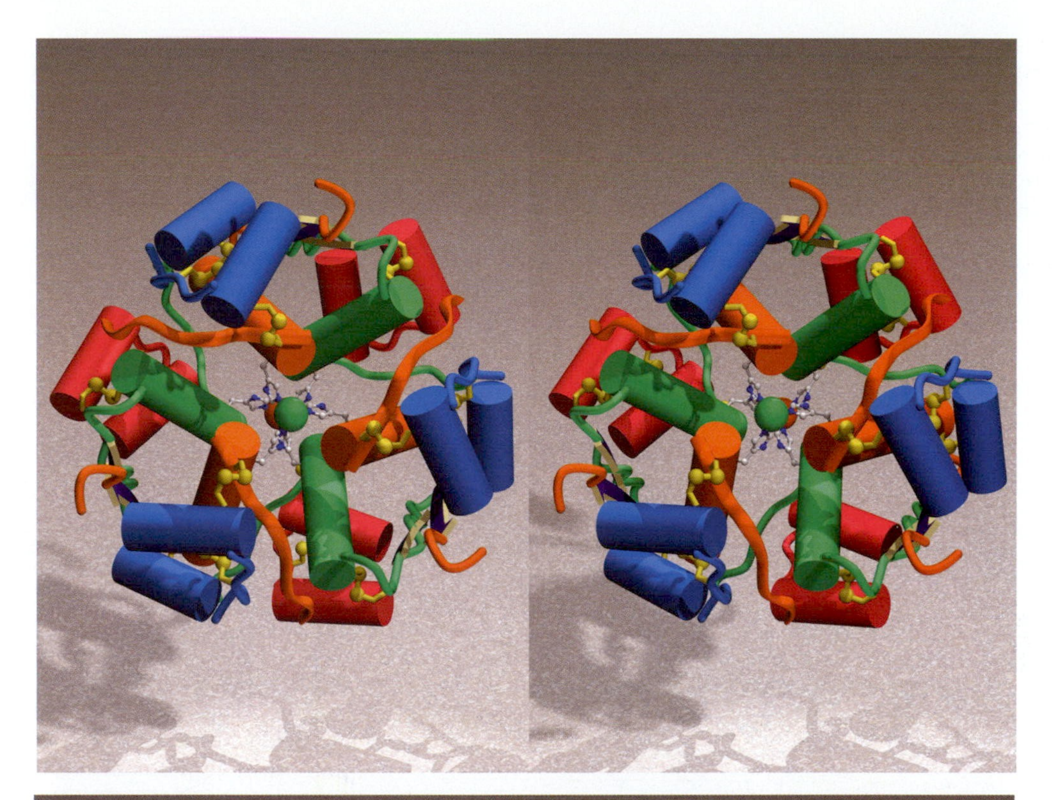

Fig. 8.6a

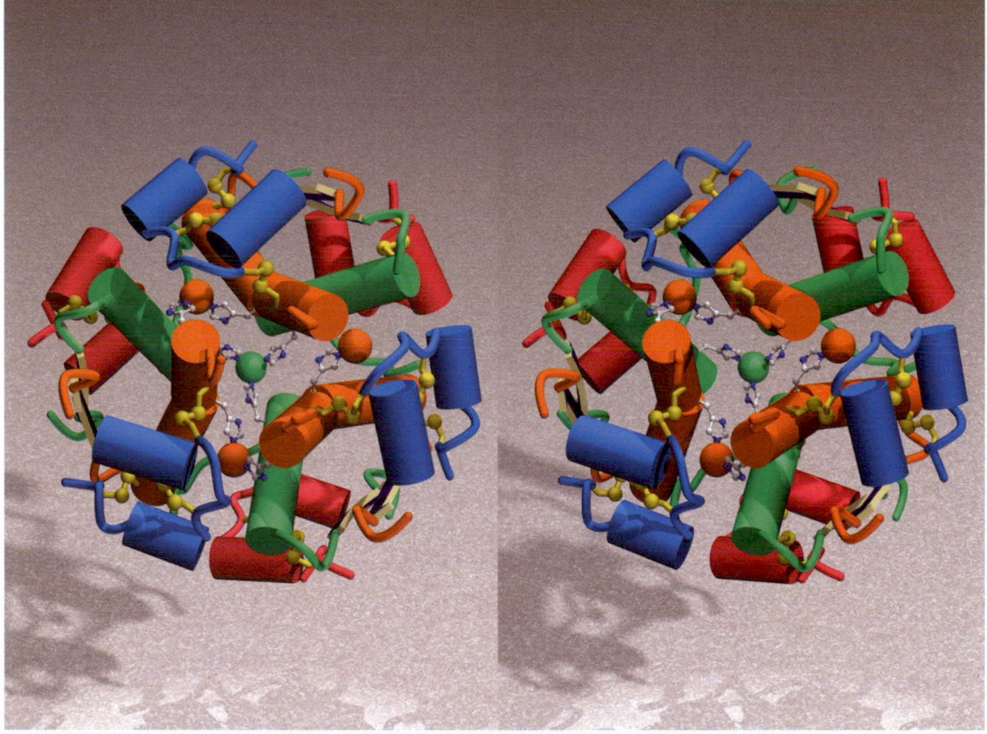

Fig. 8.6b

Fig. 8.6 The insulin hexamers: (a) 2Zn form [3INS]. (b) 4Zn form [1ZNI].

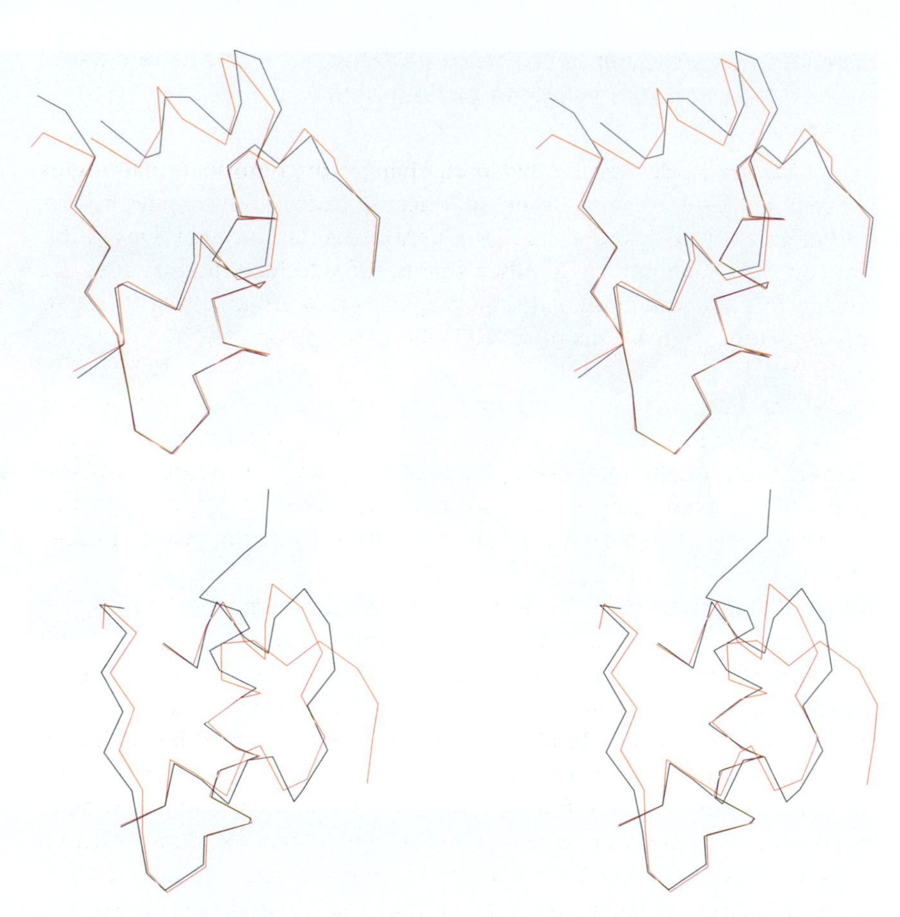

Fig. 8.7 Comparison of main chain conformation of the two monomers of (a) 2Zn insulin [3INS], (b) 4Zn insulin [1ZNI].

conformations. The only sizeable segment that exhibits a gross change in folding is the N-terminus of the B chain, to form the new off-axial Zn binding site, in one of the molecules of the 4Zn form. The well-fitting regions can be superposed on the homologous regions of the reference molecule with r.m.s. deviations of main chain atoms of 0.11–0.25 Å. However, these regions of locally-conserved structure have different mutual relative geometry in the reference structure and the second monomers. They are shifted and rotated with respect to one another. The magnitudes of the movements are in most cases no more than about 1.5 Å.

These shifts are similar in nature although smaller in magnitude than the shifts seen in the evolution of homologous proteins (Chapters 5 and 6). In evolution, the impulse comes from mutations within the protein. In insulin, as we shall see, the impulse is external.

How are these movements accommodated: why does the close-packing of interfaces between packed helices not hinder them severely?

The shifts are made possible by small changes in conformational angles which allow the sidechains in the interfaces to shift. These changes in conformation are usually quite small, not permitting the torsion angles to flip from one local minimum to another; that is, the sidechain rotamer does not change. It is this condition of being trapped in local minima that seems to impose a limit on the magnitude of the shifts of one helix relative to another.

The mechanism of conformational change in insulin appears then as follows: shifts in packed helices are facilitated by small conformational changes within helix interfaces, which permit finite but limited displacements of the mainchain atoms. This 'helix interface shear' mechanism of conformational change allows a maximum displacement of packed helices of about 1.5–2.0 Å, which appears to represent the limit of plastic deformation of the helix interface. These shifts can be seen in Figure 8.7, using the 1.5 Å rise per residue in an α-helix as a measure of distance.

This mechanism has interesting general implications for the long-range transmission of conformational change. Consider two extremes: if proteins were infinitely 'soft', any local conformational change would be dissipated in the immediate vicinity of the perturbation, and could not be transmitted farther; if proteins were infinitely 'hard', any local conformational change could cause only global movements of a rigid unit, and not allow deformations of the type required by 'induced fit' of enzymes to substrates, or alterations of tertiary structures in allosteric changes. The observation that proteins do have a potential for deformation, but a limited one, shows how conformational change can be transmitted over long distances, and even amplified.

These observations explain an old puzzle about the insulin structures: the difference in conformation of Phe B25 between the two monomers in both 2Zn and 4Zn insulin (see Figure 8.8). In one molecule in both 2Zn and 4Zn insulin, the ring of Phe B25 packs into its own monomer; in the other the ring points out across the dimer interface to pack against the other monomer. This change takes place because the pocket occupied by Phe B25 in the reference structure is deformed in each of the other monomers. The deformation is the result of the transmission of conformational change from a perturbation arising from crystal packing forces, initiated at a site 20 Å away at the surface of the hexamer.

Citrate synthase

Many enzymes change conformation in response to the binding of substrates and cofactors. Often the active site occupies a cleft between two domains, and binding of cofactors is accompanied by the closing of the cleft over the ligands. Functional reasons for such a conformational change may include the necessity to orient catalytic groups around the substrate,

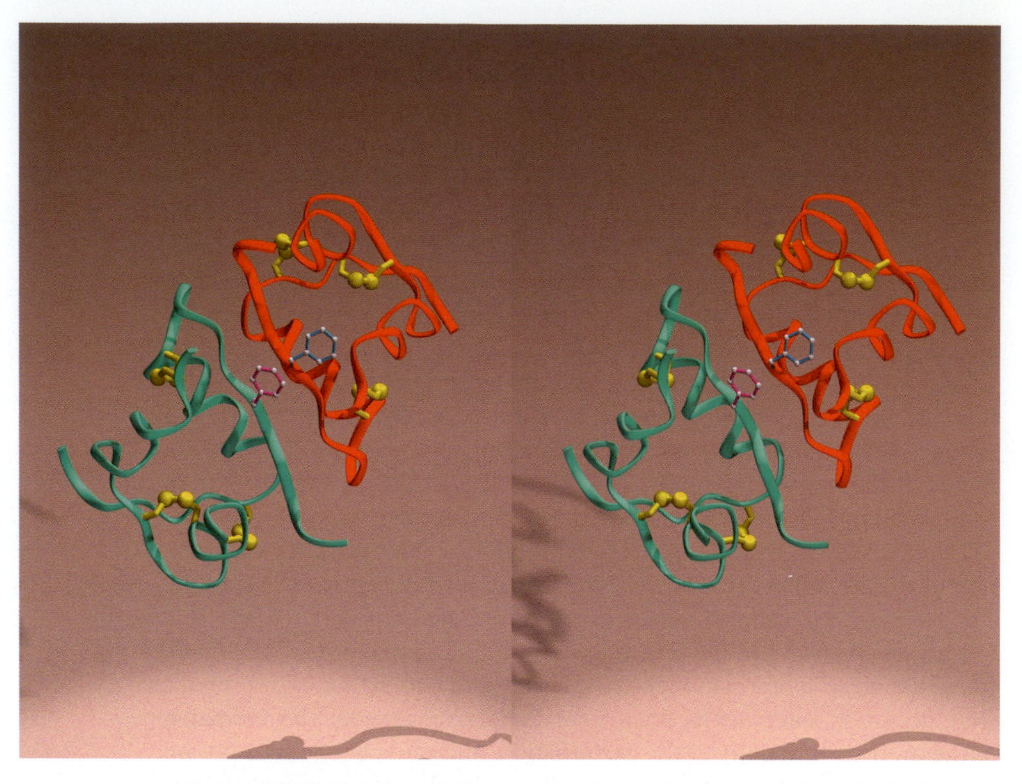

Fig. 8.8 The 2Zn insulin dimer, showing the different positions of Phe B25 in the two monomers [3INS].

discrimination against unwanted competitive ligands, and the exclusion of water from the active site. These conformational changes occur in some proteins that show no allosteric binding effects.

An example of such closure of an interdomain cleft occurs in the enzyme citrate synthase, a large dimeric protein. The monomer (of about 540 residues) contains 20 helices that form a structure containing two domains of unequal size (Figure 8.9). One crystal structure contains the product, oxalacetate, but not the cofactor coenzyme A; in the other the protein binds oxalacetate and coenzyme A. In the absence of coenzyme A the cleft between the two domains is open. With product and coenzyme A bound, the molecule has changed conformation so as to bury the product and cofactor almost completely. A loop between two helices moves by 6 Å and rotates by 28° to cover the ligands and to form hydrogen bonds to them. Some atoms move as much as 10 Å.

Citrate synthase has an extensive interdomain interface, which is incompatible with purely 'rigid-body' hinge movements of the domains. Indeed, in citrate synthase there is considerable conformational change within the domains, which contributes to the observed closure of the cleft. A group of seven helices of the large domain has a very similar structure in both forms of the molecule. These form a semi-rigid kernel containing only about one-third of the helices and limited to one of the domains. The other 13 helices

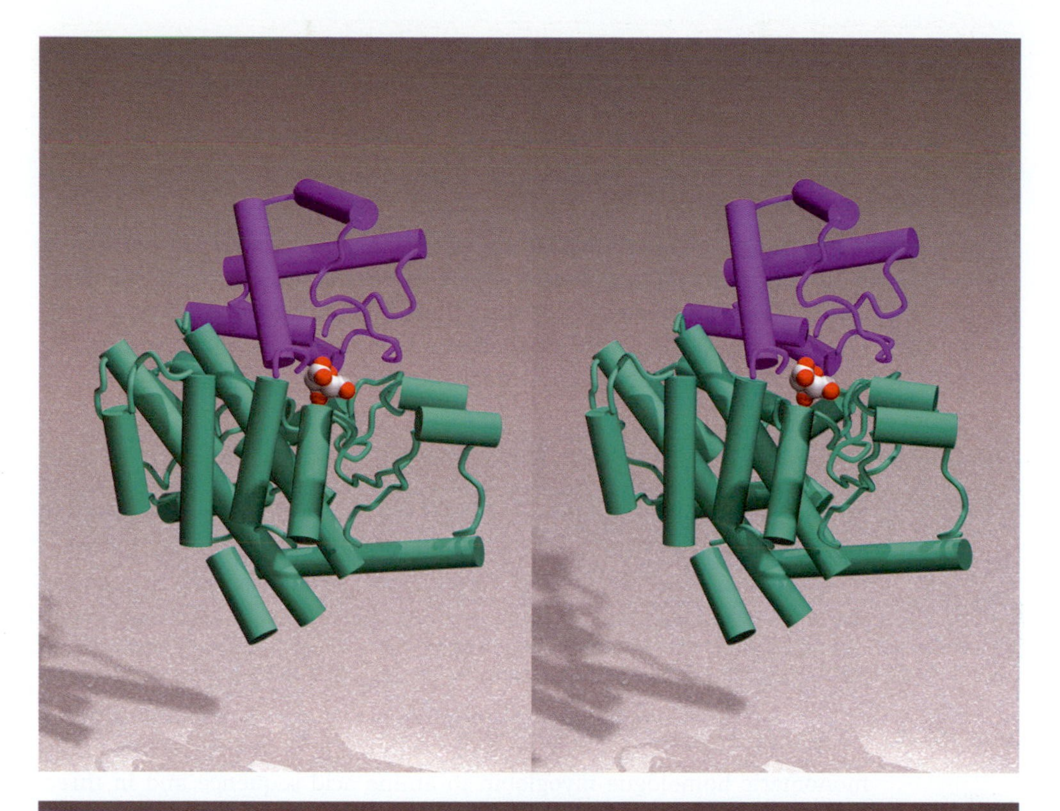

Fig. 8.9a

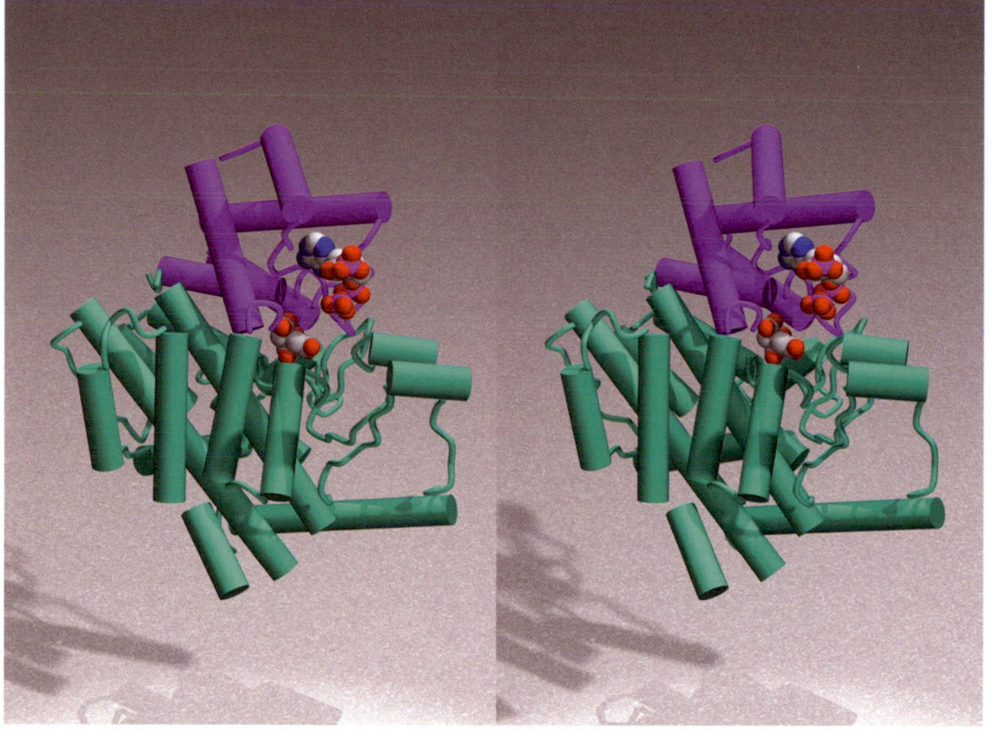

Fig. 8.9b

Fig. 8.9 (a) Open and (b) closed forms of citrate synthase [3CTS, 5CTS].

reorganize their relative spatial disposition, both with respect to the set of seven and in most cases with respect to one another. At the domain interface, the relative movements of packed helices vary from 0.2 Å displacement and 4° of rotation, up to 1.8 Å displacement and 11° degrees of rotation.

How then do the large motions required to close the interdomain cleft occur? The helix movements in citrate synthase are consistent with the 'helix interface shear' mechanism observed in insulin. Limited displacements of individual pairs of packed helices are coupled to produce the large shifts as a cumulative effect. Because there is a limit to the excursion of any single pair of packed helices, the large conformational changes must be built up by adding the effects of smaller ones.

The allosteric change in haemoglobin

To play its physiological role in oxygen distribution effectively, haemoglobin must capture oxygen in the lungs as efficiently as possible, and release as much as possible to other tissues. To achieve this 'take from the rich, give to the poor' effect, haemoglobin has a high oxygen affinity at high oxygen partial pressure (pO_2) and a low affinity at low pO_2. The binding is *cooperative*, in that binding of some oxygen enhances binding of additional oxygen.

The vertebrate haemoglobin molecule is a tetramer containing two α chains and two β chains. (Figure 8.10). Both α and β chains resemble their monomeric homologue myoglobin in amino acid sequence and in three-dimensional structure. These (and other) globins diverged from each other after gene duplication, myoglobin splitting from haemoglobin about 600–800 million years ago and the α and β strands of haemoglobin splitting about 500 million years ago.

In vivo, the haemoglobin molecule can have either of two structural forms. One is characteristic of deoxy (unligated) haemoglobin, and has low oxygen affinity; the other is characteristic of oxyhaemoglobin (four oxygen molecules bound) and has high oxygen affinity. The difference in colour between arterial and venous blood reveals the different electronic state of the iron. In the erythrocyte, haemoglobin is an equilibrium mixture of deoxy and oxy forms; the concentration of partially-ligated forms is tiny. (However, the structures of partially-ligated haemoglobin can be observed in crystal structures.) Binding of oxygen induces structural changes that alter the relative free energy of the two forms, shifting the equilibrium towards the high-affinity form. Starting from the deoxy state, partial ligation (between two and three oxygens) is enough to shift the equilibrium to the oxy state, which will then pick up the remaining oxygens with greater affinity. Conversely, starting with the fully-ligated oxy structure, partial loss of oxygen will shift the equilibrium in the opposite direction, stimulating release of the remaining oxygen. Other molecules that modify oxygen affinity, such as diphosphoglycerate, a natural allosteric effector, operate in part by shifting this equilibrium, by preferentially stabilizing one of the two forms.

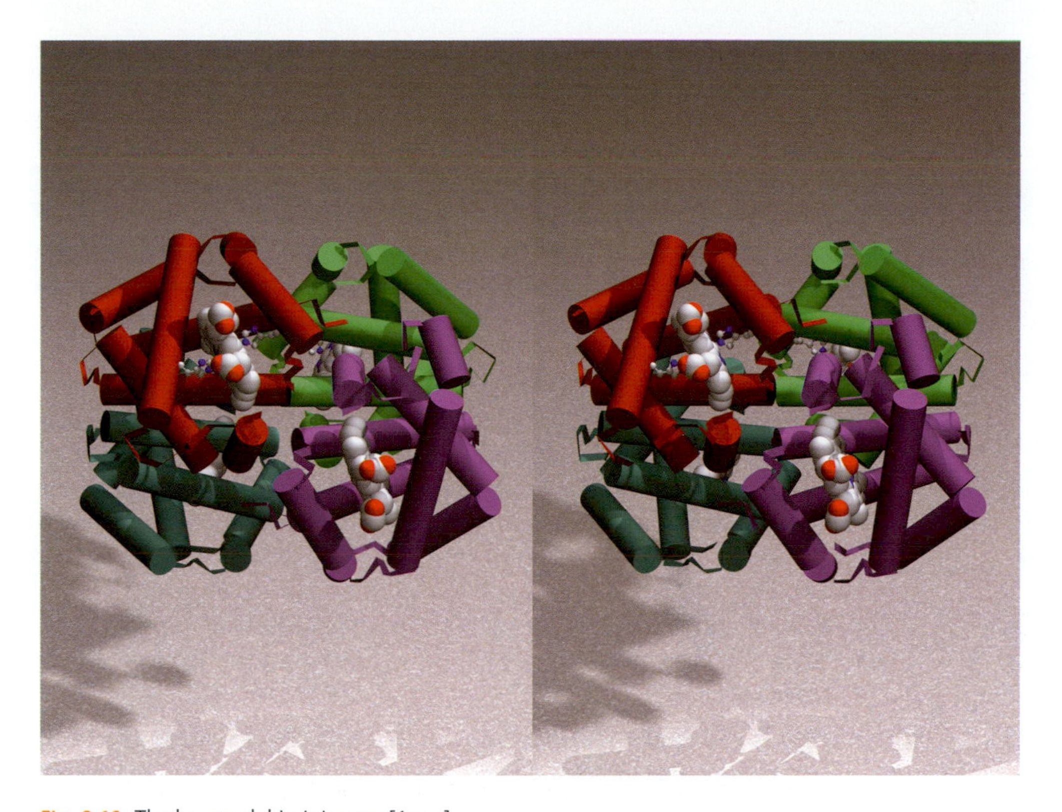

Fig. 8.10 The haemoglobin tetramer [1HHO].

The oxygen affinity of the oxy form of haemoglobin is similar in magnitude to that of isolated α and β subunits, and to that of myoglobin. The oxygen affinity of the deoxy form is much less: The ratio of the binding constants for the first and fourth oxygens is 1:150–300, depending on conditions. Therefore it is the deoxy form which is special, which has had its oxygen affinity 'artificially' reduced. In the terminology of Monod, Wyman and Changeux, in their general theory of allosteric change, the reduced oxygen affinity of the deoxy form of haemoglobin arises from structural constraints that hold the structure in a 'tense' (T), internally inhibited state; while the oxy form is in a 'relaxed' (R) state, as free to bind oxygen as the isolated monomer.

Oxy and deoxy forms of human haemoglobin

Property	Oxy	Deoxy
Oxygen affinity	High (K_4 = 0.17 mm Hg)	Low (K_1 = 26 mm Hg)
Spin state of iron	Low spin	High spin
$r(Fe^{2+})$	1.94 Å	2.06 Å
Monod, Wyman and Changeux state	Relaxed (R)	Tense (T)

The binding of oxygen is accompanied by a change in the state of the iron. In the deoxy state, the iron has five ligands—the four pyrrole nitrogens of the haem group and the proximal histidine—and is in a high-spin Fe (II) state with an ionic radius of 2.06 Å. In the oxy state, the iron has six ligands—oxygen being the sixth—and is in a low-spin Fe (II) state with an ionic radius of 1.94 Å. These radii are important because the distance from the pyrrole nitrogens to the centre of the haem is 2.03 Å; this implies that the iron will fit in the plane of the pyrrole nitrogens in the oxy state but not in the deoxy state.

Structural differences between deoxy- and oxyhaemoglobin

The structures of haemoglobin in different states of ligation have been studied with intense interest, partly because of their physiological and medical importance, and partly because they were thought to offer a paradigm of the mechanism of allosteric change. (As the structures of other proteins showing allosteric changes have been determined, it is becoming apparent that different systems achieve cooperativity by different mechanisms.)

The two crucial questions to ask of the haemoglobin structures are:

1. What is the mechanism by which the oxygen affinity of the deoxy form is reduced?
2. How is the equilibrium between low- and high-affinity states altered by oxygen binding and release?

Comparison of the oxy and deoxy structures has defined the changes in tertiary structures of individual subunits; and in the quaternary structure, or the relative geometry of the subunits and the interactions at their interfaces. Here is a simple description of how these changes are coupled: the quaternary structure is determined by the way the subunits fit together. This fit depends on the shapes of their surfaces. The tertiary structural changes alter the shapes of the surfaces of the subunits, changing the way they fit together.

Tertiary structural changes between deoxy- and oxyhaemoglobin

The tertiary structural changes in α and β subunits are similar but not identical.

At the haem group itself, in the deoxy form the iron atom is out of the plane of the four pyrrole nitrogens of the haem group. There are two reasons for this: the larger radius of the iron in its high-spin state, and the orientation of the proximal histidine, which produces steric repulsions between the Nε of the proximal histidine and the pyrrole nitrogens. The haem group is 'domed'; that is, the iron-bound nitrogens of the pyrrole rings are out of the plane of the carbon atoms of the porphyrin ring of the haem, by 0.16 Å in the α subunit and 0.10 Å in the β subunit.

Forming the link between iron and oxygen would, in the absence of tertiary structural change, create strain in the structure. For, without constraint the haem would become planar, and the iron would move into this plane. But these changes are resisted by the steric interactions between the proximal histidine and the haem group, and by the packing of the FG corner against the haem group (the FG corner is the region between the C-terminus of the F helix and the N-terminus of the G helix). In the β subunit, an additional barrier to the binding of oxygen without tertiary structural change is the position of Val E11 in the region of space to be occupied by the oxygen itself.

In the oxy form, these impediments are relieved by changes in tertiary structure. Describing these changes locally, relative to the haem group, there is in both subunits a shift of the F helix across the haem plane (Figure 8.11) by about 1 Å, and a rotation relative to the haem plane. The effect is to permit a reorientation of the proximal histidine so that the iron atom can enter the haem plane. Associated with this shift in the F helix, there are conformational changes in the FG corner.

To make the connection between tertiary and quarternary structural changes, we must describe the tertiary structural changes in terms of their effects on the shape of the entire subunit. The purely local description of what happens around the haem group is important to rationalize the energetics of ligation but is the wrong frame of reference for giving an account of the change in quaternary structure.

Intersubunit interactions in haemoglobin

J.M. Baldwin and C. Chothia presented a detailed analysis of the allosteric change in haemoglobin.

The haemoglobin tetramer can be thought of as a pair of dimers: $\alpha_1\beta_1$ and $\alpha_2\beta_2$. In the allosteric change, the $\alpha_1\beta_1$ and $\alpha_2\beta_2$ interfaces retain their structure, as does a portion of the molecule adjoining these interfaces, including the B,C,G,H regions of both subunits and the D helix of the β subunit. The overall allosteric change involves a rotation of 15° of the $\alpha_1\beta_1$ dimer with respect to the $\alpha_2\beta_2$, around an axis approximately perpendicular to their interface. (The motion is like that of a pair of shears with α_1 and α_2 = the blades and β_1 and β_2 = the handles.)

Given that the structure of the $\alpha_1\beta_1$ interface is conserved, it provides the appropriate frame of reference for describing the tertiary structural changes in a way that relates them to the changes in surface topography that account for the quaternary structural change.

With respect to the $\alpha_1\beta_1$ interface, the tertiary structural changes appear as follows (Figure 8.12a). In both subunits, in going from deoxy to oxy structures the haem groups move into the haem pockets, ending up 2 Å closer together in oxy than in deoxy structures. The backbone of the F helix moves with the haem (it also moves relative to the haem as discussed above). The FG corner also shifts. These tertiary structural changes of the F helix and FG corner do not extend beyond the EF corner on the N-terminal side and the beginning of the G helix on the C-terminal side. Note that this is consistent with the statement that the C and G regions are in the part of the structure that is conserved in the allosteric change.

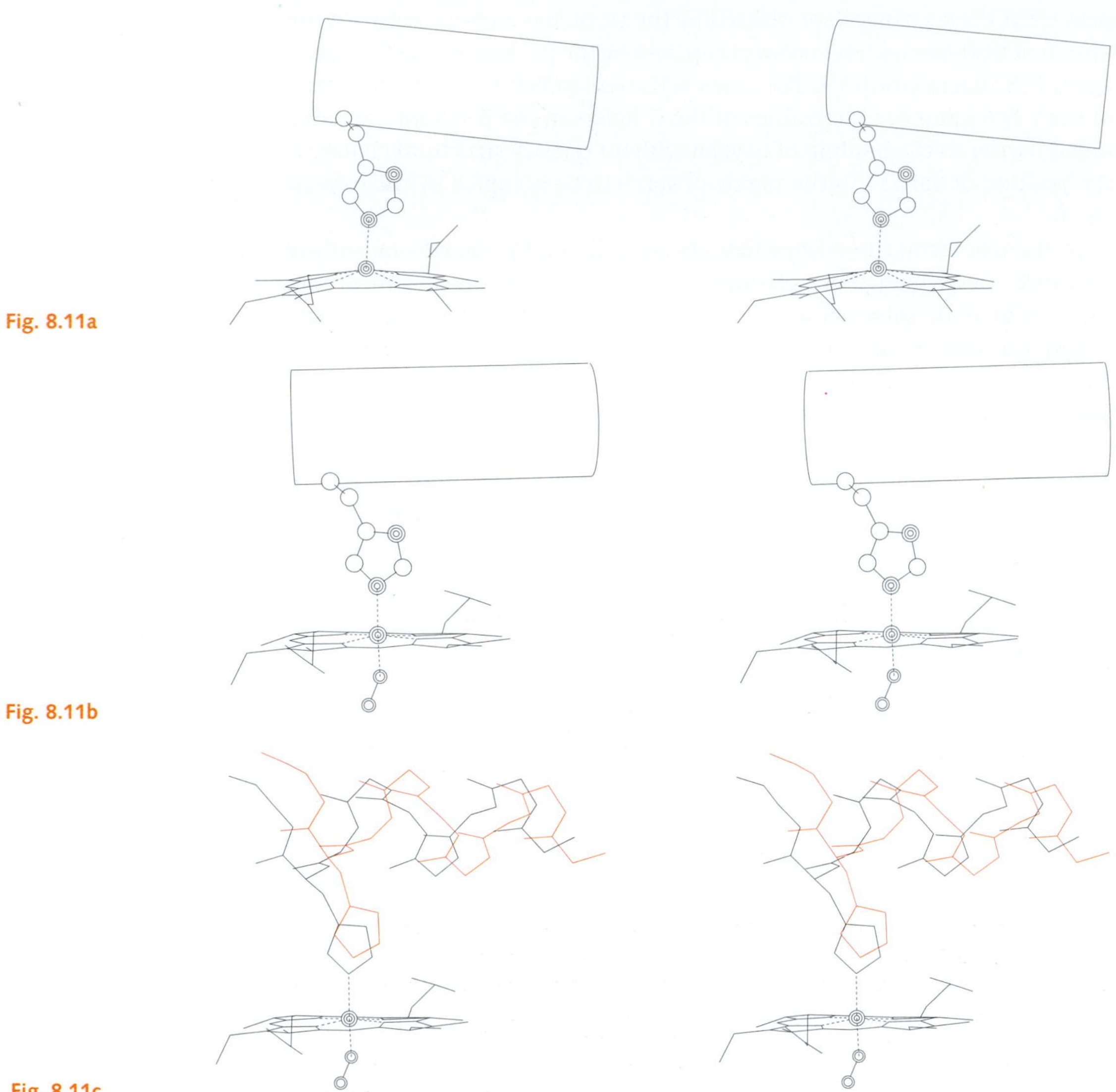

Fig. 8.11 Shifts in human haemoglobin as a result in change in state of ligation. This figure shows the F helix, proximal histidine, and haem group of the a chain in the (a) deoxy [2HHB] (observe the out-of-plane position of the iron), and (b) oxy [1HHO] forms of human haemoglobin. (c) Superposition of oxy (black) and deoxy (red); only the oxy haem is shown. The structures were superposed on the haem group. (Contrast Fig. 8.1.)

To understand the quaternary structural change, we must analyze how the interface between the $\alpha_1\beta_1$ and $\alpha_2\beta_2$ dimers changes. The most important intersubunit contacts are between the α_1–β_2 and α_2–β_1 subunits. (In the open-shears image, the important variable contacts are between each blade and the opposite handle. The contacts between each blade and its own handle are—in haemoglobin as in shears—rigid.)

The interacting regions are the α_1 FG corner–β_2 C helix and β_2 FG corner–α_1

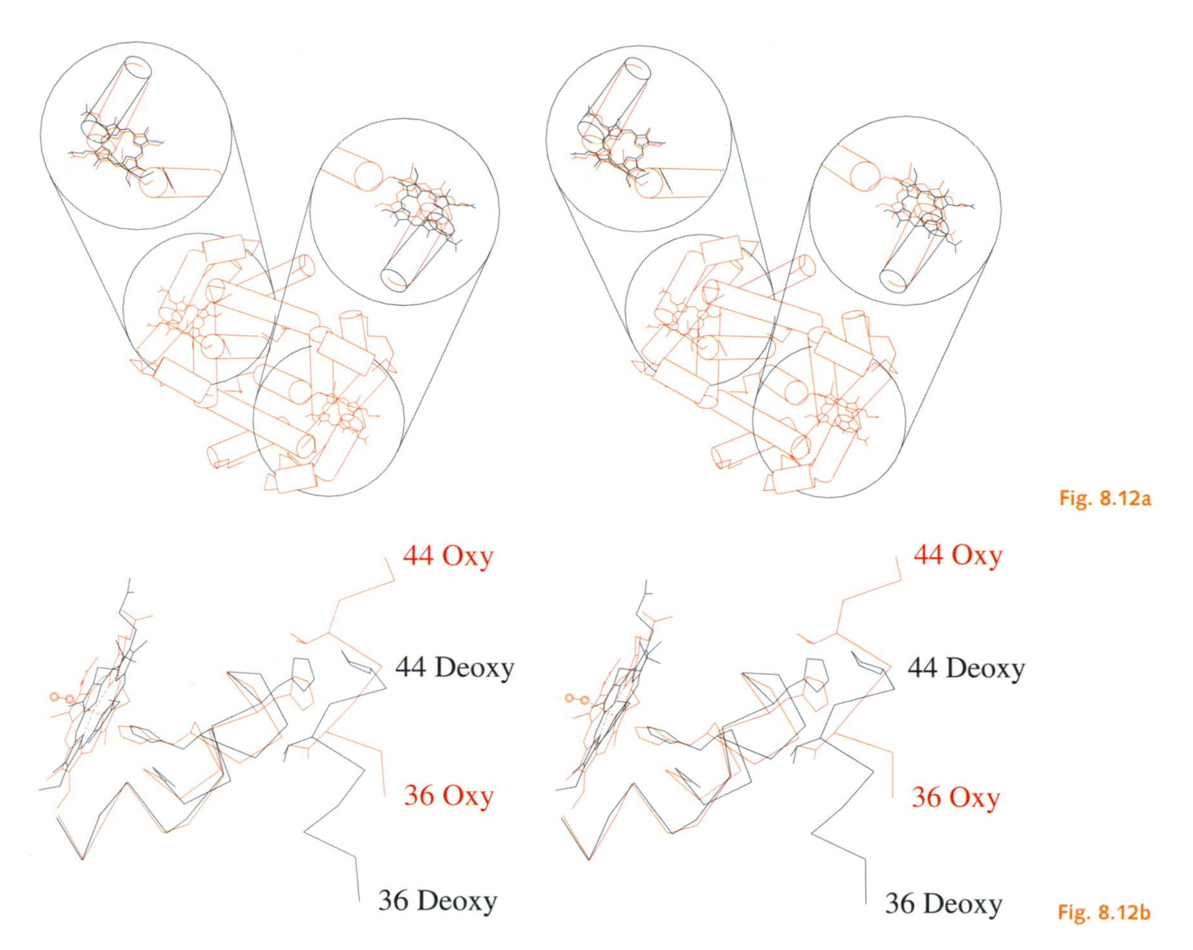

Fig. 8.12 Some important structural differences between oxy- and deoxyhaemoglobin [1HHO, 2HHB]. (a) The $\alpha_1\beta_1$ dimer in oxy (red) and deoxy (black, in blown-up regions only) forms. In the blown-up regions only the F helix, FG corner, G helix, and haem group are shown. The oxy and deoxy $\alpha_1\beta_1$ dimers have been superposed on their interface; in this frame of reference there is a small shift in the haem groups, and a shift and conformational change in the FG corners. (b) Alternative packings of α_1 and β_2 subunits in oxyhaemoglobin (red) and deoxyhaemoglobin (black). The oxy and deoxy structures have been superposed on the F and G helices of the β_2 monomer. Although for purposes of this illustration we have regarded the β_2 subunit as fixed and the α_1 subunit as mobile, only the relative motion is significant.

C helix. These are not identical. (However, these interactions *are* the same, by symmetry, as those of the α_2 FG corner–β_1 C helix and β_1 FG corner–α_2 C helix, respectively.)

The α_1 FG–β_2 C interaction is very similar in oxy and deoxy structures. Residues Arg 92 FG4, Asp 94 G1 and Pro 95 G2 of the α_1 subunit are in contact with Tyr 37 C3 and Arg 40 C6 in the β_2 subunit. In comparing oxy and deoxy structures there are small conformational changes in these residues but the pattern of interactions is retained.

The other region of contact, β_2 FG corner–α_1 C helix, differs substantially between oxy and deoxy structures (Figure 8-12b). In the deoxy structure, His β_2 97 (FG4) packs between Thr α_1 41 (C6) and Pro α_1 (44) (CD2), and there is a hydrogen bond between the sidechains of Asp β_2 99 (G1) and Tyr α_1 42 (C7). In the oxy structure, His β_2 97 packs between Thr α_1 38 (C3) and Thr α_1 41 (C6). (Because the C helix is a 3_{10} helix, the *three*-residue skip corresponds to a jump of one turn relative to the His β_2 97 against which it packs. The shift in β_2 FG relative to α_1 C is approximately 6 Å.) The Asp–Tyr hydrogen bond is not made in the oxy structure.

This explains the two discrete quaternary states: the β_2 FG corner–α_1 C helix contact has two possible states, depending on the subunit shape presented by the tertiary-structural state. The other contact, α_1 FG–β_2 C changes only slightly.

The requirement for the quaternary structural change arises from the tertiary structural changes; in particular, from the shifts that bring the FG corners of the α_1 and β_1 subunits in the $\alpha_1\beta_1$ dimer 2.5 Å closer together in the oxy structure relative to the deoxy structure. One tertiary structural state of the $\alpha_1\beta_1$ and $\alpha_2\beta_2$ dimers can form a tetramer with one state of packing at the interface; the other tertiary structural state is compatible with the alternative packing.

In conclusion, it may be useful to trace the logical connection between the structural changes. Starting from the deoxy structure, ligation of oxygen requires relief of strain around the haem group by shifting the F helix and the FG corner. To accommodate these shifts there must be a set of tertiary-structural changes, which alter the overall shape of the $\alpha_1\beta_1$ and $\alpha_2\beta_2$ dimers; notably the shifting of the relative positions of the FG corners. In consequence, the deoxy quaternary structure is destabilized because the dimers no longer fit together properly (having changed their shape). Adopting the alternative quaternary structure requires the tertiary structural changes to take place even in subunits not yet liganded. As a result of the quaternary structural change, these unligated subunits have been brought to a state of enhanced oxygen affinity. It is important to emphasize that this is a sequence of steps in a logical process and not a description of a temporal pathway of a conformational change.

Serpins: SERine Proteinase INhibitors

Some proteins undergo very large conformational changes, some so far as alterations in folding topology itself. These include systems of medical importance—prion and amyloid proteins, and serpins (SERine Proteinase INhibitors).

The serpins are a family of proteins with a variety of biological roles, including but not limited to inhibition of chymotrypsin-like serine proteinases. They are of interest both for their medical importance—clinical consequences of serpin dysfunction include blood clotting disorders, emphysema, cirrhosis, and mental disorder—and for the very unusual mechanism of their function, involving a conformational change known as the stressed → relaxed (S → R) transition, between conformational states of different folding topology.

That serpins undergo dramatic conformational changes when cleaved by proteinases was first recognized in 1983 when H. Loebermann, F. Lottspeich, W. Bode, and R. Huber, interpreting the electron-density map of α_1-antitrypsin cleaved at the scissile bond, found the residues adjacent to the cleaved peptide bond to be 65 Å apart! The region N-terminal to the scissile bond, called the reactive centre loop, formed a strand within a large β-sheet (Figure 8.13).

The structure of cleaved α_1-antitrypsin implied that the mechanism of proteinase inhibition by serpins must differ fundamentally from that of other inhibitors. In the more typical case of bovine pancreatic trypsin inhibitor, a loop binds specifically to the active site of the proteinase, and the inhibitor is not cleaved more than transiently (see Figure 3.22f). Serpins, in contrast, are 'suicide inhibitors', i.e. inhibition produces a relatively stable proteinase + cleaved inhibitor (or enzyme–product) complex in which the proteinase is distorted.

The initial interaction with the proteinase must occur with an uncleaved serpin, but knowing only the structure of cleaved α_1-antitrypsin it was unclear what the structure of the active intact molecule would be. However, Loebermann and colleagues proposed the correct answer: The removal from the central β-sheet of the strand comprising the reactive centre loop, and a subsequent rearrangement in this β-sheet and an adjacent α-helix, could bring together the two residues linked by the scissile bond. Crystal structures of uncleaved serpins confirmed this hypothesis. The first uncleaved serpin structure was that of the non-inhibitory serpin ovalbumin. In ovalbumin, the region corresponding to the reactive centre loop in inhibitory serpins forms an α-helix, and is not part of the central β-sheet (Figure 8.14a). Subsequently, structures of uncleaved inhibitory serpins (Figure 8.14b–d) revealed the structure of the reactive centre in inhibitors that actually interact with proteinases.

Structural states of serpins

The secondary structure of the serpin fold includes three β-sheets and nine α-helices. The reactive centre contains the residues that, in inhibitory serpins, interact directly with the cognate proteinase. In the active or 'native' form of typical serpins, the molecule is intact and the reactive centre is free to interact with proteinases. In the cleaved form the conformational change integrates the reactive-centre loop into the body of the molecule; a free cleaved serpin cannot inhibit proteinase. The mechanism of inhibition involves initial formation of a complex between an uncleaved serpin in its native conformation and the proteinase, followed by cleavage of the serpin.

Crystallography has revealed three different classes of conformations of serpins.

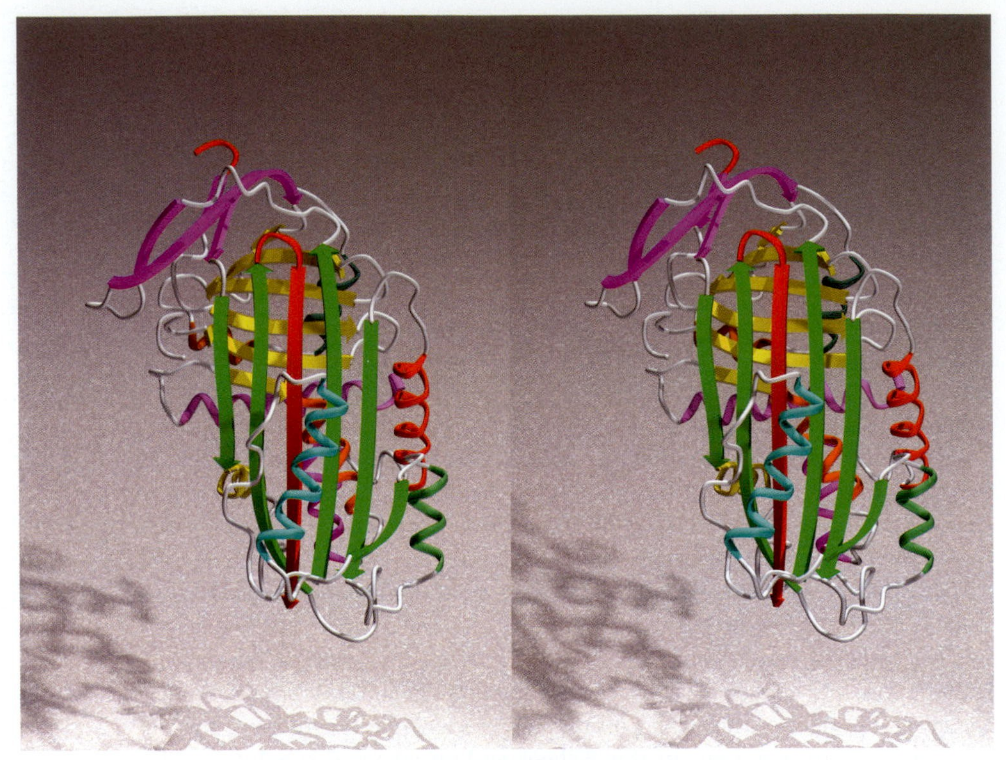

Fig. 8.13 Cleaved α_1-antitrypsin [7API]. Reactive centre loop in red.

1. *The native or S state* is the active form that can interact with the cognate proteinase. In the native state, the reactive centre is uncleaved and outside the β-sheet, free to bind to a proteinase.
2. Cutting serpins at the scissile bond produces the *cleaved* or *R* conformation, in which the reactive centre loop is inserted into the main β-sheet. During inhibition, cleavage of the serpin scissile bond and its change to the R conformation take place while in complex with the proteinase. The scenario comprises an initial association between the proteinase and inhibitor, followed by cleavage to a stable enzyme–product complex, and insertion of the loop into the β-sheet, dragging the proteinase from one pole of the serpin to the other.
3. *The latent state* is another uncleaved form, in which the reactive centre loop is inserted into the main β-sheet as in the cleaved form (Figure 8.15). This state should be considered an alternative R state.

Different folding topologies in one protein family

The rule is that homologous proteins have structures containing a core with the same folding pattern. Serpins are an exception. Now, the observation that native and cleaved α_1-antitrypsin and other inhibitory serpins have different folding topologies is not difficult to reconcile with the idea that amino acid sequence determines protein structure, because native and

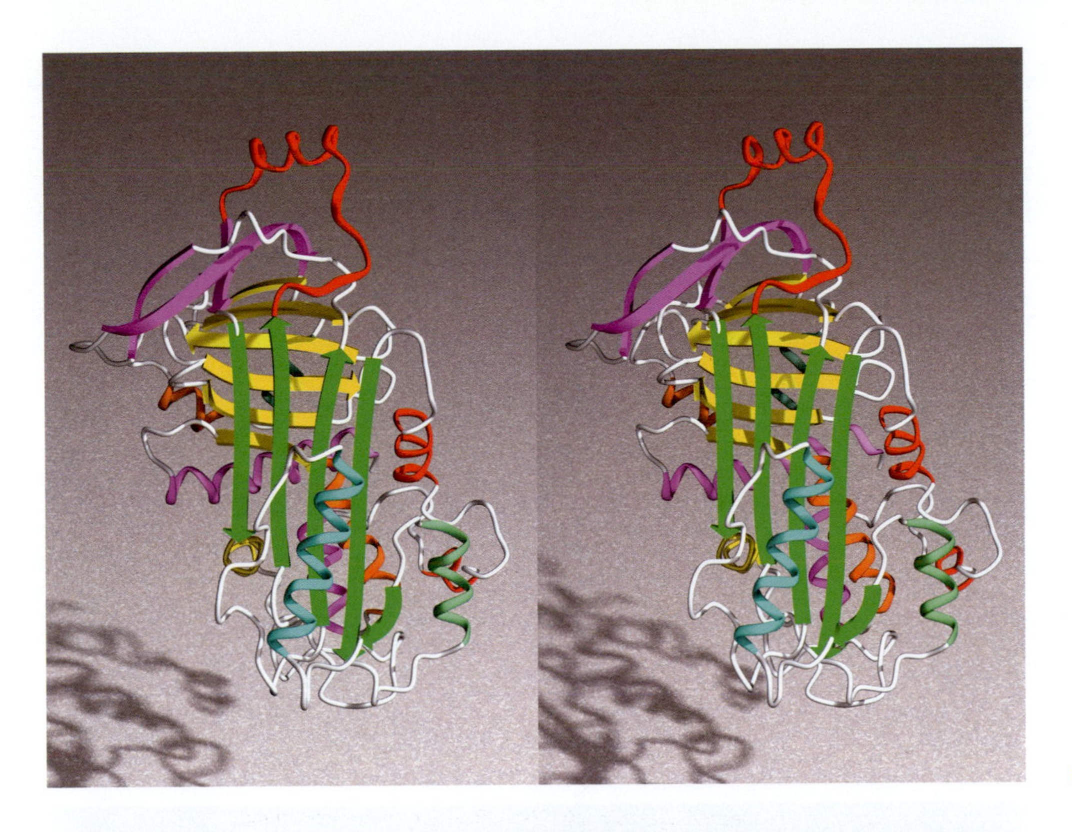

Fig. 8.14a

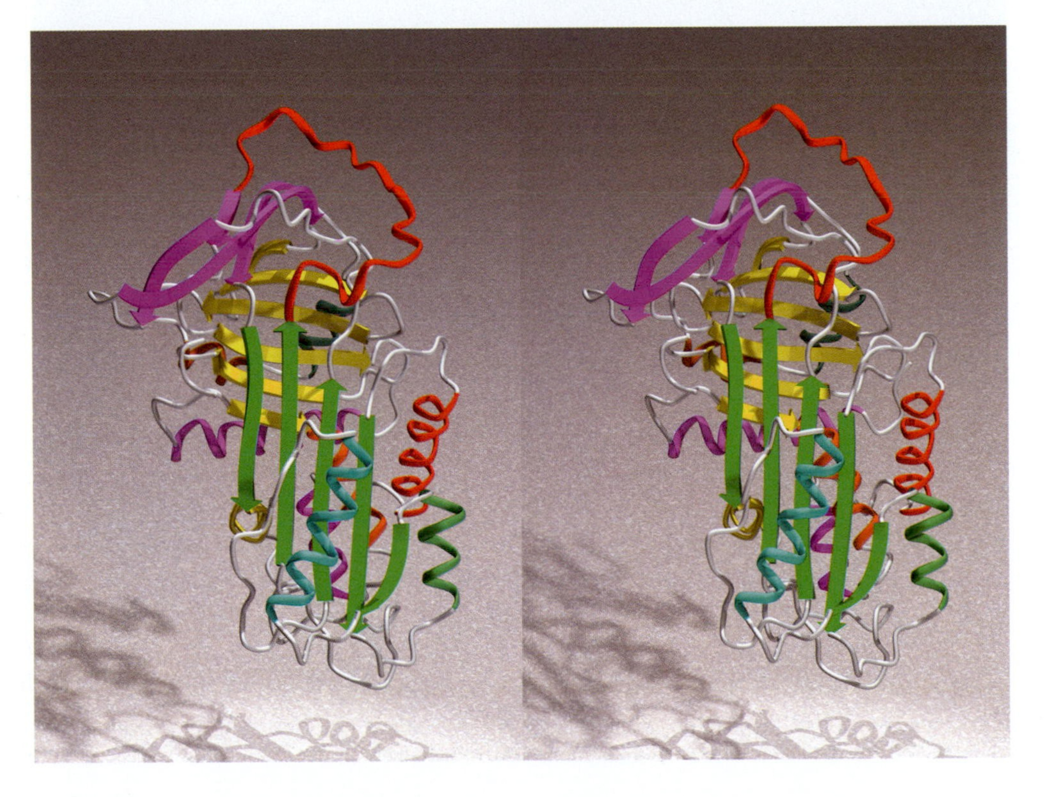

Fig. 8.14b

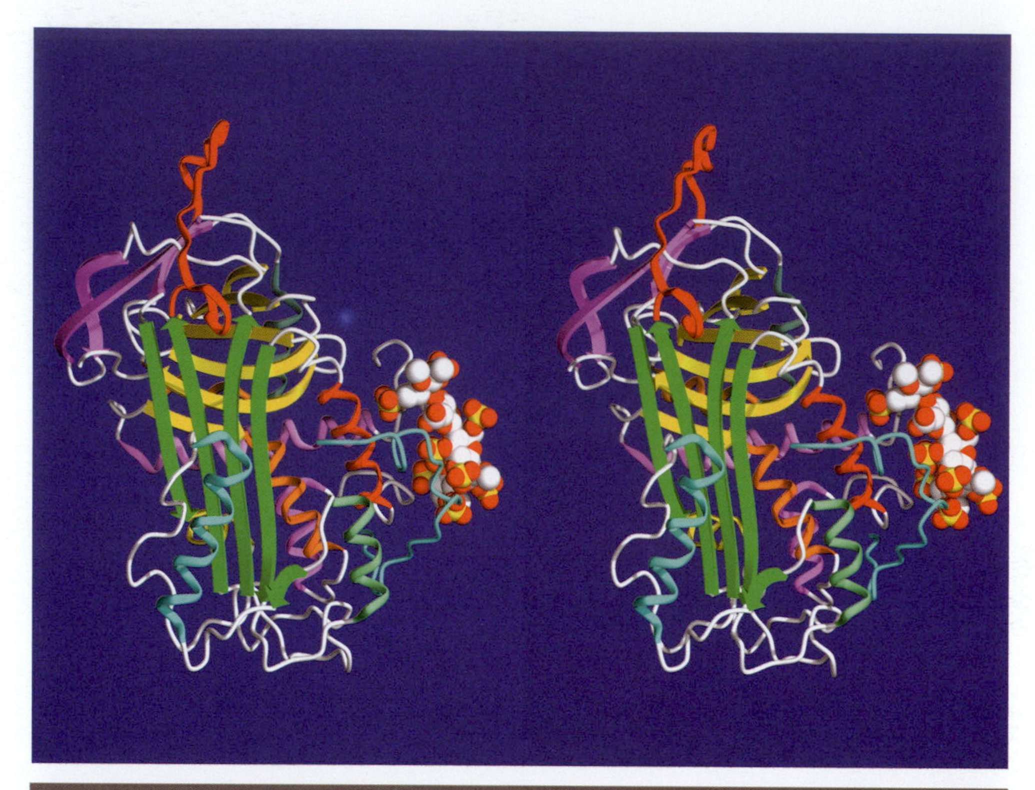

Fig. 8.14c

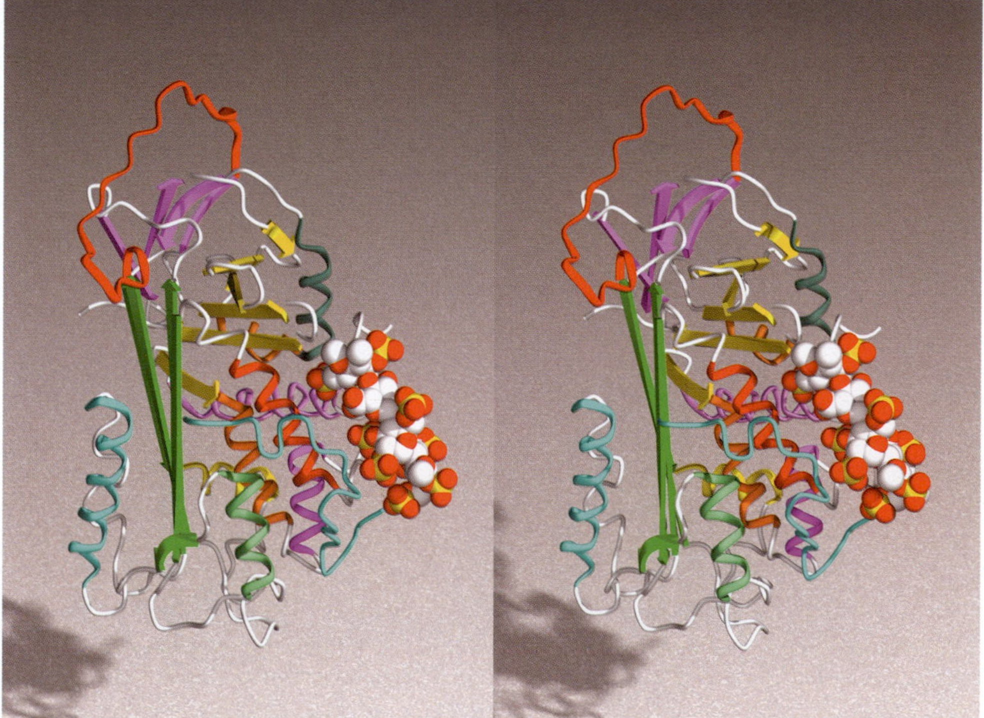

Fig. 8.14d

Fig. 8.14 The 'native' state of uncleared serpins. (a) Ovalbumin, the first structure of an uncleaved serpin [1OVA]. (b) α_1-antitrypsin [1QPL]. (c, d) The heparin-activated from of antithrombin [1AZX] (two views).

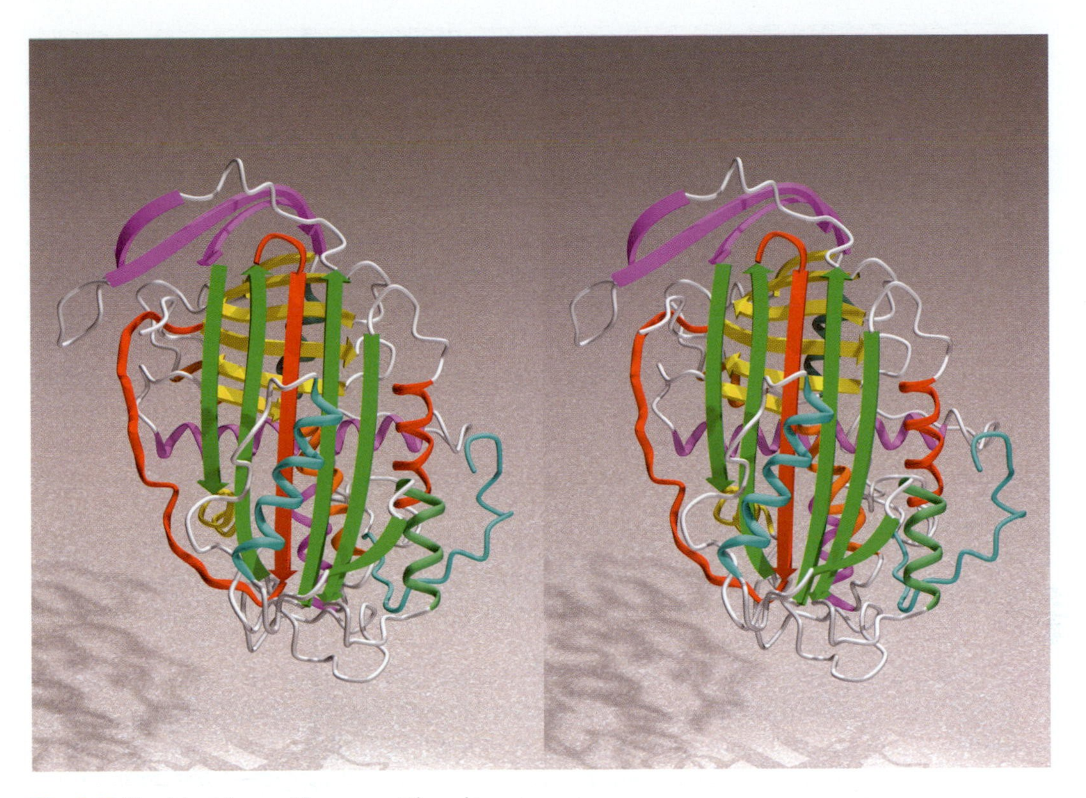

Fig. 8.15 The latent form of human antithrombin.

cleaved states do not have the same primary chemical bonds. The native → latent transition is more surprising. Under physiological conditions, the native states of inhibitory serpins are metastable, converting spontaneously to the latent state at different rates (within about 2 h for Plasminogen activator inhibitor 1). This is a challenge to the idea that there is a large energy gap between the native ground state and all other conformations. The native state, which is in fact the active state, is a long-lived intermediate in the folding of uncleaved inhibitory serpins to their final, lowest-energy, latent conformation.

It is the different folding pattern of ovalbumin, which does not form the latent state (nor the R state when cleaved), compared with inhibitory serpins which do, that is an exception to the rule that folding topology is conserved during protein evolution.

Mechanism of the S → R transition

P.E. Stein and C. Chothia showed that insertion of the reactive-centre loop into the central β-sheet is made possible by an opening of the β-sheet, based on a sliding motion of the three rightmost strands across the hydrophobic core beneath them. This movement is accompanied by shifts in three helices. The insertion of the reactive centre loop into the main β-sheet is accompanied by structural changes in the rest of the molecule that allow the β-sheet to open.

A large fragment of the serpins, including three helices and two β-sheets (not including the main β-sheet that is involved in the large conformational change), stays rigid during the conformational change, and can be seen as acting as a fulcrum for it. Figure 8.16 shows native and cleaved α_1-antitrypsin superposed on this large rigid fragment. To allow insertion of the reactive centre loop, the three rightmost strands of the main β-sheet shift to the right. The positions, relative to the main rigid fragment, that these three strands occupy in the native form are occupied in the cleaved form by other strands. In the cleaved state the reactive centre loop forms the third strand from the left. The fourth and fifth strands from the left in the cleaved form are the third and fourth strands from the left in the native form, shifted right by one strand position. Note that the third strand from the left—the reactive-centre loop in the cleaved form—has not only a different sequence, but different sense (antiparallel rather than parallel to its neighbours), in the two conformations.

Does the new set of strands provide an equivalent interface to pack against the large common fragment? What would be the constraints on the sequence for the change in the main β-sheet to leave its contribution to the interface unchanged?

Because of the geometry of a β-sheet, *alternate* residues from strands of the main β-sheet pack *into* the molecule, and the others point out into solution and do not contribute to the interface. The inserted reactive-centre loop in the cleaved form is antiparallel to the strand it replaces. Therefore, for the interface to remain constant, the amino acid sequence of the reactive-centre loop would have to match the sequence of the strand it replaces at *alternate* positions, and in the *reverse* (N to C) direction. α_1-antitrypsin employs this puzzle-like solution partially but not exclusively.

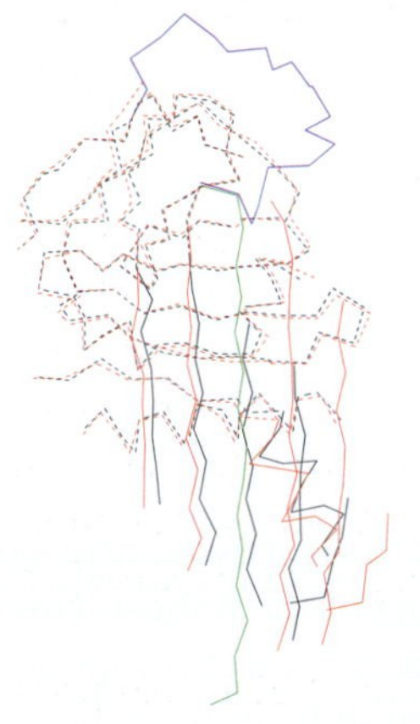
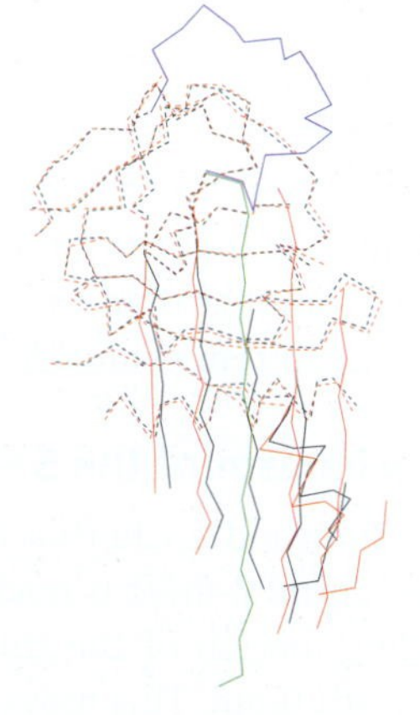

Fig. 8.16 The native [2PSI] (black) and cleaved [7API] (red) forms of α_1-antitrypsin, superposed on the largest rigid substructure (broken lines).

Higher-level structural changes

So far we have discussed conformational changes within individual proteins. There is some structural information available about larger-scale changes.

The GroEL–GroES chaperonin complex

Proteins fold spontaneously to their native states, based only on information contained in the amino acid sequence . . . but sometimes they need a little help. After all, the dilute salt solution of the physical chemistry laboratory is one thing, the intracellular medium is quite another, with concentrations of macromolecules in the range of 0.3–0.4 $g \cdot L^{-1}$, comparable to concentrations in protein crystals. The danger is that partially folded or misfolded proteins may form aggregates.

Cells have therefore developed molecules that catalyse protein folding, called chaperones. The name is apt: molecular chaperones supervise the states of nascent proteins, hold them to the proper pathway of folding, and keep them apart from improper influences that might lead to incorrect assembly or non-specific aggregation. Heat shock or viral infection enhance the danger. These conditions induce overexpression of chaperone proteins, which is how some of them were originally discovered and why they are often called heat-shock proteins.

The existence of chaperones does not contradict the basic tenet that the three-dimensional structures of proteins are dictated by their amino acid sequences. Chaperones act *catalytically* to speed up the process of protein folding, by lowering the activation barrier between misfolded and native states. They do not alter the result. Indeed, the observation that a chaperone can catalyse the folding of many proteins, with very different secondary and tertiary structures, is consistent with the idea that chaperones themselves contain no information about particular folding patterns. They anneal misfolded proteins, and allow them, rather than direct them, to find the native state.

The chaperonin system GroEL–GroES of *E. coli* contains two products of the GroE operon, GroEL (L for large, M_r = 58 000) and GroES (S for small, M_r = 10 000.) The active complex contains 14 copies of GroEL and seven copies of GroES, for a total M_r of almost 10^6.

In the absence of GroES, 14 GroEL molecules form two sevenfold rings packed back-to-back (Figure 8.17). Each ring surrounds a cavity open at one end to receive substrate (i.e. misfolded protein). The cavity is closed at the bottom (by a wall between the two rings), so that bound protein cannot pass internally from one ring to the other. Nevertheless, the two rings are not independent; they communicate via allosteric structural changes that are an essential component of the mechanism of action. Mutants containing only a single GroEL ring form complexes with substrate, ATP and GroES, but cannot release them.

With substrate in the cavity, binding of ATP and GroES enlarges the cavity, closes it off and changes its structure. The GroES subunits form another seven-membered ring that caps the GroEL ring. The GroEL ring capped by

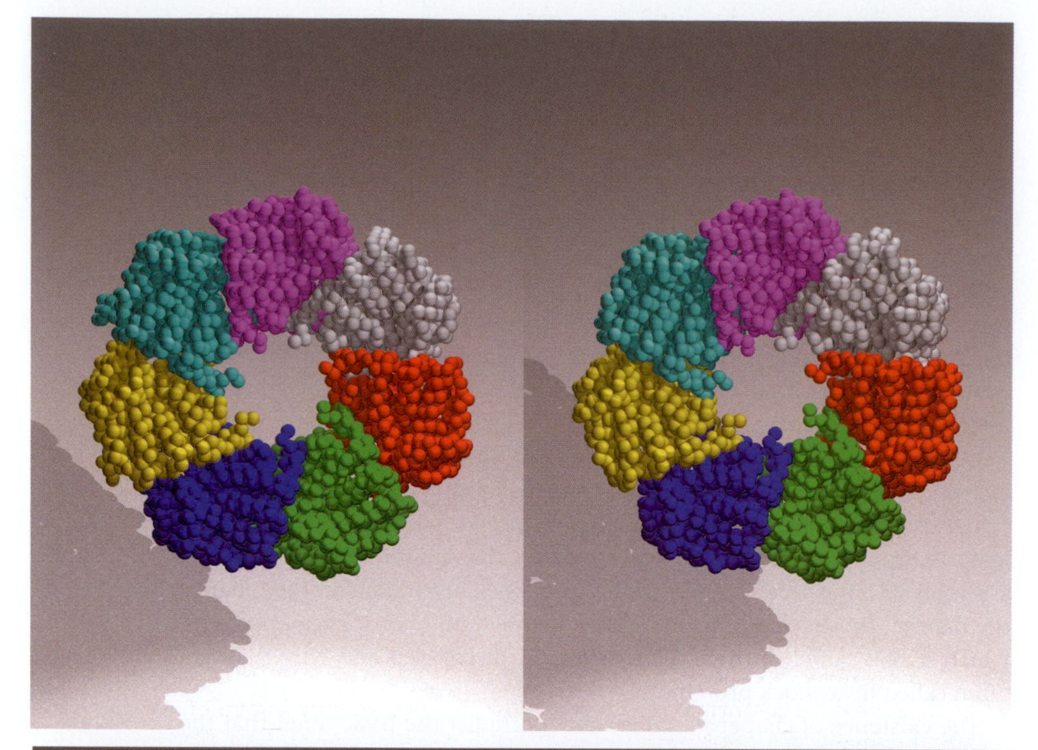

Fig. 8.17a

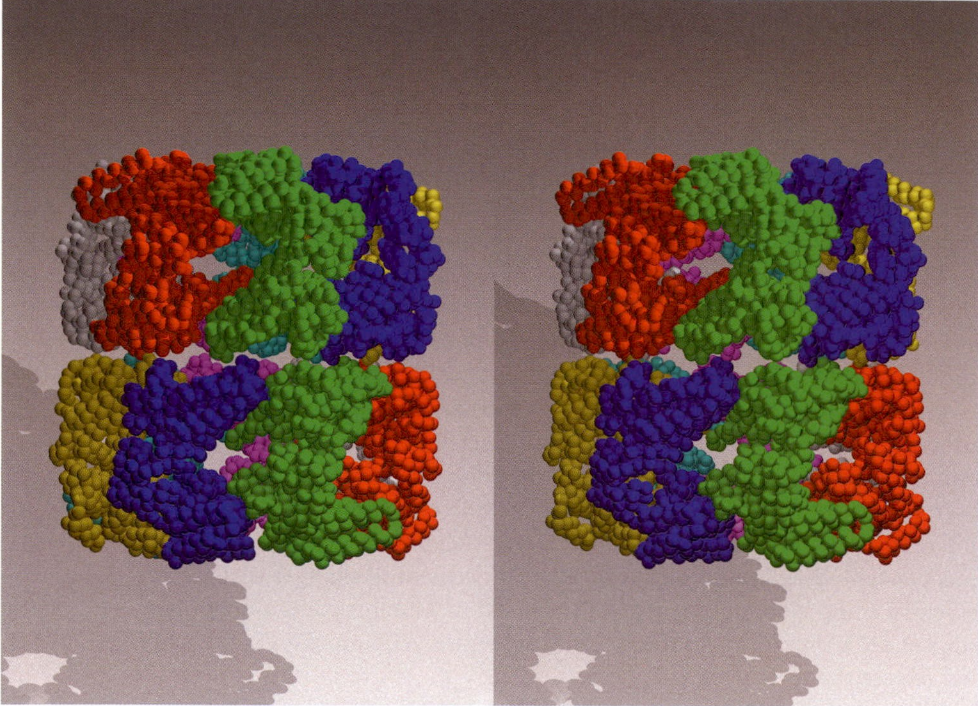

Fig. 8.17b

Fig. 8.17 The structure of GroEL in the absence of substrate and GroES [1DER]. Fourteen subunits assemble into two seven-membered rings packed back to back. What are the symmetry elements of this structure? (a) View down axis. (b) View perpendicular to axis.

GroES is called the *cis* ring, and the noncapped ring is the *trans* ring. Formation of the GroEL–GroES complex requires a large and remarkable conformational change in the *cis* GroEL ring, changing the interior surface of the cavity from hydrophobic to hydrophilic, and breaking the symmetry between the two GroEL rings (Figure 8.18). This is an allosteric change that mediates a negative cooperativity between the rings, precluding *both* rings from forming the GroES-capped structure simultaneously.

The enclosed cavity is the site of protein folding. Misfolded proteins in the cavity are given a chance to refold. After ~20 s they are expelled, either folded successfully, or, if not, with the chance to enter the same or another chaperonin complex to try again. Misfolded proteins are treated as juvenile offenders—they are caught, incarcerated, kept in solitary confinement, serve their time, and—having been given an opportunity to reform—they are released.

The GroEL–GroES conformational change

The GroEL monomer contains three domains (Figure 8.19), an apical domain (red), a hinge domain (green), and an equatorial domain (blue). Figure 8.19 compares the structures of the unbound form of GroEL, and the bound form, with one subunit of GroES bound. The viewpoint is perpendicular to the seven-fold axis, that is, tangential to the ring. Figure 8.20 shows the complex, cut away to expose the cavity.

The most obvious conformational change in GroEL is the swinging up of the apical domain by a ~60° hinge motion. A second hinge, near the ATP-binding site in the equatorial domain, rotates to lock in the ATP, and to expose a critical residue for its hydrolysis. The most remarkable feature of the conformational change is the ~90° rotation of the apical domain to expose a different surface to the interior. In the unbound form the residues lining the GroEL cavity are hydrophobic. In the bound form they are hydrophilic, with the hydrophobic residues that formed the lining now taking part in intersubunit contacts.

This is inspired, but of course perfectly logical. Recall that typical globular proteins have preferentially hydrophobic interiors and charged/polar residues on the exterior. The characteristic of misfolded proteins, that renders them subject to non-specific aggregation, is the surface exposure of hydrophobic residues that are buried in the native state. Proteins in such states bind to the open form of GroEL, with its channel lined with hydrophobic residues. What would one want a chaperone to do to such a misfolded protein, once it has it in its clutches? Altering the interior surface of the chaperone from hydrophobic to hydrophilic encourages the misfolded protein to turn itself right side out.

Operational cycle

The assembly functions like a two-state motor. Each of the seven-member GroEL rings may be in one of two states: open, ready to receive misfolded proteins, or closed, containing misfolded proteins and capped by the GroES ring.

1. In the unbound state the GroEL ring is open to allow protein to enter. The interior presents a flexible hydrophobic lining, suitable to bind

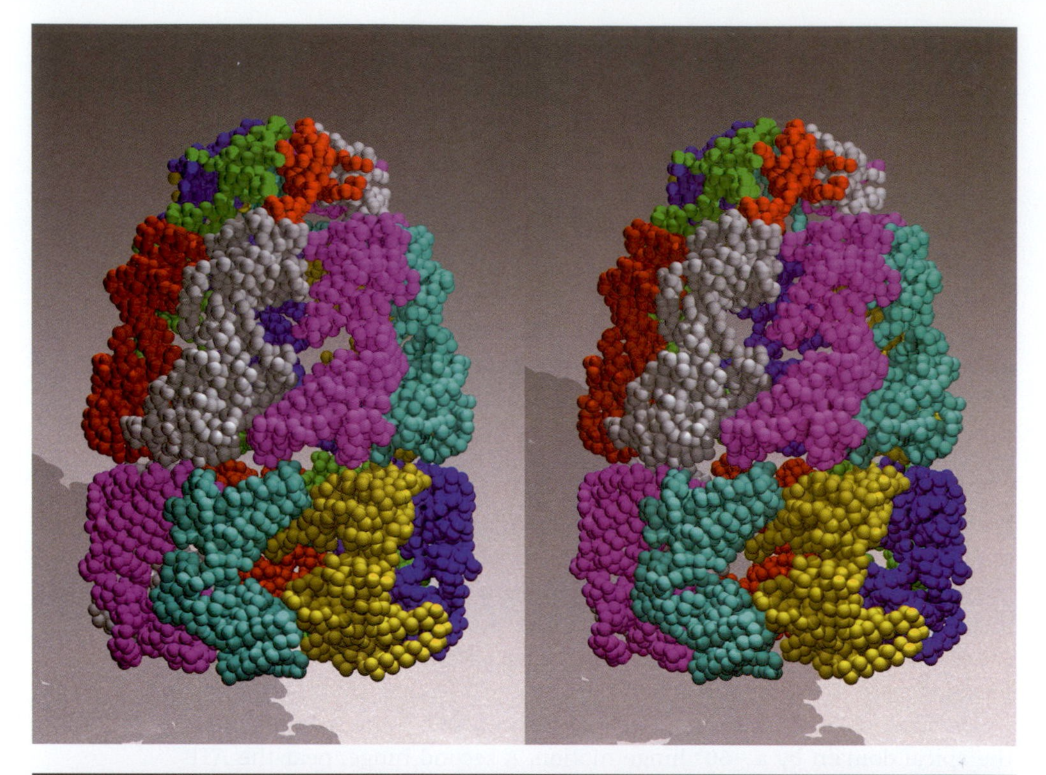

Fig. 8.18a

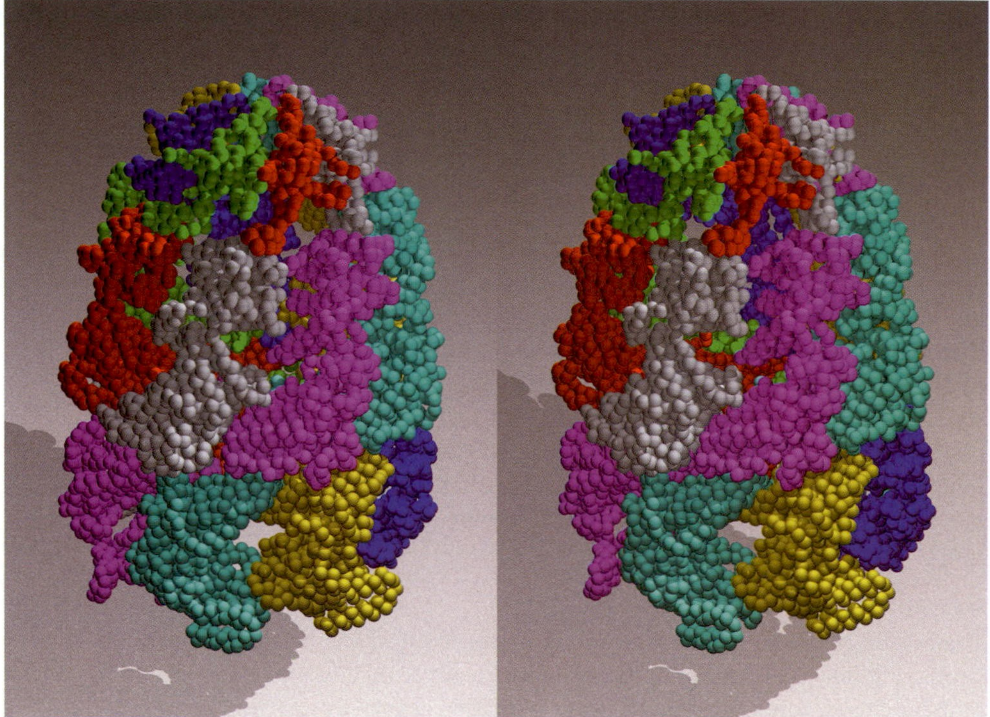

Fig. 8.18a

Fig. 8.18 The GroEL–GroES complex [1AON]. One of the GroEL rings is 'capped' by a third seven-membered ring of GroES subunits. Two slightly different orientations.

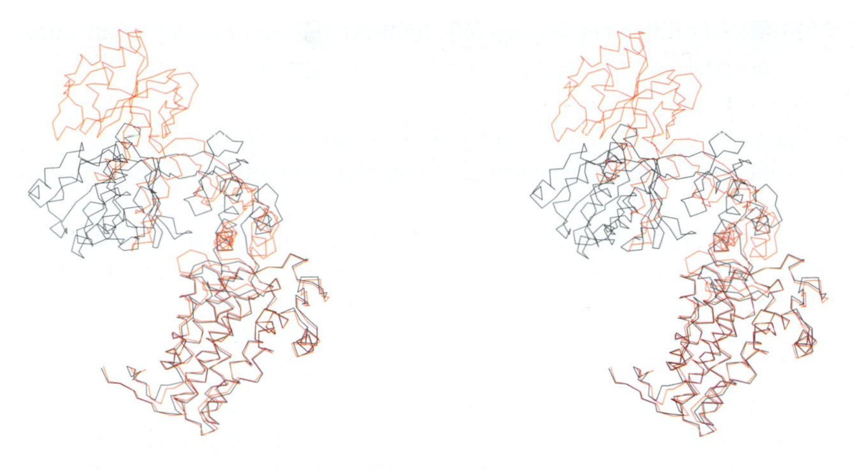

Fig. 8.19 The conformational change in one GroEL subunit. The apical domain hinges up, and rotates by ~90° to change the interior surface of the cavity. A second hinge motion locks in the ATP. [1DER] and [1AON].

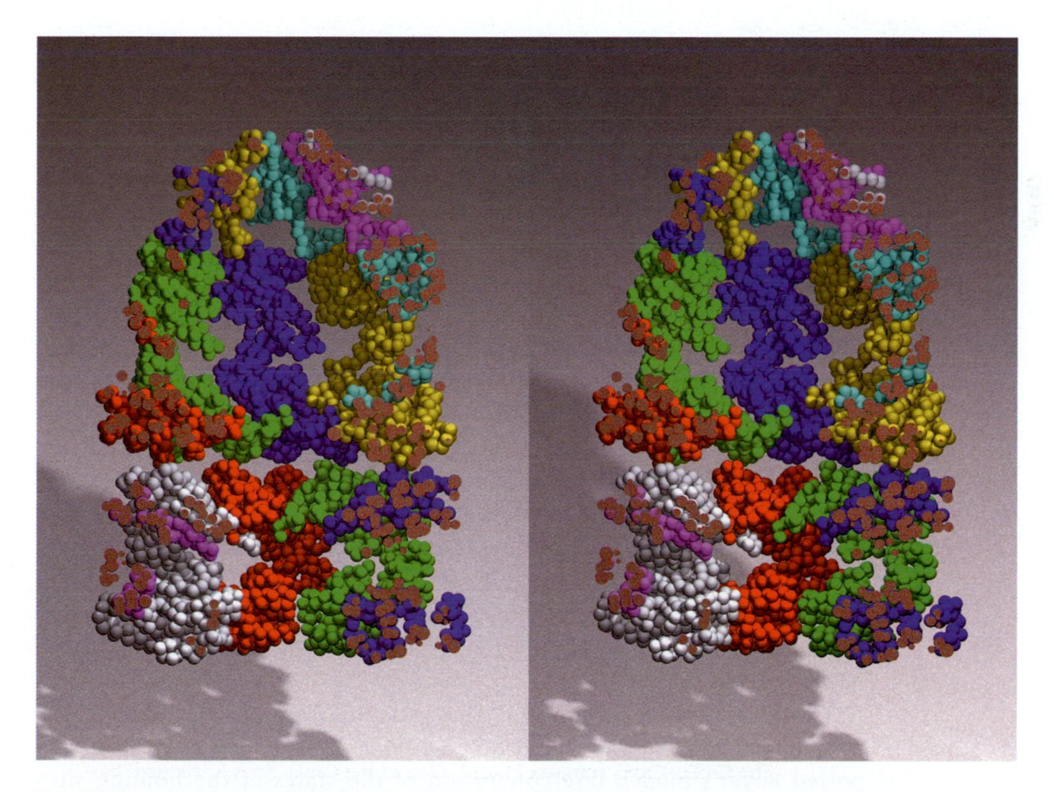

Fig. 8.20 A cutaway view of the GroEL–GroES complex, showing the cavity [1AON].

misfolded proteins by non-specific hydrophobic and van der Waals interactions. Indeed the binding process may even partially unfold proteins in incorrect states.

2. The binding of ATP and GroES, and the conformational change in the *cis* GroEL ring, creates the closed cavity in which the substrate protein, once released from the apical domains, can refold, sequestered away from potential aggregation partners. The conformational change more than doubles the volume of the cavity, to accommodate less compact unfolding/refolding transition states. The interior surface changes from hydrophobic to hydrophilic, peeling the bound misfolded protein off the surface and unfolding it even further. The burial of the original interior GroEL surface in intersubunit contacts within the GroEL–GroES complex itself breaks the binding of the protein to the original hydrophobic internal surface, leaving a macroclathrate complex.
3. Hydrolysis of ATP in the *cis* ring weakens the structure of the *cis* ring/GroES complex. Binding of ATP (but not necessarily its hydrolysis) in the *trans* ring triggers the disassembly of the *cis* assembly and release of GroES and substrate protein, restoring the ring to its original state.

Each cycle of the engine requires hydrolysis of seven or even 14 ATP molecules. The cost is much larger than the energy of unfolding of a protein, but small compared to synthesis of the polypeptide chain, and very small compared to the death of the cell.

ATPase

ATPase, the subject of this final section, forms a nice bracket with the reaction centre, the structure that began Chapter 1. The reaction centre uses light to pump protons out of the cell, converting radiant energy to electrochemical potential. ATPase couples the protons' 'return home' with synthesis of ATP. It converts the free energy *localized* in the potential gradient across a membrane to the high-energy phosphate bond of ATP, which is then distributed and used to drive all manner of life processes.

ATPase comprises a membrane-bound complex F_0 and a soluble complex, F_1. Separation of the two components revealed that F_0 contains the proton transport machinery, and F_1 the catalytic sites that synthesize ATP. In cow mitochondria, F_1 is a complex of nine polypeptides, containing copies of five different molecules, with the subunit composition: $\alpha_3\beta_3\gamma\delta\varepsilon$.

P.D. Boyer proposed the 'binding change' mechanism of ATPase. The enzyme contains three catalytic sites that cycle among three different states (Figure 8.21). The *loose* state, L, binds substrates (ADP and P_i) weakly but does not synthesize ATP. The *tight* state, T, binds substrates strongly and converts them to ATP. In the *open* state, O, the affinity for ligands is small, and bound ATP is released.

The crystal structure of the F_1 ATPase subunit from cow heart mitochondria, solved by J.P. Abrahams, A.G.W. Leslie, R. Lutter and J.E. Walker, supported Boyer's model. Interconversion of the states of the binding sites

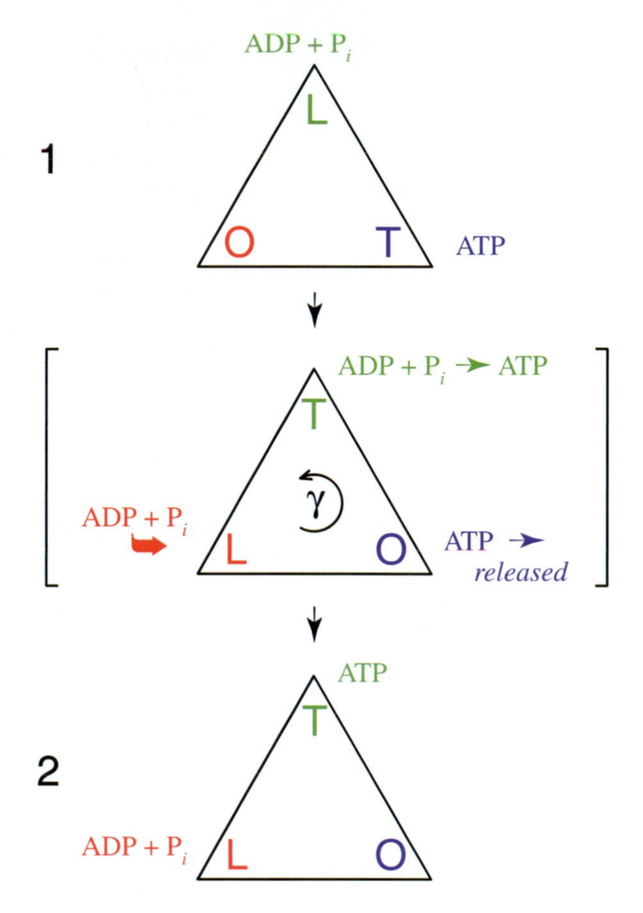

Fig. 8.21 The scheme of P.D. Boyer, confirmed by X-ray crystallography, for the mechanism of ATPase. The molecule contains three binding sites, which interconvert between three conformational states.

This diagram shows one stage of the active cycle. The three αβ dimers have three different states. In 1, the red subunit is in the open state, and is empty; the green subunit is in the loose state L, and contains substrates ADP + P_i, and the blue subunit is in the tight state T, and contains the product ATP. Rotation of the γ, δ and ε subunits (not shown) within the $(\alpha\beta)_3$ hexamer converts the red subunit from the O state to the L state, the blue subunit from T to O, and the green subunit from L to T. The L state accepts a charge of substrates. Concomitantly, the new T state forms ATP, and the new O state releases ATP. Note that bracketed state is presented solely for explanation, and *does not* represent a physical trappable intermediate.

At stage 2, the ATP has fallen out of the O state, new ADP + P_i have bound to the L state, and the T state has synthesized ATP.

Comparison of states 1 and 2 shows that the molecule has returned to its initial form with *different* subunits in the L, O and T states.

occurs during the operation of a microscopic motor. The translocation of protons is coupled to synthesis of ATP, through the generation of mechanical energy within the enzyme.

The F_1 subunit has the general shape of a mushroom (Figure 8.22). The cap of the mushroom contains a ring of three αβ dimers, forming an approximately spherical assembly ~90 Å in diameter. The binding sites are at the α–β interfaces, with the catalytic sites primarily in the β subunits. The γ subunit, containing long coiled helices, is the stem of the mushroom. It penetrates the sphere along a diameter and extends below it to form part of a stalk connecting the F_1 and F_0 subunits (Figure 8.23). The δ and ε subunits do not appear in the crystal structure. They are probably disordered in the absence of contacts with the F_0 subunit.

The symmetry of the $(\alpha\beta)_3$ subunits is broken by interactions with the asymmetric γ subunit (Figure 8.23). The enzyme operates as a microscopic rotory motor: changing patterns of interaction of the rotating γ, δ and ε subunits with the $(\alpha\beta)_3$ hexamer, which remains fixed to the membrane-bound F_0 portion, produce conformational changes, interconverting the states cyclically as in the Boyer model.

The interface between the $(\alpha\beta)_3$ and γ subunits has features that would be

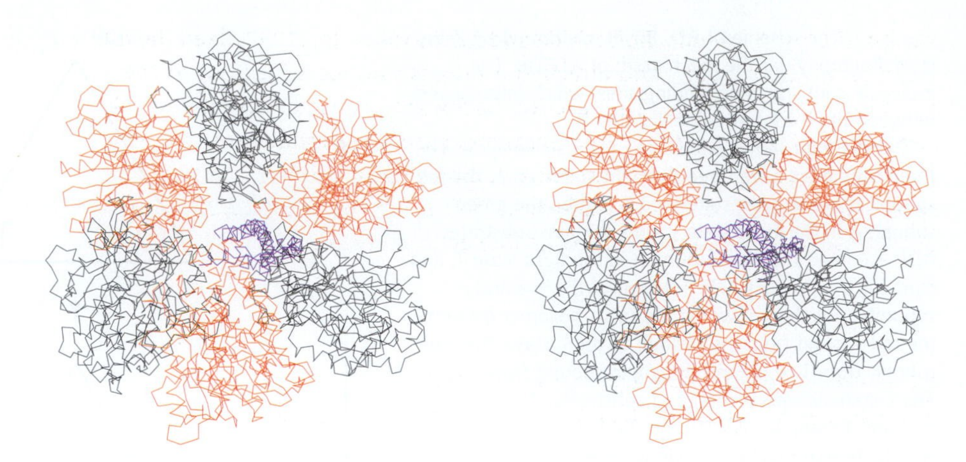

Fig. 8.22 Top view of cow mitochondrial ATPase [1BMF]. α subunits black, β subunits red, γ subunits blue, α and β subunits alternate like segments of an orange. View looking down on membrane.

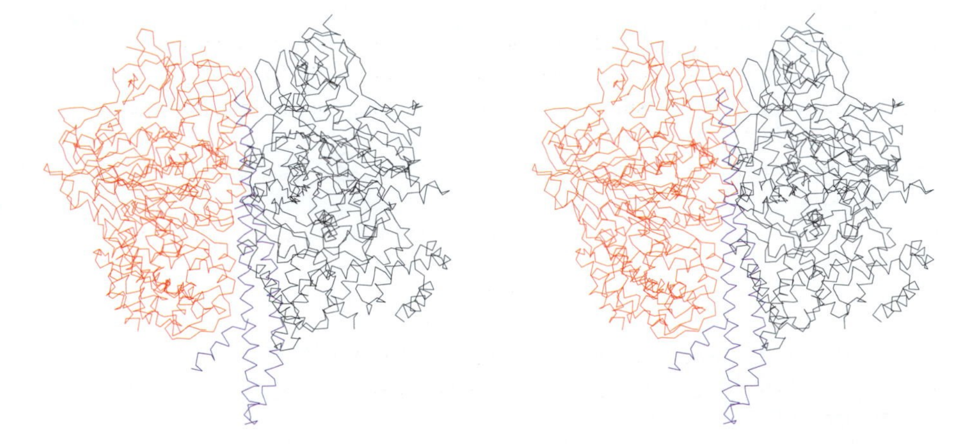

Fig. 8.23 Penetration of the $(\alpha\beta)_3$ hexamer by the γ subunit [1BMF]. The γ subunit (blue) breaks the symmetry of the $(\alpha\beta)_3$ hexamer, causing the three αβ subunits to adopt three different conformations as in the Boyer model. Rotation of the γ subunit inside the $(\alpha\beta)_3$ hexamer cyclically interconverts the states. The system converts redox energy from the proton gradient across the membrane, to mechanical energy of rotation within the enzyme, to the chemical bond energy in ATP. View looking parallel to membrane.

expected from a rotary mechanism. The internal surface of the $(\alpha\beta)_3$ (the 'bearing') forms a hydrophobic sleeve around the γ subunit (the 'axle'). The coiled-coil structure of the γ subunit is typical of proteins that require mechanical rigidity. More subtly, the high overall similarity of the three αβ dimers suggests that there are no high barriers to the interconversion of the states.

This idea that ATPase functions as a motor, made plausible by the crystal structure, and supported by observations such as inactivation of the enzyme by crosslinking of β and γ subunits, was dramatically proved by direct observation. H. Noji, R. Yasuda, M. Yoshida and K. Kinoshita, Jr. (1997) fixed the αβ units to a solid support, and attached a fluorescent actin filament to the γ subunit. Adding ATP runs the reaction backwards, coupling ATP hydrolysis to rotation of the enzyme. No membrane, no proton pumping; the energy is lost as heat. Rotation of the actin during activity of the enzyme was seen and recorded. This movie certainly deserves an Oscar in the natural history category, starring ATPase (and with some fine actin' in the cast . . .).

Useful web sites

Database of conformational changes in proteins:
http://bioinfo.mbb.yale.edu/MolMovDB/db/ProtMotDB.main.html

Recommended reading and references

Abrahams, J.P., Leslie, A.G.W., Lutter, R. and Walker, J.E. (1994). The structure of F1-ATPase from bovine heart mitochondria determined at 2.8 Å resolution. *Nature* **386**, 299–302.

Baldwin, J. and Chothia, C. (1979). Haemoglobin: the structural changes related to ligand binding and its allosteric mechanism. *J. Mol. Biol.* **129**, 175–220.

Buckle, A.M., Zahn, R. and Fersht, A.R. (1997). A structural model for GroEL–polypeptide recognition. *Proc. Nat. Acad. Sci. U.S.A.* **94**, 3571–5.

Bukau, B. and Horwich, A.L. (1998). The Hsp70 and Hsp60 chaperone machines. *Cell* **92**, 351–66.

Gerstein, M. and Krebs, W. (1998). A data base of molecular motions. *Nucl. Acid. Res.* **26**, 4280–90.

Gerstein, M., Lesk, A.M. and Chothia, C. (1994). Structural mechanisms for domain movements in proteins. *Biochemistry* **33**, 6739–49.

Holmes, K.C. (1998). A powerful stroke. *Nat Struc. Biol.* **5**, 940–2.

Holmes, K.C. (1998). A molecular model for muscle contraction. *Acta Cryst.* **A54**, 789–97.

Mattevi, A, Rizzi, M. and Bolognesi, M. (1996). New structures of allosteric proteins revealing remarkable conformational changes. *Current Opin. Struct. Biol.* **6**, 824–9.

Noji, H., Yasuda, R., Yoshida, M. and Kinoshita, K., Jr. (1997). Direct observation of the rotation of F1-ATPase. *Nature* **386**, 299–302.

Perutz, M. F. (1990). *Mechanisms of cooperativity and allosteric regulation in proteins.* Cambridge University Press, Cambridge.

Perutz, M.F., Wilkinson, A.J., Paoli, M. and Dodson, G.G. (1998). The stereochemical mechanism of hemoglobin revisited. *Ann. Revs. Biophys. Biomol. Str.* **27**, 1–34.

Ranson, N.A., White, H.E. and Saibil, H.R. (1998). Chaperonins. *Biochem. J.* **333**, 233–42.

Rye, H.S., Burston, S.G., Fenton, W.A., Beechem, J.M., Xu, Z., Sigler, P.B. and Horwich, A.L. (1997). Distinct actions of *cis* and *trans* ATP within the double ring of the chaperonin GroEL. *Nature* **388**, 792–8.

Sigler, P.B., Xu, Z., Rye, H.S., Burston, S.G., Fenton, W.A. and Horwich, A.L. (1998). Structure and function in GroEL-mediated protein folding. *Ann. Revs. Biochem.* **67**, 581–608.

Whisstock, J.C., Skinner, R. and Lesk, A.M. (1998). An atlas of serpin conformations. *Trends in Biochem. Sci.* **23**, 63–7.

Walker, J.E. (1998). ATP synthesis by rotary catalysis. *Angew. Chem. Int. Ed.* **37**, 2308–19.

Xu, Z. and Sigler, P.B. (1998). GroEL/GroES: structure and function of a two-stroke folding engine. *J. Struct. Biol.* **124**, 129–41.

Xu, Z., Horwich, A.L. and Sigler, P.B. (1997). The crystal structure of the asymmetric GroEL–GroES–$(ADP)_7$ chaperonin complex. *Nature* **388**, 741–50.

Exercises and problems

Exercises

8.1. The axis of rotation of the hinge motion of lactoferrin is approximately perpendicular to the page. On a photocopy of Figure 8.3, draw a * where you estimate that the axis intersects the page.

8.2. From inspection of Figure 8.5, decide which Zn-binding site—the upper or the lower—of the 2Zn form (Figure 8.5b) will be preserved in the 4Zn form (Figure 8.5c, in which the Zn ions are not shown).

8.3. A mutant haemoglobin (Hb Barcelona β94Asp → His) lacks the salt bridge between β94Asp and β146His that in normal haemoglobin occurs in the T (deoxy) state but not the R (oxy) state. The effect of the mutation is therefore to destabilize the deoxy state relative to the oxy state. How would you expect the oxygen affinity to be affected?

8.4. Sickle-cell anaemia is caused by a single mutation on the surface of the β chain of haemoglobin 6Glu → Val, which creates additional hydrophobic surface and leads to the aggregation of deoxyhaemoglobin within the erythrocyte. Suppose you could arrange for GroEL and GroES to be expressed in the erythrocyte. Would this be a potentially useful approach to therapy for sickle-cell disease? Why or why not?

8.5. Draw the next intermediate and next stage, following Figure 8.22, in the activity of ATPase.

8.6. For bovine mitochondrial ATPase, $V_{max} \sim 400\ s^{-1}$. What rate of rotation does this imply?

8.7. On a photocopy of Figure 8.13, indicate the two residues that are connected in the uncleaved form.

Problems

8.1. Estimate the number of additional backbone hydrogen bonds formed in the cleaved form of α_1-antitrypsin over the native form.

8.2. The volume of the cavity in the open form of GroEL is ~85 000 Å^3. How large a protein (number of residues) would it be possible to fit into this cavity? A coat protein of bacteriophage T4 is too large to fit in the cavity. How might the phage adapt to take advangage of the GroEL–GroES system to assist the folding of its coat protein? (When Einstein said,

'Nature is subtle but not malicious,' he obviously did not know about viral mechanisms at the molecular level!)

8.3. Figure 8.4c shows the two states of myosin looking approximately perpendicular to the axis of rotation. (a) What is the angle of rotation of the long helix? (b) What is the approximate displacement of the end of the long helix? Compare with the experimental estimate of 10 nm.

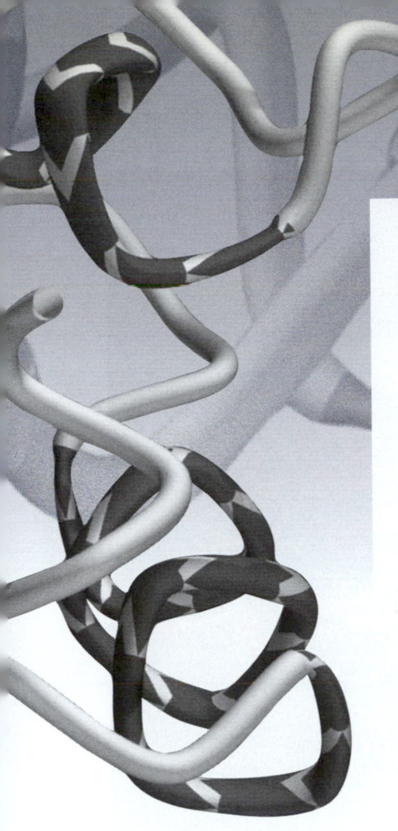

APPENDIX 1

An album for browsing

In the first edition of this book, it was possible to provide a reasonably complete atlas of the contents of the Protein Data Bank at that time. Now, with over 10 000 entries, a complete atlas, even of unique folding patterns, is out of the question. Nevertheless, this section contains pictures of proteins not illustrated in the preceding chapters. They are intended to give the reader a chance to practise the skills at visual analysis that it has been the goal of this book to develop, to give a somewhat fuller idea of the kinds of variety in folding patterns that appear in nature, and to provide the basis for additional problems that individual instructors may wish to devise.

All or mostly α-helical proteins

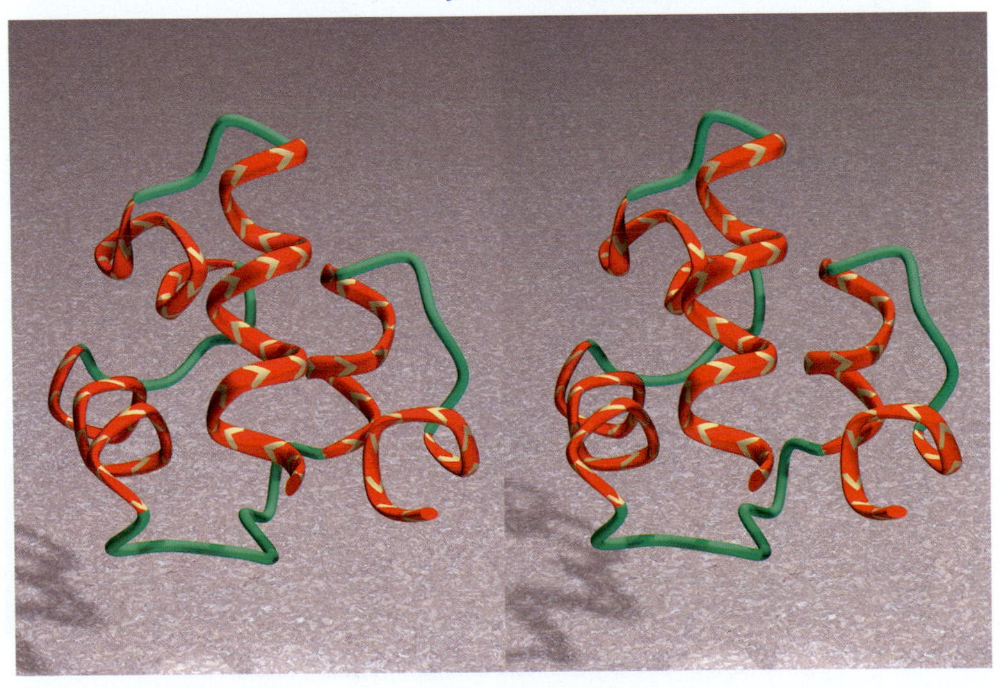

Fig. A.1 434 repressor, N–terminal domain (phage 434) [1R69].

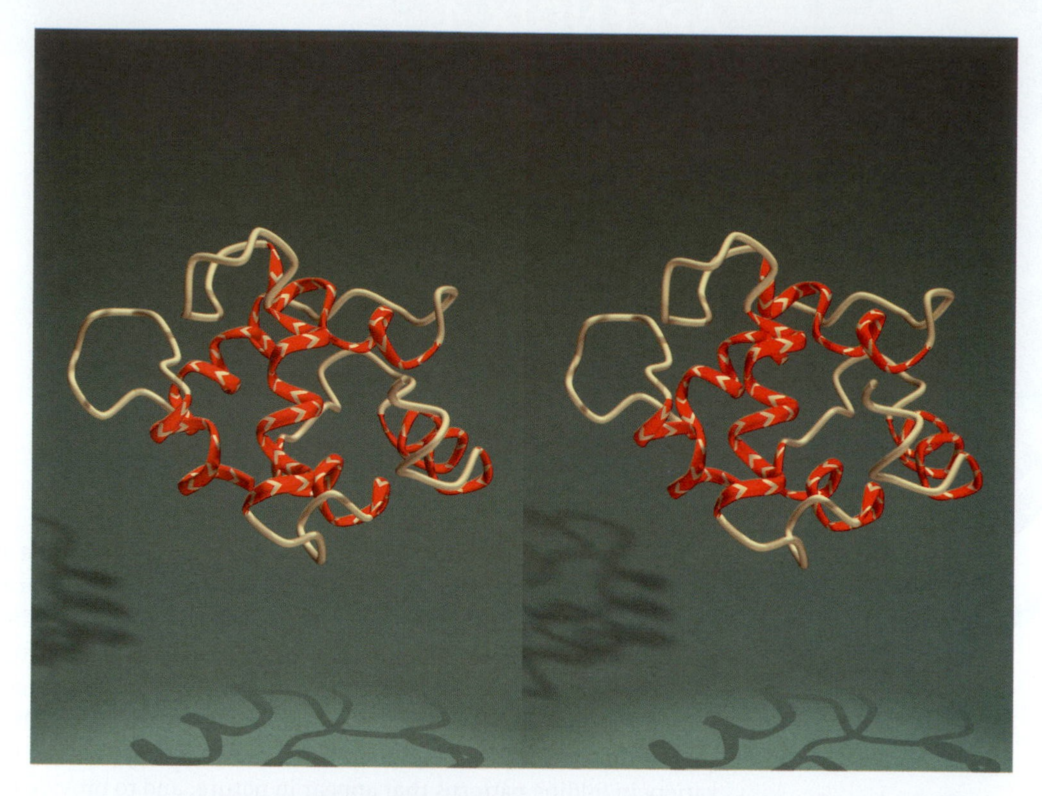

Fig. A.2 Oncomodulin (rat) [1RRO].

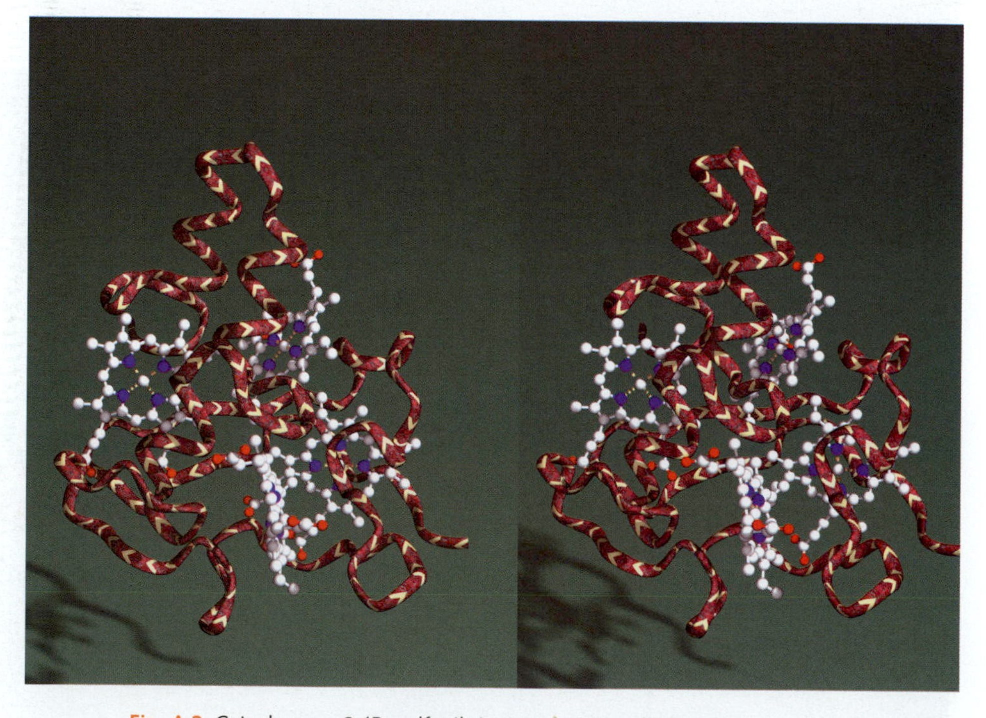

Fig. A.3 Cytochrome c3 (*Desulfovibrio vugaris Miyazaki*) [2CDV].

Fig. A.4 Trp repressor (*E. coli*) [3WRP].

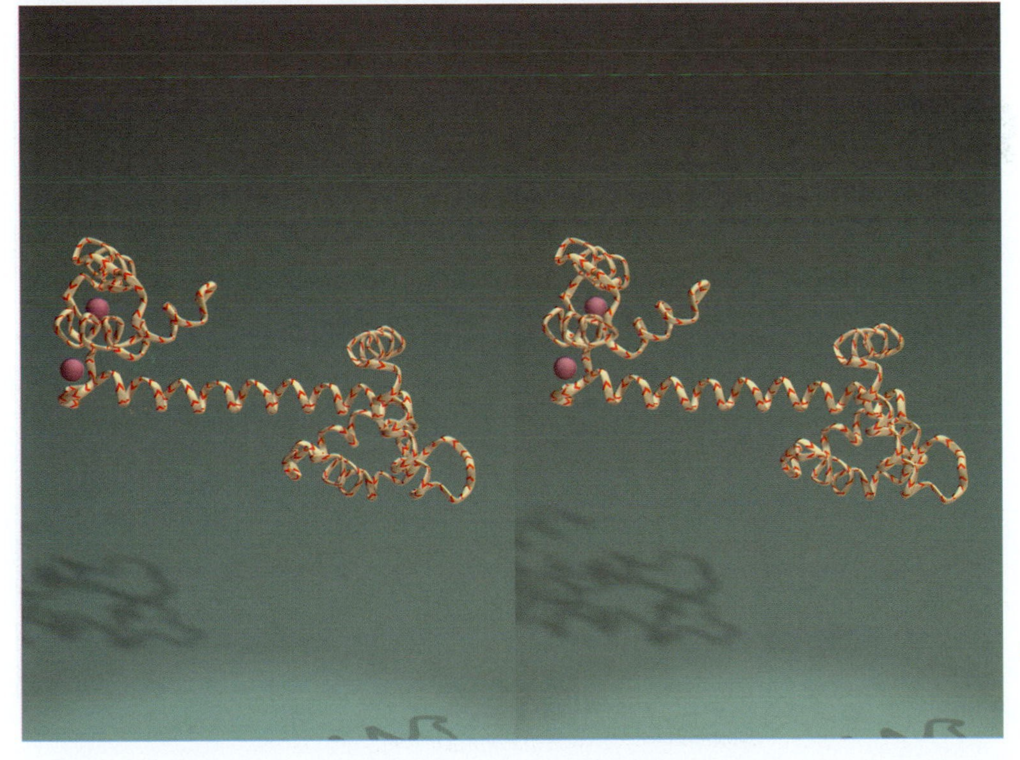

Fig. A.5 Troponin C (chicken) [4TNC].

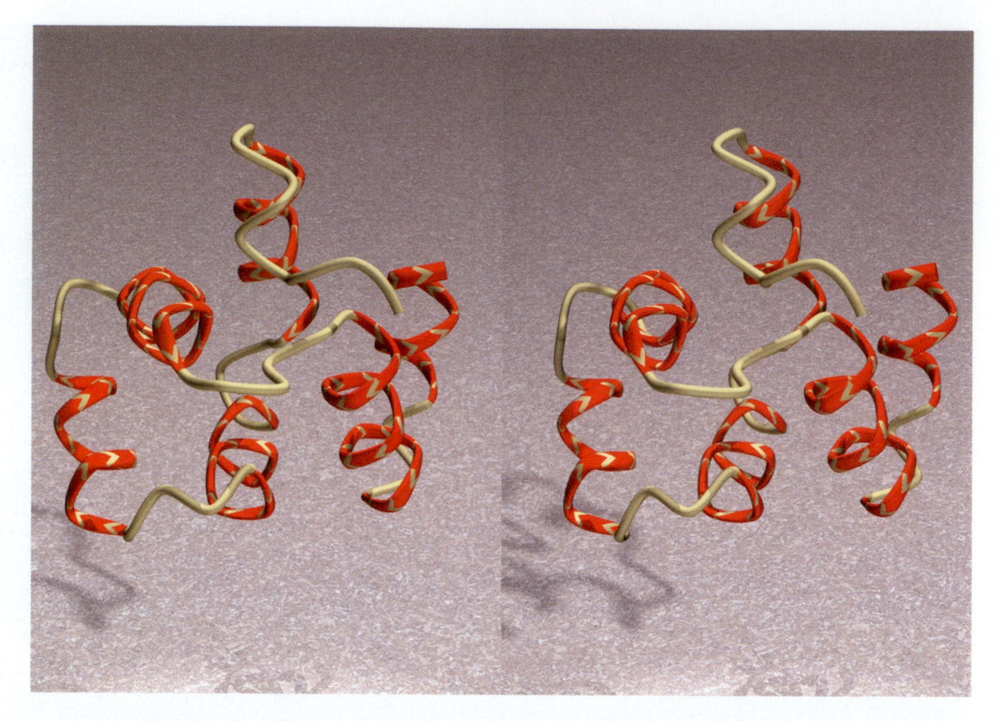

Fig. A.6 Death domain of p75 low affinity neurotrophin receptor (rat) [1NGR].

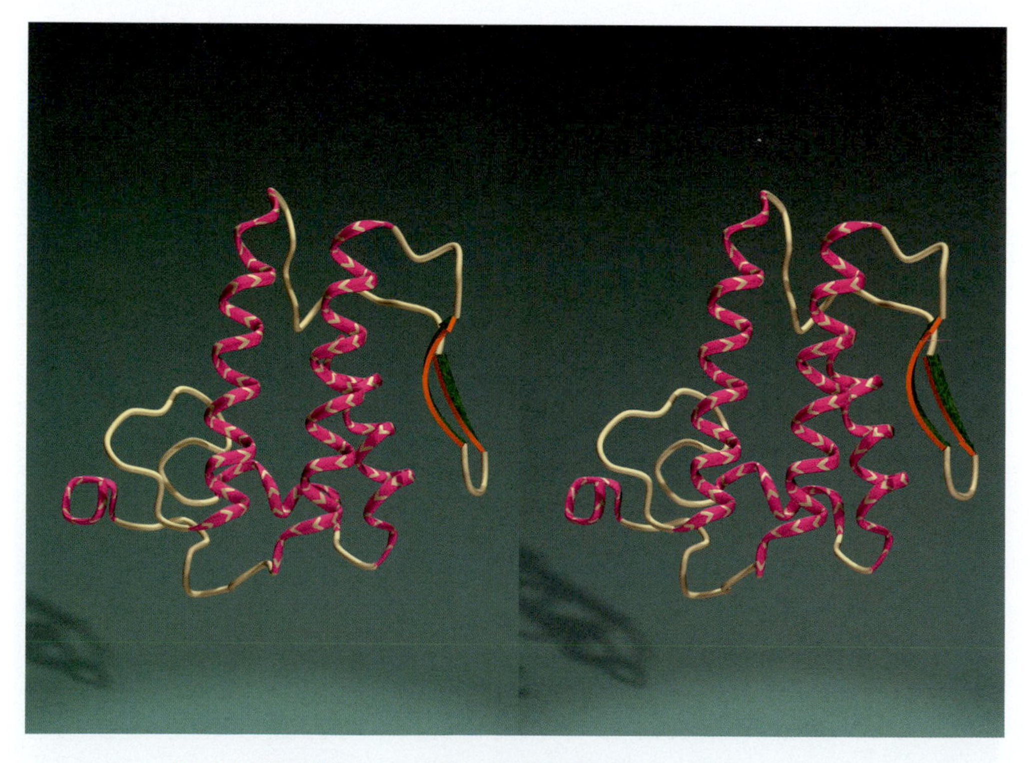

Fig. A.7 Phospholipase A_2 (Taiwan cobra) [1POA].

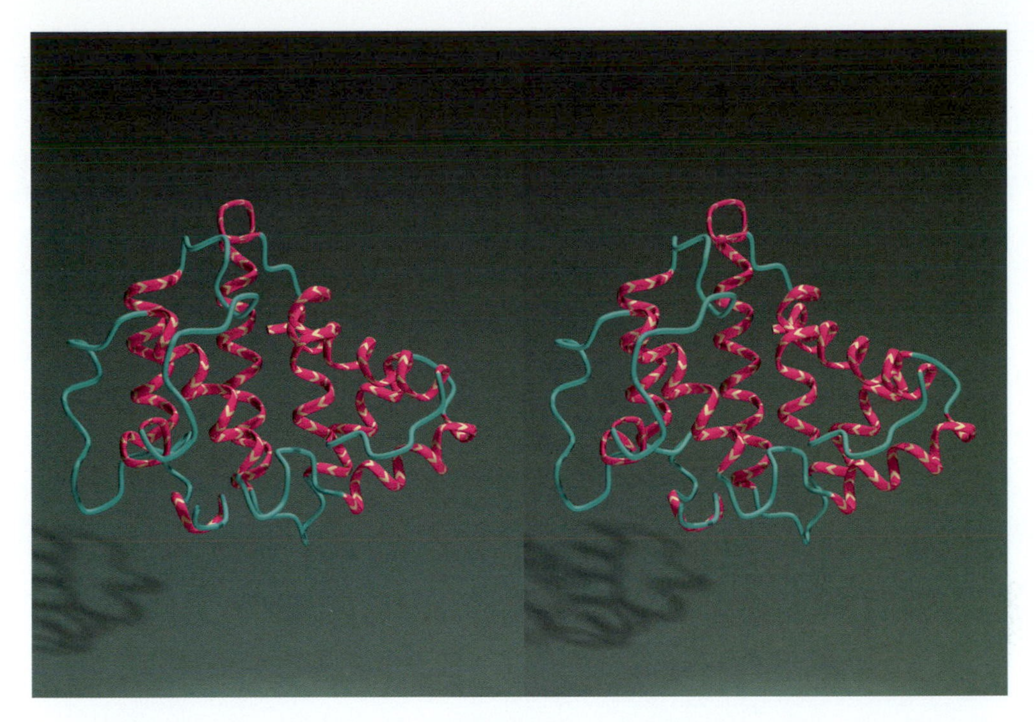

Fig. A.8 Phosphatidylinositol 3–kinase (human) [1PBW].

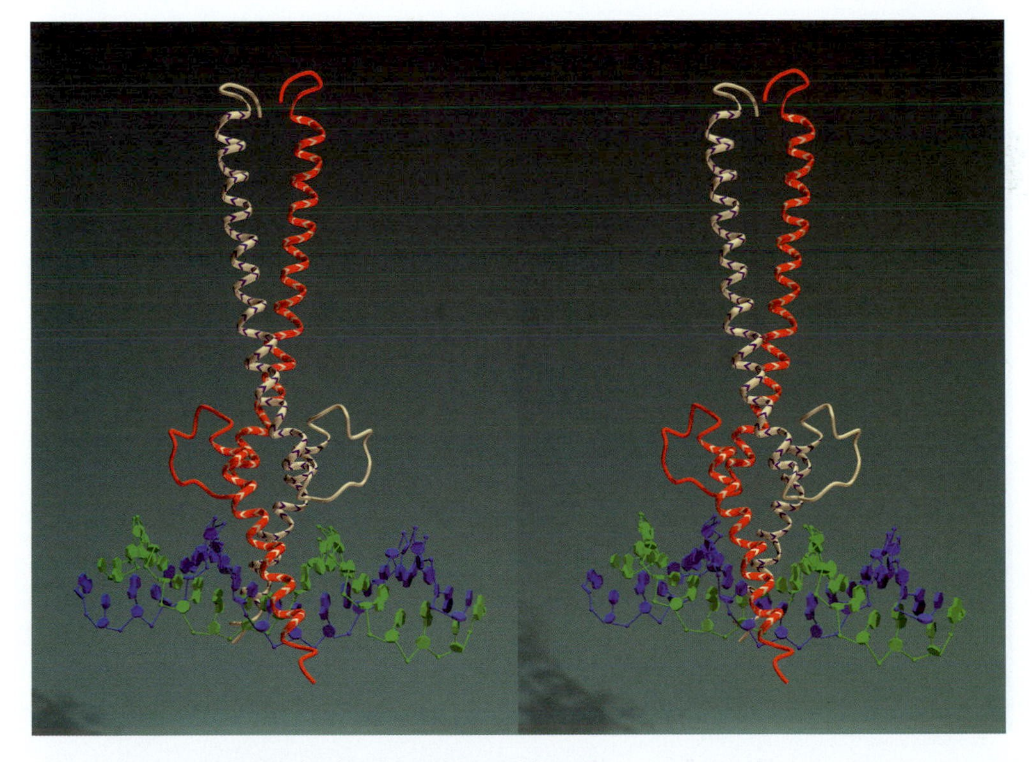

Fig. A.9 Max protein, DNA–binding domain (mouse) [1AN2].

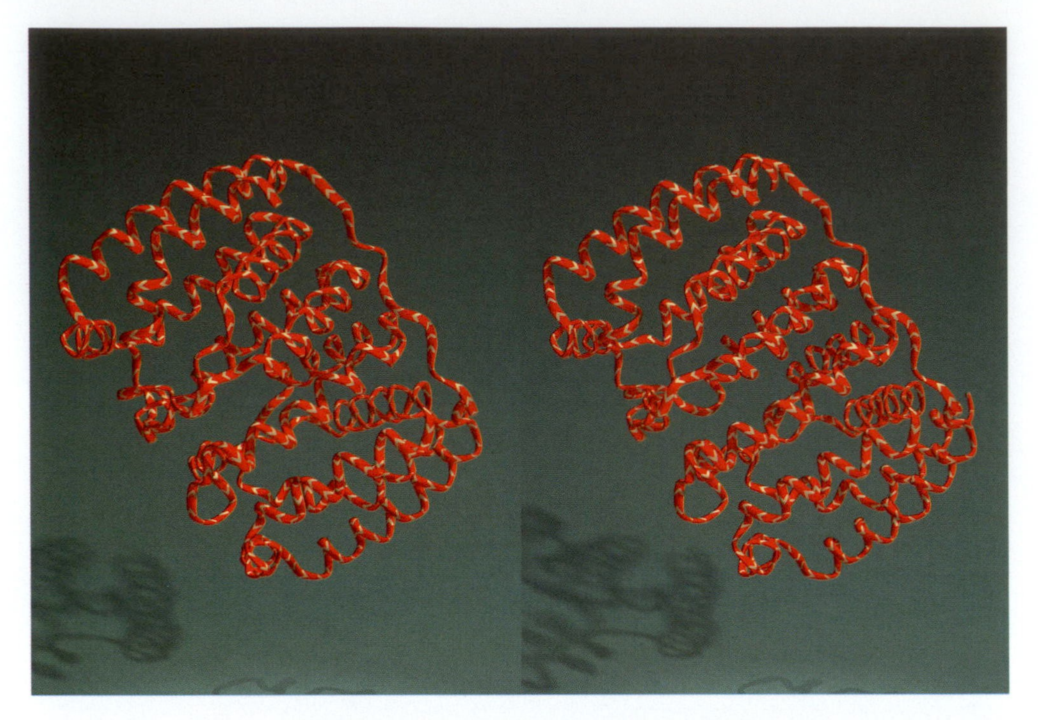

Fig. A.10 Light-harvesting protein: peridinin–chlorophyll protein (*Amphidinium carterae*) [1PPR].

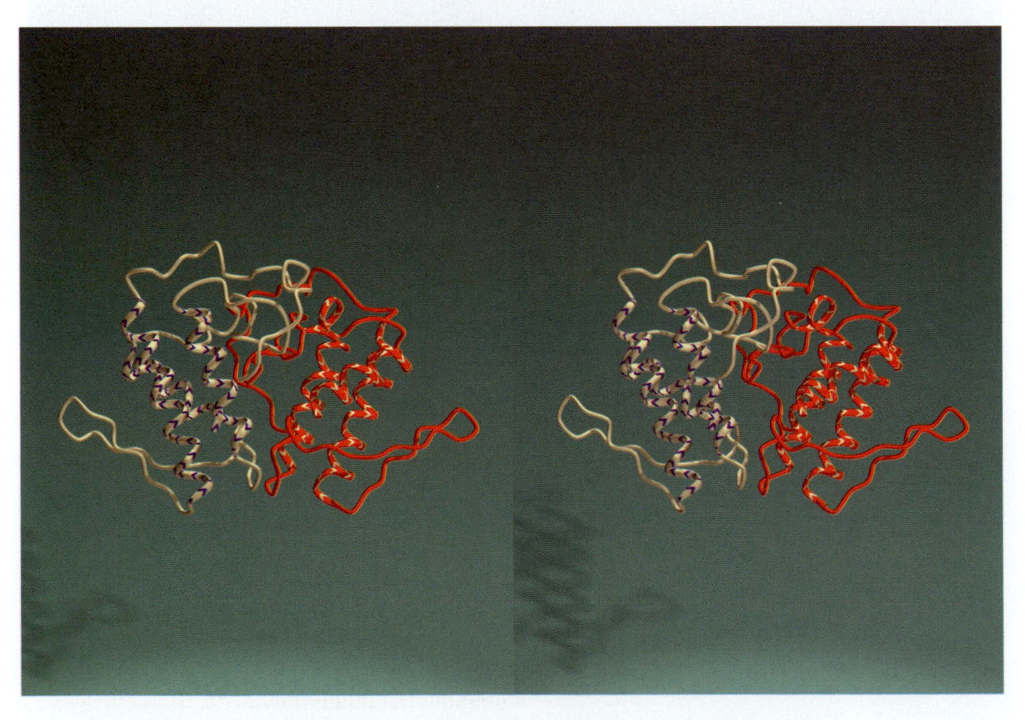

Fig. A.11 Phospholipase A_2 (western diamondback rattlesnake) [1PP2].

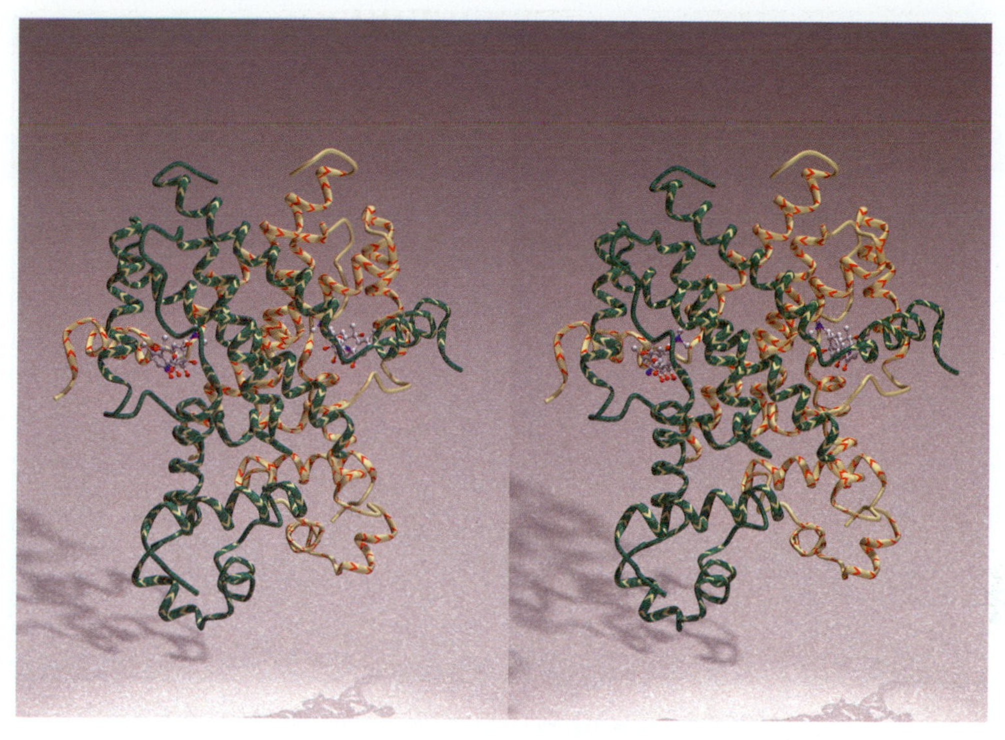

Fig. A.12 Tetracycline repressor (*E. coli*) [2TCT].

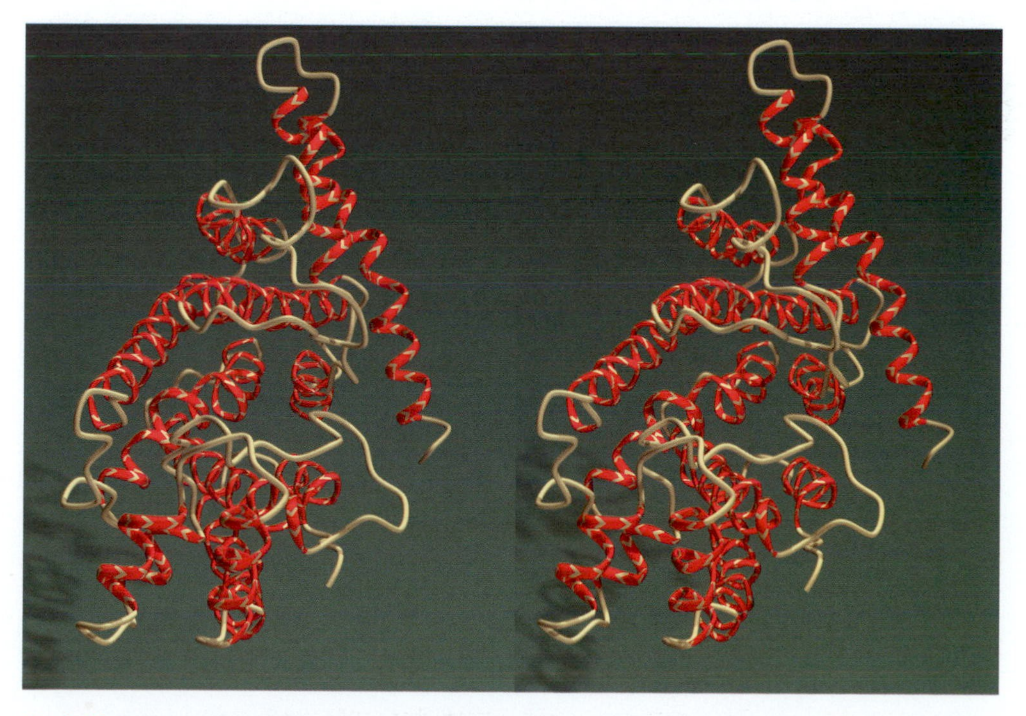

Fig. A.13 Farnesyl diphosphate synthase (chicken) [1FPS].

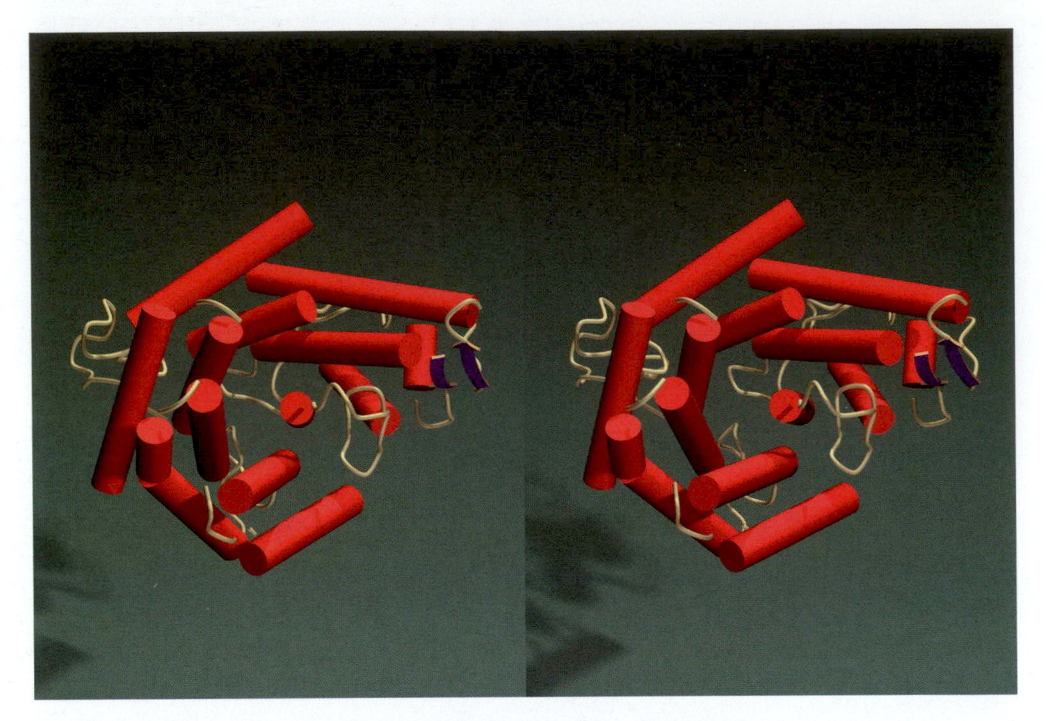

Fig. A.14 Glucoamylase fragment (*Aspergillus awamori*) [1GAI].

All or mostly β-sheet proteins

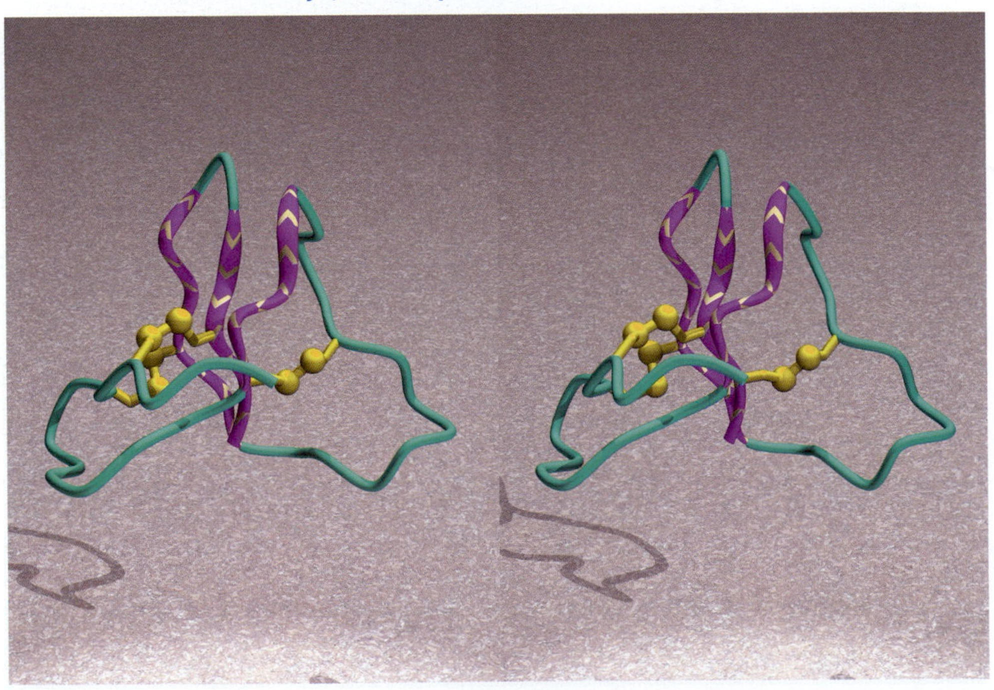

Fig. A.15 Anti-hypertensive, anti-viral protein (sea anemone) [1BDS].

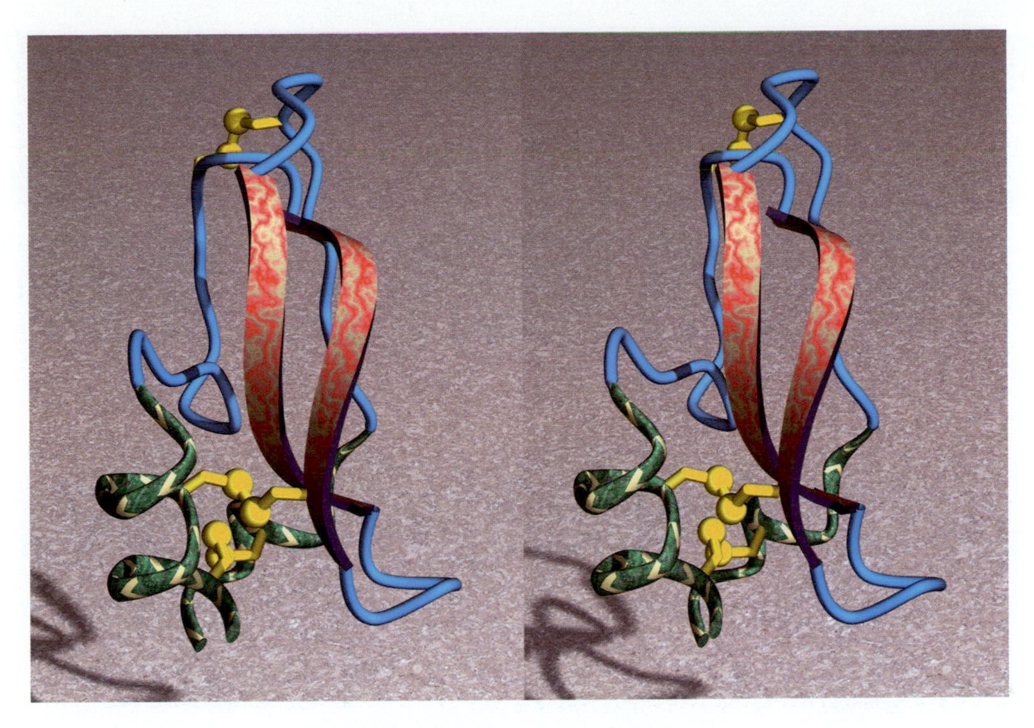

Fig. A.16 Pancreatic trypsin inhibitor (cow) [5PTI].

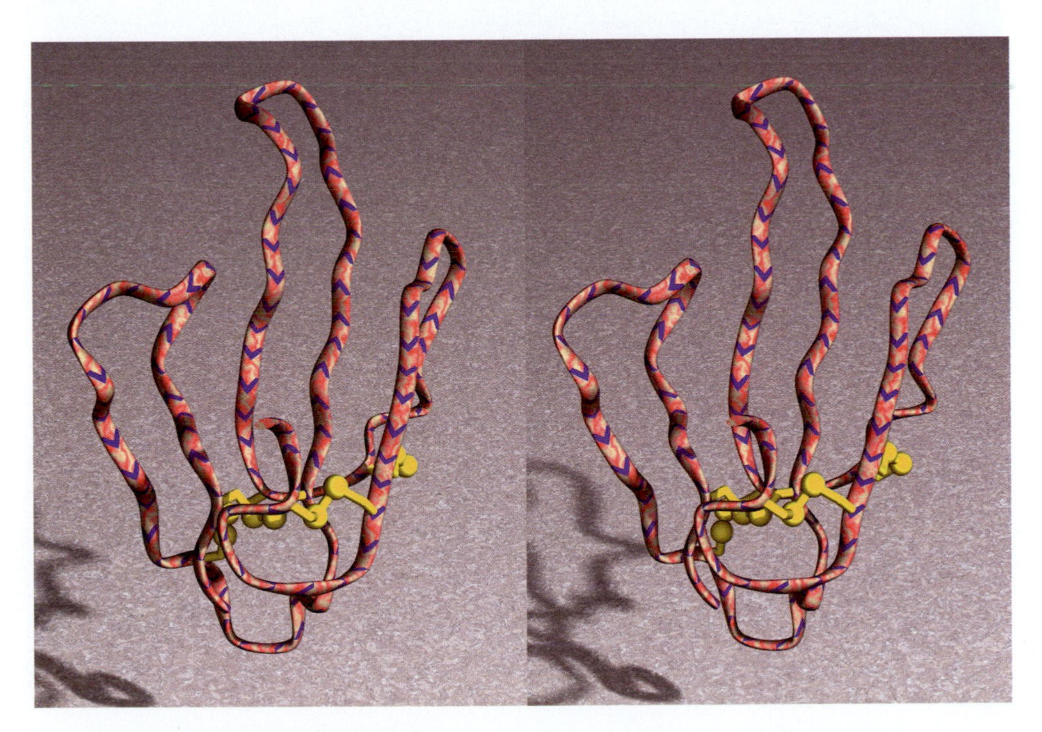

Fig. A.17 Neurotoxin B (sea snake) [1NXB].

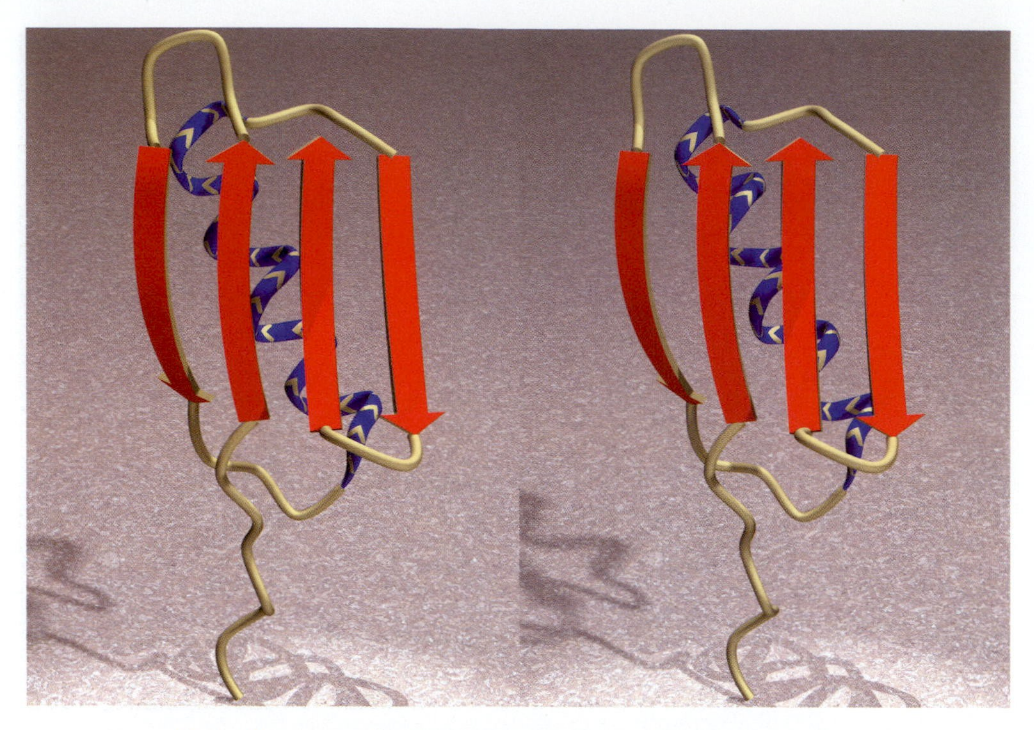

Fig. A.18 Protein G: third IgG-binding domain (*Streptococcus*) [1IGD].

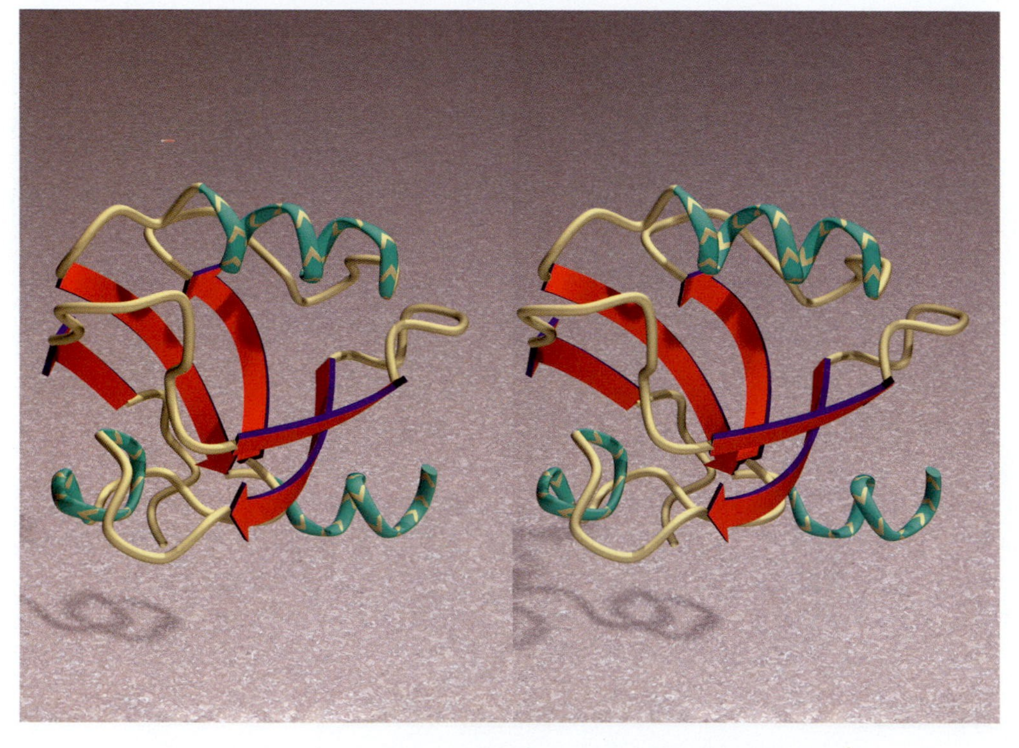

Fig. A.19 Discs large protein, PDZ3 domain, DHR3 domain (human) [1PDR].

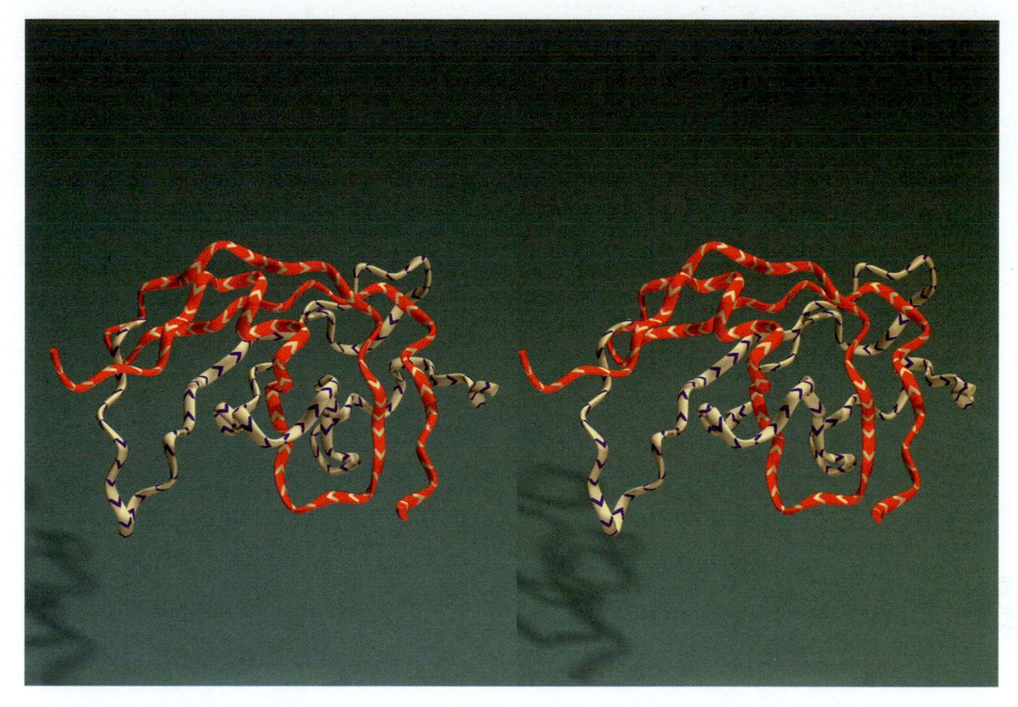

Fig. A.20 Protease (Rous sarcoma virus) [2RSP].

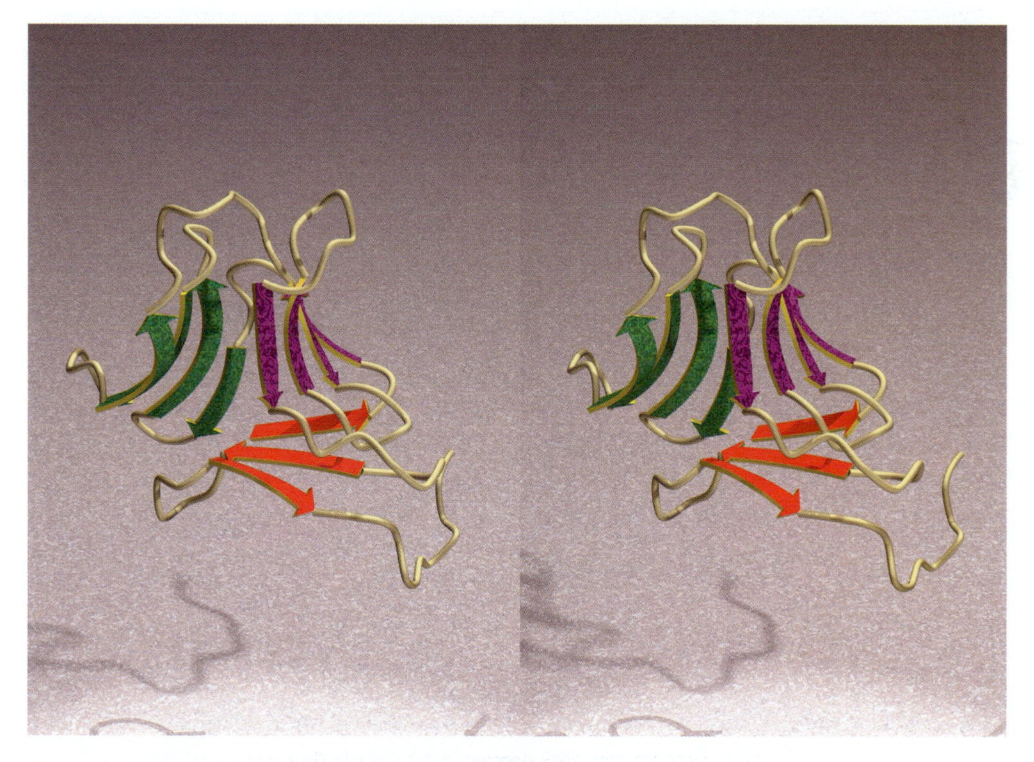

Fig. A.21 Mannose-specific agglutinin (snowdrop) [1JPC].

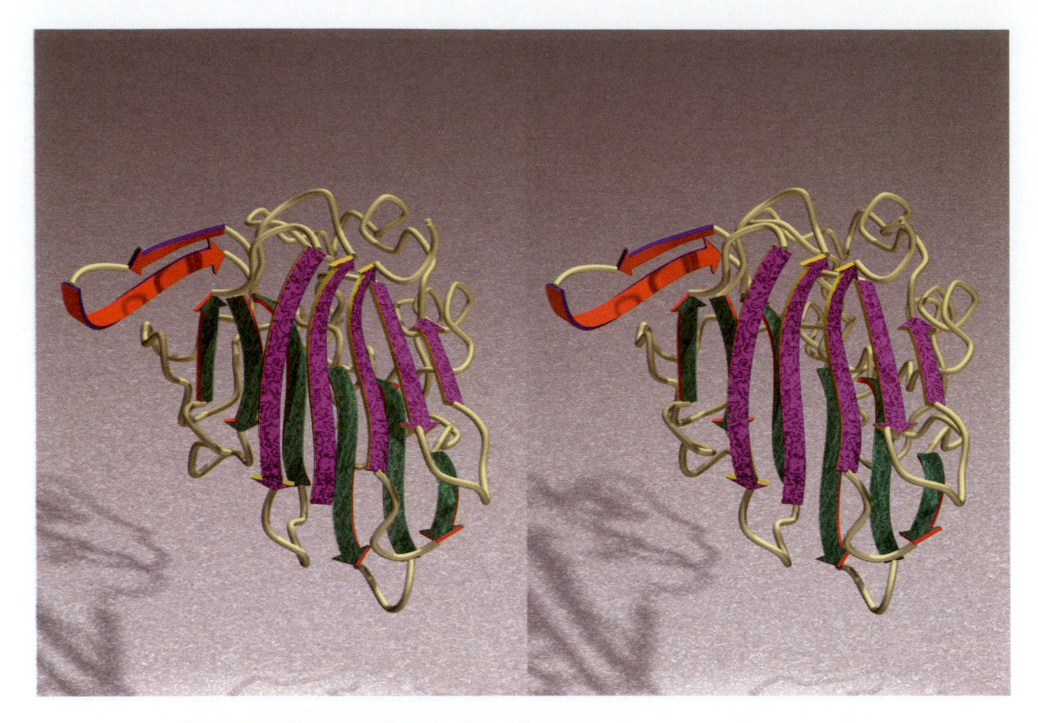

Fig. A.22 Thaumatin (African berry) [1THW].

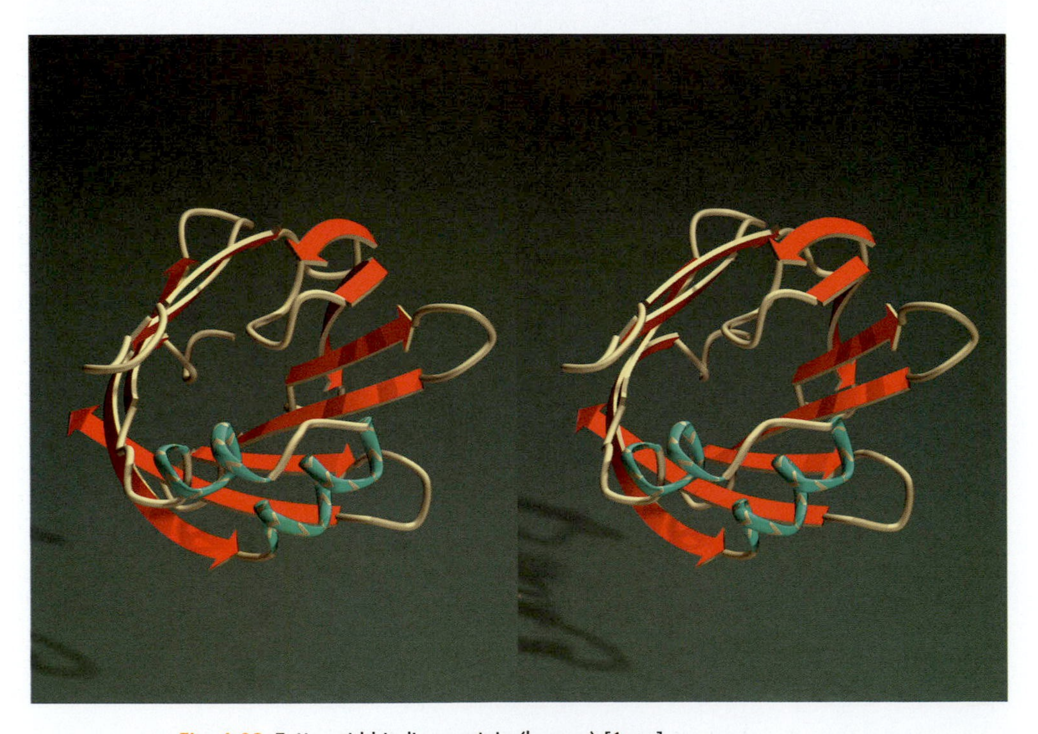

Fig. A.23 Fatty acid binding protein (human) [1HMS].

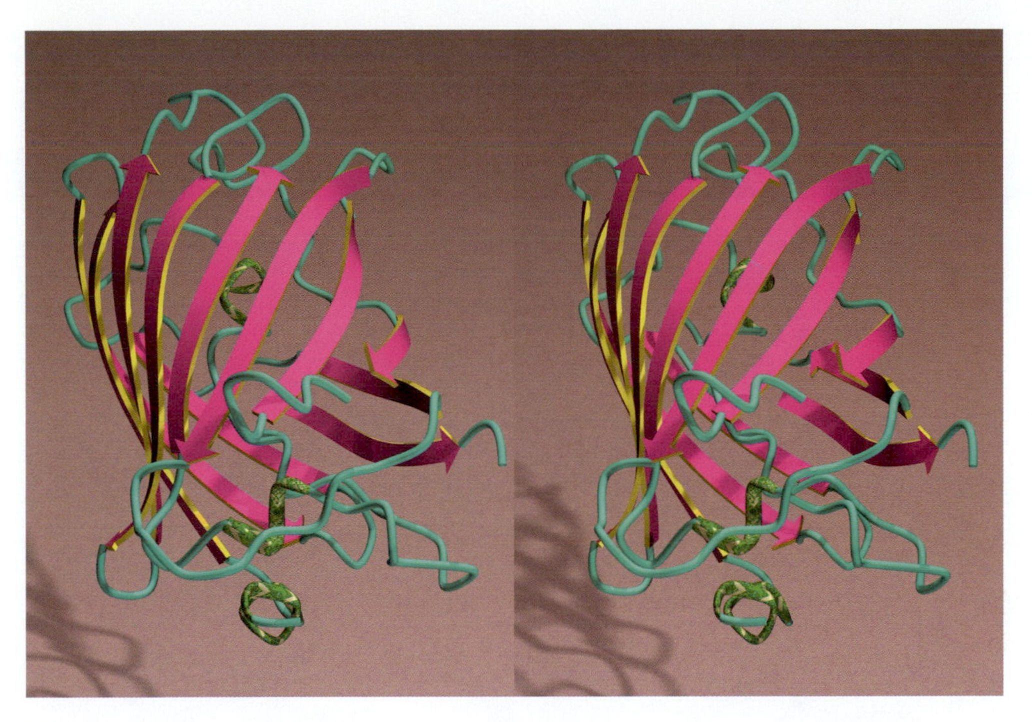

Fig. A.24 Green fluorescent protein (*Aequorea victoria*) [1EMA].

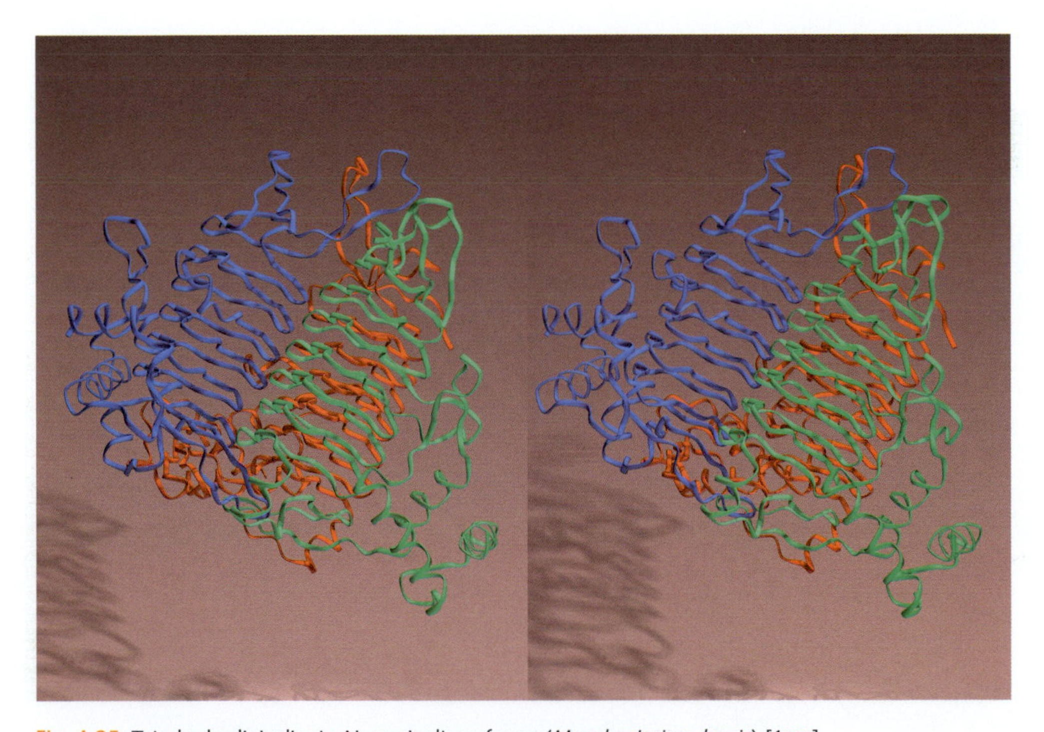

Fig. A.25 Tetrahydrodipicolinate-N-succinyltransferase (*Mycobacterium bovis*) [1TDT].

$\alpha + \beta$ proteins

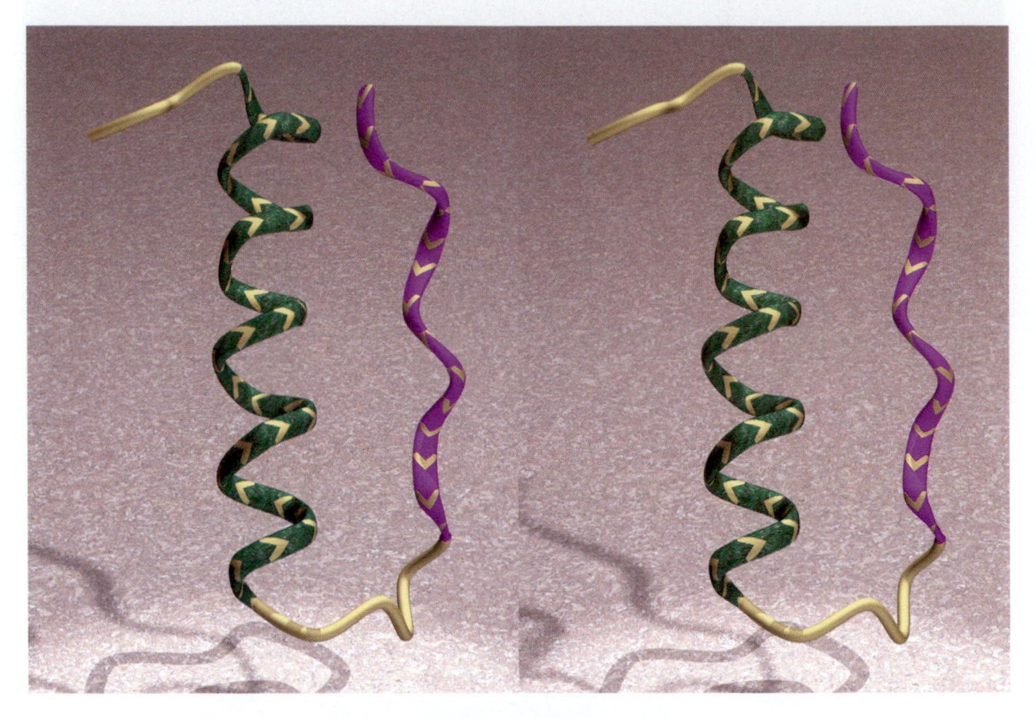

Fig. A.26 Avian pancreatic polypeptide (turkey) [1PPT].

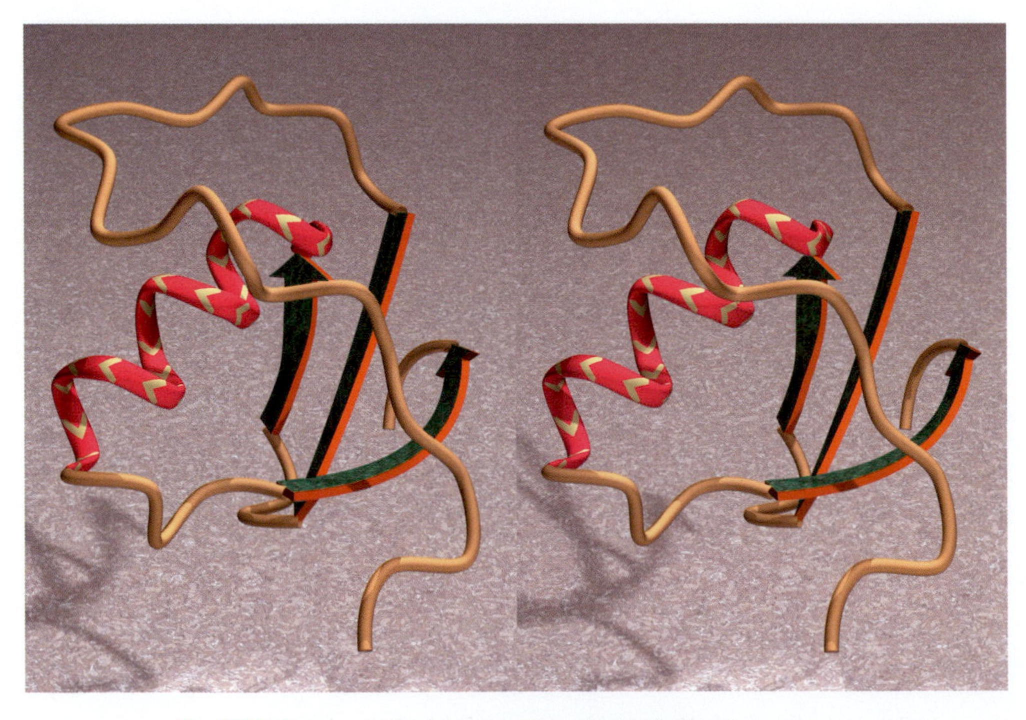

Fig. A.27 Ovomucoid third domain (silver pheasant) [2OVO].

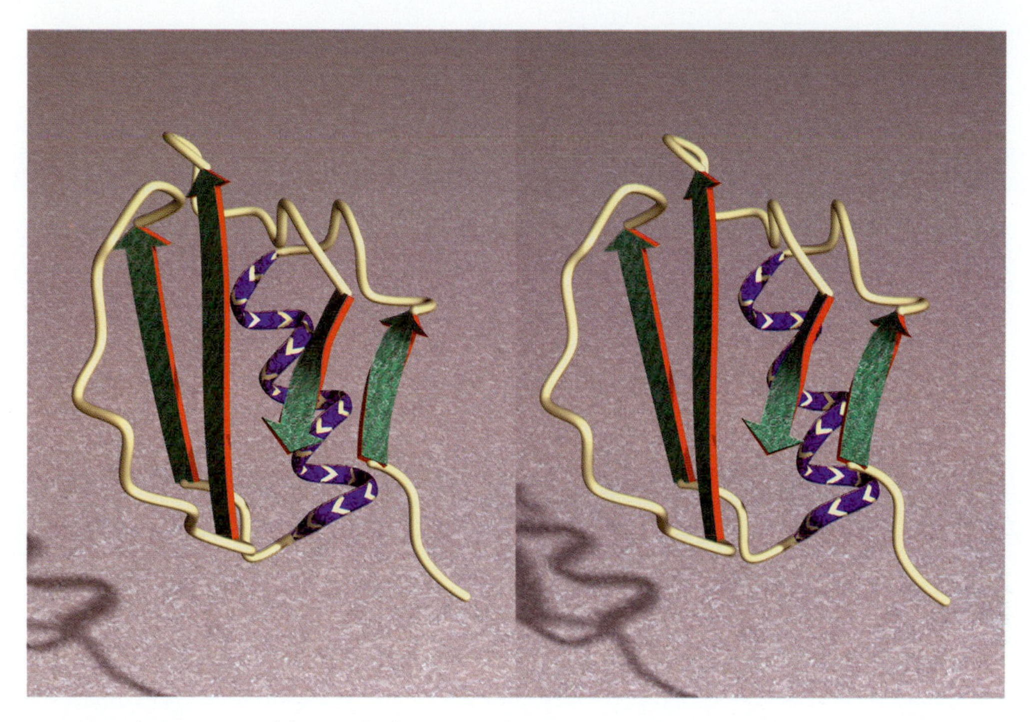

Fig. A.28 Chymotrypsin inhibitor 2 (barley) [2CI2].

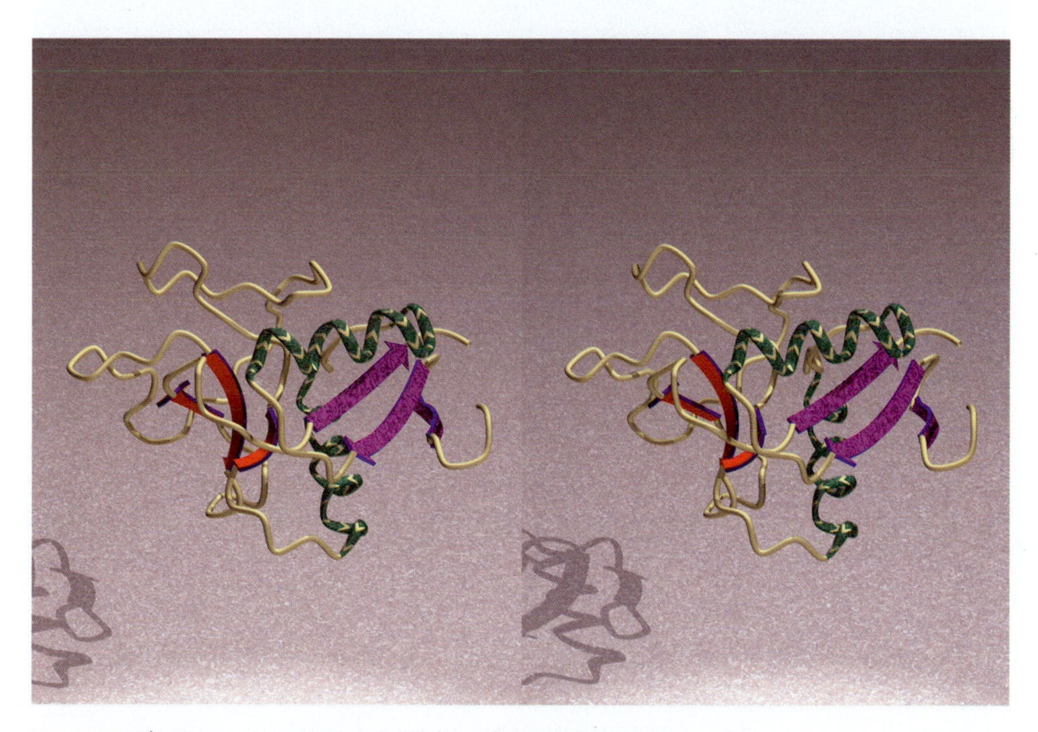

Fig. A.29 Lithostatine: pancreatic stone inhibitor (human) [1LIT].

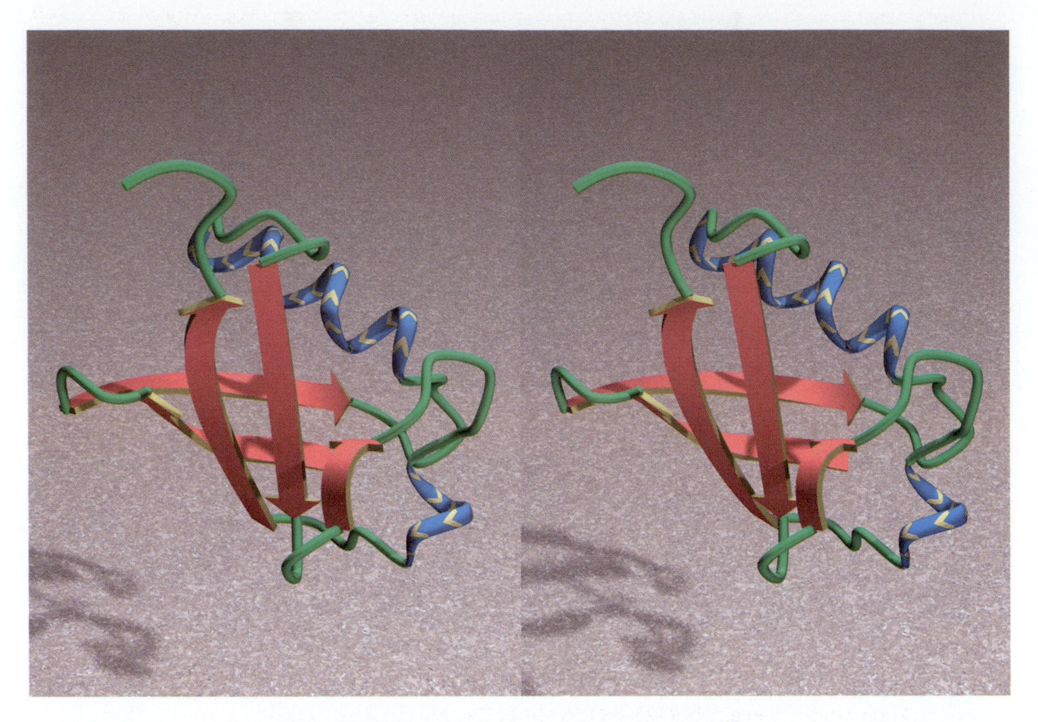

Fig. A.30 Ubiquitin (human) [1UBQ].

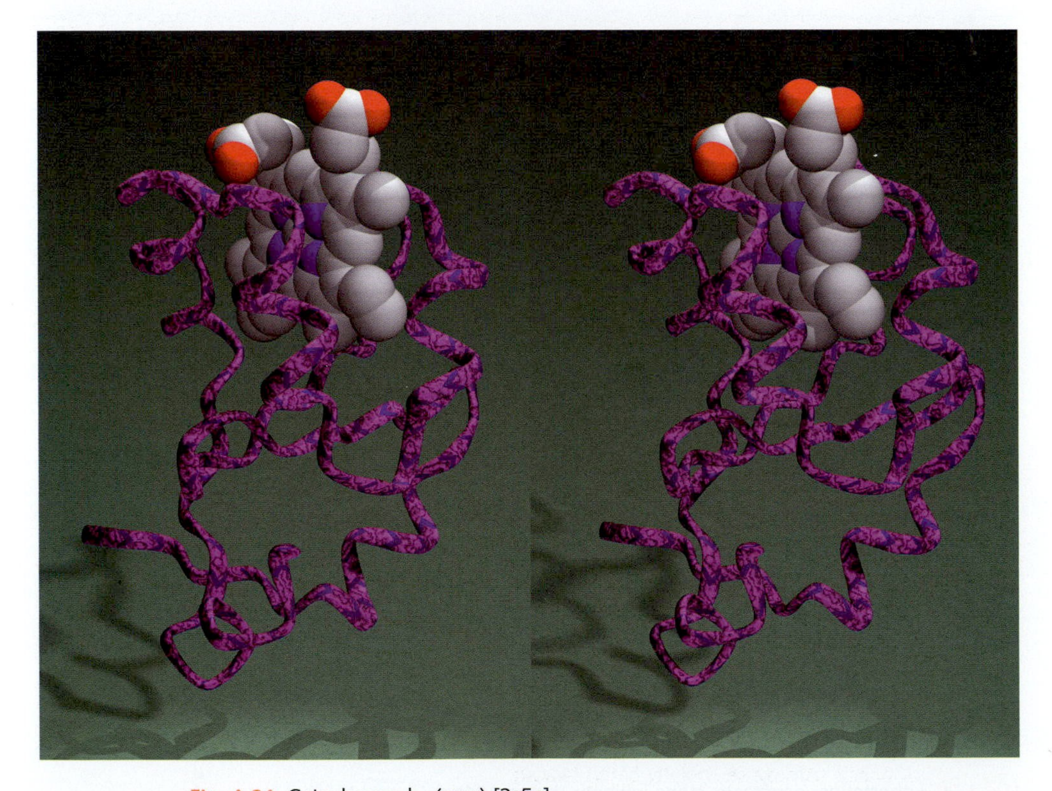

Fig. A.31 Cytochrome b_5 (cow) [3B5C].

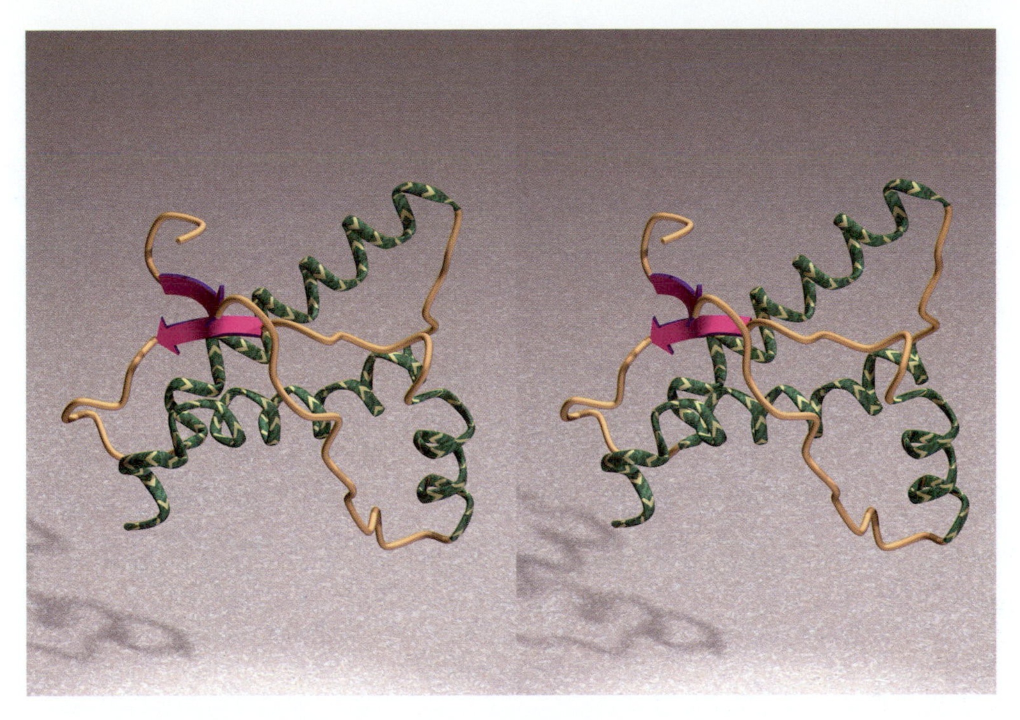

Fig. A.32 Prion protein domain prp(121–231) (mouse) [1AG2].

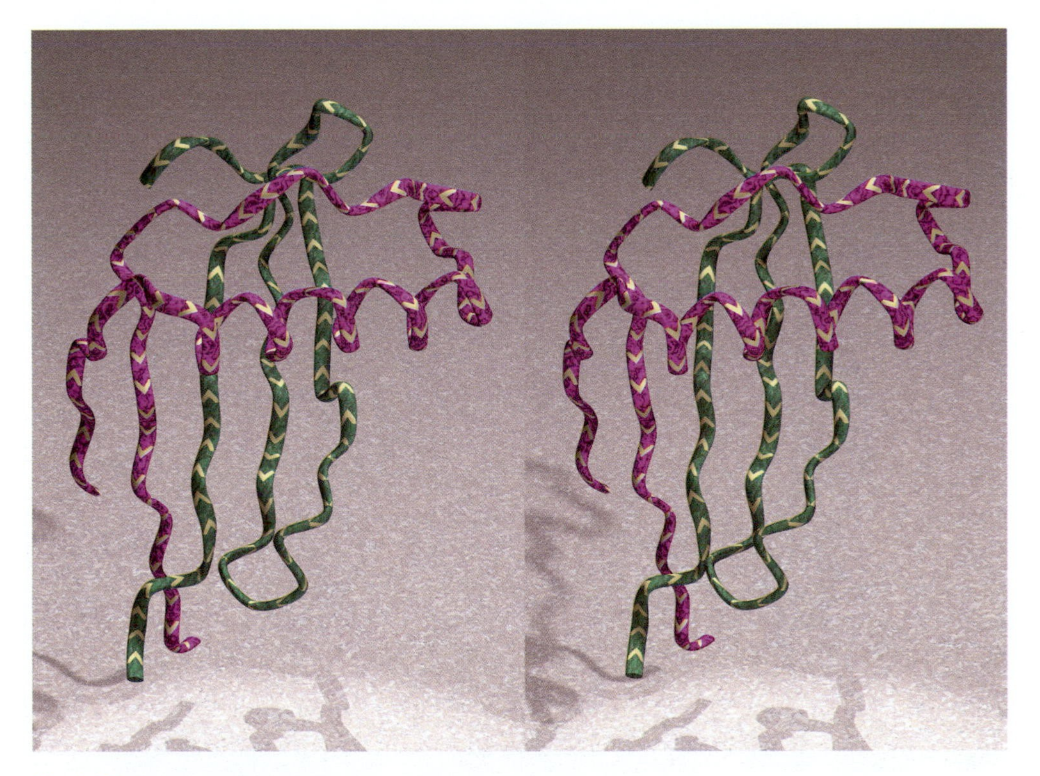

Fig. A.33 Monellin (African serendipity berry) [4MON].

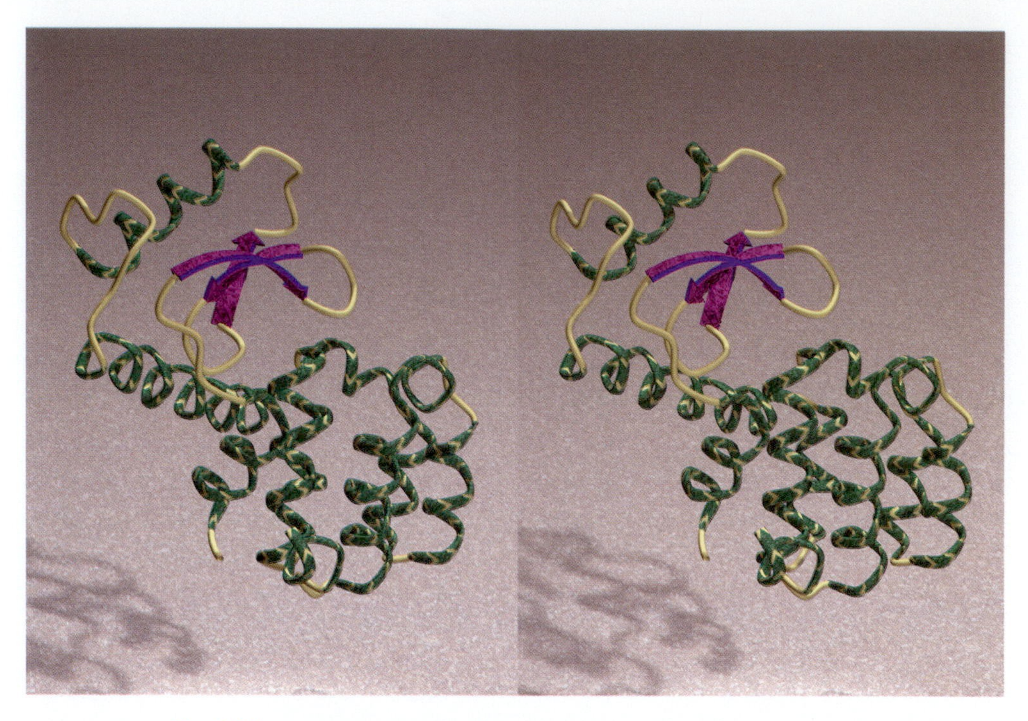

Fig. A.34 Lysozyme (phage T_4) [3LZM].

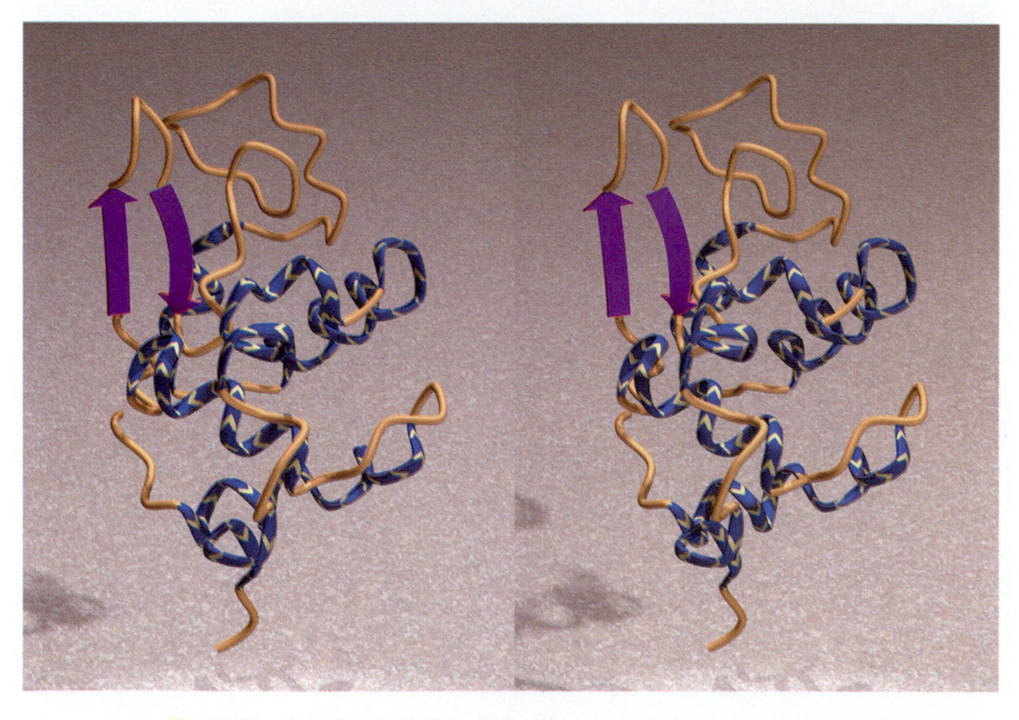

Fig. A.35 α-lactalbumin (baboon) [1ALC].

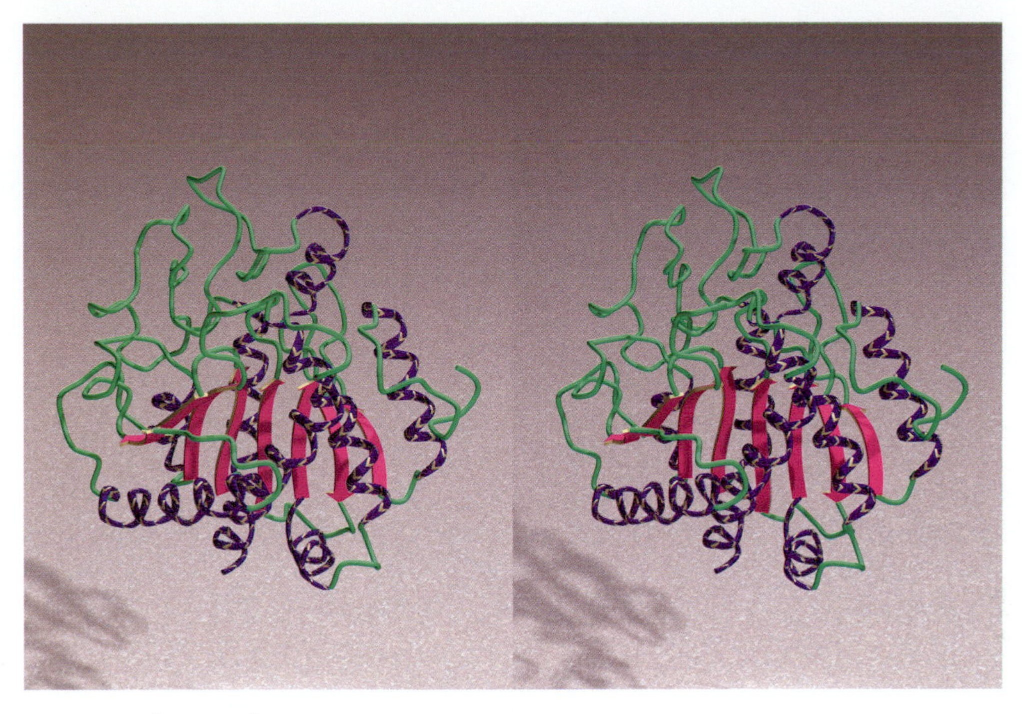

Fig. A.36 Carboxypeptidase a (cow) [5CPA].

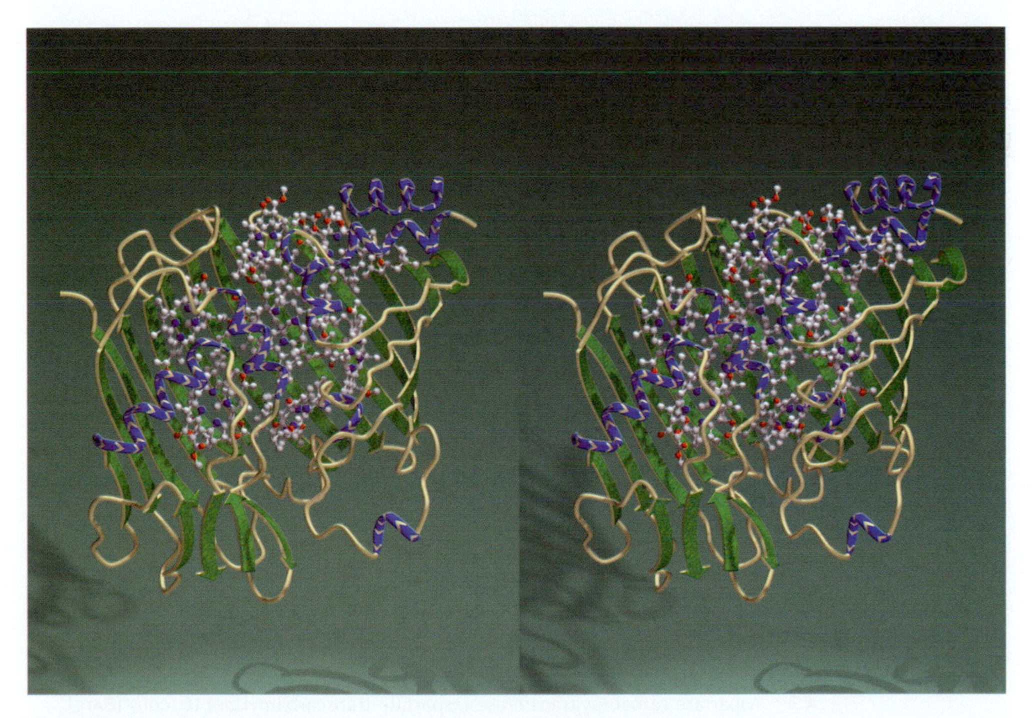

Fig. A.37 Bacteriochlorophyll a protein (*Prosthecochloris aestuarii*) [4BCL].

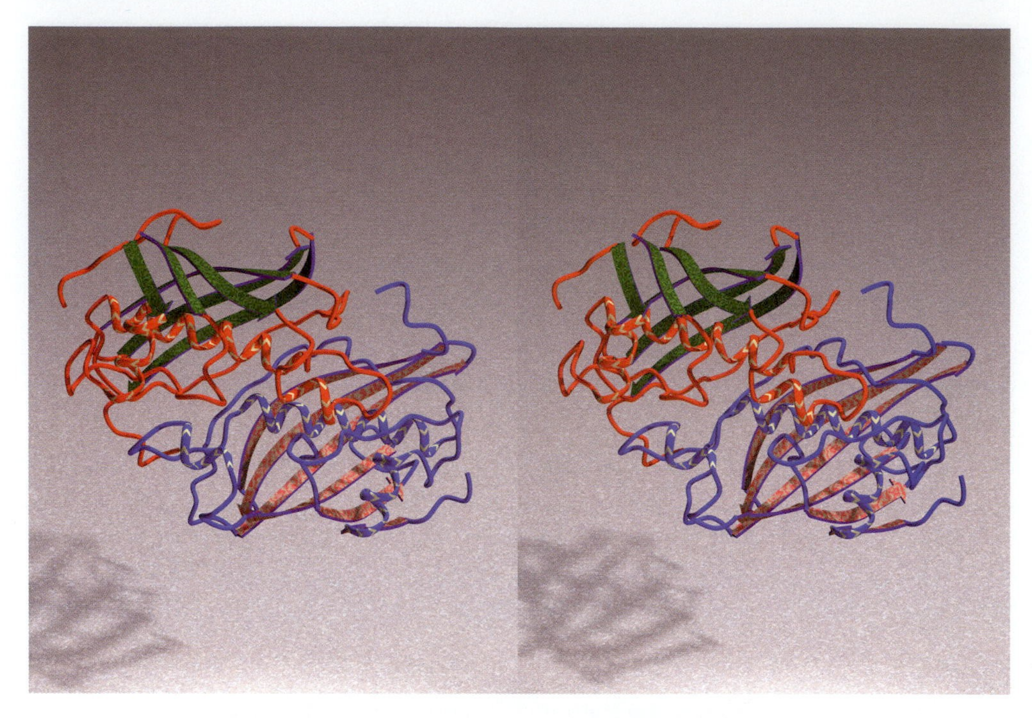

Fig. A.38 β-hydroxydecanoyl thiol ester dehydrase (*E. coli*) [1MKA].

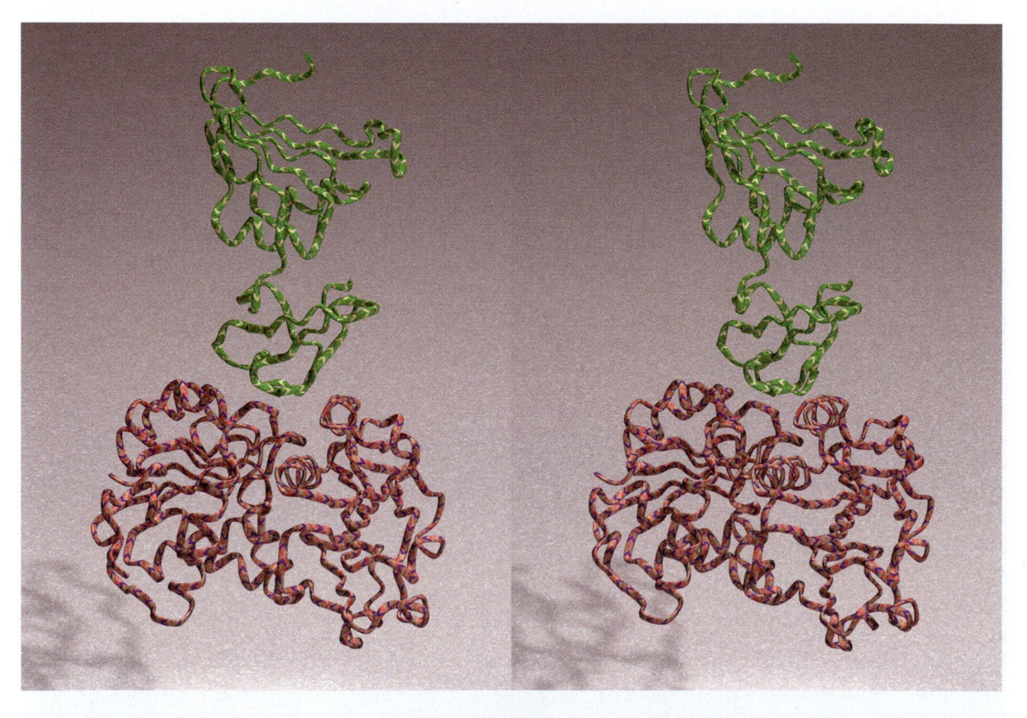

Fig. A.39 Aspartate carbamoyltransferase (aspartate transcarbamylase) (*E. coli*) [8ATC]. Catalytic subunit red; regulatory subunit green.

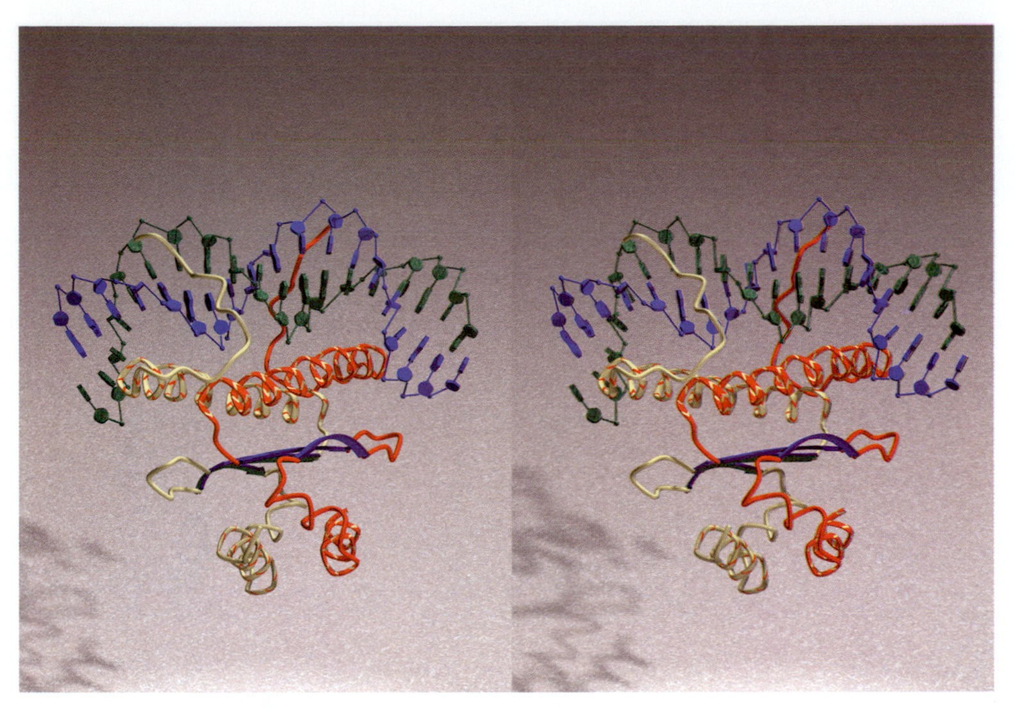

Fig. A.40 Serum response factor, core fragment (human) [1SRS].

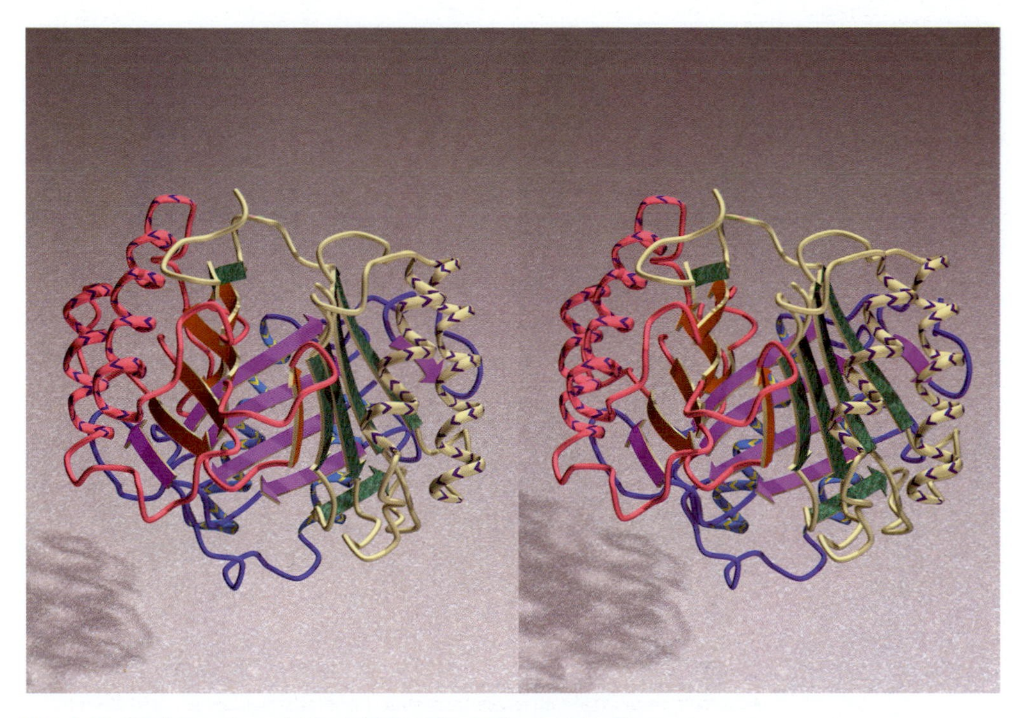

Fig. A.41 Cytokine: glycosylation-inhibiting factor (human) [1GIF].

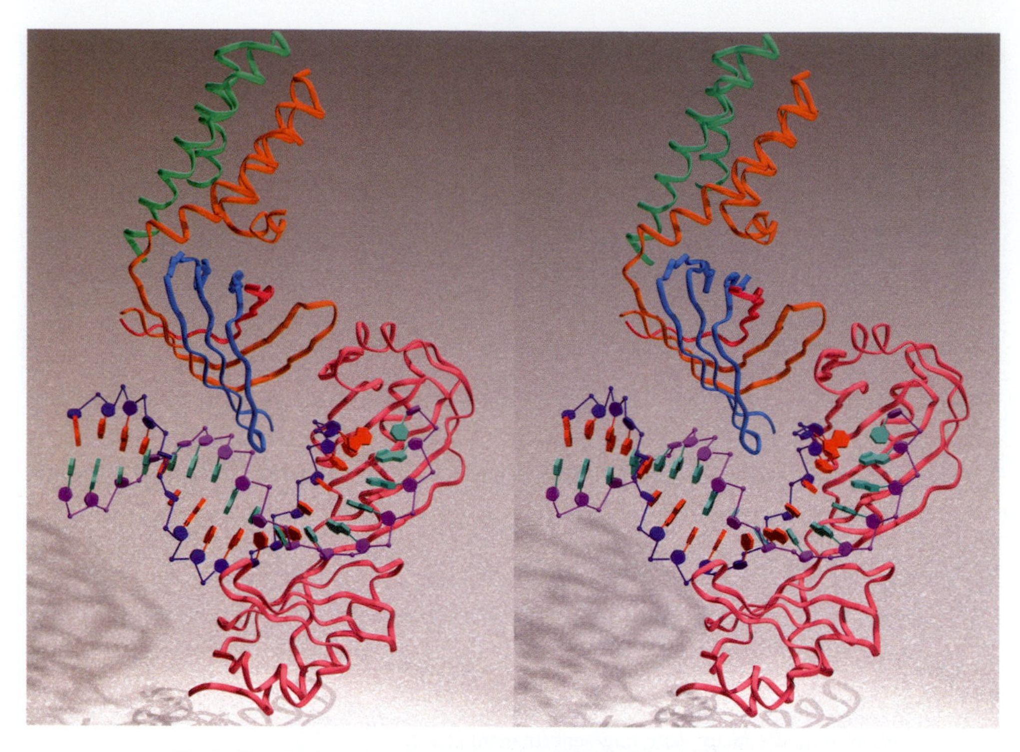

Fig. A.42 TATA-box binding protein (yeast) [1YTF] α/β proteins.

$\alpha + \beta$ proteins

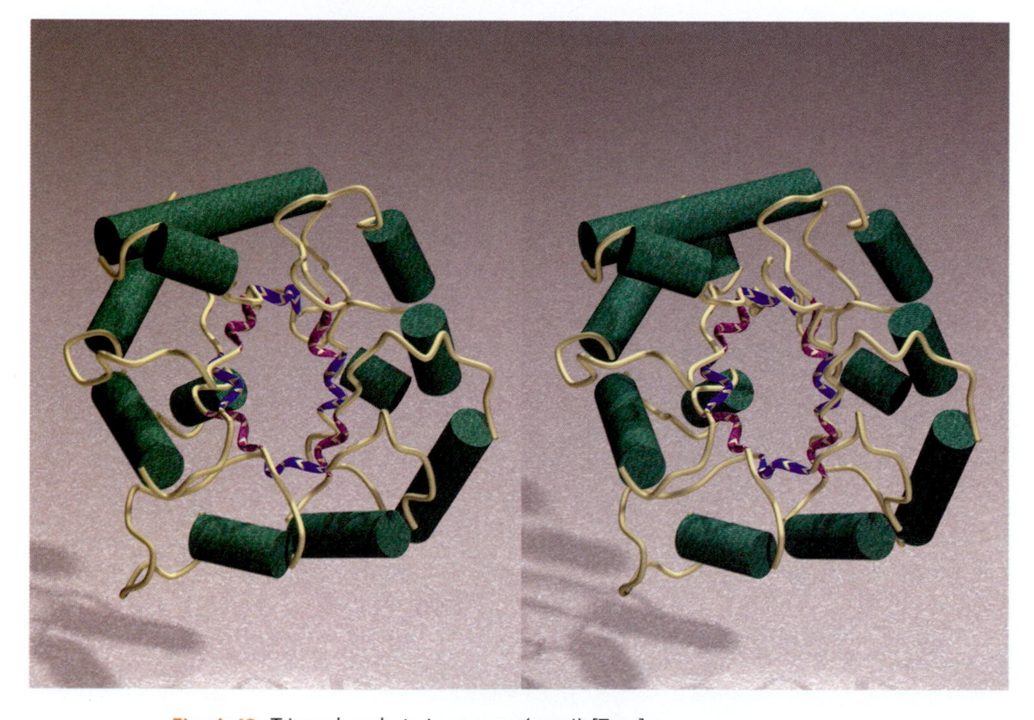

Fig. A.43 Triosephosphate isomerase (yeast) [7TIM].

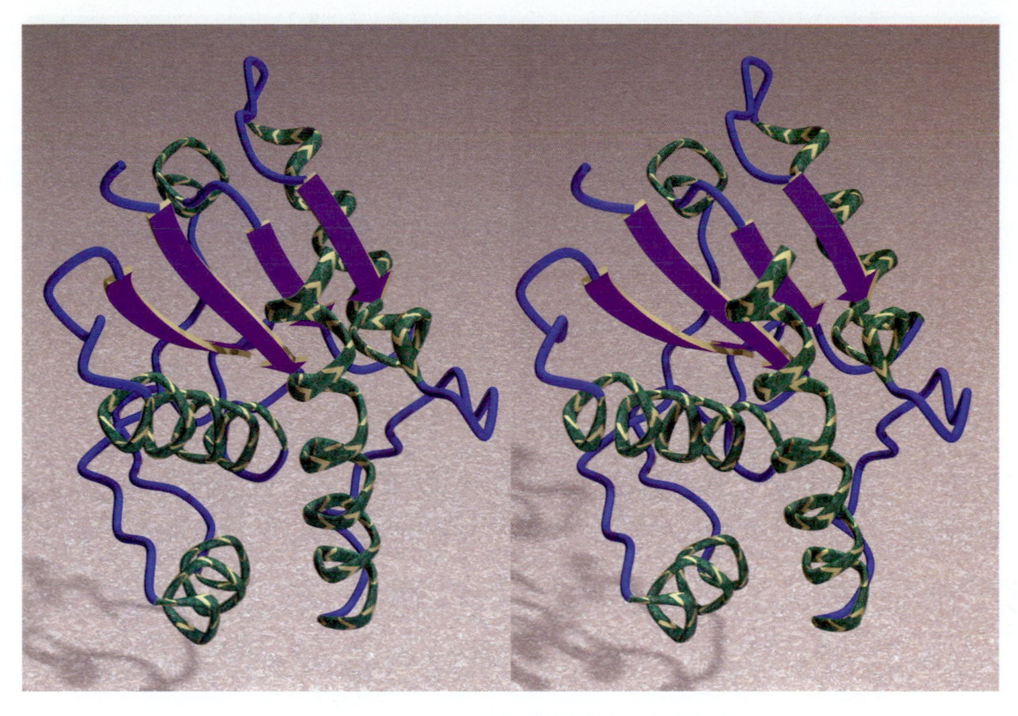

Fig. A.44 Phosphotyrosine protein phosphatase (cow) [1PHR].

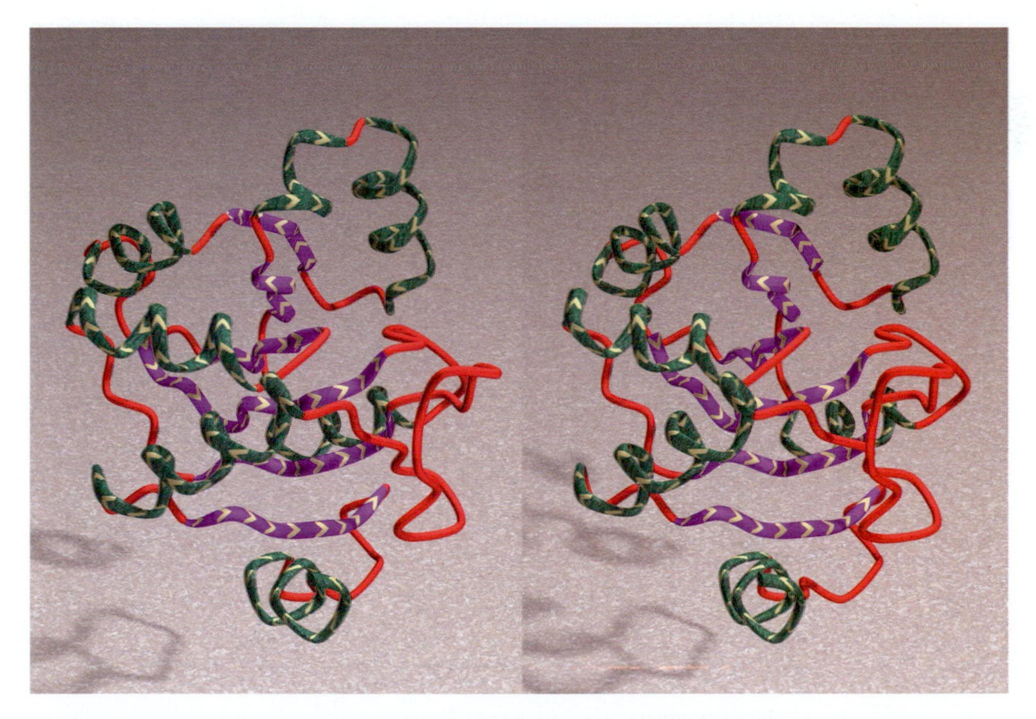

Fig. A.45 Fructose permease subunit IIb (*E. coli*) [1BLE].

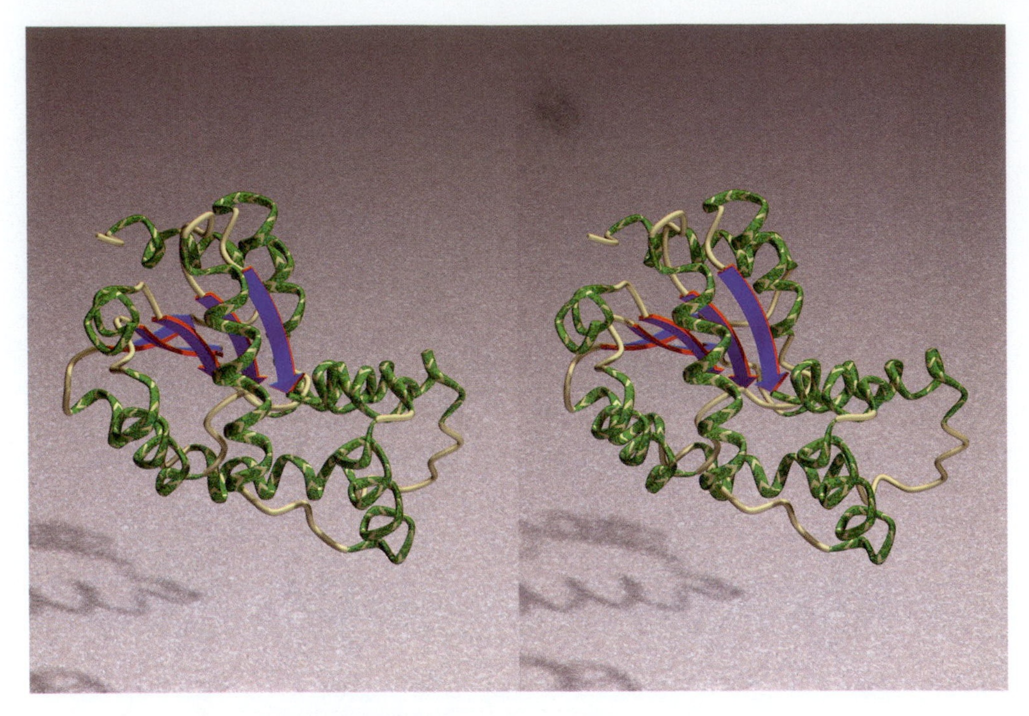

Fig. A.46 Adenylate kinase (pig) [3ADK].

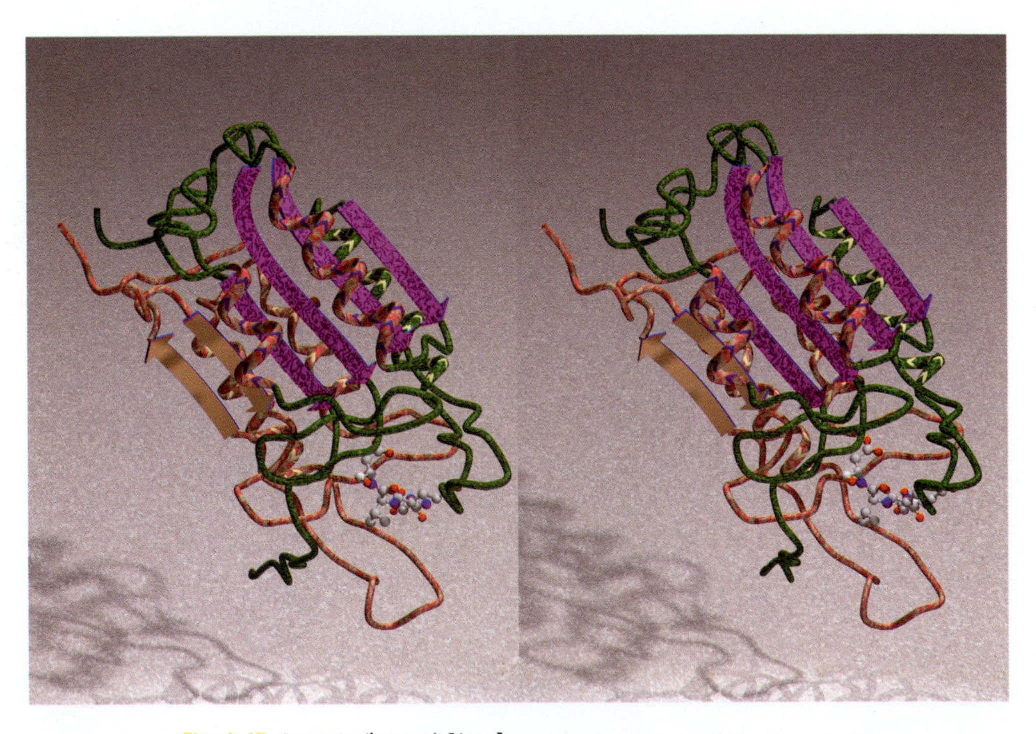

Fig. A.47 Apopain (human) [1PAU].

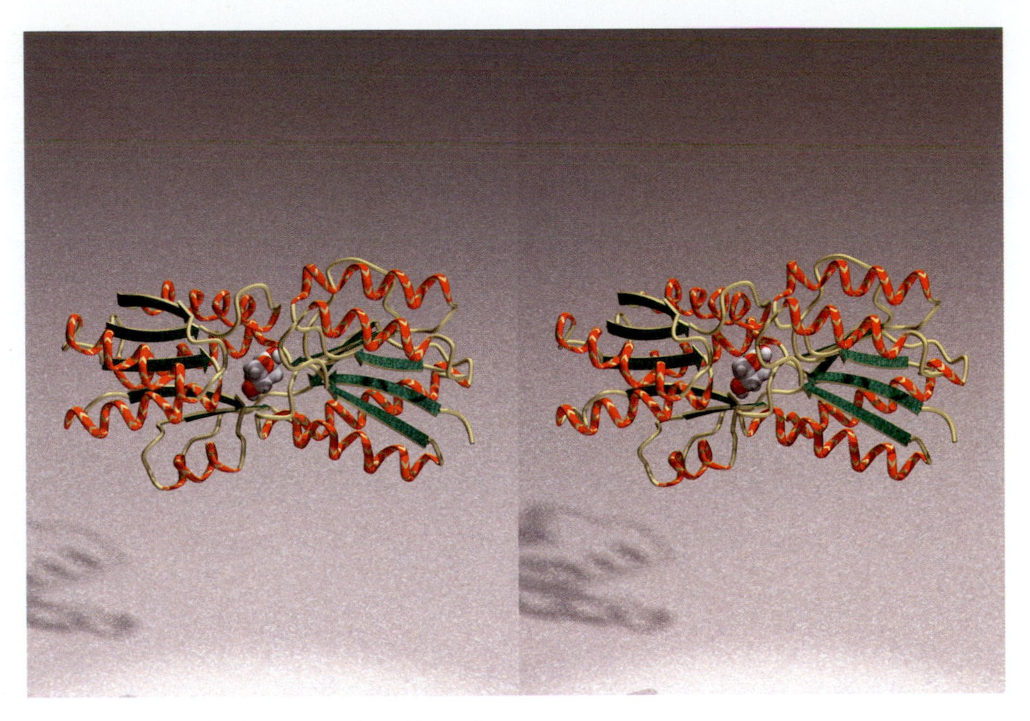

Fig. A.48 D-ribose-binding protein (*E. coli*) [2DRI].

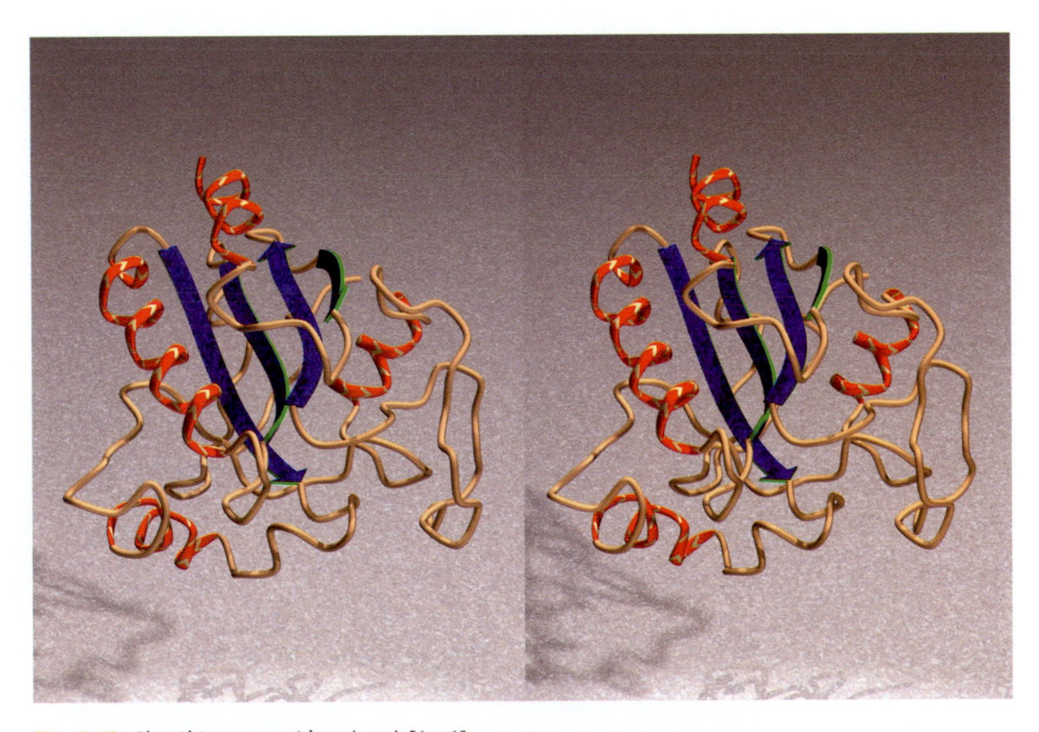

Fig. A.49 Glutathione peroxidase (cow) [1GP1].

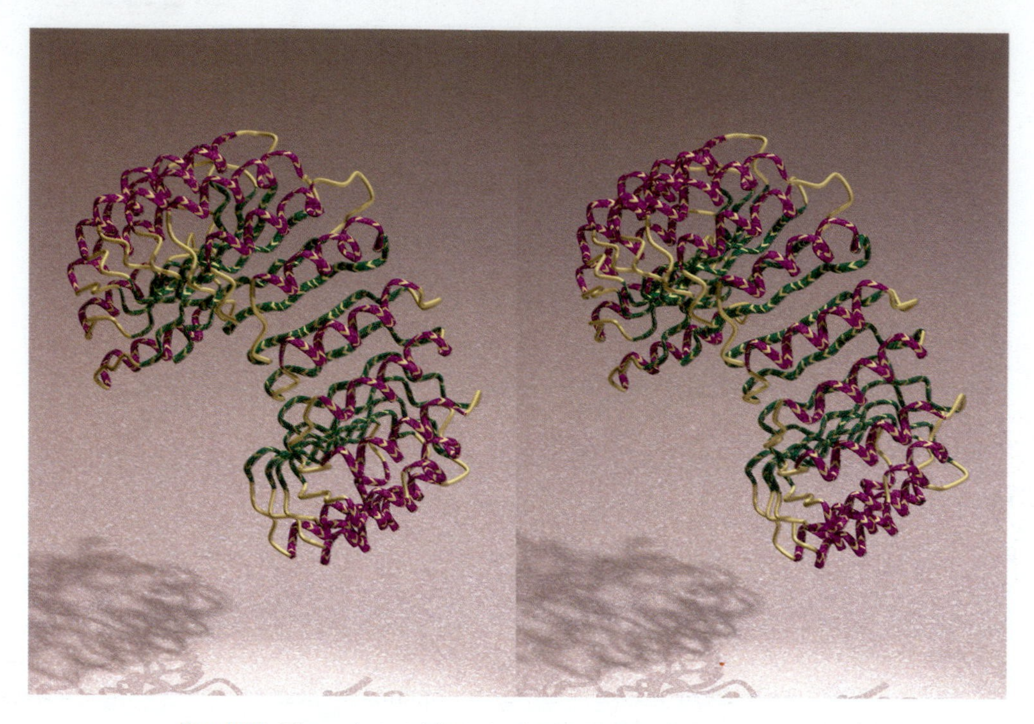

Fig. A.50 Ribonuclease inhibitor (pig) [2BNH].

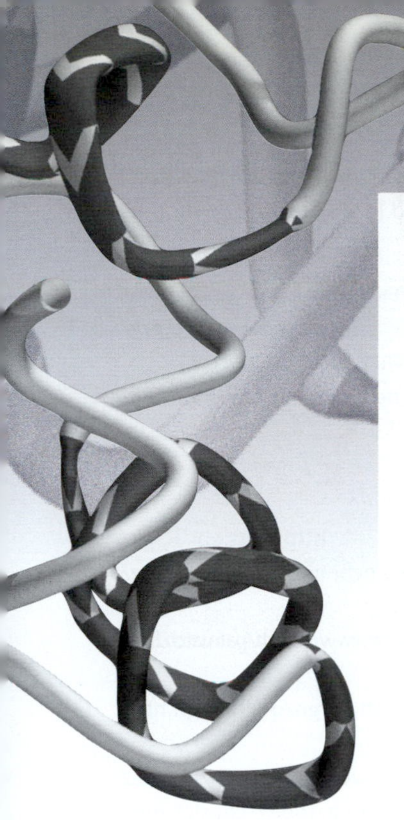

APPENDIX 2

Useful web sites

Good starting points for 'surfing' in molecular biology:

National Center for Biotechnology Information
http://www.ncbi.nlm.nih.gov/

European Bioinformatics Institute http://www.ebi.ac.uk

Expasy (Swiss Institute of Bioinformatics) http://www.expasy.ch/

GenomeNew (Kyoto University and University of Tokyo)
http://www.genome.ad.jp/

Collection of on-line analysis tools, including database searches
http://www-biol.univ-mrs.fr/english/logligne.html
http://www.fccc.edu/research/labs/dunbrack/molecularmodeling.html

Index of web sites in molecular biology, including specialized databases
http://www.cbs.dtu.dk/biolink.html

BCM search launcher—various database searches and associated tools
http://kiwi.imgen.bcm.tmc.edu:8088/search-launcher/launcher.html

Collections of protein analysis tools
http://www.graylab.ac.uk/cancerweb/research/protanal.html
http://www-biol.univ-mrs.fr/english/logligne.html

General information:

Primer on molecular genetics
http://www.ornl.gov/TechResources/Human_Genome/publicat/primer/prim1.html

Human genome project information
http://www.ornl.gov/hgmis/

Genome sequencing project information:
http://www.mcs.anl.gov/home/gaasterl/genomes.html
http://www-biol.univ-mrs.fr/english/genome.html
(Organelles): http://megasun.bch.umontreal.ca/ogmpproj.html

Growth of databanks http://www.genome.ad.jp/dbget/db_growth.gif
http://www.genome.ad.jp/dbget/db_growth.html

Database of metabolic pathways
http://www-c.mcs.anl.gov/home/compbio/PUMA/Production/puma_graphics.html
Electronic scholarly publishing http://www.esp.org/
Bibliographic database: PubMed (U.S. National Library of Medicine)
http://www.ncbi.nlm.nih.gov/PubMed/

Primarily sequence information:

Sequence retrieval:
SRS http://srs.ebi.ac.uk/
Entrez http://www.ncbi.nlm.nih.gov/Entrez/
Oligopeptide dictionary at PIR
http://www-nbrf.georgetown.edu/pirwww/search/patmatch.html
Calculation of multiple sequence alignments
http://dot.imgen.bcm.tmc.edu:9331/multi-align/multi-align.html
Prosite database of sequence motifs:
http://www.expasy.ch/prosite/
Scan Prosite: http://www.expasy.ch/tools/scnpsit1.html
Collections of multiple sequence alignments:
http://www2.ebi.ac.uk/dali/fssp/fssp.html
http://www.sanger.ac.uk/Pfam or http://pfam.wustl.edu/
Analysis of genome sequences:
PEDANT http://pedant.mips.biochem.mpg.de

Primarily structure information:

Polypeptide conformation http://www.chem.qmw.ac.uk/iupac/misc/biop.html
Protein Data Bank http://www.rcsb.org
ReLiBase (receptor–ligand complexes)
http://www.pdb.bnl.gov:8081/home.html
Promise: prosthetic groups and metal ions in protein active sites
http://bmbsgi11.leeds.ac.uk/promise/MAIN.html
Secondary structure assignments:
DSSP http://ara.ebi.ac.uk/dssp/
STRIDE http://www.emblheidelberg.de/cgi/stride_serv

Protein loop classification http://bonsai.lif.icnet.uk/bmm/loop/
Protein modules http://www.bork.embl-heidelberg.de/Modules/

Classifications of protein structures:
SCOP http://scop.mrc-lmb.cam.ac.uk/scop/
CATH http://www.biochem.ucl.ac.uk/bsm/cath/
DALI http://www2.embl-ebi.ac.uk/dali/
FSSP (Fold classification based on Structure–Structure alignment of
Proteins) http://www2.embl-ebi.ac.uk/dali/fssp/

Indices of other protein structure classifications
http://www.bioscience.org/urllists/protdb.htm
http://msd.ebi.ac.uk/add/Links/fold.shtml

Databases of protein sequences homologous to those of known structures:
FSSP (Fold classification based on Structure–Structure alignment of Proteins)
http://www2.ebi.ac.uk/dali/fssp/

HSSP (Homology-derived secondary structure)
http://www.sander.embl-heidelberg.de/hssp/

Access to structural databases at University College London
http://www.biochem.ucl.ac.uk/bsm/biocomp/index.html\#data bases

Calculation of accessible surface area
http://www.bork.embl-heidelberg.de/ASC/scr1-form.html

Calculation of hydrophobicity profile
http://bmbsgi11.leeds.ac.uk/bmb5dp/profiles.html

Sites specialized to specific protein families:

Index: http://msd.ebi.ac.uk/add/Links/family.html
Globins: http://bmbsgi11.leeds.ac.uk/promise/GLOBINS.html
Protein kinases http://www.sdsc.edu/kinases/
MEROPS database of peptidases:
http://www.bi.bbsrc.ac.uk/Merops/Merops.htm
IUBMB EC list for peptidases:
http://www.chem.qmw.ac.uk/iubmb/enzyme/EC34
AIDS-related information http://www-fbsc.ncifcrf.gov/HIVdb/

Proteins of the immune system:

General antibody-related material: http://www.antibodyresource.com
Searching the Kabat databank of antibody sequences and specificities
http://immuno.bme.nwu.edu/
Compilation of links to on-line databases and resources of immunological interest, (as well as much other material related to computational molecular biology)
http://www.infobiogen.fr/services/deambulum/english/db5.html
Immunogenetics database, plus links to sequence analysis tools
http://www.genetik.uni-koeln.de/dnaplot/

Conformational changes:

Database of conformational changes in proteins:
http://bioinfo.mbb.yale.edu/MolMovDB/db/ProtMotDB.main.html

Protein structure prediction:

Protein structure prediction centre http://predictioncenter.llnl.gov/

Homology modelling server
http://www.expasy.ch/swissmod/SWISS-MODEL.html
Threaders:
UCLA server http://www.doe-mbi.ucla.edu/people/frsvr/frsvr.html
EMBL server http://www.embl-heidelberg.de/predictprotein/predictprotein.html
Rotamer libraries: http://www.fccc.edu/research/labs/dunbrack/index.html
http://duc.urbb.jussieu.fr/rotamer.html

Major data archive projects in molecular biology

Name of data bank (home URL)	Type of data	Location
GenBank www.ncbi.nlm.nih.gov/	Nucleic acid sequences	National Library of Medicine,Washington, D.C., U.S.A.
EMBL Data Library www.ebi.ac.uk/ebi_docs/embl_db/ebi/topembl.html	Nucleic acid sequences	European Bioinformatics Institute, Hinxton, U.K.
DNA Data Bank of Japan www.ddbj.nig.ac.jp/	Nucleic acid sequences	National Institute of Genetics, Mishima, Japan
Protein Identification Resource www-nbrf.georgetown.edu/pir/	Amino acid sequences	Georgetown University, Washington, D.C., U.S.A.
Munich Information Center for Protein Sequences (MIPS) speedy.mips.biochem.mpg.de/	Amino acid sequences	Max-Planck-Institute für Biochemie, Martinsried, Germany
International Protein Information Database in Japan (JIPID)	Amino acid sequences	Science University of Tokyo Noda, Japan
SWISS–PROT www.expasy.ch/sprot/	Amino acid sequences	Geneva, Switzerland and Hinxton, U.K.
Protein Data Bank www.rcsb.org	Protein structures	Rutgers University, New Jersey, U.S.A.
Nucleic Acid Data Bank ndbserver.rutgers.edu/	Nucleic acid structures	Rutgers University, New Jersey, U.S.A
BioMagResBank www.bmrb.wisc.edu/	NMR Structure determinations	Madison, Wisconsin, U.S.A.
Cambridge Structural Database www.ccdc.cam.ac.uk/	Small-molecule crystal structures	Cambridge, U.K.

Web Sites for database searches

Task Web site	Program Name
Retrieve one sequence	
http://srs.ebi.ac.uk/	Sequence Retrieval System (SRS)
http://www.ncbi.nlm.nih.gov/Entrez/	Entrez

Retrieve one structure	Protein Data Bank (PDB)
www.rcsb.org	

Match one sequence to one sequence — ALIGN
http://www-hto.usc.edu/software/seqaln/seqaln-query.html
http://vega.igh.cnrs.fr/bin/align-guess.cgi

Multiple sequence alignment — ClustalW
http://www.ebi.ac.uk/clustalw

Probe sequence databank with sequence — PSI-BLAST
http://www.ncbi.nlm.nih.gov/blast/psiblast.cgi
or — Hidden Markov models
http://www.sanger.ac.uk/Software/Pfam/search.shtml
http://www.cse.ucsc.edu/research/compbio/HMM-apps/HMMapplications.html

Probe structure databank with structure — DALI
http://www2.embl-ebi.ac.uk/dali/

Probe structure databank with sequence (Fold recognition, threaders)
http://www.embl-heidelberg.de/predictprotein/predictprotein.html
http://www.doe-mbi.ucla.edu/people/frsvr/frsvr.html

Bibliographical search — U.S. National Library of Medicine (PubMed)
http://www.ncbi.nlm.nih.gov/

Index of structures illustrated

Subject index